Foundations and Applications of Statistics

An Introduction Using R

The Sally SERIES

Pure and Applied
UNDERGRADUATE // TEXTS · 13

Foundations and Applications of Statistics

An Introduction Using R

Randall Pruim

American Mathematical Society
Providence, Rhode Island

2010 *Mathematics Subject Classification.* Primary 62–01; Secondary 60–01.

For additional information and updates on this book, visit
www.ams.org/bookpages/amstext-13

Library of Congress Cataloging-in-Publication Data

Pruim, Randall J.
  Foundations and applications of statistics : an introduction using R / Randall Pruim.
    p. cm. — (Pure and applied undergraduate texts ; v. 13)
  Includes bibliographical references and index.
  ISBN 978-0-8218-5233-0 (alk. paper)
  1. Mathematical statistics—Data processing.   2. R (Computer program language)   I. Title.

QA276.45.R3P78   2010
519.50285—dc22

                                                                    2010041197

Contents

Preface

Intended Audience

As the title suggests, this book is intended as an introduction to both the foundations and applications of statistics. It is an introduction in the sense that it does not assume a prior statistics course. But it is not introductory in the sense of being suitable for students who have had nothing more than the usual high school mathematics preparation. The target audience is undergraduate students at the equivalent of the junior or senior year at a college or university in the United States.

Students should have had courses in differential and integral calculus, but not much more is required in terms of mathematical background. In fact, most of my students have had at least another course or two by the time they take this course, but the only courses that they have all had is the calculus sequence. The majority of my students are not mathematics majors. I have had students from biology, chemistry, computer science, economics, engineering, and psychology, and I have tried to write a book that is interesting, understandable, and useful to students with a wide range of backgrounds and career goals.

This book is suitable for what is often a two-semester sequence in "mathematical statistics", but it is different in some important ways from many of the books written for such a course. I was trained as a mathematician first, and the book is clearly mathematical at some points, but the emphasis is on the statistics. Mathematics and computation are brought in where they are useful tools. The result is a book that stretches my students in different directions at different times – sometimes statistically, sometimes mathematically, sometimes computationally.

The Approach Used in This Book

Features of this book that help distinguish it from other books available for such a course include the following:

- The use of R, a free software environment for statistical computing and graphics, throughout the text.

  Many books claim to integrate technology, but often technology appears to be more of an afterthought. In this book, topics are selected, ordered, and discussed in light of the current practice in statistics, where computers are an indispensable tool, not an occasional add-on.

  R was chosen because it is both powerful and available. Its "market share" is increasing rapidly, so experience with R is likely to serve students well in their future careers in industry or academics. A large collection of add-on packages are available, and new statistical methods are often available in R before they are available anywhere else.

  R is open source and is available at the Comprehensive R Archive Network (CRAN, http://cran.r-project.org) for a wide variety of computing platforms at no cost. This allows students to obtain the software for their personal computers – an essential ingredient if computation is to be used throughout the course.

  The R code in this book was executed on a 2.66 GHz Intel Core 2 Duo MacBook Pro running OS X (version 10.5.8) and the current version of R (version 2.12). Results using a different computing platform or different version of R should be similar.

- An emphasis on practical statistical reasoning.

  The idea of a statistical study is introduced early on using Fisher's famous example of the lady tasting tea. Numerical and graphical summaries of data are introduced early to give students experience with R and to allow them to begin formulating statistical questions about data sets even before formal inference is available to help answer those questions.

- Probability *for* statistics.

  One model for the undergraduate mathematical statistics sequence presents a semester of probability followed by a semester of statistics. In this book, I take a different approach and get to statistics early, developing the necessary probability as we go along, motivated by questions that are primarily statistical. Hypothesis testing is introduced almost immediately, and p-value computation becomes a motivation for several probability distributions. The binomial test and Fisher's exact test are introduced formally early on, for example. Where possible, distributions are presented as statistical models first, and their properties (including the probability mass function or probability density function) derived, rather than the other way around. Joint distributions are motivated by the desire to learn about the sampling distribution of a sample mean.

  Confidence intervals and inference for means based on t-distributions must wait until a bit more machinery has been developed, but my intention is that a student who only takes the first semester of a two-semester sequence will have a solid understanding of inference for one variable – either quantitative or categorical.

- The linear algebra middle road.

    Linear models (regression and ANOVA) are treated using a geometric, vector-based approach. A more common approach at this level is to introduce these topics without referring to the underlying linear algebra. Such an approach avoids the problem of students with minimal background in linear algebra but leads to mysterious and unmotivated identities and notions.

    Here I rely on a small amount of linear algebra that can be quickly reviewed or learned and is based on geometric intuition and motivation (see Appendix C). This works well in conjunction with R since R is in many ways vector-based and facilitates vector and matrix operations. On the other hand, I avoid using an approach that is too abstract or requires too much background for the typical student in my course.

Brief Outline

The first four chapters of this book introduce important ideas in statistics (distributions, variability, hypothesis testing, confidence intervals) while developing a mathematical and computational toolkit. I cover this material in a one-semester course. Also, since some of my students only take the first semester, I wanted to be sure that they leave with a sense for statistical practice and have some useful statistical skills even if they do not continue. Interestingly, as a result of designing my course so that stopping halfway makes some sense, I am finding that more of my students are continuing on to the second semester. My sample size is still small, but I hope that the trend continues and would like to think it is due in part because the students are enjoying the course and can see "where it is going".

The last three chapters deal primarily with two important methods for handling more complex statistical models: maximum likelihood and linear models (including regression, ANOVA, and an introduction to generalized linear models). This is not a comprehensive treatment of these topics, of course, but I hope it both provides flexible, usable statistical skills and prepares students for further learning.

Chi-squared tests for goodness of fit and for two-way tables using both the Pearson and likelihood ratio test statistics are covered after first generating empirical p-values based on simulations. The use of simulations here reinforces the notion of a sampling distribution and allows for a discussion about what makes a good test statistic when multiple test statistics are available. I have also included a brief introduction to Bayesian inference, some examples that use simulations to investigate robustness, a few examples of permutation tests, and a discussion of Bradley-Terry models. The latter topic is one that I cover between Selection Sunday and the beginning of the NCAA Division I Basketball Tournament each year. An application of the method to the 2009–2010 season is included.

Various R functions and methods are described as we go along, and Appendix A provides an introduction to R focusing on the way R is used in the rest of the book. I recommend working through Appendix A simultaneously with the first chapter – especially if you are unfamiliar with programming or with R.

Some of my students enter the course unfamiliar with the notation for things like sets, functions, and summation, so Appendix B contains a brief tour of the basic mathematical results and notation that are needed. The linear algebra required for parts of Chapter 4 and again in Chapters 6 and 7 is covered in Appendix C. These can be covered as needed or used as a quick reference. Appendix D is a review of the first four chapters in outline form. It is intended to prepare students for the remainder of the book after a semester break, but it could also be used as an end of term review.

Access to R Code and Data Sets

All of the data sets and code fragments used in this book are available for use in R on your own computer. Data sets and other utilities that are not provided by R packages in CRAN are available in the `fastR` package. This package can be obtained from CRAN, from the companion web site for this book, or from the author's web site.

Among the utility functions in `fastR` is the function `snippet()`, which provides easy access to the code fragments that appear in this book. The names of the code fragments in this book appear in boxes at the right margin where code output is displayed. Once `fastR` has been installed and loaded,

```
snippet('snippet')
```

will both display and execute the code named "snippet", and

```
snippet('snippet', exec=FALSE)
```

will display but not execute the code.

`fastR` also includes a number of additional utility functions. Several of these begin with the letter x. Examples include `xplot`, `xhistogram`, `xpnorm`, etc. These functions add extra features to the standard functions they are based on. In most cases they are identical to their x-less counterparts unless new arguments are used.

Companion Web Site

Additional material related to this book is available online at

`http://www.ams.org/bookpages/amstext-13`

Included there are

- an errata list,
- additional instructions, with links, for installing R and the R packages used in this book,
- additional examples and problems,
- additional student solutions,
- additional material – including a complete list of solutions – available only to instructors.

Acknowledgments

Every author sets out to write the perfect book. I was no different. Fortunate authors find others who are willing to point out the ways they have fallen short of their goal and suggest improvements. I have been fortunate.

Most importantly, I want to thank the students who have taken advanced undergraduate statistics courses with me over the past several years. Your questions and comments have shaped the exposition of this book in innumerable ways. Your enthusiasm for detecting my errors and your suggestions for improvements have saved me countless embarrassments. I hope that your moments of confusion have added to the clarity of the exposition.

If you look, some of you will be able to see your influence in very specific ways here and there (happy hunting). But so that you all get the credit you deserve, I want to list you all (in random order, of course): Erin Campbell, John Luidens, Kyle DenHartigh, Jessica Haveman, Nancy Campos, Matthew DeVries, Karl Stough, Heidi Benson, Kendrick Wiersma, Dale Yi, Jennifer Colosky, Tony Ditta, James Hays, Joshua Kroon, Timothy Ferdinands, Hanna Benson, Landon Kavlie, Aaron Dull, Daniel Kmetz, Caleb King, Reuben Swinkels, Michelle Medema, Sean Kidd, Leah Hoogstra, Ted Worst, David Lyzenga, Eric Barton, Paul Rupke, Alexandra Cok, Tanya Byker Phair, Nathan Wybenga, Matthew Milan, Ashley Luse, Josh Vesthouse, Jonathan Jerdan, Jamie Vande Ree, Philip Boonstra, Joe Salowitz, Elijah Jentzen, Charlie Reitsma, Andrew Warren, Lucas Van Drunen, Che-Yuan Tang, David Kaemingk, Amy Ball, Ed Smilde, Drew Griffioen, Tim Harris, Charles Blum, Robert Flikkema, Dirk Olson, Dustin Veldkamp, Josh Keilman, Eric Sloterbeek, Bradley Greco, Matt Disselkoen, Kevin VanHarn, Justin Boldt, Anthony Boorsma, Nathan Dykhuis, Brandon Van Dyk, Steve Pastoor, Micheal Petlicke, Michael Molling, Justin Slocum, Jeremy Schut, Noel Hayden, Christian Swenson, Aaron Keen, Samuel Zigterman, Kobby Appiah-Berko, Jackson Tong, William Vanden Bos, Alissa Jones, Geoffry VanLeeuwen, Tim Slager, Daniel Stahl, Kristen Vriesema, Rebecca Sheler, and Andrew Meneely.

I also want to thank various colleagues who read or class-tested some or all of this book while it was in progress. They are

Ming-Wen An
    Vassar College
	Daniel Kaplan
	    Macalester College

Alan Arnholdt
    Appalacian State University
	John Kern
	    Duquesne University

Stacey Hancock
    Clark University
	Kimberly Muller
	    Lake Superior State University

Jo Hardin
    Pomona College
	Ken Russell
	    University of Wollongong, Australia

Nicholas Horton
    Smith College
	Greg Snow
	    Intermountain Healthcare

Laura Kapitula
    Calvin College
	Nathan Tintle
	    Hope College

Interesting data make for interesting statistics, so I want to thank colleagues and students who helped me locate data for use in the examples and exercises in this book, especially those of you who made original data available. In the latter cases, specific attributions are in the documentation for the data sets in the `fastR` package.

Thanks also go to those at the American Mathematical Society who were involved in the production of this book: Edward Dunne, the acquisitions editor with whom I developed the book from concept to manuscript; Arlene O'Sean, production editor; Cristin Zannella, editorial assistant; and Barbara Beeton, who provided TeXnical support. Without their assistance and support the final product would not have been as satisfying.

Alas, despite the efforts of so many, this book is still not perfect. No books are perfect, but some books are useful. My hope is that this book is both useful and enjoyable. A list of those (I hope few) errors that have escaped detection until after the printing of this book will be maintained at

http://www.ams.org/bookpages/amstext-13

My thanks in advance to those who bring these to my attention.

What Is Statistics?

Some Definitions of Statistics

This is a course primarily about statistics, but what exactly is *statistics*? In other words, what is this course about?[1] Here are some definitions of statistics from other people:

- a collection of procedures and principles for gaining information in order to make decisions when faced with uncertainty (J. Utts [**Utt05**]),

- a way of taming uncertainty, of turning raw data into arguments that can resolve profound questions (T. Amabile [**fMA89**]),

- the science of drawing conclusions from data with the aid of the mathematics of probability (S. Garfunkel [**fMA86**]),

- the explanation of variation in the context of what remains unexplained (D. Kaplan [**Kap09**]),

- the mathematics of the collection, organization, and interpretation of numerical data, especially the analysis of a population's characteristics by inference from sampling (American Heritage Dictionary [**AmH82**]).

While not exactly the same, these definitions highlight four key elements of statistics.

Data – the raw material

Data are the raw material for doing statistics. We will learn more about different types of data, how to collect data, and how to summarize data as we go along. This will be the primary focus of Chapter 1.

---

[1] As we will see, the words *statistic* and *statistics* get used in more than one way. More on that later.

Information – the goal

The goal of doing statistics is to gain some information or to make a decision. Statistics is useful because it helps us answer questions like the following:

- Which of two treatment plans leads to the best clinical outcomes?
- How strong is an I-beam constructed according to a particular design?
- Is my cereal company complying with regulations about the amount of cereal in its cereal boxes?

In this sense, statistics is a science – a method for obtaining new knowledge.

Uncertainty – the context

The tricky thing about statistics is the uncertainty involved. If we measure one box of cereal, how do we know that all the others are similarly filled? If every box of cereal were identical and every measurement perfectly exact, then one measurement would suffice. But the boxes may differ from one another, and even if we measure the same box multiple times, we may get different answers to the question *How much cereal is in the box?*

So we need to answer questions like *How many boxes should we measure?* and *How many times should we measure each box?* Even so, there is no answer to these questions that will give us absolute certainty. So we need to answer questions like *How sure do we need to be?*

Probability – the tool

In order to answer a question like *How sure do we need to be?*, we need some way of measuring our level of certainty. This is where mathematics enters into statistics. Probability is the area of mathematics that deals with reasoning about uncertainty. So before we can answer the statistical questions we just listed, we must first develop some skill in probability. Chapter 2 provides the foundation that we need.

Once we have developed the necessary tools to deal with uncertainty, we will be able to give good answers to our statistical questions. But before we do that, let's take a bird's eye view of the processes involved in a statistical study. We'll come back and fill in the details later.

A First Example: The Lady Tasting Tea

There is a famous story about a lady who claimed that tea with milk tasted different depending on whether the milk was added to the tea or the tea added to the milk. The story is famous because of the setting in which she made this claim. She was attending a party in Cambridge, England, in the 1920s. Also in attendance were a number of university dons and their wives. The scientists in attendance scoffed at the woman and her claim. What, after all, could be the difference?

All the scientists but one, that is. Rather than simply dismiss the woman's claim, he proposed that they decide how one should *test* the claim. The tenor of the conversation changed at this suggestion, and the scientists began to discuss how the claim should be tested. Within a few minutes cups of tea with milk had been prepared and presented to the woman for tasting.

Let's take this simple example as a prototype for a statistical study. What steps are involved?

(1) Determine the question of interest.

Just what is it we want to know? It may take some effort to make a vague idea precise. The precise questions may not exactly correspond to our vague questions, and the very exercise of stating the question precisely may modify our question. Sometimes we cannot come up with any way to answer the question we really want to answer, so we have to live with some other question that is not exactly what we wanted but is something we can study and will (we hope) give us some information about our original question.

In our example this question seems fairly easy to state: Can the lady tell the difference between the two tea preparations? But we need to refine this question. For example, are we asking if she *always* correctly identifies cups of tea or merely if she does better than we could do ourselves (by guessing)?

(2) Determine the population.

Just who or what do we want to know about? Are we only interested in this one woman or women in general or only women who claim to be able to distinguish tea preparations?

(3) Select measurements.

We are going to need some data. We get our data by making some measurements. These might be physical measurements with some device (like a ruler or a scale). But there are other sorts of measurements too, like the answer to a question on a form. Sometimes it is tricky to figure out just what to measure. (How do we measure happiness or intelligence, for example?) Just how we do our measuring will have important consequences for the subsequent statistical analysis.

In our example, a measurement may consist of recording for a given cup of tea whether the woman's claim is correct or incorrect.

(4) Determine the sample.

Usually we cannot measure every individual in our population; we have to select some to measure. But how many and which ones? These are important questions that must be answered. Generally speaking, bigger is better, but it is also more expensive. Moreover, no size is large enough if the sample is selected inappropriately.

Suppose we gave the lady one cup of tea. If she correctly identifies the mixing procedure, will we be convinced of her claim? She might just be guessing; so we should probably have her taste more than one cup. Will we be convinced if she correctly identifies 5 cups? 10 cups? 50 cups?

What if she makes a mistake? If we present her with 10 cups and she correctly identifies 9 of the 10, what will we conclude? A success rate of 90% is, it seems, much better than just guessing, and anyone can make a mistake now and then. But what if she correctly identifies 8 out of 10? 80 out of 100?

And how should we prepare the cups? Should we make 5 each way? Does it matter if we tell the woman that there are 5 prepared each way? Should we flip a coin to decide even if that means we might end up with 3 prepared one way and 7 the other way? Do any of these differences matter?

(5) Make and record the measurements.

Once we have the design figured out, we have to do the legwork of data collection. This can be a time-consuming and tedious process. In the case of the lady tasting tea, the scientists decided to present her with ten cups of tea which were quickly prepared. A study of public opinion may require many thousands of phone calls or personal interviews. In a laboratory setting, each measurement might be the result of a carefully performed laboratory experiment.

(6) Organize the data.

Once the data have been collected, it is often necessary or useful to organize them. Data are typically stored in spreadsheets or in other formats that are convenient for processing with statistical packages. Very large data sets are often stored in databases.

Part of the organization of the data may involve producing graphical and numerical summaries of the data. We will discuss some of the most important of these kinds of summaries in Chapter 1. These summaries may give us initial insights into our questions or help us detect errors that may have occurred to this point.

(7) Draw conclusions from data.

Once the data have been collected, organized, and analyzed, we need to reach a conclusion. Do we believe the woman's claim? Or do we think she is merely guessing? How sure are we that this conclusion is correct?

Eventually we will learn a number of important and frequently used methods for drawing inferences from data. More importantly, we will learn the basic framework used for such procedures so that it should become easier and easier to learn new procedures as we become familiar with the framework.

(8) Produce a report.

Typically the results of a statistical study are reported in some manner. This may be as a refereed article in an academic journal, as an internal report to a company, or as a solution to a problem on a homework assignment. These reports may themselves be further distilled into press releases, newspaper articles, advertisements, and the like. The mark of a good report is that it provides the essential information about each of the steps of the study.

As we go along, we will learn some of the standard terminology and procedures that you are likely to see in basic statistical reports and will gain a framework for learning more.

At this point, you may be wondering who the innovative scientist was and what the results of the experiment were. The scientist was R. A. Fisher, who first described this situation as a pedagogical example in his 1925 book on statistical methodology [**Fis25**]. We'll return to this example in Sections 2.4.1 and 2.7.3.

## Chapter 1

Summarizing Data

It is a capital mistake to theorize before one has data. Insensibly one begins to twist facts to suit theories, instead of theories to suit facts.

Sherlock Holmes [**Doy27**]

Graphs are essential to good statistical analysis.

F. J. Anscombe [**Ans73**]

Data are the raw material of statistics.

We will organize data into a 2-dimensional schema, which we can think of as rows and columns in a spreadsheet. The rows correspond to the **individuals** (also called **cases**, **subjects**, or **units** depending on the context of the study). The columns correspond to variables. In statistics, a **variable** is one of the measurements made for each individual. Each individual has a **value** for each variable. Or at least that is our intent. Very often some of the data are **missing**, meaning that values of some variables are not available for some of the individuals.

How data are collected is critically important, and good statistical analysis requires that the data were collected in an appropriate manner. We will return to the issue of how data are (or should be) collected later. In this chapter we will focus on the data themselves. We will use R to manipulate data and to produce some of the most important numerical and graphical summaries of data. A more complete introduction to R can be found in Appendix A.

1.1. Data in R

Most data sets in R are stored in a structure called a `data frame` that reflects the 2-dimensional structure described above. A number of data sets are included with the basic installation of R. The `iris` data set, for example, is a famous data set containing a number of physical measurements of three varieties of iris. These data were published by Edgar Anderson in 1935 [**And35**] but are famous because R. A. Fisher [**Fis36**] gave a statistical analysis of these data that appeared a year later.

The `str()` function provides our first overview of the data set.

```
> str(iris)                                                              iris-str
'data.frame':           150 obs. of  5 variables:
$ Sepal.Length:num 5.1 4.9 4.7 4.6 5 5.4 4.6 5 4.4 4.9 ...
$ Sepal.Width :num 3.5 3 3.2 3.1 3.6 3.9 3.4 3.4 2.9 3.1 ...
$ Petal.Length:num 1.4 1.4 1.3 1.5 1.4 1.7 1.4 1.5 1.4 1.5 ...
$ Petal.Width :num 0.2 0.2 0.2 0.2 0.2 0.4 0.3 0.2 0.2 0.1 ...
$ Species :Factor w/ 3 levels "setosa","versicolor",..: 1 1 1 1 1 1 1
    1 1 1 ...
```

From this output we learn that our data set has 150 observations (rows) and 5 variables (columns). Also displayed is some information about the type of data stored in each variable and a few sample values.

While we could print the entire data frame to the screen, this is inconvenient for large data sets. We can look at the first few or last few rows of the data set using `head()` and `tail()`. This is enough to give us a feel for how the data look.

```
> head(iris,n=3)              # first three fows                        iris-head
  Sepal.Length Sepal.Width Petal.Length Petal.Width Species
1          5.1         3.5          1.4         0.2  setosa
2          4.9         3.0          1.4         0.2  setosa
3          4.7         3.2          1.3         0.2  setosa
```

```
> tail(iris,n=3)              # last three rows                         iris-tail
    Sepal.Length Sepal.Width Petal.Length Petal.Width   Species
148          6.5         3.0          5.2         2.0 virginica
149          6.2         3.4          5.4         2.3 virginica
150          5.9         3.0          5.1         1.8 virginica
```

We can access any subset we want by directly specifying which rows and columns are of interest to us.

```
> iris[c(1:3,148:150),3:5]  # first and last rows, only 3 columns       iris-subset
    Petal.Length Petal.Width   Species
1            1.4         0.2    setosa
2            1.4         0.2    setosa
3            1.3         0.2    setosa
148          5.2         2.0 virginica
149          5.4         2.3 virginica
150          5.1         1.8 virginica
```

It is also possible to look at just one variable using the `$` operator.

```
> iris$Sepal.Length    # get one variable and print as vector           iris-vector
 [1] 5.1 4.9 4.7 4.6 5.0 5.4 4.6 5.0 4.4 4.9 5.4 4.8 4.8 4.3 5.8 5.7
[17] 5.4 5.1 5.7 5.1 5.4 5.1 4.6 5.1 4.8 5.0 5.0 5.2 5.2 4.7 4.8 5.4
< 5 lines removed >
[113] 6.8 5.7 5.8 6.4 6.5 7.7 7.7 6.0 6.9 5.6 7.7 6.3 6.7 7.2 6.2 6.1
```

Box 1.1. Using the snippet( ) function

If you have installed the `fastR` package (and any other additional packages that may be needed for a particular example), you can execute the code from this book on your own computer using `snippet()`. For example,

```
snippet('iris-str')
```

will both display and execute the first code block on page 2, and

```
snippet('iris-str', exec=FALSE)
```

will display the code without executing it. Keep in mind that some code blocks assume that prior blocks have already been executed and will not work as expected if this is not true.

```
[129] 6.4 7.2 7.4 7.9 6.4 6.3 6.1 7.7 6.3 6.4 6.0 6.9 6.7 6.9 5.8 6.8
[145] 6.7 6.7 6.3 6.5 6.2 5.9
```

```
> iris$Species          # get one variable and print as vector    iris-vector2
  [1] setosa       setosa      setosa      setosa      setosa
  [6] setosa       setosa      setosa      setosa      setosa
 [11] setosa       setosa      setosa      setosa      setosa
< 19 lines removed >
[111] virginica  virginica  virginica  virginica  virginica
[116] virginica  virginica  virginica  virginica  virginica
[121] virginica  virginica  virginica  virginica  virginica
[126] virginica  virginica  virginica  virginica  virginica
[131] virginica  virginica  virginica  virginica  virginica
[136] virginica  virginica  virginica  virginica  virginica
[141] virginica  virginica  virginica  virginica  virginica
[146] virginica  virginica  virginica  virginica  virginica
Levels: setosa versicolor virginica
```

This is not a particularly good way to get a feel for data. There are a number of graphical and numerical summaries of a variable or set of variables that are usually preferred to merely listing all the values – especially if the data set is large. That is the topic of our next section.

It is important to note that the name `iris` is not reserved in R for this data set. There is nothing to prevent you from storing something else with that name. If you do, you will no longer have access to the `iris` data set unless you first reload it, at which point the previous contents of `iris` are lost.

```
> iris <- 'An iris is a beautiful flower.'               iris-reload
> str(iris)
 chr "An iris is a beautiful flower."
> data(iris)              # explicitly reload the data set
> str(iris)
'data.frame':      150 obs. of  5 variables:
 $ Sepal.Length:num 5.1 4.9 4.7 4.6 5 5.4 4.6 5 4.4 4.9 ...
 $ Sepal.Width :num 3.5 3 3.2 3.1 3.6 3.9 3.4 3.4 2.9 3.1 ...
```

```
$ Petal.Length:num 1.4 1.4 1.3 1.5 1.4 1.7 1.4 1.5 1.4 1.5 ...
$ Petal.Width :num 0.2 0.2 0.2 0.2 0.2 0.4 0.3 0.2 0.2 0.1 ...
$ Species :Factor w/ 3 levels "setosa","versicolor",..: 1 1 1 1 1 1 1
   1 1 1 ...
```

The `fastR` package includes data sets and other utilities to accompany this text. Instructions for installing `fastR` appear in the preface. We will use data sets from a number of other R packages as well. These include the CRAN packages `alr3`, `car`, `DAAG`, `Devore6`, `faraway`, `Hmisc`, `MASS`, and `multcomp`. Appendix A includes instructions for reading data from various file formats, for entering data manually, for obtaining documentation on R functions and data sets, and for installing packages from CRAN.

1.2. Graphical and Numerical Summaries of Univariate Data

Now that we can get our hands on some data, we would like to develop some tools to help us understand the **distribution** of a variable in a data set. By *distribution* we mean answers to two questions:

- What values does the variable take on?
- With what frequency?

Simply listing all the values of a variable is not an effective way to describe a distribution unless the data set is quite small. For larger data sets, we require some better methods of summarizing a distribution.

1.2.1. Tabulating Data

The types of summaries used for a variable depend on the kind of variable we are interested in. Some variables, like `iris$Species`, are used to put individuals into categories. Such variables are called **categorical** (or **qualitative**) variables to distinguish them from **quantitative** variables which have numerical values on some numerically meaningful scale. `iris$Sepal.Length` is an example of a quantitative variable.

Usually the categories are either given descriptive names (our preference) or numbered consecutively. In R, a categorical variable is usually stored as a factor. The possible categories of an R factor are called levels, and you can see in the output above that R not only lists out all of the values of `iris$species` but also provides a list of all the possible levels for this variable. A more useful summary of a categorical variable can be obtained using the `table()` function.

```
> table(iris$Species)  # make a table of values          iris-table

  setosa versicolor  virginica
      50         50         50
```

From this we can see that there were 50 of each of three species of iris.

Tables can be used for quantitative data as well, but often this does not work as well as it does for categorical data because there are too many categories.

```
> table(iris$Sepal.Length)   # make a table of values
```
`iris-table2`

```
4.3 4.4 4.5 4.6 4.7 4.8 4.9   5 5.1 5.2 5.3 5.4 5.5 5.6 5.7 5.8 5.9
  1   3   1   4   2   5   6  10   9   4   1   6   7   6   8   7   3
  6 6.1 6.2 6.3 6.4 6.5 6.6 6.7 6.8 6.9   7 7.1 7.2 7.3 7.4 7.6 7.7
  6   6   4   9   7   5   2   8   3   4   1   1   3   1   1   1   4
7.9
  1
```

Sometimes we may prefer to divide our quantitative data into two groups based on a threshold or some other boolean test.

```
> table(iris$Sepal.Length > 6.0)
```
`iris-logical`

```
FALSE   TRUE
   89     61
```

The `cut()` function provides a more flexible way to build a table from quantitative data.

```
> table(cut(iris$Sepal.Length,breaks=2:10))
```
`iris-cut`

```
 (2,3]  (3,4]  (4,5]  (5,6]  (6,7]  (7,8]  (8,9] (9,10]
     0      0     32     57     49     12      0      0
```

The `cut()` function partitions the data into sections, in this case with break points at each integer from 2 to 10. (The `breaks` argument can be used to set the break points wherever one likes.) The result is a categorical variable with levels describing the interval in which each original quantitative value falls. If we prefer to have the intervals closed on the other end, we can achieve this using `right=FALSE`.

```
> table(cut(iris$Sepal.Length,breaks=2:10,right=FALSE))
```
`iris-cut2`

```
 [2,3)  [3,4)  [4,5)  [5,6)  [6,7)  [7,8)  [8,9) [9,10)
     0      0     22     61     54     13      0      0
```

Notice too that it is possible to define factors in R that have levels that do not occur. This is why the 0's are listed in the output of `table()`. See `?factor` for details.

A tabular view of data like the example above can be converted into a visual representation called a histogram. There are two R functions that can be used to build a histogram: `hist()` and `histogram()`. `hist()` is part of core R. `histogram()` can only be used after first loading the `lattice` graphics package, which now comes standard with all distributions of R. Default versions of each are depicted in Figure 1.1. A number of arguments can be used to modify the resulting plot, set labels, choose break points, and the like.

Looking at the plots generated by `histogram()` and `hist()`, we see that they use different scales for the vertical axis. The default for `histogram()` is to use percentages (of the entire data set). By contrast, `hist()` uses counts. The shapes of

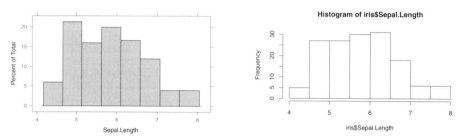

Figure 1.1. Comparing two histogram functions: `histogram()` on the left, and `hist()` on the right.

the two histograms differ because they use slightly different algorithms for choosing the default break points. The user can, of course, override the default break points (using the `breaks` argument). There is a third scale, called the density scale, that is often used for the vertical axis. This scale is designed so that the *area* of each bar is equal to the proportion of data it represents. This is especially useful for histograms that have **bins** (as the intervals between break points are typically called in the context of histograms) of different widths. Figure 1.2 shows an example of such a histogram generated using the following code:

```
> histogram(~Sepal.Length,data=iris,type="density",
+     breaks=c(4,5,5.5,6,6.5,7,8,10))
```
iris-histo-density

We will generally use the newer `histogram()` function because it has several nice features. One of these is the ability to split up a plot into subplots called panels. For example, we could build a separate panel for each species in the iris data set. Figure 1.2 suggests that part of the variation in sepal length is associated with the differences in species. Setosa are generally shorter, virginica longer, and versicolor intermediate. The right-hand plot in Figure 1.2 was created using

```
> histogram(~Sepal.Length|Species,data=iris)
```
iris-condition

If we only want to see the data from one species, we can select a subset of the data using the `subset` argument.

```
> histogram(~Sepal.Length|Species,data=iris,
+     subset=Species=="virginica")
```
iris-histo-subset

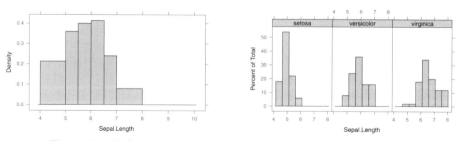

Figure 1.2. Left: A density histogram of sepal length using unequal bin widths. Right: A histogram of sepal length by species.

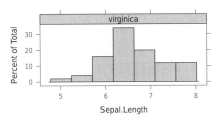

Figure 1.3. This histogram is the result of selecting a subset of the data using the `subset` argument.

By keeping the `groups` argument, our plot will continue to have a strip at the top identifying the species even though there will only be one panel in our plot (Figure 1.3).

The `lattice` graphing functions all use a similar formula interface. The generic form of a formula is

```
y ~ x | z
```

which can often be interpreted as "y modeled by x conditioned on z". For plotting, y will typically indicate a variable presented on the vertical axis, and x a variable to be plotted along the horizontal axis. In the case of a histogram, the values for the vertical axis are computed from the x variable, so y is omitted. The condition z is a variable that is used to break the data into sections which are plotted in separate panels. When z is categorical, there is one panel for each level of z. When z is quantitative, the data is divided into a number of sections based on the values of z. This works much like the `cut()` function, but some data may appear in more than one panel. In R terminology, each panel represents a shingle of the data. The term shingle is supposed to evoke an image of overlapping coverage like the shingles on a roof. Finer control over the number of panels can be obtained by using `equal.count()` or `co.intervals()` to make the shingles directly. See Figure 1.4.

1.2.2. Shapes of Distributions

A histogram gives a shape to a distribution, and distributions are often described in terms of these shapes. The exact shape depicted by a histogram will depend not only on the data but on various other choices, such as how many bins are used, whether the bins are equally spaced across the range of the variable, and just where the divisions between bins are located. But *reasonable* choices of these arguments will usually lead to histograms of similar shape, and we use these shapes to describe the underlying distribution as well as the histogram that represents it.

Some distributions are approximately **symmetrical** with the distribution of the larger values looking like a mirror image of the distribution of the smaller values. We will call a distribution **positively skewed** if the portion of the distribution with larger values (the right of the histogram) is more spread out than the other side. Similarly, a distribution is **negatively skewed** if the distribution deviates from symmetry in the opposite manner. Later we will learn a way to measure

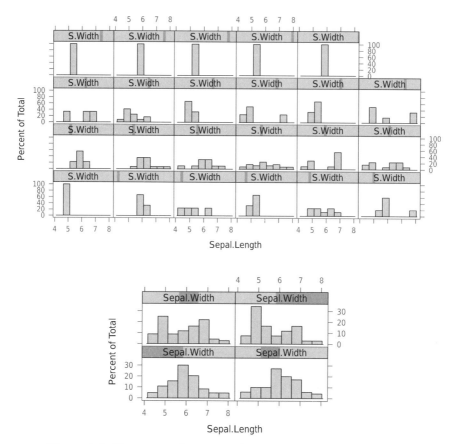

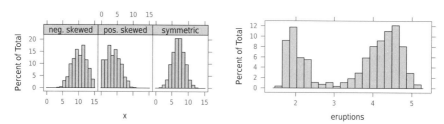

Figure 1.4. The output of `histogram(~Sepal.Length|Sepal.Width,iris)` and `histogram(~Sepal.Length|equal.count(Sepal.Width,number=4),iris)`.

Figure 1.5. Left: Skewed and symmetric distributions. Right: Old Faithful eruption times illustrate a bimodal distribution.

the degree and direction of skewness with a number; for now it is sufficient to describe distributions qualitatively as symmetric or skewed. See Figure 1.5 for some examples of symmetric and skewed distributions.

Notice that each of these distributions is clustered around a center where most of the values are located. We say that such distributions are **unimodal**. Shortly we will discuss ways to summarize the location of the "center" of unimodal distributions numerically. But first we point out that some distributions have other shapes that are not characterized by a strong central tendency. One famous example is eruption times of the Old Faithful geyser in Yellowstone National park.

```
> plot <- histogram(~eruptions,faithful,n=20)
```
faithful-histogram

produces the histogram in Figure 1.5 which shows a good example of a **bimodal** distribution. There appear to be two groups or kinds of eruptions, some lasting about 2 minutes and others lasting between 4 and 5 minutes.

1.2.3. Measures of Central Tendency

Qualitative descriptions of the shape of a distribution are important and useful. But we will often desire the precision of numerical summaries as well. Two aspects of unimodal distributions that we will often want to measure are central tendency (what is a typical value? where do the values cluster?) and the amount of variation (are the data tightly clustered around a central value, or more spread out?).

Two widely used measures of center are the **mean** and the **median**. You are probably already familiar with both. The mean is calculated by adding all the values of a variable and dividing by the number of values. Our usual notation will be to denote the n values as $x_1, x_2, \ldots x_n$, and the mean of these values as $\overline{x}$. Then the formula for the mean becomes

$$\overline{x} = \frac{\sum_{i=1}^{n} x_i}{n} \; .$$

The median is a value that splits the data in half – half of the values are smaller than the median and half are larger. By this definition, there could be more than one median (when there are an even number of values). This ambiguity is removed by taking the mean of the "two middle numbers" (after sorting the data). See the exercises for some problems that explore aspects of the mean and median that may be less familiar.

The mean and median are easily computed in R. For example,

```
> mean(iris$Sepal.Length); median(iris$Sepal.Length)
[1] 5.8433
[1] 5.8
```
iris-mean-median

Of course, we have already seen (by looking at histograms), that there are some differences in sepal length between the various species, so it would be better to compute the mean and median separately for each species. While one can use the built-in `aggregate()` function, we prefer to use the `summary()` function from the `Hmisc` package. This function uses the same kind of formula notation that the `lattice` graphics functions use.

```
> require(Hmisc) # load Hmisc package
> summary(Sepal.Length~Species,iris)          # default function is mean
Sepal.Length    N=150
```
iris-Hmisc-summary

Box 1.2. R packages used in this text

From now on we will assume that the `lattice`, `Hmisc`, and `fastR` packages have been loaded and will not show the loading of these packages in our examples. If you try an example in this book and R reports that it cannot find a function or data set, it is likely that you have failed to load one of these packages. You can set up R to automatically load these packages every time you launch R if you like. (See Appendix A for details.)

Other packages will be used from time to time as well. In this case, we will show the `require()` statement explicitly. The documentation for the `fastR` package includes a list of required and recommended packages.

```
+-------+----------+---+------------+
|       |          |N  |Sepal.Length|
+-------+----------+---+------------+
|Species|setosa    | 50|5.0060      |
|       |versicolor| 50|5.9360      |
|       |virginica | 50|6.5880      |
+-------+----------+---+------------+
|Overall|          |150|5.8433      |
+-------+----------+---+------------+
> summary(Sepal.Length~Species,iris,fun=median)  # median instead
Sepal.Length    N=150

+-------+----------+---+------------+
|       |          |N  |Sepal.Length|
+-------+----------+---+------------+
|Species|setosa    | 50|5.0         |
|       |versicolor| 50|5.9         |
|       |virginica | 50|6.5         |
+-------+----------+---+------------+
|Overall|          |150|5.8         |
+-------+----------+---+------------+
```

Comparing with the histograms in Figure 1.2, we see that these numbers are indeed good descriptions of the center of the distribution for each species.

We can also compute the mean and median of the Old Faithful eruption times.

```
> mean(faithful$eruptions)
[1] 3.4878
> median(faithful$eruptions)
[1] 4
```

faithful-mean-median

Notice, however, that in the Old Faithful eruption times histogram (Figure 1.5) there are very few eruptions that last between 3.5 and 4 minutes. So although these numbers are the mean and median, neither is a very good description of the typical eruption time(s) of Old Faithful. It will often be the case that the mean and median are not very good descriptions of a data set that is not unimodal.

```
> stem(faithful$eruptions)
```

```
  The decimal point is 1 digit(s) to the left of the |

  16 | 070355555588
  18 | 00002223333333557777777788882335777888
  20 | 00002223378800035778
  22 | 0002335578023578
  24 | 00228
  26 | 23
  28 | 080
  30 | 7
  32 | 2337
  34 | 250077
  36 | 0000823577
  38 | 2333335582225577
  40 | 000000335778888800223355555777778
  42 | 03335555778800233333555577778
  44 | 022223355577800000000023333357778888
  46 | 00002333577000000023578
  48 | 00000022335800333
  50 | 0370
```

Figure 1.6. Stemplot of Old Faithful eruption times using `stem()`.

In the case of our Old Faithful data, there seem to be two predominant peaks, but unlike in the case of the iris data, we do not have another variable in our data that lets us partition the eruption times into two corresponding groups. This observation could, however, lead to some hypotheses about Old Faithful eruption times. Perhaps eruption times at night are different from those during the day. Perhaps there are other differences in the eruptions. Subsequent data collection (and statistical analysis of the resulting data) might help us determine whether our hypotheses appear correct.

One disadvantage of a histogram is that the actual data values are lost. For a large data set, this is probably unavoidable. But for more modestly sized data sets, a stemplot can reveal the shape of a distribution without losing the actual (perhaps rounded) data values. A stemplot divides each value into a stem and a leaf at some place value. The leaf is rounded so that it requires only a single digit. The values are then recorded as in Figure 1.6.

From this output we can readily see that the shortest recorded eruption time was 1.60 minutes. The second 0 in the first row represents 1.70 minutes. Note that the output of `stem()` can be ambiguous when there are not enough data values in a row.

Comparing mean and median

Why bother with two different measures of central tendency? The short answer is that they measure different things. If a distribution is (approximately) symmetric, the mean and median will be (approximately) the same (see Exercise 1.5). If the distribution is not symmetric, however, the mean and median may be very different, and one measure may provide a more useful summary than the other.

For example, if we begin with a symmetric distribution and add in one additional value that is very much larger than the other values (an **outlier**), then the median will not change very much (if at all), but the mean will increase substantially. We say that the median is **resistant** to outliers while the mean is not. A similar thing happens with a skewed, unimodal distribution. If a distribution is positively skewed, the large values in the tail of the distribution increase the mean (as compared to a symmetric distribution) but not the median, so the mean will be larger than the median. Similarly, the mean of a negatively skewed distribution will be smaller than the median.

Whether a resistant measure is desirable or not depends on context. If we are looking at the income of employees of a local business, the median may give us a much better indication of what a typical worker earns, since there may be a few large salaries (the business owner's, for example) that inflate the mean. This is also why the government reports median household income and median housing costs.

On the other hand, if we compare the median and mean of the value of raffle prizes, the mean is probably more interesting. The median is probably 0, since typically the majority of raffle tickets do not win anything. This is independent of the values of any of the prizes. The mean will tell us something about the overall value of the prizes involved. In particular, we might want to compare the mean prize value with the cost of the raffle ticket when we decide whether or not to purchase one.

The trimmed mean compromise

There is another measure of central tendency that is less well known and represents a kind of compromise between the mean and the median. In particular, it is more sensitive to the extreme values of a distribution than the median is, but less sensitive than the mean. The idea of a **trimmed mean** is very simple. Before calculating the mean, we remove the largest and smallest values from the data. The percentage of the data removed from each end is called the trimming percentage. A 0% trimmed mean is just the mean; a 50% trimmed mean is the median; a 10% trimmed mean is the mean of the middle 80% of the data (after removing the largest and smallest 10%). A trimmed mean is calculated in R by setting the `trim` argument of `mean()`, e.g., `mean(x,trim=0.10)`. Although a trimmed mean in some sense combines the advantages of both the mean and median, it is less common than either the mean or the median. This is partly due to the mathematical theory that has been developed for working with the median and especially the mean of sample data.

1.2.4. Measures of Dispersion

It is often useful to characterize a distribution in terms of its center, but that is not the whole story. Consider the distributions depicted in the histograms in Figure 1.7. In each case the mean and median are approximately 10, but the distributions clearly have very different shapes. The difference is that distribution B is much more "spread out". "Almost all" of the data in distribution A is quite close to 10; a much larger proportion of distribution B is "far away" from 10. The intuitive (and not very precise) statement in the preceding sentence can be quantified by means of **quantiles**. The idea of quantiles is probably familiar to you since **percentiles** are a special case of quantiles.

Definition 1.2.1 (Quantile). Let $p \in [0, 1]$. A p-**quantile** of a quantitative distribution is a number q such that the (approximate) proportion of the distribution that is less than q is p.

So, for example, the 0.2-quantile divides a distribution into 20% below and 80% above. This is the same as the 20th percentile. The median is the 0.5-quantile (and the 50th percentile).

The idea of a quantile is quite straightforward. In practice there are a few wrinkles to be ironed out. Suppose your data set has 15 values. What is the 0.30-quantile? Exactly 30% of the data would be $(0.30)(15) = 4.5$ values. Of course, there is no number that has 4.5 values below it and 11.5 values above it. This is the reason for the parenthetical word *approximate* in Definition 1.2.1. Different schemes have been proposed for giving quantiles a precise value, and R implements several such methods. They are similar in many ways to the decision we had to make when computing the median of a variable with an even number of values.

Two important methods can be described by imagining that the sorted data have been placed along a ruler, one value at every unit mark and also at each end. To find the p-quantile, we simply snap the ruler so that proportion p is to the left and $1-p$ to the right. If the break point happens to fall precisely where a data value is located (i.e., at one of the unit marks of our ruler), that value is the p-quantile. If the break point is between two data values, then the p-quantile is a weighted mean of those two values.

Example 1.2.1. Suppose we have 10 data values: $1, 4, 9, 16, 25, 36, 49, 64, 81, 100$. The 0-quantile is 1, the 1-quantile is 100, the 0.5-quantile (median) is midway between 25 and 36, that is, 30.5. Since our ruler is 9 units long, the 0.25-quantile is located $9/4 = 2.25$ units from the left edge. That would be one quarter of the way from 9 to 16, which is $9 + 0.25(16 - 9) = 9 + 1.75 = 10.75$. (See Figure 1.8.) Other quantiles are found similarly. This is precisely the default method used by `quantile()`.

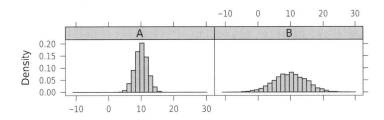

Figure 1.7. Histograms showing smaller (A) and larger (B) amounts of variation.

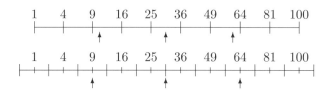

Figure 1.8. Illustrations of two methods for determining quantiles from data. Arrows indicate the locations of the 0.25-, 0.5-, and 0.75-quantiles.

```
> quantile((1:10)^2)                                     intro-quantile
    0%   25%    50%   75%   100%
  1.00 10.75  30.50 60.25 100.00                                    ◁
```

A second scheme is just like the first one except that the data values are placed midway between the unit marks. In particular, this means that the 0-quantile is not the smallest value. This could be useful, for example, if we imagined we were trying to estimate the lowest value in a population from which we only had a sample. Probably the lowest value overall is less than the lowest value in our particular sample. The only remaining question is how to extrapolate in the last half unit on either side of the ruler. If we set quantiles in that range to be the minimum or maximum, the result is another type of `quantile()`.

Example 1.2.2. The method just described is what `type=5` does.

```
> quantile((1:10)^2,type=5)                              intro-quantile05a
   0%  25%   50%  75%   100%
  1.0   9.0  30.5 64.0 100.0
```

Notice that quantiles below the 0.05-quantile are all equal to the minimum value.

```
> quantile((1:10)^2,type=5,seq(0,0.10,by=0.005))         intro-quantile05b
  0% 0.5%   1% 1.5%   2% 2.5%   3% 3.5%   4% 4.5%   5% 5.5%   6% 6.5%
1.00 1.00 1.00 1.00 1.00 1.00 1.00 1.00 1.00 1.00 1.15 1.30 1.45
  7% 7.5%   8% 8.5%   9% 9.5%  10%
1.60 1.75 1.90 2.05 2.20 2.35 2.50
```

A similar thing happens with the maximum value for the larger quantiles.   ◁

Other methods refine this idea in other ways, usually based on some assumptions about what the population of interest is like.

Fortunately, for large data sets, the differences between the different quantile methods are usually unimportant, so we will just let R compute quantiles for us using the `quantile()` function. For example, here are the **deciles** and **quartiles** of the Old Faithful eruption times.

```
> quantile(faithful$eruptions,(0:10)/10)                 faithful-quantile
    0%    10%    20%    30%    40%    50%    60%    70%    80%    90%
1.6000 1.8517 2.0034 2.3051 3.6000 4.0000 4.1670 4.3667 4.5330 4.7000
   100%
5.1000
> quantile(faithful$eruptions,(0:4)/4)
```

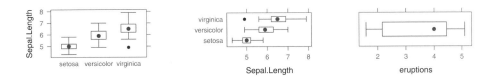

Figure 1.9. Boxplots for iris sepal length and Old Faithful eruption times.

```
     0%    25%    50%    75%   100%
1.6000 2.1627 4.0000 4.4543 5.1000
```

The latter of these provides what is commonly called the **five-number summary**. The 0-quantile and 1-quantile (at least in the default scheme) are the minimum and maximum of the data set. The 0.5-quantile gives the median, and the 0.25- and 0.75-quantiles (also called the first and third quartiles) isolate the middle 50% of the data. When these numbers are close together, then most (well, half, to be more precise) of the values are near the median. If those numbers are farther apart, then much (again, half) of the data is far from the center. The difference between the first and third quartiles is called the **interquartile range** and is abbreviated **IQR**. This is our first numerical measure of dispersion.

The five-number summary can also be presented graphically using a **boxplot** (also called box-and-whisker plot) as in Figure 1.9. These plots were generated using

iris-bwplot

```
> bwplot(Sepal.Length~Species,data=iris)
> bwplot(Species~Sepal.Length,data=iris)
> bwplot(~eruptions,faithful)
```

The size of the box reflects the IQR. If the box is small, then the middle 50% of the data are near the median, which is indicated by a dot in these plots. (Some boxplots, including those made by the `boxplot()` use a vertical line to indicate the median.) Outliers (values that seem unusually large or small) can be indicated by a special symbol. The whiskers are then drawn from the box to the largest and smallest non-outliers. One common rule for automating outlier detection for boxplots is the 1.5 **IQR rule**. Under this rule, any value that is more than 1.5 IQR away from the box is marked as an outlier. Indicating outliers in this way is useful since it allows us to see if the whisker is long only because of one extreme value.

Variance and standard deviation

Another important way to measure the dispersion of a distribution is by comparing each value with the mean of the distribution. If the distribution is spread out, these differences will tend to be large; otherwise these differences will be small. To get a single number, we could simply add up all of the **deviation from the mean**:

$$\text{total deviation from the mean} = \sum(x - \bar{x}) \,.$$

The trouble with this is that the total deviation from the mean is always 0 because the negative deviations and the positive deviations always exactly cancel out. (See Exercise 1.10).

To fix this problem, we might consider taking the absolute value of the deviations from the mean:

$$\text{total absolute deviation from the mean} = \sum |x - \overline{x}| .$$

This number will only be 0 if all of the data values are equal to the mean. Even better would be to divide by the number of data values:

$$\text{mean absolute deviation} = \frac{1}{n} \sum |x - \overline{x}| .$$

Otherwise large data sets will have large sums even if the values are all close to the mean. The mean absolute deviation is a reasonable measure of the dispersion in a distribution, but we will not use it very often. There is another measure that is much more common, namely the **variance**, which is defined by

$$\text{variance} = \text{Var}(x) = \frac{1}{n-1} \sum (x - \overline{x})^2 .$$

You will notice two differences from the mean absolute deviation. First, instead of using an absolute value to make things positive, we square the deviations from the mean. The chief advantage of squaring over the absolute value is that it is much easier to do calculus with a polynomial than with functions involving absolute values. Because the squaring changes the units of this measure, the square root of the variance, called the **standard deviation**, is commonly used in place of the variance.

The second difference is that we divide by $n-1$ instead of by n. There is a very good reason for this, even though dividing by n probably would have felt much more natural to you at this point. We'll get to that very good reason later in the course (in Section 4.6). For now, we'll settle for a less good reason. If you know the mean and all but one of the values of a variable, then you can determine the remaining value, since the sum of all the values must be the product of the number of values and the mean. So once the mean is known, there are only $n-1$ independent pieces of information remaining. That is not a particularly satisfying explanation, but it should help you remember to divide by the correct quantity.

All of these quantities are easy to compute in R.

```
> x=c(1,3,5,5,6,8,9,14,14,20)                                    intro-dispersion02
>
> mean(x)
[1] 8.5
> x - mean(x)
 [1] -7.5 -5.5 -3.5 -3.5 -2.5 -0.5  0.5  5.5  5.5 11.5
> sum(x - mean(x))
[1] 0
> abs(x - mean(x))
 [1]  7.5  5.5  3.5  3.5  2.5  0.5  0.5  5.5  5.5 11.5
> sum(abs(x - mean(x)))
[1] 46
```

```
> (x - mean(x))^2
 [1]  56.25  30.25  12.25  12.25   6.25   0.25   0.25  30.25  30.25
[10] 132.25
> sum((x - mean(x))^2)
[1] 310.5
> n= length(x)
> 1/(n-1) * sum((x - mean(x))^2)
[1] 34.5
> var(x)
[1] 34.5
> sd(x)
[1] 5.8737
> sd(x)^2
[1] 34.5
```

1.3. Graphical and Numerical Summaries of Multivariate Data

1.3.1. Side-by-Side Comparisons

Often it is useful to consider two or more variables together. In fact, we have already done some of this. For example, we looked at iris sepal length separated by species. This sort of side-by-side comparison – in graphical or tabular form – is especially useful when one variable is quantitative and the other categorical. Graphical or numerical summaries of the quantitative variable can be made separately for each group defined by the categorical variable (or by shingles of a second quantitative variable). See Appendix A for more examples.

1.3.2. Scatterplots

There is another plot that is useful for looking at the relationship between two quantitative variables. A **scatterplot** (or scattergram) is essentially the familiar Cartesian coordinate plot you learned about in school. Since each observation in a bivariate data set has two values, we can plot points on a rectangular grid representing both values simultaneously. The `lattice` function for making a scatterplot is `xyplot()`.

The scatterplot in Figure 1.10 becomes even more informative if we separate the dots of the three species. Figure 1.11 shows two ways this can be done. The first uses a conditioning variable, as we have seen before, to make separate panels for each species. The second uses the `groups` argument to plot the data in the same panel but with different symbols for each species. Each of these clearly indicates that, in general, plants with wider sepals also have longer sepals but that the typical values of and the relationship between width and length differ by species.

1.3.3. Two-Way Tables and Mosaic Plots

A 1981 paper [**Rad81**] investigating racial biases in the application of the death penalty reported on 326 cases in which the defendant was convicted of murder.

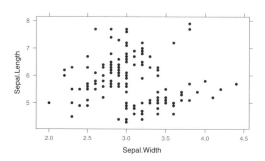

Figure 1.10. A scatterplot made with `xyplot(Sepal.Length~Sepal.Width,iris)`.

For each case they noted the race of the defendant and whether or not the death penalty was imposed. We can use R to cross tabulate this data for us:

intro-deathPenalty01

```
> xtabs(~Penalty+Victim,data=deathPenalty)
        Victim
Penalty Black White
   Death     6    30
   Not     106   184
```

Perhaps you are surprised that white defendants are more likely to receive the death penalty. It turns out that there is more to the story. The researchers also recorded the race of the victim. If we make a new table that includes this information, we see something interesting.

intro-deathPenalty02

```
> xtabs(~Penalty+Defendant+Victim,
+        data=deathPenalty)
, , Victim = Black

           Defendant
Penalty Black White
   Death     6     0
   Not      97     9
```

```
, , Victim = White

           Defendant
Penalty Black White
   Death    11    19
   Not      52   132
```

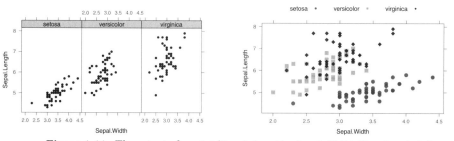

Figure 1.11. The output of `xyplot(Sepal.Length~Sepal.Width|Species,iris)` and `xyplot(Sepal.Length~Sepal.Width,groups=Species,iris,auto.key=TRUE)`.

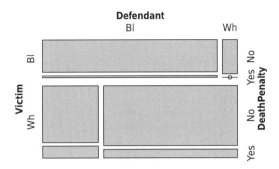

Figure 1.12. A mosaic plot of death penalty by race of defendant and victim.

It appears that black defendants are more likely to receive the death penalty when the victim is white and also when the victim is black, but not if we ignore the race of the victim. This sort of apparent contradiction is known as **Simpson's paradox**. In this case, it appears that the death penalty is more likely to be given for a white victim, and since most victims are the same race as their murderer, the result is that overall white defendants are more likely (in this data set) to receive the death penalty even though black defendants are more likely (again, in this data set) to receive the death penalty for each race of victim.

The fact that our understanding of the data is so dramatically influenced by whether or not our analysis includes the race of the victim is a warning to watch for **lurking variables** – variables that have an important effect but are not included in our analysis – in other settings as well. Part of the design of a good study is selecting the right things to measure.

These cross tables can be visualized graphically using a **mosaic plot**. Mosaic plots can be generated with the core R function `mosaicplot()` or with `mosaic()` from the `vcd` package. (`vcd` is short for visualization of categorical data.) The latter is somewhat more flexible and usually produces more esthetically pleasing output. A number of different formula formats can be supplied to `mosaic()`. The results of the following code are shown in Figure 1.12.

```
> require(vcd)                                           intro-deathPenalty03
> mosaic(~Victim+Defendant+DeathPenalty,data=deathPen)
> structable(~Victim+Defendant+DeathPenalty,data=deathPen)
                      Defendant  Bl   Wh
Victim DeathPenalty
Bl     No                       97    9
       Yes                       6    0
Wh     No                       52  132
       Yes                      11   19
```

As always, see `?mosaic` for more information. The `vcd` package also provides an alternative to `xtabs()` called `structable()`, and if you `print()` a `mosaic()`, you will get both the graph and the table.

1.4. Summary

Data can be thought of in a 2-dimensional structure in which each **variable** has a value (possibly missing) for each **observational unit**. In most statistical software, including R, columns correspond to variables and rows correspond to the observations.

The **distribution** of a variable is a description of the values obtained by a variable and the frequency with which they occur. While simply listing all the values does describe the distribution completely, it is not easy to draw conclusions from this sort of description, especially when the number of observational units is large. Instead, we will make frequent use of numerical and graphical summaries that make it easier to see what is going on and to make comparisons.

The **mean, median, standard deviation**, and **interquartile range** are among the most common numerical summaries. The mean and median give an indication of the "center" of the distribution. They are especially useful for **unimodal** distributions but may not be appropriate summaries for distributions with other shapes. When a distribution is **skewed**, the mean and median can be quite different because the extreme values of the distribution have a large effect on the mean but not on the median. A **trimmed mean** is sometimes used as a compromise between the median and the mean. Although one could imagine other measures of spread, the standard deviation is especially important because of its relationship to important theoretical results in statistics, especially the Central Limit Theorem, which we will encounter in Chapter 4.

Even as we learn formal methods of statistical analysis, we will not abandon these numerical and graphical summaries. Appendix A provides a more complete introduction to R and includes information on how to fine-tune plots. Additional examples can be found throughout the text.

1.4.1. R Commands

Here is a table of important R commands introduced in this chapter. Usage details can be found in the examples and using the R help.

`x <- c(...)`	Concatenate arguments into a single vector and store in object `x`.
`data(x)`	(Re)load the data set `x`.
`str(x)`	Print a summary of the object `x`.
`head(x,n=4)`	First four rows of the data frame `x`.
`tail(x,n=4)`	Last four rows of the data frame `x`.
`table(x)`	Table of the values in vector `x`.
`xtabs(~x+y,data)`	Cross tabulation of `x` and `y`.

`cut(x,breaks,right=TRUE)`	Divide up the range of x into intervals and code the values in x according to which interval they fall into.
`require(fastR);` `require(lattice);` `require(Hmisc)`	Load packages.
`histogram(~x|z,data,...)`	Histogram of x conditioned on z.
`bwplot(x~z,data,...)`	Boxplot of x conditioned on z.
`xyplot(y~x|z,data,...)`	Scatterplot of y by x conditioned on z.
`stem(x)`	Stemplot of x.
`sum(x); mean(x); median(x);` `var(x); sd(x); quantile(x)`	Sum, mean, median, variance, standard deviation, quantiles of x.
`summary(y~x,data,fun)`	Summarize y by computing the function `fun` on each group defined by x [Hmisc].

Exercises

1.1. Read as much of Appendix A as you need to do the exercises there.

1.2. The `pulse` variable in the `littleSurvey` data set contains self-reported pulse rates.

  a) Make a histogram of these values. What problem does this histogram reveal?

  b) Make a decision about what values should be removed from the data and make a histogram of the remaining values. (You can use the `subset` argument of the `histogram()` function to restrict the data or you can create a new vector and make a histogram from that.)

  c) Compute the mean and median of your restricted set of pulse rates.

1.3. The `pulse` variable in the `littleSurvey` data set contains self-reported pulse rates. Make a table or graph showing the distribution of the last digits of the recorded pulse rates and comment on the distribution of these digits. Any conjectures?

Note: `%%` is the modulus operator in R. So x `%%` 10 gives the remainder after dividing x by 10, which is the last digit.

1.4. Some students in introductory statistics courses were asked to select a number between 1 and 30 (inclusive). The results are in the `number` variable in the `littleSurvey` data set.

a) Make a table showing the frequency with which each number was selected using `table()`.

b) Make a histogram of these values with bins centered at the integers from 1 to 30.

c) What numbers were most frequently chosen? Can you get R to find them for you?

d) What numbers were least frequently chosen? Can you get R to find them for you?

e) Make a table showing how many students selected odd versus even numbers.

1.5. The distribution of a quantitative variable is **symmetric about** m if whenever there are k observations with value $m + d$, there are also k observations with value $m - d$. Equivalently, if the values are $x_1 \leq x_2 \leq \cdots \leq x_n$, then $x_i + x_{n+1-i} = 2m$ for all i.

a) Show that if a distribution is symmetric about m, then m is the median. (You may need to handle separately the cases where the number of values is odd and even.)

b) Show that if a distribution is symmetric about m, then m is the mean.

c) Create a small distribution such that the mean and median are equal to m but the distribution is not symmetric about m.

1.6. Describe some situations where the mean or median is clearly a better measure of central tendency than the other.

1.7. Below are histograms and boxplots from six distributions. Match each histogram (A–F) with its corresponding boxplot (U–Z).

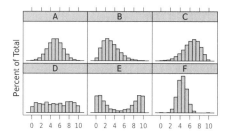 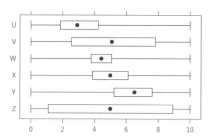

1.8. The function `bwplot()` does not use the `quantile()` function to compute its five-number summary. Instead it uses `fivenum()`. Technically, `fivenum()` computes the **hinges** of the data rather than quantiles. Sometimes `fivenum()` and `quantile()` agree:

```
> fivenum(1:11)
[1]  1.0  3.5  6.0  8.5 11.0
> quantile(1:11)
  0%  25%  50%  75% 100%
 1.0  3.5  6.0  8.5 11.0
```

But sometimes they do not:

```
> fivenum(1:10)
[1]  1.0  3.0  5.5  8.0 10.0
> quantile(1:10)
   0%   25%   50%   75%  100%
 1.00  3.25  5.50  7.75 10.00
```

Compute `fivenum()` on a number of data sets and answer the following questions:

  a) When does `fivenum()` give the same values as `quantile()`?

  b) What method is `fivenum()` using to compute the five numbers?

1.9. Design some data sets to test whether by default `bwplot()` uses the 1.5 IQR rule to determine whether it should indicate data as outliers.

1.10. Show that the total deviation from the mean, defined by

$$\text{total deviation from the mean} = \sum_{i=1}^{n} (x_i - \bar{x}) ,$$

is 0 for any distribution.

1.11. We could compute the mean absolute deviation from the *median* instead of from the mean. Show that the mean absolute deviation from the median is never larger than the mean absolute deviation from the mean.

1.12. We could compute the mean absolute deviation from any number c (c for center). Show that the mean absolute deviation from c is always at least as large as the mean absolute deviation from the median. Thus the median is a minimizer of mean absolute deviation.

1.13. Let $SS(c) = \sum (x_i - c)^2$. (SS stands for sum of squares.) Show that the smallest value of $SS(c)$ occurs when $c = \bar{x}$. This shows that the mean is a minimizer of SS.

1.14. Find a distribution with 10 values between 0 and 10 that has as large a variance as possible.

1.15. Find a distribution with 10 values between 0 and 10 that has as small a variance as possible.

1.16. The `pitching2005` data set in the `fastR` package contains 2005 season statistics for each pitcher in the major leagues. Use graphical and numerical summaries of this data set to explore whether there are differences between the two leagues, restricting your attention to pitchers that started at least 5 games (the variable `GS` stands for 'games started'). You may select the statistics that are of interest to you.

If you are not much of a baseball fan, try using `ERA` (earned run average), which is a measure of how many runs score while a pitcher is pitching. It is measured in runs per nine innings.

1.17. Repeat the previous problem using batting statistics. The `fastR` data set `batting` contains data on major league batters over a large number of years. You may want to restrict your attention to a particular year or set of years.

1.18. Have major league batting averages changed over time? If so, in what ways? Use the data in the `batting` data set to explore this question. Use graphical and numerical summaries to make your case one way or the other.

1.19. The `faithful` data set contains two variables: the duration (`eruptions`) of the eruption and the time until the next eruption (`waiting`).

a) Make a scatterplot of these two variables and comment on any patterns you see.

b) Remove the first value of `eruptions` and the last value of `waiting`. Make a scatterplot of these two vectors.

c) Which of the two scatterplots reveals a tighter relationship? What does that say about the relationship between eruption duration and the interval between eruptions?

1.20. The results of a little survey that has been given to a number of statistics students are available in the `littleSurvey` data set. Make some conjectures about the responses and use R's graphical and numerical summaries to see if there is any (informal) evidence to support your conjectures. See `?littleSurvey` for details about the questions on the survey.

1.21. The `utilities` data set contains information from utilities bills for a personal residence over a number of years. This problem explores gas usage over time.

a) Make a scatterplot of gas usage (`ccf`) vs. time. You will need to combine `month` and `year` to get a reasonable measurement for time. Such a plot is called a **time series plot**.

b) Use the `groups` argument (and perhaps `type=c('p','l')`, too) to make the different months of the year distinguishable in your scatterplot.

c) Now make a boxplot of gas usage (`ccf`) vs. `factor(month)`. Which months are most variable? Which are most consistent?

d) What patterns do you see in the data? Does there appear to be any change in gas usage over time? Which plots help you come to your conclusion?

1.22. Note that March and May of 2000 are outliers due to a bad meter reading. Utility bills come monthly, but the number of days in a billing cycle varies from month to month. Add a new variable to the `utilities` data set using

```
> utilities$ccfpday <- utilities$ccf / utilities$billingDays    utilities-ccfpday
> plot1 <- xyplot( ccfpday ~ (year + month/12), utilities, groups=month )
> plot2 <- bwplot( ccfpday ~ factor(month), utilities )
```

Repeat the previous exercise using `ccfpday` instead of `ccf`. Are there any noticeable differences between the two analyses?

1.23. The `utilities` data set contains information from utilities bills for a personal residence over a number of years. One would expect that the gas bill would be related to the average temperature for the month.

Make a scatterplot showing the relationship between `ccf` (or, better, `ccfpday`; see Exercise 1.22) and `temp`. Describe the overall pattern. Are there any outliers?

1.24. The `utilities` data set contains information from utilities bills for a personal residence over a number of years. The variables `gasbill` and `ccf` contain the gas bill (in dollars) and usage (in 100 cubic feet) for a personal residence. Use plots to explore the cost of gas over the time period covered in the `utilities` data set. Look for both seasonal variation in price and any trends over time.

1.25. The `births78` data set contains the number of births in the United States for each day of 1978.

a) Make a histogram of the number of births. You may be surprised by the shape of the distribution. (Make a stemplot too if you like.)

b) Now make a scatterplot of births vs. day of the year. What do you notice? Can you conjecture any reasons for this?

c) Can you make a plot that will help you see if your conjecture seems correct? (Hint: Use `groups`.)

## Chapter 2

Probability and Random Variables

The excitement that a gambler feels when making a bet is equal to the amount he might win times the probability of winning it.

Blaise Pascal [**Ros88**]

In this chapter we will develop the foundations of **probability**. As an area of mathematics, probability is the study of random processes.[1] Randomness, as we use it, describes a particular type of uncertainty.

Box 2.1. What is randomness?

We will say that a repeatable process is random if its outcome is

- unpredictable in the short run and
- predictable in the long run.

Note that it is the process that is random, not the outcome, even though we often speak as if it were the outcome that is random.

A good example is flipping a coin. The result of any given toss of a fair coin is unpredictable in advance. It could be heads, or it could be tails. We don't know with certainty which it will be. Nevertheless, we can say something about the long-run behavior of flipping a coin many times. This is what makes us surprised if someone flips a coin 20 times and gets heads all 20 times, but not so surprised if the result is 12 heads and 8 tails.

---

[1] It is traditional within the study of probability to refer to random processes as *random experiments*. We will avoid that usage to avoid confusion with the *randomized experiments* – statistical studies where the values of some variables are determined using randomness.

To facilitate talking about randomness more generally, it is useful to introduce some terminology. An **outcome** of a random process is one of the potential results of the process. An **event** is any set of outcomes. Two important events are the empty set (the set of no outcomes, denoted $\emptyset$) and the set of all outcomes, which is called the **sample space**. The **probability of an event** will be a number between 0 and 1 (inclusive) that indicates its relative likelihood of occurring. Things that happen most of the time will be assigned numbers near 1. Things that almost never happen "by chance" will be assigned numbers near 0.

But how do we assign these numbers? There are two important methods for assigning probabilities: the empirical method and the theoretical method. We will look at those in the next section. We will then rapidly turn our attention to a special case, namely when the outcomes of our random process are numbers or can be converted into numbers. Such a random process is called a **random variable**, and just as we did with data, we will develop a number of graphical and numerical tools for studying the distributions of random variables.

In our definitions of probability and the derivations of some basic properties of random variables, we make use of the standard mathematical notation for sets and functions. This notation is reviewed in Appendix B.

2.1. Introduction to Probability

As mentioned in the introduction to this chapter, a probability is a number assigned to an event that represents how likely it is to occur. In this section we will consider two ways of assigning these numbers.

2.1.1. Empirical Probability Calculations

The empirical method for assigning probabilities is straightforward and is based on the fact that we are interested in the long-run behavior of a random process. If we repeat the random process many times, some of the outcomes will belong to our event of interest while others will not. We could then define the probability of the event E to be

$$\mathrm{P}(E) = \frac{\text{number of times outcome was in the event } E}{\text{number of times the random process was repeated}} . \tag{2.1}$$

There is much to be said for this definition. It will certainly give numbers between 0 and 1 since the numerator is never larger than the denominator. Furthermore, events that happen frequently will be assigned large numbers and events which occur infrequently will be assigned small numbers. Nevertheless, (2.1) doesn't make a good definition, at least not in its current form. The problem with (2.1) is that if two different people each repeat the random process and calculate the probability of E, very likely they will get different numbers. So perhaps the following would be a better statement:

$$\mathrm{P}(E) \approx \frac{\text{number of times outcome was in the event } E}{\text{number of times the random process was repeated}} . \tag{2.2}$$

It would be nice if we knew something about how accurate such an approximation is. And in fact, we will be able to say something about that later. But for now, intuitively, we expect such approximations to be better when the number of repetitions is larger. A simulation of 1000 tosses of a fair coin supports this intuition. After each simulated toss, the proportion of heads to that point was calculated. The results are displayed in Figure 2.1.

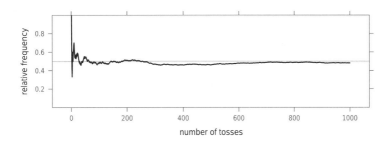

Figure 2.1. Cumulative proportion of heads in 1000 simulated coin tosses.

The observation that the relative frequency of our event appears to be converging as the number of repetitions increases might lead us to try a definition like

$$\mathrm{P}(E) = \lim_{n \to \infty} \frac{\text{number of times in } n \text{ repetitions that outcome was in the event } E}{n}.$$
(2.3)

It's not exactly clear how we would formally define such a limit, and even less clear how we would attempt to evaluate it. But the intuition is still useful, and for now we will think of the empirical probability method as an approximation method (postponing for now any formal discussion of the quality of the approximation) that estimates a probability by repeating a random process some number of times and determining what percentage of the outcomes observed were in the event of interest.

Such empirical probabilities can be very useful, especially if the process is quick and cheap to repeat. But who has time to flip a coin 10,000 or more times just to see if the coin is fair? Actually, there have been folks who have flipped a coin a large number of times and recorded the results. One such was John Kerrich, a South African mathematician who recorded 5067 heads in 10,000 flips while in a prison camp during World War II. That isn't exactly 50% heads, but it is pretty close.

Since repeatedly carrying out even a simply random process like flipping a coin can be tedious and time consuming, we will often make use of computer simulations. If we have a reasonably good model for a random event and can program it into a computer, we can let the computer repeat the simulation many times very rapidly and (hopefully) get good approximations to what would happen if we actually repeated the process many times. The histograms below show the results of 1000 simulations of flipping a fair coin 1000 times (left) and 10,000 times (right).

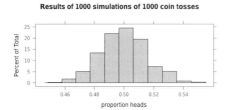
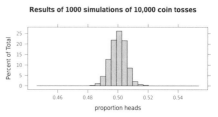

Figure 2.2. As the number of coin tosses increases, the results of the simulation become more consistent from simulation to simulation.

Notice that the simulation-based probability is closer to 0.50 more of the time when we flip 10,000 coins than when we flip only 1000 but that there is some variation in both cases. Simulations with even larger sample sizes would reveal that this variation decreases as the sample size increases but that there is always some amount of variation from sample to sample. Also notice that John Kerrich's results are quite consistent with the results of our simulations, which assumed a tossed coin has a 50% probability of being heads.

So it appears that flipping a fair coin is a 50-50 proposition. That's not too surprising; you already knew that. But *how* do you "know" this? You probably have not flipped a coin 10,000 times and carefully recorded the outcomes of each toss.[2] So you must be using some other method to derive the probability of 0.5.

2.1.2. Theoretical Probability Calculations

When you say you "know" that the toss of a fair coin has a 50% chance of being heads, you are probably reasoning something like this:

- the outcome will be either a head or a tail,
- these two outcomes are equally likely (because the coin is fair),
- the two probabilities must add to 100% because there is a 100% chance of getting one of these two outcomes.

From this we conclude that the probability of getting heads is 50%. This kind of reasoning is an example of the *theoretical method* of probability calculation. The basic idea is to combine

- some general properties that should be true of all probability situations (the **probability axioms**) and
- some additional assumptions about the situation at hand,

using deductive reasoning to reach conclusions, just as we did with the coin example.

To use this method, we first need to have some axioms. These should be statements about probability that are intuitively true and provide us with enough to get the process going. Box 2.2 contains our three probability axioms. That's it. Notice that each of these makes intuitive sense and is easily seen to be true of

---

[2]Another person who has flipped a coin a large number of times is Persi Diaconis, professor of statistics and mathematics at Stanford University – and former magician – who in 2004 wrote a paper with some colleagues demonstrating that there is actually a slight bias in a coin toss: It is more likely to land with the same side up that was up before the toss [**DHM07**].

Box 2.2. Axioms for probability

Let S be a sample space for a random process and let $E \subseteq S$ be an event. Then:

(1) $P(E) \in [0, 1]$.

(2) $P(S) = 1$ (where S is the sample space).

(3) The probability of a *disjoint* union is the sum of probabilities.
- $P(A \cup B) = P(A) + P(B)$, provided $A \cap B = \emptyset$.
- $P(A_1 \cup A_2 \cup \cdots \cup A_k) = P(A_1) + P(A_2) + \cdots + P(A_k)$, provided $A_i \cap A_j = \emptyset$ whenever $i \neq j$.
- $P\left(\bigcup_{i=1}^{\infty} A_i\right) = \sum_{i=1}^{\infty} P(A_i)$, provided $A_i \cap A_j = \emptyset$ whenever $i \neq j$.

---

our empirical probabilities (at least given a fixed set of repetitions). Despite the fact that our axioms are so few and so simple, they are quite useful. In Section 2.2 we'll see that a number of other useful general principles follow easily from our three axioms. Together these rules and the axioms form the basis of theoretical probability calculations. But before we provide examples of using these rules to calculate probabilities, we'll introduce the important notion of a random variable.

2.1.3. Random Variables

We are particularly interested in random events that produce (or can be converted to) numerical outcomes. For example:

- We may roll a die and consider the outcome as a number between 1 and 6 (inclusive).
- We may roll two dice and consider the outcome to be the sum of the numbers showing on each die.
- We may flip a coin 10 times and count the number of heads.
- We may telephone 1000 randomly selected households, ask the oldest person at home if he or she voted in the last election, and record the number (or percentage) who say yes.

In each case a number is being obtained as or from the result of a random process. We will refer to such numbers as **random variables**. More formally, we define a random variable as a function on the sample space.

Definition 2.1.1 (Random Variable). Let S be a sample space, and let $X : S \to \mathbb{R}$. Then X is called a *random variable*. $\qquad\square$

We will be concerned primarily with two types of random variables: discrete and continuous. A **discrete random variable** is one that can only take on a finite or countably infinite set of values. Often these values are a subset of the integers. A **continuous random variable** on the other hand can take on all values in some interval.

Example 2.1.1. Let X be the number of heads obtained when flipping a coin 10 times. Let Y be the length of time (in hours) a randomly selected light bulb burns continuously before burning out.

X is clearly a discrete random variable since range$(X) = \{0, 1, \ldots, 10\}$. Y is at least theoretically continuous, with a range of $[0, \infty)$. In practice we may only measure the time to the nearest minute (or second, or five minutes). Technically, this measured variable would be discrete, but it is usually more useful to model this situation by considering the underlying continuous random variable.   ◁

We will focus our attention first on discrete random variables and return to consider continuous random variables in Chapter 3, at which point we will be able to define them more carefully as well. There are, by the way, random variables that are neither discrete nor continuous. Exercise 3.3 asks you to describe one.

Example 2.1.2. If we flip a coin 3 times and record the results, our sample space is

$$S = \{\text{HHH}, \text{HHT}, \text{HTH}, \text{HTT}, \text{THH}, \text{THT}, \text{TTH}, \text{TTT}\}.$$

If we let X be the number of heads, then formally X is the following function.

$$
\begin{array}{llll}
\text{HHH} & \to & 3 & \qquad \text{THH} \to 2 \\
\text{HHT} & \to & 2 & \qquad \text{THT} \to 1 \\
\text{HTH} & \to & 2 & \qquad \text{TTH} \to 1 \\
\text{HTT} & \to & 1 & \qquad \text{TTT} \to 0
\end{array}
$$

◁

One of the advantages of introducing random variables is that it allows us to describe certain kinds of events very succinctly. In our coin flipping example, we can now write $\mathrm{P}(X \geq 2)$ instead of $\mathrm{P}(\text{we get at least 2 heads})$.

Example 2.1.3. A local charity is holding a raffle. They are selling 1000 raffle tickets for \$5 each. The owners of five of the raffle tickets will win a prize. The five prizes are valued at \$25, \$50, \$100, \$1000, and \$2000. Let X be the value of the prize associated with a random raffle ticket (0 for non-winning tickets). Then:

- $\mathrm{P}(\text{the ticket wins a prize}) = \mathrm{P}(X > 0) = 5/1000$.
- $\mathrm{P}(\text{the ticket wins the grand prize}) = \mathrm{P}(X = 2000) = 1/1000$.
- $\mathrm{P}(\text{the ticket wins a prize worth more than \$75}) = \mathrm{P}(X > 75) = 3/1000$.   ◁

You are probably already familiar with the reasoning used to determine the probabilities in the preceding example. Since there are 1000 raffle tickets and each is equally likely to be selected, we can form the probabilities by counting the number of raffle tickets in our event and dividing by 1000. This follows from what we will call the "equally likely rule". In the next section we will turn our attention to this and other methods of calculating theoretical probabilities and will show that a number of useful principles like the equally likely rule are consequences of our probability axioms.

2.2. Additional Probability Rules and Counting Methods

In this section we develop some additional probability rules and counting principles. We don't include these probability rules in our list of axioms because we can derive them from the axioms. But once we have shown this, then we can use them just like the axioms in subsequent calculations.

2.2.1. The Equally Likely Rule

Lemma 2.2.1 (Equally Likely Rule). *Suppose our sample space S consists of n equally likely outcomes. Then*

$$\mathrm{P}(E) = \frac{|E|}{|S|} = \frac{|E|}{n} = \frac{\text{size of event}}{\text{size of sample space}} .$$

Proof. This follows readily from the axioms: Let p be the probability of any outcome. $\mathrm{P}(S) = 1 = p + p + p + \cdots + p = np$. We may add because the outcomes are mutually exclusive (axiom (3)). The sum must be 1 by axiom (2). This means the probability of each outcome is $1/n$. Now we use axiom (3) again:

$$\mathrm{P}(E) = \sum_{A \in E} \mathrm{P}(A) = |E| \cdot \frac{1}{n} = \frac{|E|}{n} . \qquad \square$$

This "equally likely rule" is very useful. We already used it in our raffle example (Example 2.1.3). We can use it to determine the probability of heads when flipping a fair coin (1/2), the probability of getting a 6 when rolling a fair die (1/6), or the probability that a given Pick-3 lottery ticket is a winner (1/1000, since there are 1000 possible numbers but only 1 is a winner).

Here is a slightly more interesting example.

Example 2.2.1.
Q. Flip four coins and let X be the number of heads. What is $\mathrm{P}(X = x)$ for each value of x?

A. Since the sample space is small, we can write down all the possible outcomes:

HHHH HHHT HHTH HHTT HTHH HTHT HTTH HTTT
THHH THHT THTH THTT TTHH TTHT TTTH TTTT

Since each of these outcomes is equally likely, all we need to do is count the number of outcomes with each possible number of heads and divide by 16, the number of outcomes. We can display the results in the following probability table:

value of X	0	1	2	3	4
probability	$\frac{1}{16}$	$\frac{4}{16}$	$\frac{6}{16}$	$\frac{4}{16}$	$\frac{1}{16}$

Note that $\mathrm{P}(X = 2) < \mathrm{P}(X = 1 \text{ or } X = 3) = \frac{8}{16}$, so a 3-1 split in one direction or the other is more common than a 2-2 split. ◁

Be aware that it is not always the case that all outcomes in a sample space are equally likely, so this rule does not always apply. But when it applies, probability is as easy as counting. Of course, in complicated situations the counting itself may be challenging. If the sample space is small, we can simply make a complete list of all the outcomes, but when the sample space is large, this method is clearly unsatisfactory and we will need to develop a more systematic approach – which we will do soon. Once we have done so, we can apply the equally likely principle to determine probabilities of winning various lotteries, of getting various hands when playing cards, of rolling a Yahtzee, or of getting exactly 10 heads in 20 flips of a fair coin. It will also be important in our study of random sampling.

2.2.2. The Bijection Principle

Example 2.2.2. Suppose you have invited a number of people over for dinner. The preparations have been made, the places are set, and you are about to invite the guests to the table. Suddenly you have the sinking feeling that you may have set the wrong number of places at the table. What to do?

There are two choices. You could quickly count the people and count the chairs to see if the two counts match. But if the number of guests is large, this might not be so easy to do. A second option would be to have each person come and take a place at the table. If each person has a chair and each chair has a person, then you know that you have the correct number of places. Of course, there are other more embarrassing outcomes. There may be unused chairs or, worse, there may a person with no chair.   ◁

The preceding example gets at the heart of what it means to count. Counting is establishing a matching between two sets. When you learned to count as a child, you were matching up the set of objects being counted with a set of special number words (which you first had to memorize in the correct order). We can generalize this idea to matchings between any pair of sets. If such a matching exists, then the two sets have the same size (also called the **cardinality** of the set). The cardinality of a set A is denoted $|A|$.

In order to establish this same-size relationship, our matching must have two properties: we must not skip anything, and we must not double-match anything. Every item in one set must be matched to *exactly* one item in the other set. We will call a function that describes a matching **one-to-one** if it does no double matching and **onto** if it does no skipping. A function with both properties is called a **bijection** (or is said to be one-to-one and onto). The term bijection comes from alternate nomenclature: a one-to-one function is also called an **injection**, and an onto function is also called a **surjection**.

Definition 2.2.2. Let $f : A \to B$.

- If $f(x) = f(y)$ implies that $x = y$, then we say that f is *one-to-one*.
- If for every $b \in B$ there is an $a \in A$ such that $f(a) = b$, then we say that f is *onto*.
- A bijection is a function that is both one-to-one and onto.   □

Definition 2.2.2 is merely a formalization of the reasoning in Example 2.2.2. The key to counting in situations where making a complete list is not practical – whether we express it formally or informally – is the bijection principle, which can be summed up as

$$\boxed{\textbf{No double counting. No skipping.}}$$

Example 2.2.3.

Q. How many numbers between 10 and 100 (inclusive) are divisible by 6?

A. It is sufficient to count a set that has the same size. Here are several such sets along with a brief explanation of why each is the same size as the one before:

- Numbers between 12 and 96 (inclusive) that are divisible by 6.
    This is because 10, 11, 97, 98, 99, and 100 are all not divisible by 6.
- Numbers between 2 and 16 (inclusive).
    Match each number divisible by 6 with its quotient upon division by 6.
- Numbers between 1 and 15 (inclusive).
    These can be obtained by subtracting 1 from each of the numbers 2–16.

From this last description it is clear that there are 15 numbers between 10 and 100 that are divisible by 6.

$\triangleleft$

Example 2.2.4.

Q. How many ways are there to distribute three balls (1 red, 1 white, and 1 blue) to 10 people? (We'll allow giving one person more than one ball.)

A. Again we will think about other sets that have the same size:

- The number of ways to write down three names in order (repetitions allowed).
- The number of ways to write down three numbers between 1 and 10 (inclusive) in order (repetitions allowed).
- The number of ways to write down three numbers between 0 and 9 (inclusive) in order (repetitions allowed).
- The number of integers between 0 and 999 (inclusive).
- The number of integers between 1 and 1000 (inclusive).

From the last description it is clear that there are 1000 ways to distribute the three balls. $\triangleleft$

Here is a somewhat more challenging example of the bijection principle at work.

Example 2.2.5.

Q. How many ways are there to select 3 positive integers x, y, and z, so that $x + y + z = 20$? That is, how large is

$$A = \{\langle x, y, z \rangle \mid x, y, z \text{ are positive integers and } x + y + z = 20\}?$$

A. We could make a list, and with a computer that might not be too hard to accomplish. But it would be pretty tedious to do by hand. On the other hand, it isn't immediately clear how we should proceed. Let's begin by finding some other sets that are the same size with the hope that we will eventually find something that is easier to count. Our descriptions of the necessary bijections will be quite informal – but you should convince yourself that there is no skipping or double counting going on. In Exercise 2.4 you are asked to prove the bijections explicitly.

- $B = \{\langle x, y, z\rangle \mid x, y, \text{ and } z \text{ are non-negative integers and } x + y + z = 17\}$.

    If we subtract 1 from each of x, y, and z, then we reduce the sum by 3. It is easy to see that every solution in B is matched with exactly one solution in A.

- $C = \{\langle x, y\rangle \mid x \text{ and } y \text{ are non-negative integers and } x + y \leq 17\}$.

    If we know the first two numbers in our sum, we know the third one too because the sum must be 17.

- $D = \{\langle x, S\rangle \mid x \text{ and } S \text{ are integers and } 0 \leq x \leq S \leq 17\}$.

    Let $S = x + y$.

D is something we can count. Consider the following table:

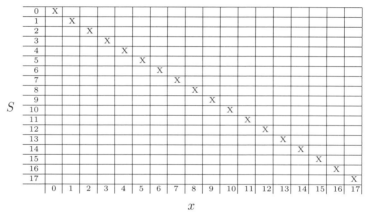

A pair $\langle x, S\rangle$ is in D if it is along the diagonal marked by 'X' or in the lower triangle of the table. So the size of D (and hence of A, B, and C as well) is

$$18 + \frac{1}{2}(18^2 - 18) = 171 \;.$$

$\triangleleft$

2.2.3. The Multiplication Principle

Where did the 18^2 come from in Example 2.2.5? From the picture it is clear; the table has 18 rows and 18 columns, so there are $18 \cdot 18$ cells in the table. This is an example of a more general principle called the multiplication principle. Before stating this principle, we define the Cartesian product of two sets.

Definition 2.2.3 (Cartesian Product). Let A and B be sets. Then the Cartesian product of A and B, denoted $A \times B$, is defined by

$$A \times B = \{\langle a, b\rangle \mid a \in A \text{ and } b \in B\} \;.$$

$\square$

If we know the sizes of A and B, then it is easy to determine the size of $A \times B$ (written $|A \times B|$).

Lemma 2.2.4 (Multiplication Rule). *For finite sets A and B*

$$|A \times B| = |A| \cdot |B| \,. \qquad \square$$

We can convince ourselves that Lemma 2.2.4 is true by making a rectangular diagram like the one in Example 2.2.5. Such a diagram will have $|A|$ rows and $|B|$ columns, giving it $|A| \cdot |B|$ cells. Lemma 2.2.4 is easily generalizable to products of three or more sets. For counting purposes, we will typically combine the multiplication rule with the bijection principle and reason as in the following example. We will refer to this type of reasoning as the multiplication principle.

Example 2.2.6.

Q. How many ways are there to distribute three balls (1 red, 1 white, and 1 blue) to 10 people if we do not allow anyone to receive more than one ball?

A. We can imagine the process of distributing the balls in three steps: (1) pick the person to get the red ball, (2) pick the person to get the white ball, and (3) pick the person to get the blue ball. There are 10 ways to do step (1). No matter how we do step (1), there will be 9 ways to do step (2). Which nine people remain after step (1) will depend on who was selected in step (1), but there will always be nine. Similarly, there will always be 8 ways to select the person to receive the blue ball (since we cannot choose the people who were given the red and white balls). By the bijection principle, the number of ways to distribute the balls is the same as the number of ways to select a number 1–10, followed by a number 1–9, followed by a number 1–8. By the multiplication rule, there are $10 \cdot 9 \cdot 8 = 720$ ways to do this.

It is important to note that what is important here is that the *number* of ways to perform each step *once the preceding steps are done* is the same no matter how the preceding steps are done. As in this example, it is usually the case that the actual sets of "next step options" differ depending on how the preceding steps are done. To be very explicit, let A, B, ..., J be the 10 people. Then any triple $\langle i, j, k \rangle$ of three integers corresponds to picking the person who is position i in the alphabetical order, followed by the person in position j *among the remaining 9 people*, followed by the person in position k among the remaining 8 people. This is a bijection because for any set of three people we can determine the appropriate coding as 3 numbers, and for any 3-number code, we can determine the three people. For example,

$$
\begin{aligned}
\langle A, B, C \rangle &\leftrightarrow \langle 1, 1, 1 \rangle \\
\langle C, B, A \rangle &\leftrightarrow \langle 3, 2, 1 \rangle \\
\langle E, G, J \rangle &\leftrightarrow \langle 5, 6, 8 \rangle \\
\langle 1, 2, 3 \rangle &\leftrightarrow \langle A, C, E \rangle \\
\langle 2, 4, 6 \rangle &\leftrightarrow \langle B, E, H \rangle
\end{aligned}
$$

This use of the multiplication rule combined with the bijection principle can be more succinctly described schematically as

$$\underbrace{10}_{\text{give red ball}} \cdot \underbrace{9}_{\text{give white ball}} \cdot \underbrace{8}_{\text{give blue ball}} = 720 \, .$$

Each blank holds the number of ways to perform the step indicated once the preceding steps have been performed.   ◁

Example 2.2.7.

Q. If you are dealt 5 cards from a standard 52-card deck (consisting of 4 suits of 13 cards each), what is the probability that you receive a flush? (A flush is a set of cards that are all from the same suit.)

A. We will use the equally likely principle, since every hand is equally likely. First we determine the number of (ordered) hands:

$$\underbrace{52}_{\text{first card}} \cdot \underbrace{51}_{\text{second card}} \cdot \underbrace{50}_{\text{third card}} \cdot \underbrace{49}_{\text{fourth card}} \cdot \underbrace{48}_{\text{fifth card}} = \frac{52!}{47!} \, .$$

Next we count the number of (ordered) flushes:

$$\underbrace{52}_{\text{first card}} \cdot \underbrace{12}_{\text{second card}} \cdot \underbrace{11}_{\text{third card}} \cdot \underbrace{10}_{\text{fourth card}} \cdot \underbrace{9}_{\text{fifth card}} = \frac{52 \cdot 12!}{8!} \, .$$

So the desired probability is

$$\frac{12 \cdot 11 \cdot 10 \cdot 9}{51 \cdot 50 \cdot 49 \cdot 48} = 0.00198 \, ,$$

which is just under once every 500 hands.   ◁

We have emphasized here that we are counting *ordered* hands. This means that we consider being dealt $K\heartsuit$ followed by $9\diamondsuit$ to be different from first getting $9\diamondsuit$ and then $K\heartsuit$. It is fine to count this way even though the order in which the cards are dealt does not matter for most card games. In Section 2.2.5 we will learn how to count the unordered hands in this situation. This will change both the numerator and the denominator, but not the probability.

Example 2.2.8. In Example 2.2.1 we calculated the probabilities of getting various numbers of heads when flipping a fair coin 4 times. To do this, we determined the size of the sample space by listing all of its elements. But we could also have done this using the multiplication principle. The process of flipping a coin four times can be broken naturally into 4 steps, one for each toss. Each step has 2 possibilities, so the total number of coin toss outcomes is

$$\underbrace{2}_{\text{first toss}} \cdot \underbrace{2}_{\text{second toss}} \cdot \underbrace{2}_{\text{third toss}} \cdot \underbrace{2}_{\text{fourth toss}} = 2^4 = 16 \, .$$

This clearly generalizes to any number of tosses. If we toss a coin n times, there are 2^n possible patterns of heads and tails.   ◁

2.2.4. The Complement Rule

The complement rule is simple, but we will see that it has many important uses. The idea is this: if we want to know the probability that something happens, it is good enough to know the probability that it *doesn't* happen. If it doesn't happen 25% of the time, then it must happen 75% of the time. That leads us to the following rule.

Lemma 2.2.5 (Complement Rule). *Let E be any event. Then*

$$P(E) = 1 - P(E^c) \, ,$$

where E^c is the event that E does not happen.

Proof. The complement rule follows easily from the probability axioms since $E \cap E^c = \emptyset$, so $1 = P(S) = P(E \cup E^c) = P(E) + P(E^c)$, from which it follows that $P(E^c) = 1 - P(E)$. □

E^c is read "the complement of E". The complement rule is useful because sometimes it is easier to determine $P(E^c)$ than to determine $P(E)$ directly. By the complement rule, every probability problem is really two problems, and we can solve whichever is easier.

Example 2.2.9.

Q. If we roll 5 standard dice, what is the probability that at least two of the numbers rolled match?

A. It is possible to attack this problem directly, but it is complicated because there are so many different ways to have at least two numbers match. It is much easier to determine the probability that none of the numbers match and then apply the complement rule.

$$P(\text{no numbers match}) = \frac{6 \cdot 5 \cdot 4 \cdot 3 \cdot 2}{6^5} = 0.09259 \, .$$

The denominator is calculated in the same manner as we did for coin tosses, but now there are 6 possibilities for each die. For the numerator, if none of the numbers match, then we can get any number on the first roll, any of five on the second (anything except what was on the first die), any of four on the third, etc. Finally,

$$P(\text{at least two numbers match}) = 1 - P(\text{no numbers match}) = 0.90741 \, . \quad \triangleleft$$

2.2.5. The Division Principle

The division principle provides us with a way to compensate for a certain type of double counting.

Example 2.2.10.

Q. 100 hundred cookies are distributed among a group of children. Each child gets 4 cookies. How many children were there?

A. We are matching cookies to kids. The matching is onto (each child receives a cookie), but it is not one-to-one (some child receives more than one cookie).

But since each child receives the same number of cookies, we can make the easy adjustment: The number of children is $100/4 = 25$. ◁

Example 2.2.11.

Q. How many ways are there to distribute three identical balls to 10 people if we do not allow anyone to receive more than one ball?

A. We have already seen in Example 2.2.6, that there are $10 \cdot 9 \cdot 8 = 720$ ways to distribute three distinguishable balls. But now the only issue is who gets a ball and who does not, not which color the balls have. It doesn't matter to us if we first select Alice, then Bob, then Claire; or first Claire, then Alice, then Bob; etc. So 720 is clearly an overcount.

The good news is that the number of different orders in which to pick three particular people is the same, no matter which three people we select. So the number we are looking for is

$$\frac{720}{\text{the number of orders in which three balls can be given to three people}}.$$

The denominator is easily calculated using the multiplication principle again: $3 \cdot 2 \cdot 1 = 6$ (because there are three choices for who gets the first ball, two choices remaining for who gets the second, and only one choice for who gets the third ball). So the total number of ways to select three of the ten people is $720/6 = 120$. ◁

The previous example is a very important one because it generalizes to many situations. It applies whenever we want to know how many ways there are to select k objects from a set of n objects. This situation arises so frequently that there is special notation for this number.

Definition 2.2.6. Let S be a set with $|S| = n$, and let k be an integer. Then

$$\binom{n}{k} = \text{the number of subsets of } S \text{ of size } k.$$

$\square$

The notation $\binom{n}{k}$ is called a **binomial coefficient** and is usually read "n choose k". Many calculators have keys that will calculate the binomial coefficients. Often they are labeled differently; one common alternative is ${}_nC_k$. The R function to compute binomial coefficients is `choose(n,k)`.

Example 2.2.12.

Q. What is $\binom{5}{2}$?

A. For this small example, we can make a list of all 2-element subsets and count them. Let $S = \{1, 2, 3, 4, 5\}$. Then the 2-element subsets of S are

$$\{1,2\}, \{1,3\}, \{1,4\}, \{1,5\}, \{2,3\}, \{2,4\}, \{2,5\}, \{3,4\}, \{3,5\}, \{4,5\}$$

so $\binom{5}{2} = 10$. We can check our counting using R.

```
> choose(5,2)
[1] 10
```
choose ◁

Theorem 2.2.7. *Let n and k be integers with $0 \le k \le n$. Then*

$$\binom{n}{k} = \frac{n!}{k!(n-k)!} .$$

Proof. We can use the same reasoning that we used in Example 2.2.11 to give a general formula for the binomial coefficients. First we count the number of ordered k-tuples from a set of size n. There are $n(n-1)(n-2)\cdots(n-k+1) = \frac{n!}{(n-k)!}$ of these. Then we count the number of ways to rearrange k selected items. There are $k(k-1)\cdots 1 = k!$ such rearrangements. So the number of subsets is

$$\frac{n!}{(n-k)!}/k! = \frac{n!}{k!(n-k)!} . \qquad \square$$

Corollary 2.2.8. *For any n and k, $\binom{n}{k} = \binom{n}{n-k}$.*

Proof. We provide two proofs of this fact. The first uses the formula in Theorem 2.2.7. We simply note that

$$\binom{n}{n-k} = \frac{n!}{(n-k)!(n-(n-k))!} = \frac{n!}{(n-k)!k!} = \binom{n}{k} .$$

This type of proof is sometimes called an *algebraic argument*.

The second proof is an example of what is called a *combinatorial argument*. $\binom{n}{k}$ is the number of ways to put k out of n items into a subset. But any time we select k items to put into the subset, we are selecting the other $n - k$ items to leave out. Thus by the bijection principle, $\binom{n}{n-k} = \binom{n}{k}$. $\qquad \square$

Example 2.2.13.

Q. Five cards are dealt from a standard deck of cards. What is the probability that they all have the same suit?

A. We already did this problem in Example 2.2.7. But we will do it a different way this time. There are 52 cards in a standard deck, and we will be dealt 5 of them. So there are $\binom{52}{5}$ different hands. Notice that we are not concerned with the order in which we receive the cards, not the order the cards are held in our hand, only with the five cards themselves.

To get a flush, we need to pick a suit and then pick 5 cards from that suit. There are $\binom{4}{1} = 4$ ways to pick one suit. There are $\binom{13}{5}$ ways to pick five cards from the selected suit. By the multiplication principle, the number of flush hands is therefore $4\binom{13}{5}$.

Since each hand is equally likely,

$$\mathrm{P}(\text{a flush is dealt}) = \frac{4 \cdot \binom{13}{5}}{\binom{52}{5}} \, .$$

R will happily compute this value for us.

```
> 4 * choose(13,5) / choose(52,5)
[1] 0.001980792
```
◁

Example 2.2.14. Michael is the chairman of his school's gaming club. Each month they have a big "game-in" for which he orders 8 pizzas. The pizza shop he orders from offers four types of pizza (pepperoni, sausage, cheese, and veggie).

Q. How many different orders can Michael place?

A. Although this is equivalent to counting the elements in

$$\{ \langle x, y, z, w \rangle \mid x, y, z, w \text{ are non-negative integers, and } x + y + z + w = 8 \} \, ,$$

we will choose a different approach from the one we used in Example 2.2.5.

This time imagine that the pizzas are placed in boxes and stacked with all the pepperoni pizzas on the bottom, followed by all the sausage pizzas, followed by all the cheese pizzas, and finally with all the veggie pizzas on top. Furthermore, let's suppose there is an empty pizza box between each type of pizza. So the pizzas are delivered in a stack of 11 pizza boxes – 8 with pizzas inside and 3 empty.

The key observation is that if we know where the empty boxes are, we can determine what the order was. For example, if the empty boxes are in positions 3, 4, and 8, then the order was 2 pepperoni, 0 sausage, 3 cheese, and 3 veggie. So the number of possible orders is $\binom{11}{3} = 165$ – one for each way of placing the empty boxes.
◁

2.2.6. Inclusion-Exclusion

Our third probability axiom tells us how to compute the probability of $A \cup B$ when A and B are mutually exclusive. But what if $A \cap B \neq \emptyset$? There is a more general formula to handle this case. It is obtained by splitting $A \cup B$ into three disjoint pieces (see Figure 2.3):

$$A \cup B = (A - B) \cup (A \cap B) \cup (B - A) \, .$$

Now we can use axiom (3) repeatedly:

- $\mathrm{P}(A) = \mathrm{P}((A - B) \cup (A \cap B)) = \mathrm{P}(A - B) + \mathrm{P}(A \cap B)$, so $\mathrm{P}(A - B) = \mathrm{P}(A) - \mathrm{P}(A \cap B)$.
- $\mathrm{P}(B) = \mathrm{P}((B - A) \cup (B \cap A)) = \mathrm{P}(B - A) + \mathrm{P}(B \cap A)$, so $\mathrm{P}(B - A) = \mathrm{P}(B) - \mathrm{P}(A \cap B)$.
- Combining these, we see that

$$\begin{aligned}
\mathrm{P}(A \cup B) &= \mathrm{P}(A - B) + \mathrm{P}(B - A) + \mathrm{P}(A \cap B) \\
&= \mathrm{P}(A) - \mathrm{P}(A \cap B) + \mathrm{P}(B) - \mathrm{P}(A \cap B) + \mathrm{P}(A \cap B) \\
&= \mathrm{P}(A) + \mathrm{P}(B) - \mathrm{P}(A \cap B) \, .
\end{aligned}$$

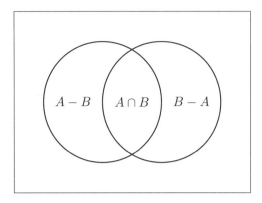

Figure 2.3. A Venn diagram illustrating inclusion-exclusion.

One useful way to think about this statement goes as follows. $P(A \cup B) \neq P(A) + P(B)$ in general. This is because the right side counts the probability of $A \cap B$ twice, once as part of $P(A)$ and again as part of $P(B)$. Subtracting $P(A \cap B)$ compensates for this "double counting".

If we replace $P(X)$ with $|X|$ in the argument above, we see that the same principle works for cardinality as well as for probability:

$$|A| + |B| - |A \cap B|.$$

The principle of inclusion-exclusion can be extended to more than two sets; see Exercise 2.16.

2.2.7. Conditional Probability and Independent Events

Example 2.2.15.

Q. Suppose a family has two children and one of them is a boy. What is the probability that the other is a girl?

A. We'll make the simplifying assumption that boys and girls are equally likely (which is not exactly true). Under that assumption, there are four equally likely families: BB, BG, GB, and GG. But only three of these have at least one boy, so our sample space is really $\{BB, BG, GB\}$. Of these, two have a girl as well as a boy. So the probability is $2/3$ (see Figure 2.4).

Figure 2.4. Illustrating the sample space for Example 2.2.15.

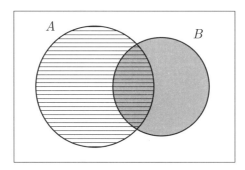

Figure 2.5. A Venn diagram illustrating the definition of conditional probability. $P(A \mid B)$ is the ratio of the area of the football shaped region that is both shaded and striped ($A \cap B$) to the area of the shaded circle (B).

We can also think of this in a different way. In our original sample space of four equally likely families,

$$P(\text{at least one girl}) = 3/4 \,,$$

$$P(\text{at least one girl } and \text{ at least one boy}) = 2/4 \,, \text{ and}$$

$$\frac{2/4}{3/4} = 2/3 \,;$$

so 2/3 of the time when there is at least one boy, there is also a girl. We will denote this probability as $P(\text{at least one girl} \mid \text{at least one boy})$. We'll read this as "the probability that there is at least one girl *given that* there is at least one boy". See Figure 2.5 and Definition 2.2.9.   ◁

Definition 2.2.9 (Conditional Probability). Let A and B be two events such that $P(B) \neq 0$. The *conditional probability* of A given B is defined by

$$P(A \mid B) = \frac{P(A \cap B)}{P(B)} \,.$$

If $P(B) = 0$, then $P(A \mid B)$ is undefined.   □

Example 2.2.16. A class of 5th graders was asked to give their favorite color. The most popular color was blue. The table below contains a summary of the students' responses:

	Favorite Color	
	Blue	Other
Girls	7	9
Boys	10	8

Q. Suppose we randomly select a student from this class. Let A be the event that a child's favorite color is blue. Let B be the event that the child is a boy, and let G be the event that the child is a girl. Express each of the following probabilities in words and determine their values:

- $P(A)$,

- P($A \mid B$),
- P($B \mid A$),
- P($A \mid G$),
- P($B \mid G$).

A. The conditional probabilities can be computed in two ways. We can use the formula from the definition of conditional probability directly, or we can consider the condition event to be a new, smaller sample space and read the conditional probability from the table.

- P(A) = 17/34 = 1/2 because 17 of the 34 kids prefer blue.
    This is the probability that a randomly selected student prefers blue.
- P($A \mid B$) = $\dfrac{10/34}{18/34} = \dfrac{10}{18}$ because 10 of the 18 boys prefer blue.
    This is the probability that a randomly selected boy prefers blue.
- P($B \mid A$) = $\dfrac{10/34}{17/34} = \dfrac{10}{17}$ because 10 of the 17 students who prefer blue are boys.
    This is the probability that a randomly selected student who prefers blue is a boy.
- P($A \mid G$) = $\dfrac{7/34}{16/34} = \dfrac{7}{16}$ because 7 of the 16 girls prefer blue.
    This is the probability that a randomly selected girl prefers blue.
- P($B \mid G$) = $\dfrac{0}{16/34} = 0$ because none of the girls are boys.
    This is the probability that a randomly selected girl is a boy.   ◁

One important use of conditional probability is as a tool to calculate the probability of an intersection.

Lemma 2.2.10. *Let A and B be events with non-zero probability. Then*
$$P(A \cap B) = P(A) \cdot P(B \mid A)$$
$$= P(B) \cdot P(A \mid B).$$

Proof. This follows directly from the definition of conditional probability by a little bit of algebra.   □

Example 2.2.17.

Q. If you roll two standard dice, what is the probability of doubles? (Doubles is when the two numbers match.)

A. Let A be the event that we get a number between 1 and 6 on the first die. So P(A) = 1. Let B be the event that the second number matches the first. Then the probability of doubles is P($A \cap B$) = P(A) $\cdot$ P($B \mid A$) = $1 \cdot \frac{1}{6} = \frac{1}{6}$ since regardless of what is rolled on the first die, 1 of the 6 possibilities for the second die will match it.   ◁

Lemma 2.2.10 can be generalized to more than two events. You will be asked to prove this in Exercise 2.19. Here is an example using the generalization.

Example 2.2.18.

Q. A 5-card hand is dealt from a standard 52-card deck. What is the probability of getting a flush (all cards the same suit)?

A. Imagine dealing the cards in order. Let A_i be the event that the ith card is the same suit as all previous cards. Then

$$
\begin{aligned}
\mathrm{P}(\text{flush}) &= \mathrm{P}(A_1 \cap A_2 \cap A_3 \cap A_4 \cap A_5) \\
&= \mathrm{P}(A_1) \cdot \mathrm{P}(A_2 \mid A_1) \cdot \mathrm{P}(A_3 \mid A_1 \cap A_2) \cdot \mathrm{P}(A_4 \mid A_1 \cap A_2 \cap A_3) \\
&\quad \cdot \mathrm{P}(A_5 \mid A_1 \cap A_2 \cap A_3 \cap A_4) \\
&= 1 \cdot \frac{12}{51} \cdot \frac{11}{50} \cdot \frac{10}{49} \cdot \frac{9}{48} \, .
\end{aligned}
$$

$\lhd$

Example 2.2.19.

Q. In the game Yahtzee, players roll 5 standard dice. A large straight occurs when 5 consecutive numbers appear on the five dice. What is the probability of getting a large straight when rolling 5 dice?

A. We'll work this out in stages.

- $\mathrm{P}(5 \text{ different numbers}) = 1 \cdot \frac{5}{6} \cdot \frac{4}{6} \cdot \frac{3}{6} \cdot \frac{2}{6} = 0.0925926$.
- $\mathrm{P}(\text{large straight} \mid 5 \text{ different numbers}) = 2/6 = 1/3$.

  This is because there are 6 possible missing numbers – all equally likely – and two of these yield a large straight.

- So

  $$
  \begin{aligned}
  \mathrm{P}(\text{large straight}) &= \mathrm{P}(5 \text{ different}) \cdot \mathrm{P}(\text{large straight} \mid 5 \text{ different}) \\
  &= \left(\frac{5}{6} \cdot \frac{4}{6} \cdot \frac{3}{6} \cdot \frac{2}{6} \right) \frac{1}{3} = 0.0308642 \, .
  \end{aligned}
  $$

  $\lhd$

Notice that the formula for $\mathrm{P}(A \cap B)$ is especially simple when $\mathrm{P}(B \mid A) = \mathrm{P}(B)$. Then we can avoid the use of conditional probability and the formula becomes $\mathrm{P}(A \cap B) = \mathrm{P}(A) \cdot \mathrm{P}(B)$. This situation is so important, it has a special name.

Definition 2.2.11 (Independent Events). If $\mathrm{P}(A \mid B) = \mathrm{P}(A)$, then we say that A and B are independent events. $\qquad\square$

Notice that this definition does not appear to be symmetric in A and B. The following lemma shows that this asymmetry is inconsequential.

Lemma 2.2.12. *If* $\mathrm{P}(A \mid B) = \mathrm{P}(A)$, *then* $\mathrm{P}(B \mid A) = \mathrm{P}(B)$.

Proof. Exercise 2.20. $\qquad\square$

Example 2.2.20. Mendelian genetics is a simple probabilistic model based on certain assumptions (which are not always true but are at least approximately true in some interesting cases, including some human diseases). The peas from one of Mendel's genetics experiments [**Men65**] could be either yellow or green. In this simple situation, color is determined by a single gene. Each organism inherits two of these genes (one from each parent), which can be one of two types (Y or y). A pea plant produces yellow peas unless *both* genes are type y.

As a first step, Mendel bred pure line pea plants of type YY and yy. The offspring of these plants are very predictable:

- $YY \times YY$: all offspring are YY (and therefore yellow),

- $yy \times yy$: all offspring are yy (and therefore green),

- $YY \times yy$: all offspring are Yy (and therefore yellow).

More interestingly, consider $Yy \times Yy$:

- $P(\text{green}) = P(yy) = P(y \text{ from 1st } Yy \text{ and } y \text{ from 2nd } Yy) = P(y \text{ from 1st}) \cdot P(y \text{ from 2nd}) = 1/2 \cdot 1/2 = 1/4,$

- $P(\text{yellow}) = 1 - P(\text{green}) = 3/4,$

- $P(\text{heterozygote}) = P(Yy \text{ or } yY) = 1/4 + 1/4 = 1/2.$

Notice that in these calculations we are repeatedly using the assumptions:

- offspring are equally likely to inherit each parental gene,

- the genes inherited from the parents are independent of each other.

Later we will learn how to test how well data fit theoretical probabilities like this. (Mendel's data fit very well.) ◁

Example 2.2.21. Mendel actually looked at several traits [**Men65**]. Among them were color (yellow or green) and shape (round or wrinkled). According to his model, yellow (Y) and round (R) were dominant. If these traits are inherited independently, then we can make predictions about a $YyRr \times YyRr$ cross:

- $P(\text{yellow}) = \frac{3}{4},$

- $P(\text{green}) = \frac{1}{4},$

- $P(\text{round}) = \frac{3}{4},$

- $P(\text{wrinkled}) = \frac{1}{4}.$

So

- $P(\text{yellow and round}) = \frac{3}{4} \cdot \frac{3}{4} = \frac{9}{16} = 0.5625,$

- $P(\text{yellow and wrinkled}) = \frac{3}{4} \cdot \frac{1}{4} = \frac{3}{16} = 0.1875,$

- $P(\text{green and round}) = \frac{1}{4} \cdot \frac{3}{4} = \frac{3}{16} = 0.1875,$

- $P(\text{green and wrinkled}) = \frac{1}{4} \cdot \frac{1}{4} = \frac{1}{16} = 0.0625.$ ◁

Example 2.2.22.

Q. Suppose a test correctly identifies diseased people 98% of the time and correctly identifies healthy people 99% of the time. Furthermore assume that in a certain population, one person in 1000 has the disease. If a random person is tested and the test comes back positive, what is the probability that the person has the disease?

A. We begin by introducing some notation. Let D be the event that a person has the disease. Let H be the event that the person is healthy. Let $+$ be the event that the test comes back positive (meaning it indicates disease – probably a negative from the perspective of the person tested). Let $-$ be the event that the test is negative.

- $P(D) = 0.001$, so $P(H) = 0.999$.
- $P(+ \mid D) = 0.98$, so $P(- \mid D) = 0.02$.

   $P(+ \mid D)$ is called the **sensitivity** of the test. (It tells how sensitive the test is to the presence of the disease.)

- $P(- \mid H) = 0.99$, so $P(+ \mid H) = 0.01$.

   $P(- \mid H)$ is called the **specificity** of the test.

- $$P(D \mid +) \quad = \quad \frac{P(D \cap +)}{P(+)}$$

   $$= \quad \frac{P(D) \cdot P(+ \mid D)}{P(D \cap +) + P(H \cap +)}$$

   $$= \quad \frac{0.001 \cdot 0.98}{0.001 \cdot 0.98 + 0.999 \cdot 0.01} \quad = \quad 0.0893.$$

This low probability surprises most people the first time they see it. This means that if the test result of a random person comes back positive, the probability that that person has the disease is less than 9%, even though the test is "highly accurate". This is one reason why we do not routinely screen an entire population for a rare disease – such screening would produce many more false positives than true positives.

Of course, if a doctor orders a test, it is usually because there are some other symptoms. This changes the *a priori* probability that the patient has the disease. Exercise 2.28 gives you a chance to explore this further. ◁

Example 2.2.23. In a simple (Mendelian) model of a recessive disease, each person has one of three genotypes (AA, Aa, aa) based on which of two forms (called alleles) each of their two genes (one from each parent) has. Those with genotype aa have the disease; the others do not.

A more realistic model assigns to each of these genotypes a probability of getting the disease. The following probabilities are said to describe the **penetrance** of the disease:

- $P(D \mid AA)$,
- $P(D \mid Aa)$,
- $P(D \mid aa)$.

Consider a model where the penetrances are

- $P(D|AA) = 0.01$,
- $P(D|Aa) = 0.05$,
- $P(D|aa) = 0.50$.

Q. Now consider an $AA \times aa$ cross. What is the probability that a child will have the disease?

A. In this case we know the child will have genotype Aa, so $\mathrm{P}(D) = \mathrm{P}(D \mid Aa) = 0.05$.

Q. What is the probability that a child will have the disease in an $AA \times Aa$ cross?

A. We can divide the event D into three mutually exclusive cases and sum their probabilities:

$$
\begin{aligned}
\mathrm{P}(D) &= \mathrm{P}(D \text{ and } AA) + \mathrm{P}(D \text{ and } Aa) + \mathrm{P}(D \text{ and } aa) \\
&= \mathrm{P}(AA) \cdot \mathrm{P}(D \mid AA) + \mathrm{P}(Aa) \cdot \mathrm{P}(D \mid Aa) + \mathrm{P}(aa) \cdot \mathrm{P}(D \mid aa) \\
&= (0.5)(0.01) + (0.5)(0.05) + (0)(0.5) = 0.03 \ .
\end{aligned}
$$
$\triangleleft$

2.3. Discrete Distributions

2.3.1. The Distribution of a Discrete Random Variable: pmfs and cdfs

Recall that the distribution of a variable in a data set described what values occurred and with what frequency. What we need now is a way to describe the distribution of a random variable. We do this somewhat differently for discrete and continuous random variables, so for the moment we will focus our attention on the discrete case.

One useful way to describe the distribution of a discrete random variable – especially one that has a finite range – is in a table like the one we used in Example 2.2.1 for the random variable X that counts the number of heads in four tosses of a fair coin:

value of X	0	1	2	3	4
probability	0.0625	0.2500	0.3750	0.2500	0.0625

Notice that the probability table allows us to assign to each possible outcome a probability. This means that the table is really describing a function. This function is called a **probability mass function** or **pmf**. A pmf can be any function that obeys the probability axioms.

Definition 2.3.1. Let $X : S \to \mathbb{R}$ be a random variable. The *probability mass function* (pmf) for X is a function $f : \mathbb{R} \to [0, 1]$ such that for all $x \in \mathbb{R}$,

$$
f(x) = \mathrm{P}(X = x) \ .
$$
$\square$

We will write f_X for the pmf of X when we want to emphasize the random variable.

Lemma 2.3.2. *Let f be the pmf for a random variable $X : S \to \mathbb{R}$. Then*

(1) $f(x) \in [0,1]$ *for all $x \in \mathbb{R}$, and*

(2) $\sum_{s \in S} f(X(s)) = 1$.

Furthermore if g is a function such that

(1) $g(x) \in [0,1]$ *for all $x \in \mathbb{R}$, and*

(2) $\sum_{g(x) \neq 0} g(x) = 1 \in [0,1]$ *for all $x \in \mathbb{R}$,*

then g is a pmf for a random variable.

Proof. Exercise 2.38.   □

As we will see, the pmf will be much more important to our study of random variables than the set S or the particular function mapping S to $\mathbb{R}$.

It is possible to write down an explicit formula for the pmf of the random variable in Example 2.2.1, namely

$$f_X(x) = \begin{cases} \frac{4!}{16x!(4-x)!} & \text{if } x \in \{0,1,2,3,4\}, \\ 0 & \text{otherwise.} \end{cases}$$

You can easily check that the values given by this formula match those in the table above. Sometimes it is not so easy to write down a formula for the pmf of a random variable, but the function exists nonetheless, since we can always define it by

$$f_X(x) = \mathrm{P}(X = x) . \tag{2.4}$$

Figure 2.6 shows three different ways that we can plot the pmf from Example 2.2.1. In the first plot, the non-zero values of this function are represented by dots placed according to the Cartesian coordinate scheme. In the second plot, vertical lines are used instead. In the third plot, lines are drawn connecting the dots. It is important to remember that although these lines help us see the shape of the distribution, the value of the pmf is zero between the dots. All of these plots can be made using `xyplot()` by setting the `type` argument to `'p'`, `'h'` (for histogram-like), or `c('p','l')`, respectively.

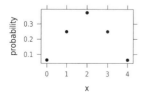 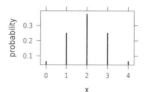 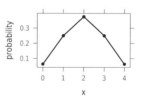

Figure 2.6. Graphs of a pmf.

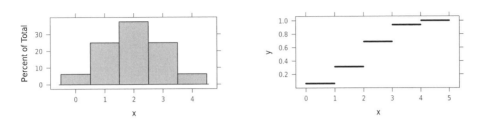

Figure 2.7. A probability histogram (left) and the graph of a cdf (right).

```
                                                                pmf-plot
# define pmf (vectorized),
# checking to be sure input is in { 0, 1, 2, 3, 4 }
> f <- function(x) {
+     sapply(x, function(x) {
+         if ( ! ( x %in% 0:4 ) ) { return(0) }
+         return( factorial(4) / ( 16 * factorial(x) * factorial(4-x) ) )
+     })
+ }
> f(0:6)
[1] 0.0625 0.2500 0.3750 0.2500 0.0625 0.0000 0.0000
> probplot1 <- xyplot(f(0:4)~0:4, xlab="x", ylab="probability")
> probplot2 <- xyplot(f(0:4)~0:4, xlab="x", ylab="probability",
+                     type="h")
> probplot3 <- xyplot(f(0:4)~0:4, xlab="x", ylab="probability",
+                     type=c("l","p"))
```

Another useful picture is a **probability histogram**. A probability histogram is made very much like the density histograms we made from data. Typically vertical bars are centered at the possible values of the random variable, but this is not required and for random variables with many possible values, it may be useful to combine some of the bins. In any case, the area of each bar represents the probability that the random variable takes on a value covered by the base of the rectangle. (If we choose boundaries that are not possible values of the random variable, we do not need to worry about whether the left or right endpoint belongs to the bin.) An example appears in Figure 2.7.

There is yet one more important way to describe the distribution of a discrete random variable, with a **cumulative distribution function** (**cdf**).

Definition 2.3.3 (Cumulative Distribution Function)**.** The *cumulative distribution function* F_X of a random variable X is defined by

$$F_X(x) = \mathrm{P}(X \le x) \,.$$ □

The graph of a cdf of a discrete random variable is a step function. See Figure 2.7. There is, of course, a connection between the pmf and cdf of a given random variable, namely

$$F_X(x) = \sum_{w \le x} f_X(w), \text{ and}$$

$$f_X(x) = F(x) - F(x^-),$$

where $F(x^-) = \max\{F(w) \mid w < x\}$.[3] The notation here is more challenging than the idea. To get the cdf from the pmf, we simply add up the probabilities for all possible values up to and including x. To get the pmf from the cdf, we look at how much the cdf has increased since its "last change".

Example 2.3.1. Suppose that the cdf for a discrete random variable is given by

$$F(x) = \begin{cases} 1 - \frac{1}{\lfloor x+1 \rfloor} & \text{if } x > 0, \\ 0 & \text{otherwise .} \end{cases}$$

Q. What is the pmf for this random variable?

A. We begin by calculating $f(3)$:

$$f(3) = F(3) - F(3^-) = F(3) - F(2) = \left(1 - \frac{1}{4}\right) - \left(1 - \frac{1}{3}\right) = \frac{1}{12}.$$

More generally,

$$f(x) = \begin{cases} \frac{1}{x(x+1)} & \text{if } x \text{ is a positive integer,} \\ 0 & \text{otherwise.} \end{cases} \qquad \triangleleft$$

In the next section we turn our attention to two important examples of discrete distributions.

2.3.2. The Binomial and Negative Binomial Distributions

In Example 2.2.1 the random variable X counted the number of heads in four tosses of a fair coin. This is an example of an important family of distributions called the **binomial distributions**. A binomial random variable arises in a situation where our random process can be divided up into a sequence of smaller random processes called trials and

(1) the number of trials (usually denoted n) is specified in advance,

(2) there are two outcomes (traditionally called success and failure) for each trial,

(3) the probability of success (frequently denoted p or π) is the same in each trial, and

(4) each trial is independent of the other trials.

---

[3]Technically, if X can take on infinitely many values, we may need to use the supremum (sup) in place of the maximum (max). The supremum of a bounded, non-empty set is the smallest number that is at least as large as all numbers in the set. This is the maximum when the maximum exists. See Exercise 2.39.

The binomial random variable counts the *number of successes*. There are actually many different binomial distributions, one for each positive integer n and probability π. Collectively, we refer to these distributions as the **binomial family**. Members of the family are distinguished by the values of the **parameters** n and π but are otherwise very similar. We will use the following notation to succinctly describe binomial random variables.

Notation 2.3.4. If X is a binomial random variable with parameters n and π, we will write $X \sim \mathsf{Binom}(n, \pi)$.

We would like to have a general formula for the pmf of a binomial distribution:

$$f_X(x; n, \pi) = \mathrm{P}(X = x) = ??? \ .$$

Notice the use of the semi-colon in this equation. The semi-colon separates the parameters of the distribution from the independent variable of the pmf.

The tools we developed in Section 2.2 make deriving a formula for the pmf of a binomial random variable straightforward.

Theorem 2.3.5 (pmf for Binomial Distributions). *Let* $X \sim \mathsf{Binom}(n, \pi)$. *Then the pmf for* X *is given by*

$$f_X(x; n, \pi) = \binom{n}{x} \pi^x (1 - \pi)^{n-x} \ .$$

Proof. For a fixed n, there are 2^n possible outcomes (see Example 2.2.8). These outcomes are not equally likely unless $\pi = 0.5$, but we can determine the probability of any particular outcome. For example, if $n = 4$, then

$$\mathrm{P}(SSFS) = \pi \cdot \pi \cdot (1 - \pi) \cdot \pi = \pi^3 (1 - \pi)^1 \ .$$

More generally, any outcome with x successes in n trials will have probability $\pi^x (1 - \pi)^{n-x}$, and the number of such outcomes is $\binom{n}{x}$ since we must select x of the n trials to be successful. So

$$\mathrm{P}(X = x) = \underbrace{\binom{n}{x}}_{\substack{\text{number of outcomes} \\ \text{with } X = x}} \cdot \underbrace{\pi^x (1 - \pi)^{n-x}}_{\substack{\text{probability of} \\ \text{each outcome}}} \ . \qquad \square$$

The cdf for a binomial random variable cannot be expressed simply in closed form, but R offers functions to compute both the pmf and the cdf, as well as a function that will make random draws from a binomial distribution. These functions are summarized in Box 2.3.

Example 2.3.2. Here are some example uses of the functions in Box 2.3.

```
> randomData <- rbinom(n=20,size=4,prob=0.5)          binom-demo01
> randomData
 [1] 3 1 2 2 3 1 3 4 2 2 0 1 1 2 2 3 2 3 4 2
> table(randomData)
randomData
0 1 2 3 4
1 4 8 5 2
```

Box 2.3. Working with Binom(size, prob) in R

The following functions are available in R for working with a binomial random variable $X \sim$ Binom(size, prob):

function (& arguments)	explanation
dbinom(x,size,prob)	returns $P(X = x)$ (the pmf)
pbinom(q,size,prob)	returns $P(X \leq q)$ (the cdf)
qbinom(p,size,prob)	returns smallest x such that $P(X \leq x) \geq p$
rbinom(n,size,prob)	makes n random draws of the random variable X and returns them in a vector
set.seed(seed)	sets the seed for the random number generator; see ?set.seed for details

---

```
> dbinom(0:4,size=4,prob=0.5)           # matches earlier example
[1] 0.0625 0.2500 0.3750 0.2500 0.0625
> dbinom(0:4,size=4,prob=0.5) * 20      # pretty close to our table above
[1] 1.25 5.00 7.50 5.00 1.25
> pbinom(0:4,size=4,prob=0.5)           # same as cumsum(dbinom(...))
[1] 0.0625 0.3125 0.6875 0.9375 1.0000                                    ◁
```

It is important to note that

- R uses size for the number of trials (n), n for the number of random draws, and prob for the probability of success (π). prob and size can be abbreviated to p and s if desired, but most often we will simply use them without names and in the required order.

- pbinom() gives the cdf not the pmf. Reasons for this naming convention will become clearer later.

- There are similar functions in R for many of the distributions we will encounter, and they all follow a similar naming scheme. We simply replace binom with the R-name for a different distribution.

Example 2.3.3.

Q. Free Throw Freddie is a good free throw shooter. Over the last few seasons he has made 80% of his free throws. Let's assume that each of Freddie's shots is independent of the others[4] and that he has an 80% probability of making each. At the end of each practice, Freddie shoots 20 free throws to keep sharp. What is the probability that he makes all 20? At least 15? Exactly 16 (80% of 20)?

A. We will model this situation as a binomial random variable. Let X be the number of made free throws in 20 attempts. Then $X \sim$ Binom(20, 0.8). We'll let

---

[4]How one might gather data to test this assumption of independence is an interesting question. But it will have to wait for another day.

> **Box 2.4. Working with NBinom(size, prob) in R**
>
> The following functions are available in R for working with a binomial random variable $X \sim$ NBinom(size, prob):
>
> | function (& arguments) | explanation |
> |---|---|
> | rnbinom(n,size,prob) | makes n random draws of the random variable X and returns them in a vector |
> | dnbinom(x,size,prob) | returns $P(X = x)$ (the pmf) |
> | qnbinom(p,size,prob) | returns smallest x such that $P(X \le x) \ge$ p |
> | pnbinom(q,size,prob) | returns $P(X \le x)$ (the cdf) |

R do the number crunching here, but you are welcome to check these using the formulas.

```
> dbinom(20,20,0.8)        # probability of making all 20      binom-freddy01
[1] 0.011529
> 1 - pbinom(14,20,0.8)    # probability of NOT making 14 or fewer
[1] 0.8042
> dbinom(16,20,0.8)        # probability of making exactly 16
[1] 0.2182                                                           ◁
```

The **negative binomial** random variables arise in a very similar situation to that of the binomial random variables. The difference is that instead of deciding in advance how many trials to perform and counting the number of successes, now we will decide how many successes there will be and repeat the trials until we have obtained the desired number of successes. The negative binomial random variable counts *the number of failures* that occur before getting the desired number of successes.[5]

Let X be negative binomial with parameters s (number of successes) and π (probability of success). We will denote this $X \sim$ NBinom(s, π). The R functions related to negative binomial distributions are similar to those for the binomial distributions (see Box 2.4). In R the number of successes is called size rather than s but can be abbreviated to s.

Example 2.3.4.

Q. Suppose you roll a pair of standard 6-sided dice until you get double sixes. How many rolls will it take? What is the probability that it will take you at least 20 rolls? At least 30? At least 40? At least 50?

A. We'll work out formulas for the negative binomial distributions shortly. For now, we'll let R do the work. Let $X \sim$ NBinom($1, 1/36$). Remember that X is

---

[5] Whether the negative binomial variable counts the number of *failures* or the number of *trials* varies in the literature, so when looking at information about this distribution, be sure to check which convention the author is using. The convention we are choosing matches what is done in R.

the number of failures. Let $Y = X + 1$ be the total number of rolls. We want to determine $P(Y \geq 20) = P(X \geq 19) = 1 - P(X \leq 18)$ and the corresponding probabilities for the other questions. We can do them all in one step in R.

```
> 1-pnbinom(c(18,28,38,48),size=1,prob=1/36)
[1] 0.58552 0.44177 0.33332 0.25148
```
nbinom-first-example01

You could be rolling quite a while if you are waiting for double sixes.   ◁

Now we want to derive the formula for the pmf of a negative binomial variable. The simplest case is when we stop after the first success, i.e., when $s = 1$. In this case

$$P(X = x) = P(x \text{ failures followed by a success})$$
$$= (1 - \pi)^x \pi .$$

When $s = 1$, a negative binomial distribution is called a **geometric distribution** because the pmf forms a geometric series with ratio $(1 - \pi)$. The sum of the series is

$$\sum_{x=0}^{\infty}(1 - \pi)^x \pi = \frac{\pi}{1 - (1 - \pi)} = \frac{\pi}{\pi} = 1 .$$

That's a good thing, since all pmfs are supposed to sum to 1. You can access the geometric distribution directly in R using `rgeom()`, `dgeom()`, and `pgeom()`.

Now let's try the case where $s > 1$. Let E be the event that there are x failures and $s - 1$ successes in the first $x + s - 1$ trials followed by a success in trial $x + s$. Then

$$P(X = x) = P(E) = \binom{x + s - 1}{x}(1 - \pi)^x \pi^{s-1} \cdot \pi$$
$$= \binom{x + s - 1}{x}(1 - \pi)^x \pi^s .$$

This proves the following theorem. Notice that when $s = 1$ we get the same expression that we just derived for the geometric distribution.

Theorem 2.3.6. *Let* $X \sim \mathsf{NBinom}(s, \pi)$. *Then* $P(X = x) = \binom{x+s-1}{x}(1 - \pi)^x \pi^s$. □

2.4. Hypothesis Tests and p-Values

We have said that a fair coin is equally likely to be heads or tails when tossed. Based on this assumption, we determined that the probability of getting heads is 50%. But now suppose we have a coin and we do not know if it is a fair coin. How can we test it? Clearly we need to flip the coin and check the outcomes. But how many times do we flip the coin? And what decision do we make? If we flip the coin 100 times, we would expect roughly 50 heads and 50 tails, but we know that it is very likely we won't get *exactly* 50 of each. At what point would we become suspicious that the coin is *biased* (more likely to give one outcome than the other)?

2.4.1. The (Exact) Binomial Test

If we flip a coin n times and let X be the number of heads, then $X \sim \mathsf{Binom}(n, \pi)$ for some *unknown* value of π. We want to know whether or not $\pi = 0.50$. For example, suppose that $n = 100$ and we get $x = 40$ heads in our sample. What do we conclude? Is this consistent with a fair coin? Or is it sufficient evidence to suggest that the coin is biased?

Well, if it really is the case that $\pi = 0.50$, then $\mathrm{P}(X \le 40) = 0.02844$, so we would only get 40 or fewer heads about 2.8% of the times that we did this test. In other words, getting only 40 heads is pretty unusual, but not extremely unusual. This gives us some evidence to suggest that the coin may be biased. After all, one of two things must be true. Either

- the coin is fair ($\pi = 0.50$) and we were just "unlucky" in our particular 100 tosses, or

- the coin is not fair, in which case the probability calculation we just did doesn't apply to the coin.

That in a nutshell is the logic of a **statistical hypothesis test**. We will learn a number of hypothesis tests, but they all follow the same basic four-step outline.

Step 1: State the null and alternative hypotheses

A **hypothesis** is a statement that can be either true or false. A **statistical hypothesis** is a hypothesis about a parameter (or parameters) of some population or process. In this example, the statistical hypothesis we are testing is

- H_0: $\pi = 0.50$

where π is the probability of obtaining a head when we flip the coin. This is called the **null hypothesis**. In some ways it is like a straw man. We will collect evidence (data) against this hypothesis. If the evidence is strong enough, we will reject the null hypothesis in favor of an **alternative hypothesis**. In our coin tossing example, the alternative hypothesis is

- H_a: $\pi \ne 0.50$

because $\pi = 0.50$ when the coin is a fair coin but will be some different value if the coin is biased.

Step 2: Calculate a test statistic

A **statistic** is a number calculated from sample data. Mathematically, a statistic is simply a function that assigns a real number to a data set:

$$f : \mathrm{DATA} \to \mathbb{R} \,.$$

In our coin tossing example, we may wish to count the number of heads obtained in 100 tosses of the coin. So our statistic is the function that takes any sequence of heads and tails (the data) and returns the number of heads (a number).

If we have a particular data set in mind, we will also refer to the numerical output of this function applied to that data set as a statistic. If we use this number to test a statistical hypothesis, we will call it a **test statistic**. In our example, the number of heads is 40, and we could denote this test statistic as $x = 40$. A test statistic should be a number that measures in some way how consistent the data are with the null hypothesis. In this case, a number near 50 is in keeping with the null hypothesis. The farther x is from 50, the stronger the evidence against the null hypothesis.

Step 3: Compute the p-value

Now we need to evaluate the evidence that our test statistic provides. To do this requires yet another way of thinking about our test statistic. Assuming our sample was obtained in some random way, we can also think about a statistic as a random variable. A random process produces a data set, from which we calculate some number. Schematically,

$$\text{random process} \rightarrow \text{sample} \rightarrow \text{DATA} \rightarrow \text{statistic}.$$

The distribution of this kind of random variable is called its **sampling distribution**. To distinguish between these two views (a particular number vs. a random variable), we will often use capital letters to denote random variables and lowercase to indicate particular values. So our random variable in this case will be called X.

Now we can ask probability questions about our test statistic. The general form of the question is, *How unusual would my test statistic be if the null hypothesis were true?* To answer this question, it is important that we know something about the distribution of X when the null hypothesis is true. In this case, $X \sim$ Binom$(100, 0.5)$. So how unusual is it to get only 40 heads? If we assume that the null hypothesis is true (i.e., that the coin is fair), then

$$P(X \leq 40) = \texttt{pbinom(40,100,0.5)} = 0.0284,$$

and since the Binom$(100, 0.5)$ is a symmetric distribution, we get the same probability for the other tail:

$$P(X \geq 60) = 1 - \texttt{pbinom(59,100,0.5)} = 0.0284.$$

So the probability of getting a test statistic at least as extreme (unusual) as 40 is 0.0568. This probability is called a **p-value**.

Step 4: Draw a conclusion

Drawing a conclusion from a p-value is a bit of a judgment call. Our p-value is 0.0568. This means that if we flipped 100 fair coins many times, between 5% and 6% of these times we would obtain fewer than 41 or more than 59 heads. So our result of 40 is a bit on the unusual side, but not extremely so. Our data provide some evidence to suggest that the coin may not be fair, but the evidence is far from conclusive. If we are really interested in the coin, we probably need to gather more data.

Other hypothesis tests will proceed in a similar fashion. The details of how to compute a test statistic and how to convert it into a p-value will change from test to test, but the interpretation of the p-value is always the same. On the other hand, this interpretation does involve some amount of judgment, whereas the computation of the p-value is more or less automatic (a computer will typically do it for us). A famous rule of thumb regarding p-values is that when the p-value is less than 0.05, then we have enough evidence to reject the null hypothesis. This is a useful rule of thumb, but it should not be taken too literally. A p-value of 0.049 is hardly different from a p-value of 0.051. Both indicate nearly the same strength of evidence against the null hypothesis even though one is less than 0.05 and the other greater. Furthermore, when interpreting p-values, we must take into consideration the consequences of making a mistake. We'll return to the topic of errors in a moment, but first let's do some more examples.

Example 2.4.1.

Q. Let's return to our example of the lady tasting tea. Suppose we decide to test whether the lady can tell the difference between tea poured into milk and milk poured into tea by preparing 10 cups of tea. We will flip a coin to decide how each is prepared. Then we present the ten cups to the lady and have her state which ones she thinks were prepared each way. If she gets 9 out of 10 correct, what do we conclude?

A. The null hypothesis is that she is just guessing, i.e.,

$$H_0 : \pi = 0.5 .$$

Under that assumption, $P(X \geq 9) = 0.0107$, so the chances of getting 9 or 10 correct just by guessing is just over 1%. This test can be conducted easily in R:

```
> 1-pbinom(8,10,0.5);
[1] 0.010742
> binom.test(9,10);

        Exact binomial test

data:  9 and 10
number of successes = 9, number of trials = 10, p-value = 0.02148
alternative hypothesis: true probability of success is not equal to 0.5
95 percent confidence interval:
 0.55498 0.99747
sample estimates:
probability of success
                  0.9
```

binomtest-lady-tea01

There is more in this output than we have discussed to this point, but it is easy to find the p-value. Notice that it is twice the probability that we just calculated. This is because by default, `binom.test()` does a **two-sided test**. In this case the p-value computed by `binom.test()` is $P(X \geq 9 \text{ or } X \leq 1)$. It is possible to have R compute a one-sided test instead:

```
> binom.test(9,10,alternative="greater");          binomtest-lady-tea02

        Exact binomial test

data:  9 and 10
number of successes = 9, number of trials = 10, p-value = 0.01074
alternative hypothesis: true probability of success is greater than 0.5
95 percent confidence interval:
 0.60584 1.00000
sample estimates:
probability of success
                  0.9
```

The output indicates which kind of test is being done by reporting which of the following alternative hypotheses is being used:

$$H_a: \quad \pi \neq 0.5 \quad \text{(two-sided alternative),}$$
$$H_a: \quad \pi < 0.5 \quad \text{(one-sided alternative),}$$
$$H_a: \quad \pi > 0.5 \quad \text{(one-sided alternative).}$$

The fact that R by default computes a two-sided p-value can serve as a reminder that one must offer additional justification to legitimately use a one-sided test.   ◁

There is a certain attraction to one-sided tests. Since the p-value for a one-sided test is always smaller than the p-value for a two-sided test (unless the sample proportion is on the "wrong side" of the hypothesized proportion), a one-sided test appears to give stronger evidence against the null hypothesis. But it is not appropriate to use a one-sided test simply because you want a smaller p-value. There are differing opinions about when a one-sided test is appropriate. At one extreme are those who say one should *never* do a one-sided test. In any case one thing is certain: the decision to do a one-sided test must be something that can be *defended without referring to the data*. That is, it must be based on some *a priori* knowledge about the situation.

In the coin tossing example, it is clear that a two-sided alternative is the appropriate choice. We have no reason to expect the coin, if biased, to be biased in a particular direction. So without looking at data, we wouldn't even know which of the two possible one-sided tests to do. For the lady tasting tea, we can at least identify a difference between the two possible alternatives *before collecting any data*: in one case she is correct more often than expected and in the other case she is *wrong* too often. What will we conclude if she gets 9 out of 10 wrong? Should we consider that to be evidence that she can indeed tell the difference between the two tea preparations? The answer to that question will essentially answer the question of whether to use a one-sided or two-sided alternative.

Example 2.4.2.

Q. A children's game uses a die that has a picture of a ghost named Hugo on one side and numbers on the other sides. If the die is fair, the ghost should be rolled 1 time in 6. You test the die by rolling 50 times, and the ghost is rolled 16 times. Is there any reason to be concerned that the die is not fair?

A. We can perform a binomial test with the following hypotheses:

- $H_0 : \pi = 1/6$,
- $H_a : \pi \neq 1/6$.

Not having any prior knowledge about the die, a two-sided alternative is appropriate.

Now we need to convert our test statistic, $x = 16$, into a p-value. We can do this directly with `binom.test()`.

binomtest-hugo01

```
> binom.test(16,50,1/6)

        Exact binomial test

data:  16 and 50
number of successes = 16, number of trials = 50, p-value =
0.006943
alternative hypothesis: true probability of success is not equal to 0.16667
95 percent confidence interval:
 0.19520 0.46699
sample estimates:
probability of success
              0.32
```

It is interesting to do this manually as well. Let's start with a one-sided test and compare to the results of `binom.test()`.

binomtest-hugo02

```
# one-sided test manually and using binom.test()
> 1-pbinom(15,50,1/6);
[1] 0.0057345
> binom.test(16,50,1/6,alternative="greater");

        Exact binomial test

data:  16 and 50
number of successes = 16, number of trials = 50, p-value =
0.005734
alternative hypothesis: true probability of success is greater than 0.16667
95 percent confidence interval:
 0.21210 1.00000
sample estimates:
probability of success
              0.32
```

Obtaining a two-sided p-value is a bit more challenging this time. Since $X \sim$ Binom$(50, 1/6)$, the p-value should be $\mathrm{P}(X \geq 16) + \mathrm{P}(X \leq k)$ for some number k. But what number k do we use? The distribution of X is not symmetric in this case, and the mean ($50/6 = 8.33$) is not an integer, so it isn't so easy to simply take the "mirror image" like we did when the null hypothesis was $\pi = 0.5$. The usual solution is quite clever. We will add up $\mathrm{P}(X = x)$ for all values of x with

$P(X = x) \leq P(X = 16)$:

$$\text{p-value} = \sum_{P(X=x) \leq P(X=16)} P(X = x) \, .$$

That is, we add the probabilities for all values that are at least as unusual as the value obtained from our data.

```
# finding the "other side" by inspection:                          binomtest-hugo03
> dbinom(16,50,1/6);
[1] 0.0035458
> rbind(0:4,dbinom(0:4,50,1/6));
           [,1]      [,2]       [,3]    [,4]     [,5]
[1,] 0.00000000 1.0000000 2.0000000 3.00000 4.00000
[2,] 0.00010988 0.0010988 0.0053844 0.01723 0.04049
>
# this should match the p-value from binom.test()
> pbinom(1,50,1/6) + 1 - pbinom(15,50,1/6);
[1] 0.0069432
# letting R automate finding the interval too:
> probs <- dbinom(0:50,50,1/6); sum(probs[probs <= dbinom(16,50,1/6)])
[1] 0.0069432
```

Since $P(X = x) \leq P(X = 16)$ when $x = 0$ or $x = 1$, we get the left tail probability from `pbinom(1,50,1/6)`.

Note: This situation is based on a game the author once played with his children. Basing his strategy on the expected number of ghosts that would be rolled, he lost badly and became suspicious of the die. In fact, the die had *two* ghosts (on opposite sides).   ◁

2.4.2. Types of Error and Statistical Power

When we carry out a hypothesis test, there are two kinds of mistakes we could make. It could be that the null hypothesis is true but that we reject it (because the p-value is small by random chance). This is called **type I error**. If we decide in advance what amount of type I error we can live with, that amount is called the significance level of the test. Usually it is denoted by α. You may see the result of a hypothesis test reported in terms of α instead of with a p-value. In our example above, we could say that "our results were not significant at the $\alpha = 0.05$ level" (or that they were significant at the $\alpha = 0.10$ level). This style of reporting used to be especially common when calculating p-values was more cumbersome than it is with today's computers and is equivalent to saying whether our p-value was above (not significant) or below (significant) our pre-specified threshold α. If we pick a significance level α in advance and the null hypothesis is true, then the probability of type I error is α.

On the other hand, it could be that the null hypothesis is false but that we do not reject it. This is called **type II error**. The probability of type II error is usually denoted β, but it is not as straightforward to calculate. This is because the probability of type II error depends on two things:

	reject H_0	don't reject H_0
H_0 is true	type I error	☺
H_0 is false	☺	type II error

Figure 2.8. Types of error when conducting hypothesis tests.

- the value of α that will be used and
- "just how wrong" the null hypothesis is.

Suppose that we choose $\alpha = 0.05$. Then we will reject the null hypothesis if the number of heads in our sample is less than 40 or greater than 60:

binomtest-power01

```
> qbinom(0.025,100,0.5)        # find q with pbinom(q,100,0.5) >= 0.025
[1] 40
> pbinom(39:40,100,0.5)        # double checking
[1] 0.017600 0.028444
```

On the other hand, if the coin is biased but $40 \le x \le 60$, then we will make a type II error. The values of our test statistic that lead to rejection of H_0 ($[0, 39] \cup [61, 100]$ in our example) are called the **rejection region** for the test. The boundary values of the rejection region are called **critical values**. (Note: This has nothing to do with critical values from calculus.)

If our coin is biased so that heads actually occurs 95% of the time, (i.e., $\pi = 0.95$ is our particular alternative), then we will be very likely to reject the null hypothesis and β will be small. In fact, we can easily calculate it:

binomtest-power02

```
> pbinom(60,100,0.95) - pbinom(39,100,0.95);
[1] 6.2386e-26
```

The chances of making the wrong decision – type II error – are very small in this case.

But if the bias is smaller and the coin comes up heads 55% of the time instead, then the probability of making a type II error is quite large:

binomtest-power03

```
> pbinom(60,100,0.55) - pbinom(39,100,0.55);
[1] 0.8648
```

This shows that 100 coin tosses isn't very likely to catch a coin with only a modest bias (differing from a fair coin by 5% or less).

Similarly, for any particular alternative value of π, we can calculate the probability of making a type II error. The **power** of the test against that alternative is $1 - \beta$, which is the probability that we will make the correct decision (reject the null hypothesis) when that alternative is true. That is, β and power are really functions of π_a, the probability of getting heads in some particular alternative. These functions can be plotted (by calculating a number of values and "connecting the dots").

binomtest-power05

```
> p <- seq(0,1,by=0.02);
> power <- 1 - ( pbinom(60,100,p) - pbinom(39,100,p) );
```

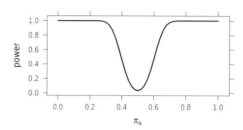

Figure 2.9. The power of a binomial test depends on π_a, the alternative value of π.

```
> myplot <- xyplot(power~p,ylab="power",xlab=expression(pi[a]),
+           type='l', lwd=2);
>
```

Figure 2.9 shows that the power is very low when π_a is near 0.5 but is nearly 1 when π_a is sufficiently far from 0.5.

Another way to explore power is to fix π_a and let n vary. This can help us determine how many coins to toss based on our desired power (see Figure 2.10).

binomtest-power06

```
> p <- rep(c(0.52,0.55,0.60), each=2000);
> plab <- paste("alt prob =", as.character(p));
> n <- rep(1:2000,times=3);
> critical <- qbinom(0.025,size=n,prob=p);
> power <- 1 - ( pbinom(n-critical+1,n,p) - pbinom(critical-1,n,p) );
> myplot <- xyplot(power~n|plab,ylab="power",xlab="number of coin tosses",
+           ylim=c(0,1.1), type='l', lwd=2);
```

A test is said to be **under-powered** if we collect too little data to have much chance of detecting an effect of some desired magnitude and **over-powered** if we collect more data than were necessary. The design of a good statistical study will include a **power analysis** that attempts to determine a reasonable sample size given some assumptions about effect size.

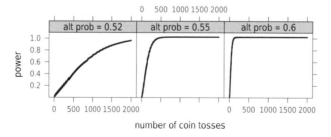

Figure 2.10. Power curves show how power depends on sample size for various values of π_a.

2.5. Mean and Variance of a Discrete Random Variable

Just as numerical summaries of a data set can help us understand our data, numerical summaries of the distribution of a random variable can help us understand the behavior of that random variable. In this section we develop two of the most important numerical summaries of random variables: mean and variance. In each case, we will use our experience with data to help us develop a definition.

2.5.1. The Mean of a Random Variable

Example 2.5.1.

Q. Let's begin with a motivating example. Suppose a student has taken 10 courses and received 5 A's, 4 B's, and 1 C. Using the traditional numerical scale where an A is worth 4, a B is worth 3, and a C is worth 2, what is this student's GPA (grade point average)?

A. The first thing to notice is that $\frac{4+3+2}{3} = 3$ is *not* correct. We cannot simply add up the values and divide by the number of values. Clearly this student should have a GPA that is higher than 3.0, since there were more A's than C's.

Consider now a correct way to do this calculation:

$$\text{GPA} = \frac{4+4+4+4+4+3+3+3+3+2}{10}$$

$$= \frac{5 \cdot 4 + 4 \cdot 3 + 1 \cdot 2}{10}$$

$$= \frac{5}{10} \cdot 4 + \frac{4}{10} \cdot 3 + \frac{1}{10} \cdot 2$$

$$= 4 \cdot \frac{5}{10} + 3 \cdot \frac{4}{10} + 2 \cdot \frac{1}{10}$$

$$= 3.4 \ . \qquad \qquad \triangleleft$$

Our definition of the mean of a random variable follows the example above. Notice that we can think of the GPA as a sum of products:

$$\text{GPA} = \sum (\text{grade})(\text{probability of getting that grade}) \ .$$

Such a sum is often called a weighted sum or weighted average of the grades (the probabilities are the weights). The expected value of a discrete random variable is a similar weighted average of its possible values.

Definition 2.5.1. Let X be a discrete random variable with pmf f. The *mean* (also called *expected value*) of X is denoted as μ_X or $\mathrm{E}(X)$ and is defined by

$$\mu_X = \mathrm{E}(X) = \sum_x x \cdot f(x) \ .$$

The sum is taken over all possible values of X. □

Example 2.5.2.

Q. If we flip four fair coins and let X count the number of heads, what is $\mathrm{E}(X)$?

A. Recall that if we flip four fair coins and let X count the number of heads, then the distribution of X is described by the following table:

value of X	0	1	2	3	4
probability	$\dfrac{1}{16}$	$\dfrac{4}{16}$	$\dfrac{6}{16}$	$\dfrac{4}{16}$	$\dfrac{1}{16}$

So the expected value is

$$0 \cdot \frac{1}{16} + 1 \cdot \frac{4}{16} + 2 \cdot \frac{6}{16} + 3 \cdot \frac{4}{16} + 4 \cdot \frac{1}{16} = 2 \,.$$

On average we get 2 heads in 4 tosses.

We can also calculate this using R.

```
> vals <- 0:4                                         discDists-mean-coins01
> probs <- c(1,4,6,4,1) / 16      # providing probabilities directly
> sum(vals * probs)
[1] 2
> sum(0:4 * dbinom(0:4,4,0.5))     # using the fact that X is binomial
[1] 2                                                                    ◁
```

Example 2.5.3.

Q. If we toss a fair coin until we get 3 heads, on average how many tosses will that be?

A. We can use the negative binomial distribution for this. Let $X \sim \mathsf{NBinom}(3, 0.5)$. The quantity we really want is $Y = X + 3$. (Remember, X only counts the tails, but we are asked about the number of *tosses*.) We can easily approximate $\mathrm{E}(X)$ using R:

```
> sum(0:100 * dnbinom(0:100,3,0.5))                   discDists-mean-coins02
[1] 3
```

While it is possible that $X > 100$, it is extremely unlikely, and indeed 3 is the exact value. Rather than using a series to show that this is true, we will derive general formulas for the expected value of any binomial or negative binomial random variable shortly.

It seems plausible that $\mathrm{E}(Y) = \mathrm{E}(X) + 3$ since $Y = X + 3$. This is also true and follows from Lemma 2.5.3. So pending a couple steps that we will justify momentarily, we have determined that the average number of flips until we obtain 3 heads is 6.   ◁

Before we can show that $\mathrm{E}(X + 3) = \mathrm{E}(X) + 3$, we first need to state precisely what $X + 3$ means. Intuitively, whatever random process generates X could be executed and then 3 added to the result. The definition below formalizes this idea and allows for an arbitrary transformation t.

Definition 2.5.2. Let X be a discrete random variable with pmf f. Let $t : \mathbb{R} \to \mathbb{R}$. Then $t(X)$ is a discrete random variable and

$$\mathrm{P}(Y = y) = \sum_{t(x)=y} \mathrm{P}(X = x) = \sum_{t(x)=y} f(x) \,.$$

□

Example 2.5.4.

Q. The pmf for X is given by the following table:

value of X	-2	-1	0	1	2
probability	0.05	0.10	0.35	0.30	0.20

Determine the pmf for X^2 and calculate $E(X^2)$. Compare this with $E(X)^2$.

A. The possible values of X^2 are 0, 1, and 4. The probabilities are easily computed. For example, $P(X^2 = 4) = P(X = 2) + P(X = -2) = 0.25$ because $2^2 = 4$ and $(-2)^2 = 4$. The rest of the table is determined similarly:

value of X	0	1	4
probability	0.35	0.40	0.25

From this it follows that

$$E(X^2) = 0(0.35) + 1(0.40) + 4(0.25) = 1.4 .$$

Note that

$$E(X)^2 = [(-2)(0.05) + (-1)(0.10) + 0(0.35) + 1(0.30) + 2(0.20)]^2 = (0.50)^2 = 0.25 ,$$

so $E(X^2) \neq E(X)^2$. ◁

Example 2.5.5.

Q. Let $X \sim \text{Binom}(n, \pi)$. What is the pmf for $X + 3$?

A. Let $Y = X + 3$. Then $P(Y = y) = P(X = y - 3)$, so the pmf for Y is

$$f_Y(y) = f_X(y - 3) = \binom{n}{y-3} \pi^{y-3} (1 - \pi)^{n-y+3}$$

provided $3 \leq y \leq n + 3$. ◁

As Example 2.5.4 shows, it is not always the case that $E(t(x)) = t(E(X))$ (see Exercise 2.58). But if the transformation is linear, then we have this desirable equality.

Lemma 2.5.3. *Let X be a discrete random variable, let a and b be constants, and let $Y = aX + b$. Then*

$$E(Y) = a E(X) + b .$$

Proof. First note that if $a = 0$, then the result is easy since in that case $Y = b$ and

$$E(b) = b \cdot 1 = b .$$

Now suppose $a \neq 0$, and let f be the pmf of X. Because t is linear and $a \neq 0$, t is one-to-one. Therefore, $P(X = x) = P(Y = t(x))$. Now we calculate the expected value of Y from the definition:

$$
\begin{aligned}
E(Y) &= \sum (ax + b) f(x) \\
&= \sum [axf(x) + bf(x)] \\
&= a \sum xf(x) + b \sum f(x) \\
&= a\,E(X) + b \cdot 1 \\
&= a\,E(X) + b \, . \qquad \qquad \square
\end{aligned}
$$

The expected value of a binomial random variable is just what you would expect.

Theorem 2.5.4. *Let $X \sim \mathsf{Binom}(n, \pi)$. Then $E(X) = n\pi$.*

Proof. We will learn other easier ways to calculate the expected value later, but for the moment our only tool is the definition:

$$
E(X) = \sum_{x=0}^{n} x \binom{n}{x} \pi^x (1 - \pi)^{n-x} \tag{2.5}
$$

$$
= \sum_{x=1}^{n} x \binom{n}{x} \pi^x (1 - \pi)^{n-x} \tag{2.6}
$$

$$
= n\pi \sum_{x=1}^{n} \binom{n-1}{x-1} \pi^{x-1} (1 - \pi)^{n-x} \tag{2.7}
$$

$$
= n\pi \sum_{y=0}^{n-1} \binom{n-1}{y} \pi^y (1 - \pi)^{(n-1)-y} \tag{2.8}
$$

$$
= n\pi \, . \tag{2.9}
$$

The justification of each step above is left as Exercise 2.59. $\qquad \square$

Theorem 2.5.5. *Let $X \sim \mathsf{NBinom}(s, \pi)$. Then $E(X) = \frac{s}{\pi} - s$.*

Proof. Again we will learn other easier ways to calculate this expected value later, but for the moment we must rely on algebraic manipulation. We will only prove the case when $s = 1$, that is, when $X \sim \mathsf{Geom}(\pi)$. A full proof will wait until we have a bit more machinery (moment generating functions) at our disposal. (See Exercise 3.27.)

$$q^1$$

$$q^2 \qquad q^2$$

$$q^3 \qquad q^3 \qquad q^3$$

$$q^4 \qquad q^4 \qquad q^4 \qquad q^4$$

$$\vdots \qquad \vdots \qquad \vdots \qquad \vdots \qquad \ddots$$

$$\downarrow \qquad \downarrow \qquad \downarrow \qquad \downarrow$$

$$\frac{q^1}{1-q} \quad \frac{q^2}{1-q} \quad \frac{q^3}{1-q} \quad \frac{q^4}{1-q} \quad \cdots \quad \rightarrow \quad \frac{q}{1-q} \cdot \frac{1}{1-q}$$

Figure 2.11. An illustration of the proof of Theorem 2.5.5.

Let $X \sim \mathsf{Geom}(\pi) = \mathsf{NBinom}(1, \pi)$, and let $q = 1 - \pi$. Then

$$
\begin{aligned}
\mathrm{E}(X) &= \sum_{x=0}^{\infty} x(1-\pi)^x \pi \\
&= \pi \sum_{x=1}^{\infty} x q^x \\
&= \pi \left[\sum_{x=1}^{\infty} q^x + \sum_{x=2}^{\infty} q^x + \sum_{x=3}^{\infty} q^x + \cdots \right] \\
&= \pi \left[\frac{q}{1-q} + \frac{q^2}{1-q} + \frac{q^3}{1-q} + \cdots \right] \\
&= \pi \left[\frac{\frac{q}{1-q}}{1-q} \right] \\
&= \pi \frac{q/\pi}{\pi} \\
&= \frac{q}{\pi} = \frac{1-\pi}{\pi} = \frac{1}{\pi} - 1 \, .
\end{aligned}
$$

Figure 2.11 illustrates the reasoning used above. This proof does not generalize easily to negative binomial distributions with $s > 1$ because the binomial coefficient $\binom{x+s-1}{x} \neq 1$ when $s > 1$. $\qquad\square$

2.5.2. The Variance of a Random Variable

Variance is a measure of spread; it is the mean squared deviation from the mean.

Definition 2.5.6. Let X be a discrete random variable. The *variance* of X is defined by

$$\sigma_X^2 = \mathrm{Var}(X) = \mathrm{E}((X - \mu_X)^2) \, .$$

The *standard deviation* is the square root of the variance. $\qquad\square$

Example 2.5.6. Let $X \sim \mathrm{Binom}(2, 0.5)$. Then $\mathrm{E}(X) = 1$, and

$$
\begin{aligned}
\mathrm{Var}(X) &= (0-1)^2 \cdot 0.25 &+& (1-1)^2 \cdot 0.5 &+& (2-1)^2 \cdot 0.25 \\
&= 0.25 &+& 0 &+& 0.25 \\
&= 0.5 \; .
\end{aligned}
$$

Shortly, we will learn a formula for the variance of a general binomial random variable. For now, the variance of binomial random variables with small values of n can be easily obtained in R. Here are some examples:

```
> x <- 0:2                                    discDists-var-example01
> sum( (x - 1)^2 * dbinom(x,2,0.5) )   # same as above
[1] 0.5
> n <- 5; p <- 0.2; x <- 0:n
> sum( (x - n*p)^2 * dbinom(x,n,p) )   # X ~ Binom(5,0.2)
[1] 0.8
> n <- 5; p <- 0.8; x <- 0:n
> sum( (x - n*p)^2 * dbinom(x,n,p) )   # X ~ Binom(5,0.8)
[1] 0.8
> n <- 10; p <- 0.8; x <- 0:n
> sum( (x - n*p)^2 * dbinom(x,n,p) )   # X ~ Binom(10,0.8)
[1] 1.6
> n <- 20; p <- 0.8; x <- 0:n
> sum( (x - n*p)^2 * dbinom(x,n,p) )   # X ~ Binom(20,0.8)
[1] 3.2
```

You may observe some interesting patterns in these variances. Try some additional examples to see if the patterns hold.   ◁

There is another equivalent formula for the variance that is not as intuitive as the definition but is often easier to use in practice.

Theorem 2.5.7. *Let X be a discrete random variable. Then*

$$
\mathrm{Var}(X) = \mathrm{E}(X^2) - [\mathrm{E}(X)]^2 \; .
$$

Proof. Let f be the pmf of X and let $\mu = \mathrm{E}(X)$. Then

$$
\mathrm{Var}(X) = \sum (x - \mu)^2 f(x) \tag{2.10}
$$

$$
= \sum (x^2 - 2x\mu + \mu^2) f(x) \tag{2.11}
$$

$$
= \sum x^2 f(x) - \sum 2x\mu f(x) + \sum \mu^2 f(x) \tag{2.12}
$$

$$
= \mathrm{E}(X^2) - 2\mu \sum x f(x) + \mu^2 \sum f(x) \tag{2.13}
$$

$$
= \mathrm{E}(X^2) - 2\mu \cdot \mu + \mu^2 \cdot 1 \tag{2.14}
$$

$$
= \mathrm{E}(X^2) - \mu^2 = \mathrm{E}(X^2) - \mathrm{E}(X)^2 \; . \tag{2.15}
$$

$\square$

An analog of Lemma 2.5.3 for variance is given in the following lemma.

Lemma 2.5.8. *Let X be a discrete random variable, and let a and b be constants. Then*

$$\text{Var}(aX + b) = a^2 \text{Var}(X) \,.$$

Proof. Exercise 2.61. $\qquad\qquad\qquad\qquad\qquad\qquad\qquad\qquad\qquad\qquad$ $\square$

Example 2.5.7.

Q. Find the mean and variance of the random variable Y described below:

value of Y	1	2	3	4
probability	0.05	0.20	0.40	0.35

A. The following R code does the necessary calculations. The variance can be calculated directly from the definition or using Theorem 2.5.7.

```
                                                          varExampleFromTable01
> y <- 1:4
> prob <- c(0.05,0.20,0.40,0.35)
> mean.y <- sum(y*prob); mean.y         # E(Y)
[1] 3.05
> sum((y-mean.y)^2 * prob)              # Var(Y)
[1] 0.7475
> sum(y^2 *prob) - mean.y^2             # Var(Y) again
[1] 0.7475                                                              ◁
```

Example 2.5.8.

Q. If $X \sim \text{Binom}(1, \pi)$, then X is called a **Bernoulli random variable**. For a Bernoulli random variable X, what are $\text{E}(X)$ and $\text{Var}(X)$?

A. We already know that $\text{E}(X) = n\pi = 1\pi = \pi$. We can calculate the variance directly from the definition:

$$\begin{aligned}
\text{Var}(X) &= (0 - \pi)^2(1 - \pi) + (1 - \pi)^2\pi \\
&= \pi^2(1 - \pi) + \pi(1 - \pi)^2 \\
&= \pi(1 - \pi)\left[(1 - \pi) + \pi\right] \\
&= \pi(1 - \pi) \,.
\end{aligned}$$

We could also calculate $\text{Var}(X)$ using Theorem 2.5.7. First note that $X^2 = X$ for a Bernoulli random variable since $0^2 = 0$ and $1^2 = 1$. Therefore,

$$\begin{aligned}
\text{Var}(X) &= \text{E}(X^2) - \text{E}(X)^2 \\
&= \text{E}(X) - \text{E}(X)^2 \\
&= \pi - \pi^2 \\
&= \pi(1 - \pi) \,.
\end{aligned}$$

As a function of π, $\text{Var}(X)$ is quadratic. Its graph is a parabola that opens downward, and since $\text{Var}(X) = 0$ when $\pi = 0$ and when $\pi = 1$, the largest variance occurs when $\pi = \frac{1}{2}$. (See Figure 2.12.) $\qquad\qquad\qquad\qquad\qquad$ ◁

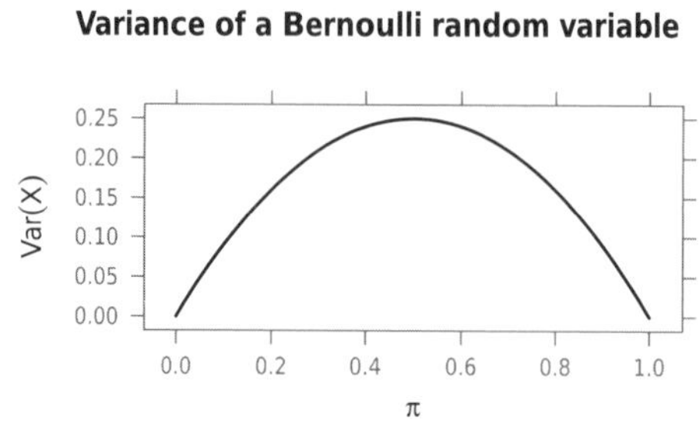

Variance of a Bernoulli random variable

Figure 2.12. The variance of a $\mathsf{Binom}(1, \pi)$ random variable depends on π. The maximum variance occurs when $\pi = 0.5$.

Any random variable that only takes on the values 0 and 1 is a Bernoulli random variable. Bernoulli random variables will play an important role in determining the variance of a general binomial random variable. But first we need to learn how to compute the mean and variance of a sum. This, in turn, requires that we learn about joint distributions.

2.6. Joint Distributions

In this section we will develop formulas for $\mathrm{E}(X + Y)$ and $\mathrm{Var}(X + Y)$ and use them to more simply compute the means and variances of random variables that can be expressed as sums (as can the binomial and negative binomial random variables). But first we need to talk about the **joint distribution** of two random variables.

2.6.1. Joint Probability Mass Functions

Just as the pmf of a discrete random variable X must describe the probability of each value of X, the **joint pmf** of X and Y must describe the probability of any *combination* of values that X and Y could have. We could describe the joint pmf of random variables X and Y using a table like the one in the following example.

Example 2.6.1. The joint distribution of X and Y is described by the following table:

		value of X		
		1	2	3
	1	0.17	0.15	0.08
value of Y	2	0.00	0.10	0.10
	3	0.08	0.20	0.12

We can now calculate the probabilities of a number of events:

- $\mathrm{P}(X = 2) = 0.15 + 0.10 + 0.20 = 0.45$,
- $\mathrm{P}(Y = 2) = 0.00 + 0.10 + 0.10 = 0.20$,
- $\mathrm{P}(X = Y) = 0.17 + 0.10 + 0.12 = 0.39$,

- $P(X > Y) = 0.15 + 0.08 + 0.10 = 0.33$.

The first two probabilities show that we can recover the pmfs for X and for Y from the joint distribution. These are known as the **marginal distributions** of X and Y because they can be obtained by summing across rows or down columns, and the natural place to record those values is in the margins of the table:

		value of X			
		1	2	3	total
	1	0.17	0.15	0.08	0.40
value of Y	2	0.00	0.10	0.10	0.20
	3	0.08	0.20	0.12	0.40
	total	0.25	0.45	0.30	1.00

We can also compute **conditional distributions**. Conditional distributions give probabilities for one random variable for a given value of the other. For example,

$$P(Y = 2 \mid X = 2) = \frac{P(Y = 2 \text{ and } X = 2)}{P(X = 2)} = \frac{0.10}{0.45} = \frac{2}{9} = 0.2222 \,.$$

Notice the general form that applies here:

$$\text{conditional} = \frac{\text{joint}}{\text{marginal}} \,. \qquad \triangleleft$$

Tables like the ones in Example 2.6.1 are really just descriptions of functions, so our formal definitions are the following.

Definition 2.6.1. The *joint pmf* of a pair of discrete random variables $\langle X, Y \rangle$ is a function $f : \mathbb{R} \times \mathbb{R} \to \mathbb{R}$ such that for all x and y,

$$P(X = x \text{ and } Y = y) = f(x, y) \,. \qquad \square$$

Definition 2.6.2. If f is the joint pmf for X and Y, then the *marginal distributions* of X and Y are defined by

- $f_X(x) = \sum_y f(x, y)$, and

- $f_Y(y) = \sum_x f(x, y).$ $\qquad \square$

Definition 2.6.3. If f is the joint pmf for X and Y, then the *conditional distributions* $f_{X|Y=y}$ and $f_{Y|X=x}$ are defined by

- $f_{X|Y=y}(x) = \dfrac{f(x, y)}{f_Y(y)}$, and

- $f_{Y|X=x}(y) = \dfrac{f(x, y)}{f_X(x)}.$ $\qquad \square$

An important (and generally easier) special case of joint distributions is the distribution of independent random variables.

Definition 2.6.4. Random variables X and Y are *independent* if for every x and y

$$f(x,y) = f_X(x) \cdot f_Y(y) \,,$$

that is, if

$$\mathrm{P}(X = x \text{ and } Y = y) = \mathrm{P}(X = x) \cdot \mathrm{P}(Y = y) \,. \qquad \square$$

Example 2.6.2. The variables X and Y from Example 2.6.1 are not independent. This can be seen by observing that

- $\mathrm{P}(X = 2) = 0.45$, and
- $\mathrm{P}(Y = 2) = 0.20$, but
- $\mathrm{P}(X = 2 \text{ and } Y = 2) = 0.10 \neq 0.45 \cdot 0.20 = 0.09$

or that

- $\mathrm{P}(X = 1) = 0.25$, and
- $\mathrm{P}(Y = 2) = 0.20$, but
- $\mathrm{P}(X = 1 \text{ and } Y = 2) = 0.00 \neq 0.25 \cdot 0.20.$

The fact that

- $\mathrm{P}(X = 3) = 0.30$,
- $\mathrm{P}(Y = 3) = 0.40$, and
- $\mathrm{P}(X = 3 \text{ and } Y = 3) = 0.12 = 0.30 \cdot 0.40$

is not enough to make the variables independent. $\triangleleft$

Just as we defined the transformation of a single random variable, we can now define functions of jointly distributed random variables.

Definition 2.6.5. Let f be the joint pmf of discrete random variables X and Y and let $t : \mathbb{R} \times \mathbb{R} \to \mathbb{R}$. Then $t(X, Y)$ is a discrete random variable with pmf given by

$$\mathrm{P}\left(t(X,Y) = k\right) = \sum_{t(x,y)=k} \mathrm{P}(X = x \text{ and } Y = y) = \sum_{t(x,y)=k} f(x,y) \,. \qquad \square$$

Example 2.6.3. Continuing with Example 2.6.1, we can determine the pmf for $X + Y$. The possible values of $X + Y$ are from 2 to 6. The table below is formed by adding the appropriate cell probabilities from the table on page 72:

value of $X + Y$	2	3	4	5	6
probability	0.17	0.15	0.26	0.30	0.12

$\triangleleft$

2.6.2. More Than Two Variables

All the definitions of the previous section can be extended to handle the case of $k \geq 3$ random variables. A joint pmf is then a function $f : \mathbb{R}^k \to \mathbb{R}$ such that

$$f(x_1, \ldots, x_k) = \mathrm{P}(X_1 = x_1 \text{ and } \cdots \text{ and } X_k = x_k) \,.$$

Marginal distributions (for any subset of the variables) are obtained by summing over all values of the other variables, and conditional distributions follow the same pattern as before:

$$\text{conditional} = \frac{\text{joint}}{\text{marginal}} .$$

The only definition that is somewhat tricky is the definition of independence.

Definition 2.6.6. Let $X_1, \ldots, X_k$ be k jointly distributed discrete random variables with joint pmf $f : \mathbb{R}^k \to \mathbb{R}$. Then:

(1) $X_1, \ldots, X_k$ are *pairwise independent* if X_i and X_j are independent whenever $i \neq j$.

(2) $X_1, \ldots, X_k$ are *independent* if the joint pmf factors into a product of marginals:

$$f(x_1, \ldots x_k) = f_{X_1}(x_1) \cdot f_{X_2}(x_2) \cdots f_{X_k}(x_k) . \qquad \square$$

Independence is stronger than pairwise independence. In Exercise 2.70 you are asked to find an example that shows this.

2.6.3. The Mean and Variance of a Sum

Example 2.6.3 is especially important because a number of important situations can be expressed as sums of random variables.

- A binomial random variable is the sum of Bernoulli random variables.
  If $X \sim \mathsf{Binom}(n, \pi)$, then $X = X_1 + X_2 + \cdots X_n$ where

$$X_i = \begin{cases} 0 & \text{if the } i\text{th trial is a failure,} \\ 1 & \text{if the } i\text{th trial is a success.} \end{cases}$$

Each $X_i \sim \mathsf{Binom}(1, \pi)$, and the X_i's are independent by the definition of the binomial distributions.

- A negative binomial random variable is the sum of geometric random variables.
  If $X \sim \mathsf{NBinom}(r, \pi)$, then $X = X_1 + X_2 + \cdots + X_r$ where each $X_i \sim \mathsf{NBinom}(1, \pi)$. Here X_i is the number of failures between success $i - 1$ and success i. Once again, the X_i's are independent of each other since the trials are independent.

- When we compute a sample mean, we first add all of the data values.
  If we have an independent random sample from some population, then the sample mean is a random variable and

$$\overline{X} = \frac{1}{n} \left(X_1 + X_2 + \cdots + X_n \right) ,$$

where X_i is the value of the ith individual in the sample. This means that once we understand sums of random variables, we can reduce the study of sample means to the study of a sample of size 1.

For these reasons and more, the distribution of sums of random variables will arise again and again. Fortunately, the next theorem makes dealing with these sums very convenient.

Theorem 2.6.7. *Let X and Y be discrete random variables. Then*

(1) $E(X + Y) = E(X) + E(Y)$,

(2) $E(XY) = E(X) \cdot E(Y)$, *provided X and Y are independent, and*

(3) $\mathrm{Var}(X + Y) = \mathrm{Var}(X) + \mathrm{Var}(Y)$, *provided X and Y are independent.*

Proof. All three parts of this theorem are proved by algebraic manipulation of the definitions involved. For the second and third parts, it will be important that X and Y be independent – otherwise the argument fails. Independence is not required for the first part.

Let f and g be the (marginal) pmfs of X and Y, respectively, and let h be the joint pmf. Then

$$E(X + Y) = \sum_{x,y}(x + y)P(X = x \ \text{and} \ Y = y)$$

$$= \sum_{x,y} x\, h(x,y) + y\, h(x,y)$$

$$= \sum_{x}\sum_{y} x\, h(x,y) + \sum_{y}\sum_{x} y\, h(x,y)$$

$$= \sum_{x} x \sum_{y} h(x,y) + \sum_{y} y \sum_{x} h(x,y)$$

$$= \sum_{x} x \cdot P(X = x) + \sum_{y} y \cdot P(Y = y)$$

$$= E(X) + E(Y)\,.$$

Similarly, for part (2),

$$E(XY) = \sum_{x,y} xy\, P(X = x \ \text{and} \ Y = y) \tag{2.16}$$

$$= \sum_{x,y} xy\, f(x)g(y) \tag{2.17}$$

$$= \sum_{x}\sum_{y} xy\, f(x)g(y)$$

$$= \sum_{x} xf(x) \sum_{y} y\, g(y)$$

$$= \sum_{x} xf(x)\, E(Y)$$

$$= E(X) \cdot E(Y)\,.$$

In the argument above, we needed independence to conclude that (2.16) and (2.17) are equivalent.

For the variance, we use part (2) of the theorem and our shortcut method for calculating the variance:

$$\begin{aligned}
\mathrm{Var}(X+Y) &= \mathrm{E}((X+Y)^2) - [\mathrm{E}(X+Y)]^2 \\
&= \mathrm{E}(X^2 + 2XY + Y^2) - [\mathrm{E}(X) + \mathrm{E}(Y)]^2 \\
&= \mathrm{E}(X^2) + \mathrm{E}(2XY) + \mathrm{E}(Y^2) - [(\mathrm{E}(X))^2 + 2\,\mathrm{E}(X)\,\mathrm{E}(Y) + (\mathrm{E}(Y))^2] \\
&= \mathrm{E}(X^2) + 2\,\mathrm{E}(X)\,\mathrm{E}(Y) + \mathrm{E}(Y^2) - [\mathrm{E}(X)^2 + 2\,\mathrm{E}(X)\,\mathrm{E}(Y) + \mathrm{E}(Y)^2] \\
& \hspace{9cm} (2.18) \\
&= \mathrm{E}(X^2) + \mathrm{E}(Y^2) - \mathrm{E}(X)^2 - \mathrm{E}(Y)^2 \\
&= \mathrm{E}(X^2) - \mathrm{E}(X)^2 + \mathrm{E}(Y^2) - \mathrm{E}(Y)^2 \\
&= \mathrm{Var}(X) + \mathrm{Var}(Y) \,. \hspace{6cm} (2.19)
\end{aligned}$$

$\square$

Induction allows us to easily extend these results from sums of two random variables to sums of three or more random variables. (See Exercise 2.73.)

We now have the tools to compute the means and variances of binomial distributions.

Theorem 2.6.8. *If $X \sim \mathsf{Binom}(n, \pi)$, then $\mathrm{E}(X) = n\pi$ and $\mathrm{Var}(X) = n\pi(1-\pi)$.*

Proof. If $X \sim \mathsf{Binom}(n, \pi)$, then we can write $X = X_1 + X_2 + \cdots + X_n$ where $X_i \sim \mathsf{Binom}(1, \pi)$. We already know that $\mathrm{E}(X_i) = \pi$, that $\mathrm{Var}(X_i) = \pi(1-\pi)$, and that the X_i's are independent. So

$$\mathrm{E}(X) = \sum_{i=1}^{n} \mathrm{E}(X_i) = n\pi \,, \text{ and}$$

$$\mathrm{Var}(X) = \sum_{i=1}^{n} \mathrm{Var}(X_i) = n\pi(1-\pi) \,. \hspace{3cm} \square$$

This method will work for finding the mean of a negative binomial random variable as well by expressing a negative binomial random variable as the sum of geometric random variables. (See Exercise 2.74.) This will not yet enable us to determine the variance of a negative binomial distribution, however, because we still have not determined a formula for the variance of a geometric distribution. In Exercise 3.27 you are asked to use moment generating functions to prove the following theorem.

Theorem 2.6.9. *If $X \sim \mathsf{NBinom}(s, \pi)$, then $\mathrm{E}(X) = \dfrac{s}{\pi} - s$ and $\mathrm{Var}(X) = s\dfrac{1-\pi}{\pi^2}$.*

$\square$

2.6.4. Covariance

Let's take another look at the proof of Theorem 2.6.7. If X and Y are independent, then $\mathrm{E}(XY) = \mathrm{E}(X)\,\mathrm{E}(Y)$, so those terms cancel in (2.18). But what if X and Y are not independent? Then we get the following equation:

$$\mathrm{Var}(X+Y) = \mathrm{Var}(X) + \mathrm{Var}(Y) + 2\,\mathrm{E}(XY) - 2\,\mathrm{E}(X)\,\mathrm{E}(Y) \,.$$

Notice that if $X = Y$ (i.e., $P(X = Y) = 1$), then

$$\mathrm{E}(XY) - \mathrm{E}(X)\,\mathrm{E}(Y) = \mathrm{E}(X^2) - \mathrm{E}(X)^2 = \mathrm{Var}(X) \ .$$

This leads us to make the following definition and theorem.

Definition 2.6.10. Let X and Y be jointly distributed random variables. Then the *covariance* of X and Y is defined by

$$\mathrm{Cov}(X, Y) = \mathrm{E}(XY) - \mathrm{E}(X)\,\mathrm{E}(Y) \ . \qquad \Box$$

An immediate consequence of this definition is the following lemma.

Lemma 2.6.11. *Let X and Y be jointly distributed random variables (possibly dependent). Then*

$$\mathrm{Var}(X + Y) = \mathrm{Var}(X) + \mathrm{Var}(Y) + 2\,\mathrm{Cov}(X, Y) \ . \qquad \Box$$

Lemma 2.6.11 can be generalized to more than two random variables. You are asked to do this in Exercise 2.75.

The intuition for the name *covariance* – besides its connection to the variance – is easier to explain using the following lemma.

Lemma 2.6.12. *Let X and Y be jointly distributed random variables. Then*

$$\mathrm{Cov}(X, Y) = \mathrm{E}\left[(X - \mu_X)(Y - \mu_Y)\right] \ .$$

Proof. Exercise 2.76. $\qquad \Box$

The expression $(X - \mu_X)(Y - \mu_Y)$ is positive when X and Y are either both greater than their means or both less than their means; it is negative when one is larger and the other smaller. So $\mathrm{Cov}(X, Y)$ will be positive if X and Y are usually large together or small together – that is, if they *vary together*. Similarly, $\mathrm{Cov}(X, Y)$ will be negative when X and Y vary in opposite directions, and $\mathrm{Cov}(X, Y)$ will be near 0 when large values of X occur about equally with large or small values of Y. In particular, as we have already seen,

Lemma 2.6.13. *If X and Y are independent, then $\mathrm{Cov}(X, Y) = 0$.* $\qquad \Box$

Unfortunately, the converse of Lemma 2.6.13 is not true. (See Exercise 2.77.) Computation of covariance is often aided by the following lemma.

Lemma 2.6.14. *Let X and Y be jointly distributed random variables. Then:*

(1) $\mathrm{Cov}(X, Y) = \mathrm{Cov}(Y, X)$.

(2) $\mathrm{Cov}(a + X, Y) = \mathrm{Cov}(X, Y)$.

(3) $\mathrm{Cov}(aX, bY) = ab\,\mathrm{Cov}(X, Y)$.

(4) $\mathrm{Cov}(X, Y + Z) = \mathrm{Cov}(X, Y) + \mathrm{Cov}(X, Z)$.

(5) $\mathrm{Cov}(aW + bX, cY + dZ) = ac\,\mathrm{Cov}(W, Y) + ad\,\mathrm{Cov}(W, Z) + bc\,\mathrm{Cov}(X, Y) + bd\,\mathrm{Cov}(X, Z)$.

Proof. Each of these can be proved by straightforward algebraic manipulations using the previously established properties of expected value. See Exercise 2.78. $\quad \Box$

Lemma 2.6.14 allows us to calculate the variance of sums of more than two random variables.

Lemma 2.6.15. *Let* $X_1, X_2, \ldots, X_k$ *be jointly distributed random variables. Then*

$$\operatorname{Var}(X_1 + X_2 + \cdots + X_k) = \sum_i \operatorname{Var}(X_i) + 2 \sum_{i<j} \operatorname{Cov}(X_i, X_j)$$

$$= \sum_{i,j} \operatorname{Cov}(X_i, X_j) \,.$$

Proof. Exercise 2.75. □

Lemma 2.6.12 also motivates the definition of a "unitless" version of covariance, called the correlation coefficient.

Definition 2.6.16. The correlation coefficient ρ of two random variables X and Y is defined by

$$\rho = \frac{\operatorname{Cov}(X,Y)}{\sqrt{\operatorname{Var}(X) \cdot \operatorname{Var}(Y)}} \,.$$

If we let $\sigma_{XY} = \operatorname{Cov}(X,Y)$, then this is equivalent to

$$\rho = \frac{\sigma_{XY}}{\sigma_X \sigma_Y} \,. \qquad \square$$

Lemma 2.6.17. *Let* ρ *be the correlation coefficient of random variables* X *and* Y. *Then*

$$-1 \leq \rho \leq 1 \,.$$

Proof.

$$\begin{aligned}
0 \leq \operatorname{Var}\left(\frac{X}{\sigma_X} + \frac{Y}{\sigma_Y}\right) &= \operatorname{Var}\left(\frac{X}{\sigma_X}\right) + \operatorname{Var}\left(\frac{Y}{\sigma_Y}\right) + 2\operatorname{Cov}\left(\frac{X}{\sigma_X}, \frac{Y}{\sigma_Y}\right) \\
&= \frac{1}{\sigma_X^2} \operatorname{Var}(X) + \frac{1}{\sigma_Y^2} \operatorname{Var}(Y) + \frac{1}{\sigma_X \sigma_Y} 2\operatorname{Cov}(X,Y) \\
&= 1 + 1 + 2\rho \,,
\end{aligned}$$

from which it follows that $\rho \geq -1$.

The other inequality is proved similarly by considering $\operatorname{Var}(\frac{X}{\sigma_X} - \frac{Y}{\sigma_Y})$. □

Lemma 2.6.18. *Let* X *and* Y *be jointly distributed variables such that* $\rho = \pm 1$. *Then there are constants* a *and* b *such that*

$$\mathrm{P}(Y = a + bX) = 1 \,.$$

Proof. If $\rho = -1$, then $\operatorname{Var}(\frac{X}{\sigma_X} + \frac{Y}{\sigma_Y}) = 0$, so for some constant c,

$$\mathrm{P}\left(\frac{X}{\sigma_X} + \frac{Y}{\sigma_Y} = c\right) = 1 \,.$$

Algebraic rearrangement gives

$$\mathrm{P}\left(Y = c\sigma_Y - \frac{\sigma_Y}{\sigma_X} X\right) = 1 \,.$$

Note that the slope of this linear transformation is negative.

The proof in the case that $\rho = 1$ is similar. When $\rho = 1$, the slope will be positive. □

2.7. Other Discrete Distributions

2.7.1. Poisson Distributions

We will motivate the Poisson random variable with a situation in which it arises and derive its pmf. Let's suppose the following:

- Some event is happening "at random times".
- The probability of an event occurring in any small time interval depends only on (and is proportional to) the length of the interval, not on when it occurs.

The Poisson random variables X counts the number of occurrences in a fixed amount of time. In Chapter 3 we will learn that in this situation, the random variable Y that measures the amount of time until the next occurrence is an exponential random variable and, therefore, that there is an important connection between Poisson distributions and exponential distributions.

Our derivation of the pmf for X begins by approximating X with something we already know, namely the binomial distributions, using the following chain of reasoning.

(1) If we divide our time into n pieces, then the probability of an occurrence in one of these sub-intervals is proportional to $1/n$ – so let's call the probability λ/n.

(2) If n is large, then the probability of having two occurrences in one sub-interval is very small – so let's just pretend it can't happen.

(3) Since the number of occurrences in each interval is independent of the number in the other sub-intervals, a good approximation for X is

$$X \approx \mathsf{Binom}(n, \lambda/n)$$

because we have n independent sub-intervals with probability λ/n of an occurrence in each one.

(4) This approximation should get better and better as $n \to \infty$. When n is very large, then

$$\begin{aligned}
\mathrm{P}(X = x) &\approx \binom{n}{x}\left(\frac{\lambda}{n}\right)^x \left(1 - \frac{\lambda}{n}\right)^{n-x} \\
&= \frac{n!}{x!\,(n-x)!} \cdot \frac{\lambda^x}{n^x} \cdot \left(1 - \frac{\lambda}{n}\right)^n \cdot \left(1 - \frac{\lambda}{n}\right)^{-x} \\
&\approx \frac{n^x}{x!} \cdot \frac{\lambda^x}{n^x} \cdot e^{-\lambda} \cdot 1 \\
&= e^{-\lambda}\frac{\lambda^x}{x!}\,.
\end{aligned}$$

(5) So we let $f(x; \lambda) = e^{-\lambda} \dfrac{\lambda^x}{x!}$ be the pmf for a Poisson random variable with **rate parameter** λ. (We'll see why it is called a rate parameter shortly.)

(6) We should check that this definition is legitimate.

- Recall from calculus (Taylor series) that $\displaystyle\sum_{x=0}^{\infty} \frac{\lambda^x}{x!} = e^{\lambda}$.

- So $\displaystyle\sum_{x=0}^{\infty} e^{-\lambda} \frac{\lambda^x}{x!} = 1$, which means that we have defined a legitimate pmf.

Definition 2.7.1. The pmf for a **Poisson random variable** with rate parameter λ is

$$f(x; \lambda) = e^{-\lambda} \frac{\lambda^x}{x!} \ .$$

We let $X \sim \mathsf{Pois}(\lambda)$ denote that X is a Poisson random variable with rate parameter λ. $\qquad\square$

Lemma 2.7.2. *Let $X \sim \mathsf{Pois}(\lambda)$. Then*

- $\mathrm{E}(X) = \lambda$,
- $\mathrm{Var}(X) = \lambda$.

Proof. First we compute the expected value:

$$\mathrm{E}(X) = \sum_{x=0}^{\infty} x e^{-\lambda} \frac{\lambda^x}{x!} = \sum_{x=1}^{\infty} x e^{-\lambda} \frac{\lambda^x}{x!}$$

$$= \lambda \sum_{x=1}^{\infty} e^{-\lambda} \frac{\lambda^{x-1}}{(x-1)!}$$

$$= \lambda \sum_{x=0}^{\infty} e^{-\lambda} \frac{\lambda^x}{(x)!}$$

$$= \lambda \ .$$

Similarly,

$$\mathrm{E}(X^2) = \sum_{x=0}^{\infty} x^2 e^{-\lambda} \frac{\lambda^x}{x!} = \sum_{x=1}^{\infty} x^2 e^{-\lambda} \frac{\lambda^x}{x!}$$

$$= \lambda \sum_{x=1}^{\infty} x e^{-\lambda} \frac{\lambda^{x-1}}{(x-1)!}$$

$$= \lambda \sum_{x=0}^{\infty} (x+1) e^{-\lambda} \frac{\lambda^x}{(x)!}$$

$$= \lambda \left(\sum_{x=0}^{\infty} x e^{-\lambda} \frac{\lambda^x}{(x)!} + \sum_{x=0}^{\infty} e^{-\lambda} \frac{\lambda^x}{(x)!} \right)$$

$$= \lambda(\lambda + 1) \ .$$

So $\mathrm{Var}(X) = \mathrm{E}(X^2) - \mathrm{E}(X)^2 = \lambda(\lambda + 1) - \lambda^2 = \lambda$.   □

Note that if $X \sim \mathsf{Pois}(\lambda)$, then $\mathrm{E}(X) = \lambda$, so λ is the expected number of occurrences per observed time.

Example 2.7.1.

Q. Customers come to a small business at an average rate of 6 per hour. Let's assume that a Poisson model is a good model for customer arrivals.

(1) How unusual is it to go 20 minutes without any customers?

(2) How unusual is it to have 10 or more customers in an hour?

A. Let X be the number of customers arriving in 20 minutes. Since the rate of customer arrivals is 6 per hour, the expected number of customers in 20 minutes is 2, and $X \sim \mathsf{Pois}(2)$. So

$$\mathrm{P}(X = 0) = e^{-2} \frac{2^0}{0!} = e^{-2} = 0.1353 \ .$$

We can also use `dpois()` to determine the probability that no customers arrive in 20 minutes.

```
> dpois(0,2)
[1] 0.13534
```
poisson-customers-1

Similarly, we can use `ppois()` to determine the probability that 10 or more customers arrive in an hour.

```
> 1- ppois(9,6)
[1] 0.083924
```
poisson-customers-2

◁

Example 2.7.2.

Q. Is the Poisson model a good model for the number of fumbles in NCAA football games?

A. The `fumbles` data set includes the the number of fumbles by each NCAA FBS football team in the first three weeks of November, 2010. The distribution of fumbles for teams playing on November 6 and some summary statistics are given below:

```
> m <- max(fumbles$week1)
> table(factor(fumbles$week1,levels=0:m))

 0  1  2  3  4  5  6  7
22 36 29 23  5  4  0  1
> favstats(fumbles$week1)
    0%    25%    50%    75%   100%   mean     sd    var
0.0000 1.0000 2.0000 3.0000 7.0000 1.7500 1.3612 1.8529
```
fumbles

Table 2.5 compares this distribution to a Poisson distribution with the same mean. By overlaying the Poisson densities on a histogram of the data, we obtain a visual comparison shown in Figure 2.13. The actual and theoretical frequencies match quite well.

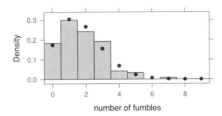

Figure 2.13. The distribution of the number of fumbles by NCAA football teams on November 6, 2010. The densities of a Poisson distribution are overlaid for comparison.

As a further check, we note that the variance of our data (1.85) is close to the mean (1.75), as we would expect for data sampled from a Poisson distribution. ◁

2.7.2. The Hypergeometric Distributions

The urn model

It is traditional to introduce the **hypergeometric distributions** in terms of an **urn model**, but we shall see that this distribution has important statistical applications as well. Suppose we have a large urn with m white balls and n black balls. If we randomly select k balls from the urn so that every subset of k balls is equally likely to be selected and let X count the number of white balls selected, then $X \sim \mathsf{Hyper}(m, n, k)$.

The pmf for X is easily computed:

$$f_X(x) = \frac{\binom{m}{x}\binom{n}{k-x}}{\binom{m+n}{k}} .$$

The denominator counts the number of ways to select k balls from the urn, and the numerator counts the number of ways to select x white balls and $k - x$ black balls.

Of course, the hypergeometric distributions have many applications that do not mention balls or urns.

Table 2.5. A comparison of football fumbles to a Poisson model.

fumbles	observed count	model count	observed %	model %
0	22	20.9	18.3	17.4
1	36	36.5	30.0	30.4
2	29	31.9	24.2	26.6
3	23	18.6	19.2	15.5
4	5	8.15	4.2	6.8
5	4	2.85	3.3	2.4
6	0	0.83	0.0	0.7
7	1	0.21	0.8	0.2

Example 2.7.3.

Q. A youth soccer team plays six-on-six. The team has 7 girls and 5 boys. If the coach randomly selects the starting line-up, what is the probability that more girls than boys start?

A. Let X be the number of girls selected to start. Then $X \sim \mathsf{Hyper}(7,5,6)$, and $\mathrm{P}(X \geq 4)$ can be calculated from the pmf or using R:

```
> 1-phyper(3,m=7,n=5,k=6)     # from "girls perspective"      youthSoccer01
[1] 0.5
> phyper(2,m=5,n=7,k=6)       # redone from "boys perspective"
[1] 0.5
```
◁

The mean and variance of a hypergeometric random variable are given by the following theorem.

Theorem 2.7.3. *Let $X \sim \mathsf{Hyper}(m,n,k)$ and let $\pi = \frac{m}{m+n}$. Then*

- $\mathrm{E}(X) = k\pi$, *and*
- $\mathrm{Var}(X) = k\pi(1-\pi) \cdot \dfrac{m+n-k}{m+n-1}.$

Proof. We can express X as a sum: $X = X_1 + X_2 + \cdots + X_k$ where

$$X_i = \begin{cases} 0 & \text{if ball } i \text{ is black,} \\ 1 & \text{if ball } i \text{ is white.} \end{cases}$$

Each X_i is a Bernoulli random variable with $\pi = \frac{m}{m+n}$, so $\mathrm{E}(X_i) = \pi$ and $\mathrm{Var}(X_i) = \pi(1-\pi)$. By Theorem 2.6.7, $\mathrm{E}(X) = k\pi$.

Since the X_i's are not independent, we need to use Lemma 2.6.11 to compute the variance. If $i \neq j$, then $X_i X_j$ is a Bernoulli random variable and

$$\mathrm{P}(X_i \cdot X_j = 1) = \mathrm{P}(X_i = 1 \text{ and } X_j = 1) = \frac{m}{m+n} \cdot \frac{m-1}{m+n-1},$$

so

$$\begin{aligned}
\mathrm{Cov}(X_i, X_j) &= \mathrm{E}(X_i Y_i) - \mathrm{E}(X_i)\,\mathrm{E}(X_j) \\
&= \frac{m}{m+n} \cdot \frac{m-1}{m+n-1} - \left(\frac{m}{m+n}\right)^2 \\
&= \pi \cdot \left[\frac{m-1}{m+n-1} - \frac{m}{m+n} \right] \\
&= \pi \cdot \left[\frac{(m-1)(m+n) - (m)(m+n-1)}{(m+n-1)(m+n)} \right] \\
&= \pi \cdot \left[\frac{-n}{(m+n-1)(m+n)} \right] \\
&= \pi(1-\pi) \cdot \left[\frac{-1}{(m+n-1)} \right],
\end{aligned}$$

and

$$\mathrm{Var}(X) = \sum_{i=1}^{k} \mathrm{Var}(X_i) + 2 \sum_{1 \leq i < j \leq k} \mathrm{Cov}(X_i, X_j)$$

$$= k \cdot \pi(1-\pi) - 2\binom{k}{2}\pi(1-\pi) \cdot \frac{1}{m+n-1}$$

$$= k \cdot \pi(1-\pi) - 2\frac{k(k-1)}{2}\pi(1-\pi) \cdot \frac{1}{m+n-1}$$

$$= k \cdot \pi(1-\pi)\left[1 - \frac{k-1}{m+n-1}\right]$$

$$= k \cdot \pi(1-\pi) \cdot \frac{m+n-k}{m+n-1} . \qquad \Box$$

Comparing the binomial and the hypergeometric distributions

Looking at the results of the previous section, we see that the means of binomial and hypergeometric distributions (with the same value of π) are the same. The variance of the hypergeometric distribution is smaller by a factor of

$$\frac{m+n-k}{m+n-1} .$$

In particular, if k is small relative to $m + n$, the two variances are nearly identical.

We can also describe binomial distributions using an urn model. If our urn has m white balls and n black balls, then the proportion of white balls in the urn is $\pi = \frac{m}{m+n}$. If we select balls one at a time *and return them to the urn* (and mix the urn thoroughly) after each selection, then the number of white balls selected will be $X \sim \mathsf{Binom}(k, \pi)$. This is often referred to as **sampling with replacement**. Typically in statistical studies we sample *without replacement*, so a hypergeometric distribution is a more accurate model.

That does not make the hypergeometric model better, however. The problem is that the increased "accuracy" comes at a cost. In the binomial distribution, there is only one unknown parameter, namely π. In particular, it does not matter how many balls are in the urn: As long as the ratio of white to black is the same, we get the same distribution. For the hypergeometric distribution, however, the number of balls in the urn does matter. In essence, this gives us one additional parameter. (We can think of the two unknown parameters as being m and n, or $\pi = \frac{m}{m+n}$ and $m + n$.) If we are primarily interested in π, then the additional parameter is a **nuisance parameter**. A nuisance parameter is a parameter that is unknown, is not of direct interest, but is nevertheless needed to do the calculations at hand. Eliminating or dealing with nuisance parameters is an important theme in statistics.

Suppose we want to know public opinion on some matter for which there are only two choices. For example, we may ask, "Do you prefer the Republican or the Democrat in the upcoming election?" (Ignore, for the moment, that some people may refuse to respond or say they don't know or care.) There is some proportion π of people that would prefer the Democrat. If we knew the value of π (among the population of actual voters), we could predict the outcome of the election. If we can get a independent sample of our desired size, then we have something very much like a binomial situation: n independent trials with probability π of "success" (preferring the Democrat) each time. It's not quite right because if the first person prefers the Democrat, then the probability that second person also prefers the Democrat goes down a bit (since one Democrat has been "used up").

We could model this situation as a hypergeometric distribution instead: The population consists of d Democrats and r Republicans. We sample n people. Let X be the number in our sample who prefer the Democrat. Then $X \sim \mathsf{Hyper}(d, r, n)$. Now suppose we wanted to do a hypothesis test with the null hypothesis

$$H_0 : \pi = \pi_0 \ .$$

To do this based on the hypergeometric distribution instead of the binomial distribution, we would need to know (or make an additional hypothesis about) the size of the population $(m + n)$ we are sampling from.

Fortunately, when $N = d + r$ is large, the difference between these two models is very small. This is why we use `binom.test()` (and why there is no such thing as `hyper.test()`). Furthermore, the binomial test is **conservative**: Because the variance of the binomial distribution is larger than the variance of the hypergeometric distribution, our p-values will have a tendency to be slightly too large, so they slightly understate the strength of the evidence against the null hypothesis. Said another way, we are less likely to be as far from the mean when sampling from the hypergeometric distribution as when sampling from the corresponding binomial distribution.

2.7.3. Fisher's Exact Test

Let's return to our story about the lady tasting tea. Suppose we design our experiment a little differently this time. We will once again prepare ten cups of tea, but this time we will randomly select 5 to prepare with tea first and 5 with milk first. Suppose further that the lady knows this. In fact, we give her five cards and ask her to place them by the five cups that had milk poured in first.

How does this new design change our hypothesis test? Our null hypothesis is still that she is merely guessing. In that case she is randomly selecting $k = 5$ cups from $m = 5$ prepared one way and $n = 5$ prepared the other way. So if she is guessing, the number of cards that are correctly placed will be $X \sim \mathsf{Hyper}(5, 5, 5)$. The one-sided p-value for a particular outcome x is then $\mathrm{P}(X \geq x)$. The table below gives the p-values corresponding to correctly identifying $x = 0, \ldots, 5$ cups correctly.

```
> cbind(0:5, 1-phyper(-1:4,5,5,5))
     [,1]       [,2]
[1,]    0 1.0000000
[2,]    1 0.9960317
[3,]    2 0.8968254
[4,]    3 0.5000000
```
ladyTeaHyper01

```
[5,]    4 0.1031746
[6,]    5 0.0039683
```

Let's compare this to the p-values for a binomial test.

```
> cbind(0:10, 1-pbinom(-1:9,10,0.5))        [6,]     5 0.62304687
         [,1]        [,2]                    [7,]     6 0.37695313
  [1,]    0 1.00000000                       [8,]     7 0.17187500
  [2,]    1 0.99902344                       [9,]     8 0.05468750
  [3,]    2 0.98925781                      [10,]     9 0.01074219
  [4,]    3 0.94531250                      [11,]    10 0.00097656
  [5,]    4 0.82812500
```

So which test is better? The p-values show that it is harder to be perfect in the original design. This pattern continues for all other situations where she does better than guessing. (Keep in mind that placing four cards correctly means that 8 of the 10 cups were correctly identified, etc., so we compare 5 with 10, 4 with 8, and 3 with 6.) The evidence is stronger in the design analyzed by the binomial test than in our new design analyzed using the hypergeometric distribution. This makes sense since our new design provides the lady with extra information (the *number* of cups that had milk poured first).

Lest you think there is no reason to study the hypergeometric distribution at all, let's consider a situation where the hypergeometric distribution really is useful.

Example 2.7.4. Here is an example that was treated by R. A. Fisher in an article [**Fis62**] and again later in a book [**Fis70**]. The data come from a study of same-sex twins where one twin had had a criminal conviction. Two pieces of information were recorded: whether the sibling had also had a criminal conviction and whether the twins were monozygotic (identical) twins and dizygotic (nonidentical) twins. Dizygotic twins are no more genetically related than any other siblings, but monozygotic twins have essentially identical DNA. Twin studies like this have often used to investigate the effects of "nature vs. nurture".

Here are the data from the study Fisher analyzed:

	Convicted	Not convicted
Dizygotic	2	15
Monozygotic	10	3

The question is whether this gives evidence for a genetic influence on criminal convictions. If genetics had nothing to do with convictions, we would expect conviction rates to be the same for monozygotic and dizygotic twins since other factors (parents, schooling, living conditions, etc.) should be very similar for twins of either sort. So our null hypothesis is that the conviction rates are the same for monozygotic and dizygotic twins. That is,

- $H_0 : \pi_1 = \pi_2$.

Given that there were 17 dizygotic twins and 13 monozygotic twins, of whom 12 had been convicted, under the null hypothesis, the number of dizygotic twins with convictions should have a hypergeometric distribution:

- $X \sim \mathsf{Hyper}(17, 13, 12)$.

Our test statistic is $x = 2$, and our one-sided p-value is

- p-value $= \mathrm{P}(X \leq 2)$ where $X \sim \mathsf{Hyper}(17, 13, 12)$.

We can, of course, do these calculations in R.

fisher-twins1

```
> phyper(2,17,13,12)
[1] 0.00046518
> convictions <- rbind(dizygotic=c(2,15), monozygotic=c(10,3))
> colnames(convictions) <- c('convicted','not convicted')
> convictions
            convicted not convicted
dizygotic           2            15
monozygotic        10             3
> fisher.test(convictions, alternative = "less")

        Fisher's Exact Test for Count Data

data:  convictions
p-value = 0.0004652
alternative hypothesis: true odds ratio is less than 1
95 percent confidence interval:
 0.00000 0.28496
sample estimates:
odds ratio
  0.046937
```

If we use any other cell from the tabulated data as our reference cell, we will need to work with a different hypergeometric distribution but will get the same p-value. You are asked to do this in Exercise 2.88.

For a two-sided p-value we will once again use the strategy employed for the binomial test. That is, we add $\mathrm{P}(X = k)$ for all values of k with $\mathrm{P}(X = k) \leq \mathrm{P}(X = 2)$. In other words, the two-sided p-value is the sum of all smaller probabilities in the pmf for the appropriate hypergeometric distribution. The R function fisher.test() can automate the entire process.

fisher-twins2

```
> fisher.test(convictions)

        Fisher's Exact Test for Count Data

data:  convictions
p-value = 0.0005367
alternative hypothesis: true odds ratio is not equal to 1
95 percent confidence interval:
 0.0033258 0.3631823
sample estimates:
```

```
odds ratio
   0.046937
```
◁

Relative risk and odds ratio

There is more information in the output of `fisher.test()` than we currently know how to interpret. We'll wait until Chapter 4 to introduce confidence intervals, but we'll say something about odds ratios now. If we look at our data, we see that the proportion of monozygotic twins with convictions is $\hat{\pi}_1 = \frac{10}{13} = 0.7692$. Similarly, the proportion of dizygotic twins with convictions is $\hat{\pi}_2 = \frac{2}{17} = 0.1176$. (We use $\hat{\pi}_i$ to indicate that we are estimating π_i.) These numbers can be interpreted as (estimates of) the **risk** of having a conviction for the two groups. The **relative risk** is the ratio $\hat{\pi}_1/\hat{\pi}_2 = 6.5385$. (We could also form the relative risk $\hat{\pi}_2/\hat{\pi}_1 = 1/6.5385 = 0.1529$.) This says that the (estimated) risk is more than 6 times greater for monozygotic twins than for dizygotic twins.

Relative risk is a convenient way to compare two groups because it is easily interpretable. It is especially useful when the two proportions being compared are both quite small, in which case the absolute difference might be small but the relative risk much larger.

Example 2.7.5. Suppose we are comparing two medical treatments and the death rate under one treatment plan is $1/100$ and in the other is $1/500$. Then the relative risk is 5, but the absolute difference is only $4/500 = 0.008$.

One downside to relative risk is its sensitivity to certain decisions made in forming it. Continuing our example, if we compute the relative risk based on survival rate instead of death rate, we get $\frac{99}{100}/\frac{499}{500} = 0.991984$, which is not the reciprocal of 5. In fact it is close to 1, indicating that the two proportions being compared are nearly the same. So one way around, there seems to be a 5-fold difference and the other way around, there is almost no difference. There are other downsides too having to do with statistical inference procedures we have not yet discussed.   ◁

The odds ratio avoids many of the inconveniences of relative risk. If π is the probability of an event, then $\frac{\pi}{1-\pi}$ is the odds of that event. We can form an **odds ratio** from odds just as we formed the relative risk from risk:

$$\text{odds ratio} = \frac{\pi_1/(1-\pi_1)}{\pi_2/(1-\pi_2)} .$$

Notice that the odds ratio satisfies

$$\frac{\pi_1/(1-\pi_1)}{\pi_2/(1-\pi_2)} = \frac{\pi_1}{\pi_2} \cdot \frac{1-\pi_2}{1-\pi_1} .$$

So the relative risk and odds ratio are nearly the same when $\frac{1-\pi_1}{1-\pi_2} \approx 1$. This is often the case when π_1 and π_2 are small – rates of diseases, for example. Since relative risk is easier to interpret (I'm 3 times as likely ...) than an odds ratio (my odds are 3 times greater than ...), we can use relative risk to help us get a feeling for odds ratios in some situations.

Also, if we flip the roles of success and failure, the odds ratio becomes

$$\frac{\frac{1-\pi_1}{\pi_1}}{\frac{1-\pi_2}{\pi_2}} = \frac{1-\pi_1}{1-\pi_2} \cdot \frac{\pi_2}{\pi_1} \,,$$

which is the reciprocal of our original odds ratio. In our previous example we have

relative risk of survival: 5.000,
relative risk of death: 0.9920,
odds ratio for survival: 5.0404,
odds ratio for death: 0.1984.

Returning to our twins data set, the odds ratio reported by R is a "best guess" at the actual odds ratio using a method called conditional maximum likelihood estimation. (See [**Agr90**] for details.) This is gives a slightly different answer from the odds ratio computed directly from the table:

$$\frac{2/15}{10/3} = \frac{2 \cdot 3}{10 \cdot 15} = 0.04 \neq 0.04694 \,.$$

2.7.4. Benford's Law: The Distribution of Leading Digits

Consider the first non-zero digit of a "random number". It seems obvious that each digit should be equally likely. This means that if we let the random variable X be the first digit of a random number, then the distribution of X should be

Value of X	1	2	3	4	5	6	7	8	9
Probability	1/9	1/9	1/9	1/9	1/9	1/9	1/9	1/9	1/9

In the late 1800s Newcomb [**New81**] noticed that some pages of the table of logarithms book he used were dirtier than others – presumably because they were used more frequently than other pages. Which page is used depends on the first non-zero digit in a number. That is, Newcomb observed data that seemed to suggest that the obvious conjecture that all first digits should be equally likely was wrong, because **our theoretical distribution and the empirical distribution do not seem to match.**

Later Benford [**Ben38**] and others came up with other probability models to explain this situation. The most famous of these is now called Benford's Law. **Benford's Law** says that under certain assumptions, the probability distribution should obey the following formula:

$$\mathrm{P}(X = k) = \log_{10}(k+1) - \log_{10}(k) \,.$$

Here is Benford's distribution described in table form (rounded to three digits):

leading digit	1	2	3	4	5	6	7	8	9
probability	0.301	0.176	0.125	0.097	0.079	0.067	0.058	0.051	0.046

Notice how much better it matches the empirical distributions in Table 2.6.

The fact that Benford's Law is not immediately intuitive has a useful side effect. It can be used for fraud detection. Most people who make up numbers do not follow Benford's Law, so one way to check for phony numbers (say, on a tax return) is to compare the distribution of leading digits to the predictions of Benford's Law [**LBC06**].

Benford's Law has been discussed in many places, and various explanations have been proposed. Part of the difficulty in explaining Benford's Law comes from the difficulty in saying what it means to select a (positive) number at random. (See, for example, the discussion in the second volume of Donald Knuth's *Art of Computer Programming* [**Knu97**].) Other discussions (among many) appear in articles by Persi Diaconis [**Dia77**] and Ralph Raimi [**Rai76, Rai85**]. The first rigorous formulation and derivation of Benford's Law appears to have been given by Theodore P. Hill in the 1990s [**Hil95a, Hil95b**]. A discussion of Benford's Law has even appeared in the *New York Times* [**Bro98**].

Table 2.6. Empirical data collected by Benford [**Ben38**].

	Title	1	2	3	4	5	6	7	8	9	N
A	Rivers, Area	31.0	16.4	10.7	11.3	7.2	8.6	5.5	4.2	5.1	335
B	Population	33.9	20.4	14.2	8.1	7.2	6.2	4.1	3.7	2.2	3259
C	Constants	41.3	14.4	4.8	8.6	10.6	5.8	1.0	2.9	10.6	104
D	Newspapers	30.0	18.0	12.0	10.0	8.0	6.0	6.0	5.0	5.0	100
E	Specific Heat	24.0	18.4	16.2	14.6	10.6	4.1	3.2	4.8	4.1	1389
F	Pressure	29.6	18.3	12.8	9.8	8.3	6.4	5.7	4.4	4.7	703
G	H.P. Lost	30.0	18.4	11.9	10.8	8.1	7.0	5.1	5.1	3.6	690
H	Mol. Wgt.	26.7	25.2	15.4	10.8	6.7	5.1	4.1	2.8	3.2	1800
I	Drainage	27.1	23.9	13.8	12.6	8.2	5.0	5.0	2.5	1.9	159
J	Atomic Wgt.	47.2	18.7	5.5	4.4	6.6	4.4	3.3	4.4	5.5	91
K	$n^{-1}, n^{-1/2}, \ldots$	25.7	20.3	9.7	6.8	6.6	6.8	7.2	8.0	8.9	5000
L	Design	26.8	14.8	14.3	7.5	8.3	8.4	7.0	7.3	5.6	560
M	Reader's Digest	33.4	18.5	12.4	7.5	7.1	6.5	5.5	4.9	4.2	308
N	Cost Data	32.4	18.8	10.1	10.1	9.8	5.5	4.7	5.5	3.1	741
O	X-Ray Volts	27.9	17.5	14.4	9.0	8.1	7.4	5.1	5.8	4.8	707
P	Am. League	32.7	17.6	12.6	9.8	7.4	6.4	4.9	5.6	3.0	1458
Q	Blackbody	31.0	17.3	14.1	8.7	6.6	7.0	5.2	4.7	5.4	1165
R	Addresses	28.9	19.2	12.6	8.8	8.5	6.4	5.6	5.0	5.0	342
S	Powers	25.3	16.0	12.0	10.0	8.5	8.8	6.8	7.1	5.5	900
T	Death Rate	27.0	18.6	15.7	9.4	6.7	6.5	7.2	4.8	4.1	418
	Average	30.6	18.5	12.4	9.4	8.0	6.4	5.1	4.9	4.7	1011

2.8. Summary

Probability is a number assigned to an **event** (a set of **outcomes**) of a random process that represents how likely that event is to occur. There are two general approaches to calculating probabilities that are important in statistics. Probabilities can be estimated by repeating a random process many times and determining the relative frequency of the event. Such a probability is called an **empirical probability**. The precision of an empirical probability improves with the number of repetitions, but the empirical method never allows us to determine probabilities exactly.

Probability can also be calculated using an axiomatic approach. A **theoretical probability** calculation combines a set of **probability axioms** (statements that are true about all probabilities) with some assumptions about the random process to derive a numerical probability. Theoretical probabilities are exact, but they may be incorrect if they are based on invalid assumptions.

A **random variable** is a random process that produces a numerical outcome. A random variable that can take on only finitely or countably many different values is said to be **discrete**. Probabilities for discrete random variables are typically obtained by combining some probability rules (derived from the probability axioms) and counting techniques. (See Sections 2.8.1 and 2.8.2.)

Discrete distributions are described by providing either a **probability mass function** (pmf) or a **cumulative distribution function** (cdf). For each possible value x of a discrete random variable X, the pmf f_X gives the probability of obtaining the value:

$$f_X(x) = \mathrm{P}(X = x) \,.$$

Notice the conventional use of capitalization (for random variables) and lowercase (for possible values). The cdf F_X gives the probability of X being at most a specified value:

$$F_X(x) = \mathrm{P}(X \le x) \,.$$

The pmf and cdf can each be derived from the other. A **family** of distributions is a set of distributions that arise in similar situations but differ in the values of certain **parameters**. Some important families of discrete distributions are summarized in Section 1.

One important application of random variables in statistics is to conduct a **hypothesis test** (also called a test of significance). Hypothesis tests attempt to answer a particular type of question based on sample data. There are many such hypothesis tests, but they generally follow the same four-step outline.

(1) State the **null and alternative hypotheses**.
These are statements about the parameters of a distribution.

(2) Compute a **test statistic** from the sample data.
A test statistic is a numerical value that can be used to measure how well the data match what would be expected if the null hypothesis were true.

(3) Compute a **p-value**.

   The p-value is the probability of obtaining a test statistic at least as extreme as the one computed from the sample data, assuming that the null hypothesis is indeed true.

(4) Draw a conclusion.

   A small p-value is an indication that the data would be very unusual if the null hypothesis were true. This is taken as evidence *against* the null hypothesis.

Because of the variation from sample to sample, it is possible to draw incorrect conclusions based on a hypothesis test. **Type I error** arises when we reject a null hypothesis that is actually true. We can control type I error by deciding how small p-values must be before we will reject the null hypothesis. This threshold is called the **significance level** of the test and is denoted α.

Type II error arises when we fail to reject the null hypothesis even though it is false. The probability of type II error (denoted β) depends on what the true situation is and is often plotted as a function. The **power** of a statistical test (given a specific alternative hypothesis) is $1 - \beta$, the probability of correctly rejecting a false null hypothesis. Generally, the power increases as the sample size increases and as the alternative hypothesis differs more from the null. Power analyses are an important part of designing a statistical study. Underpowered tests are unlikely to detect departures from the null hypothesis, even if they exist. Over-powered tests are an inefficient use of resources.

As with data, we can compute the **mean** (also called **expected value**) and **variance** of a discrete random variable:

$$\mu_X = \mathrm{E}(X) = \sum x f_X(x) \,,$$

$$\sigma_X^2 = \mathrm{Var}(X) = \sum (x - \mu_x)^2 f_X(x) = \mathrm{E}(X^2) - \mathrm{E}(X)^2 \,.$$

The expected values and variances of some important discrete random variables are summarized in Section 1. The expected value and variance of a linear transformation of a random variable are easily computed:

* $\mathrm{E}(aX + b) = a \, \mathrm{E}(X) + b$,
* $\mathrm{Var}(aX + b) = a^2 \, \mathrm{Var}(X)$.

These simple rules have many important applications.

A **joint pmf** describes the distribution of two (or more) random variables simultaneously:

$$f_{X,Y}(x, y) = \mathrm{P}(X = x \text{ and } Y = y) \,.$$

From a joint pmf we can compute the **marginal** and **conditional** distributions:

$$\text{marginal:} \qquad f_X(x) \;=\; \sum_y f_{X,Y}(x, y) \,,$$

$$\text{conditional:} \qquad f_{X|Y}(x|y) \;=\; \frac{\text{joint}}{\text{marginal}} = \frac{f_{X,Y}(x, y)}{f_Y(y)} \,.$$

When the marginal and conditional distributions for X are the same, we say that X and Y are **independent**. This is equivalent to the statement

$$f_{X,Y}(x,y) = f_X(x) \cdot f_Y(y) \,.$$

From a joint distribution we can compute the sum or product of X and Y. The sum of two or more jointly distributed random variables is an especially important random variable for statistics. Four important rules often make working with sums and products easier:

- $\mathrm{E}(X+Y) = \mathrm{E}(X) + \mathrm{E}(Y)$ for any random variables X and Y,
- $\mathrm{Var}(X+Y) = \mathrm{Var}(X) + \mathrm{Var}(Y) + 2\,\mathrm{Cov}(X,Y)$, where $\mathrm{Cov}(X,Y) = \mathrm{E}(XY) - \mathrm{E}(X)\,\mathrm{E}(Y)$,
- $\mathrm{E}(X \cdot Y) = \mathrm{E}(X) \cdot \mathrm{E}(Y)$, provided X and Y are independent, so
- $\mathrm{Var}(X+Y) = \mathrm{Var}(X) + \mathrm{Var}(Y)$, provided X and Y are independent.

These rules extend naturally to more than two variables. This will be important later when we want to derive formulas for the expected value and variance of a sample mean.

2.8.1. Rules for Counting

(1) Bijection Rule: Don't double count; don't skip.

  If there is a one-to-one and onto function $f : A \to B$, then $|A| = |B|$. More informally, if every person has a chair and no chairs are empty, then the number of people and the number of chairs must be equal. This is the essence of what it means to count.

(2) Sum Rule: Divide and Conquer.

$$\text{If } A \cap B = \emptyset, \text{ then } |A \cup B| = |A| + |B| \,.$$

  If you can divide a counting problem into disjoint (non-overlapping) pieces, then you can count each piece separately and combine the results by adding.

(3) Difference Rule: Compensate for double counting.

$$|A \cup B| = |A| + |B| - |A \cap B| \,.$$

  If you know how many things were double-counted, you can "uncount" them to get the correct total.

(4) Product Rule: One stage at a time.

$$|A \times B| = |A| \cdot |B| \,.$$

  If you can divide a counting problem into two stages and for each possible way to do stage 1 there are an equal number of ways to do stage 2, then multiplication will give you the number of ways to do stage 1 followed by stage 2.

(5) Division Rule: A way to compensate for uniform "double" counting.

  If you count everything an equal number of times, dividing the overcount by that number gives the correct count.

2.8.2. Probability Axioms and Rules

Axioms

Let S be a sample space for a random process and let $E \subseteq S$ be an event. Then:

(1) $\mathrm{P}(E) \in [0, 1]$.

(2) $\mathrm{P}(S) = 1$.

(3) The probability of a *disjoint* union is the sum of probabilities.
- $\mathrm{P}(A \cup B) = \mathrm{P}(A) + \mathrm{P}(B)$, provided $A \cap B = \emptyset$.
- $\mathrm{P}(A_1 \cup A_2 \cup \cdots \cup A_k) = \mathrm{P}(A_1) + \mathrm{P}(A_2) + \cdots + \mathrm{P}(A_k)$, provided $A_i \cap A_j = \emptyset$ whenever $i \neq j$.
- $\mathrm{P}\left(\bigcup_{i=1}^{\infty} A_i\right) = \sum_{i=1}^{\infty} \mathrm{P}(A_i)$, provided $A_i \cap A_j = \emptyset$ whenever $i \neq j$.

Additional probability rules

Let S be a sample space for a random process and let $A, B \subseteq S$ be events. Then:

(1) Complement Rule.
$$\mathrm{P}(A) = 1 - \mathrm{P}(A^c).$$

(2) Equally Likely Rule.
Suppose our sample space S consists of n equally likely outcomes. Then
$$\mathrm{P}(A) = \frac{|A|}{|S|} = \frac{|A|}{n} = \frac{\text{size of event}}{\text{size of sample space}}.$$

(3) Inclusion-Exclusion (a.k.a. Generalized Sum Rule).
$$\mathrm{P}(A \cup B) = \mathrm{P}(A) + \mathrm{P}(B) - \mathrm{P}(A \cap B).$$

(4) Intersection Rule.
$$\mathrm{P}(A \cap B) = \mathrm{P}(A) \cdot \mathrm{P}(B \mid A) = \mathrm{P}(B) \cdot \mathrm{P}(A \mid B).$$

2.8.3. Important Discrete Distributions

Table 2.7 summarizes key features of several important discrete distributions. Note:

- For the hypergeometric distribution, $\pi = \frac{m}{n+m}$ (proportion of white balls in the urn).

- In our definitions of the geometric and negative binomial distributions, the random variable counts the number of failures. Some authors choose instead to count the number of trials (successes + failures). This modifies the expected value (in the obvious way) but leaves the variance unchanged.

Table 2.7. Important Discrete Distributions.

family (R pmf)	pmf formula	expected value	variance
Bernoulli dbinom(x,1,π)	$f(1) = \pi,\ f(0) = 1 - \pi$	π	$\pi(1 - \pi)$
binomial dbinom(x,n,π)	$\binom{n}{x} \pi^x (1 - \pi)^{n-x}$	$n\pi$	$n\pi(1 - \pi)$
geometric dgeom(x,π)	$(1 - \pi)^x \pi$	$\dfrac{1}{\pi} - 1$	$\dfrac{(1 - \pi)}{\pi^2}$
negative binomial dnbinom(x,s,π)	$\binom{x+s-1}{s-1}(1 - \pi)^x \pi^s$	$s \cdot \left(\dfrac{1}{\pi} - 1 \right)$	$s\dfrac{(1 - \pi)}{\pi^2}$
hypergeometric dhyper(x,m,n,k)	$\dfrac{\binom{m}{x}\binom{n}{k-x}}{\binom{n+m}{k}}$	$k\pi$	$\dfrac{m+n-k}{m+n-1} \cdot k\pi(1 - \pi)$
Poisson dpois(x,λ)	$e^{-\lambda}\dfrac{\lambda^x}{x!}$	λ	λ

2.8.4. R Commands

Here is a table of important R commands introduced in this chapter. Usage details can be found in the examples and using the R help.

choose(n,k)	$\binom{n}{k} = \dfrac{n!}{(n-k)!k!}$.
dbinom(x,size,prob)	$P(X = \text{x})$ for $X \sim \text{Binom}(\text{size}, \text{prob})$.
pbinom(q,size,prob)	$P(X \leq \text{q})$ for $X \sim \text{Binom}(\text{size}, \text{prob})$.
qbinom(p,size,prob)	Smallest x such that $P(X \leq x) \geq \text{p}$ for $X \sim \text{Binom}(\text{size}, \text{prob})$.
rbinom(n,size,prob)	Simulate n random draws from a Binom(size, prob)-distribution.

`dpois(...); ppois(...);`
`qpois(...); rpois(...);`
`dnbinom(...); pnbinom(...);`
`qnbinom(...); rnbinom(...);`
`dhyper(...); phyper(...);`
`qhyper(...); rhyper(...)`
	Similar to the functions above but for Poisson, negative binomial, and hypergeometric distributions.

`rep(values,...)`	Create a vector of repeated values.
`sum(x,...); prod(x,...)`	Compute the sum or product of values in the vector x.
`binom.test(x,n,p,...)`	Conduct a binomial test of $H_0 : \pi = p$ from a data set with x successes in n tries.
`fisher.test(` `    rbind(c(x,y),c(z,w)),...)`	Conduct Fisher's exact test with data summarized in the table | x	y
z	w	

Exercises

2.1. A coin is tossed 3 times and the sequence of heads and tails is recorded.

a) List the elements of the sample space.

b) List the outcomes in each of the following events: A = at least two tails, B = last two tosses are tails, C = first toss is a tail.

c) List the elements in the following events: A^c, $A \cap B$, $A \cup C$.

2.2. A red die and a blue die are rolled and the number on each is recorded.

a) List the elements of the sample space.

b) List the elements in the following events: A = the sum of the two numbers is at least 9, B = the blue die has a larger value than the red die, C = the blue die is a 5.

c) List the elements of the following events: $A \cap B$, $B \cup C$, $A \cap (B \cup C)$.

2.3. Write an R function that prints out all 171 sums from Example 2.2.5.

2.4. Let A, B, C, and D be as in Example 2.2.5. The bijection $f : A \to B$ described there is

$$f(\langle x, y, z \rangle) = \langle x - 1, y - 1, z - 1 \rangle .$$

a) Use the definition of bijection to prove that f is indeed a bijection.

b) Provide explicit bijections from B to C and from C to D as well.

2.5. A donut store is offering a special price on a dozen (12) donuts. There are three varieties of donuts available: plain, glazed, and chocolate covered. How many different dozens can you order?

(One example order would be 3 plain, 5 glazed, and 4 chocolate.)

2.6. Five cards are dealt from a standard 52-card deck. What is the probability that the hand is a full house?

(A full house is 3 cards of one denomination and 2 of another. Example: 2♣, 2◇, 2♡, 10♣, 10♠.)

2.7. Five cards are dealt from a standard 52-card deck. What is the probability that the hand contains two pairs?

(This means there are 2 cards of one denomination, 2 of another, 1 more card of a third denomination. Example: 2♣, 2◇, 8♡, 10♣, 10♠.)

2.8. Five cards are dealt from a standard 52-card deck. What is the probability that the hand contains three of a kind?

(Three of a kind means that there are 3 cards of one denomination, 1 of another, and 1 of a third. Example: 2♣, 2◇, 2♡, 7♣, 10♠. Note that by this definition a full house is *not* three of a kind.)

2.9. Let X count the number of suits in a 5-card hand dealt from a standard 52-card deck. Complete the following table:

value of X	0	1	2	3	4
probability		0.00198			

2.10. Assume that there are 365 days in a year and that all birthdays are equally likely.

a) If ten people are selected at random, what is the probability that (at least) two of them have the same birthday?

b) Find the smallest number n such that a random group of n people has a greater than 50% chance that two (or more) people in the group share a common birthday.

2.11. In Exercise 2.10, we assume that all birthdays are equally likely. As the `births78` data set shows, this is not actually true. There are both weekly and seasonal fluctuations to the number of births.

a) Prove that if at least n birthdays have non-zero probability, then the probability of n matching birthdays is lowest when the birthdays are equally likely. [Hint: Show that if two birthdays have probability p and q, then matches are less likely if we replace both of those probabilities with $\frac{p+q}{2}$.]

b) Use the `births78` data set and simulations to estimate the probability that two (or more) people in a random group of n people have matching birthdays for several values of n. Assume that birthday frequencies are as they were in the United States in 1978.

Further discussion of the birthday problem without the assumption of a uniform distribution of birthdays can be found in [**Ber80**].

2.12. Redo Exercise 2.5 using the method of Example 2.2.14.

2.13. Redo Example 2.2.5 using the method of Example 2.2.14.

2.14. Use Venn diagrams to show De Morgan's Laws:

a) $(A \cup B)^c = A^c \cap B^c$,

b) $(A \cap B)^c = A^c \cup B^c$.

2.15. Prove Bonferroni's Inequality:

$$P(A \cap B) \geq P(A) + P(B) - 1 .$$

2.16. Derive an inclusion-exclusion rule that works for three sets. That is, find a general formula for $P(A \cup B \cup C)$, and explain why it is valid.

2.17. Let X be the number of girls in a family and let Y be the number of boys. Using these random variables, express all of the probabilities in Example 2.2.15 in random variable notation.

2.18. Suppose items are produced on two assembly lines. Some are good and some are defective (bad). One day the production data were as follows:

	Bad	**Good**
Assembly Line 1	2	6
Assembly Line 2	1	9

Determine the following probabilities:

a) the probability that an item is bad,

b) the probability that an item is bad given that it was produced on assembly line 1,

c) the probability that an item was produced on assembly line 1 given that the item is bad.

2.19. Let A_i be any events.

a) Show that

$$P(A_1 \cap A_2 \cap A_3) = P(A_1 \cap A_2 \cap A_3)$$
$$= P(A_1) \cdot P(A_2 \mid A_1) \cdot P(A_3 \mid A_1 \cap A_2) .$$

b) Generalize this to give a formula for the intersection of any number of events. (Hint: Use induction.)

2.20. Prove Lemma 2.2.12.

2.21. Use the definition of independent events to show that

a) if A and B are independent, then A and B^c are also independent;

b) if A and B are independent, then A^c and B^c are also independent.

2.22. A survey was given to a class of students. One of the questions had two versions – a) and b) below – which were assigned to the students randomly:

> Social science researchers have conducted extensive empirical studies and concluded that the expression
>
> **a)** "absence makes the heart grow fonder"
>
> **b)** "out of sight out of mind"
>
> is generally true. Do you find this result surprising or not surprising?

Here are the results.

	A absence	A^c sight
surprised (S)	6	5
not surprised (S^c)	44	36

Is "being surprised" independent of the form of the question asked?

2.23. In Example 2.2.23, what is the probability that a child will have the disease in an $Aa \times Aa$ cross?

2.24. DMD is a serious form of Muscular Dystrophy, a sex-linked recessive disease.

- If a woman is a carrier, her sons have a 50% chance of being affected; daughters have a 50% chance of being carriers.
- 2/3 of DMD cases are inherited (from Mom); 1/3 of DMD cases are due to spontaneous mutations.

Now suppose there is a screening test to check whether a woman is a carrier such that

- $P(T + |C+) = 0.7$, and
- $P(T - |C-) = 0.9$,

where $T+$ and $T-$ are the events that the test is positive and negative, respectively, and $C+$ and $C-$ are the events that a woman is and is not a carrier, respectively.

A woman has a child with DMD and then has a screening test done to see if she is a carrier. If the test comes back positive, what is the probability that she is a carrier? What if the screening test is negative?

2.25. Show that for any events A and B with non-zero probability, $P(A \mid B) = \frac{P(B|A)\,P(A)}{P(B)}$. (This result is sometimes referred to as Bayes' Theorem.)

2.26. Suppose you roll 5 standard dice. Determine the probability that all 5 numbers are the same.

2.27. Suppose you deal 5 cards from a standard deck. Determine the probability that all 5 cards are the same color.

2.28. Return to the setting of Example 2.2.22.

a) Suppose the rate of the disease is 1 in 100 instead of 1 in 1000. How does that affect the interpretation of the test results?

b) What if the rate of disease is 1 in 10? (This would also be the situation if an individual is tested after the doctors estimate that the chances of having the disease is 10%.)

2.29. How many ways can you arrange the letters in the word STATISTICS? [Hint: Combine the division principle with the multiplication principle.]

2.30. Acceptance sampling is a procedure that tests some of the items in a lot or shipment and decides to accept or reject the entire lot based on the results of testing the sample. In a simple case, each tested item is determined to be either acceptable or defective. Suppose a purchaser orders 100 items, tests 4 of them, and rejects the lot if one or more of those four are found to be defective.

a) If 10% of the items in the lot of 100 are defective, what is the probability that the purchaser will reject the shipment?

b) Make a graph showing how the probability of rejection is related to the percent of defective items in the shipment.

2.31. A fair coin is tossed five times. What are the chances that there is a run of three consecutive heads?

2.32. In the game of Mastermind, player 1 makes a hidden code by placing colored pegs into holes behind a plastic shield.

a) In the original Mastermind, there are four holes and 6 colors available. How many secret codes are there?

b) In Super Mastermind, there are 5 holes and 8 colors. How many secret codes are there now?

2.33. A poker player is dealt three spades and two diamonds. She discards the two diamonds and is dealt two more cards. What is the probability that she is dealt two more spades, making a flush?

2.34. A gentleman claims he can distinguish between four vintages of a particular wine. His friends, assuming he has probably just had too much of each, decide to test him. They prepare one glass of each vintage and present the gentleman with four unlabeled glasses of wine. What is the probability that the gentleman correctly identifies all four simply by guessing?

2.35. Prove the following identities:

a) $\binom{n}{k} = \binom{n}{n-k}$.

b) $\binom{n}{k} = \binom{n-1}{k} + \binom{n-1}{k-1}$.

Each identity can be proved two ways: either by algebraic manipulation or by interpreting the meaning of each side combinatorially.

2.36. Fred's sock drawer has 8 black socks, 5 blue socks, and 4 brown socks. To avoid waking his roommate, Fred grabs 3 socks in the dark and brings them to another room to see what they are.

  a) What is the probability that he has a pair of matching socks?

  b) What is the probability that he has a pair of black socks?

2.37. A box has three coins in it. One has heads on both sides, one has tails on both sides, and the third is a fair coin with heads on one side and tails on the other. A coin is selected at random and flipped.

  a) If the result of the flip is heads, what is the probability that the coin is two-headed?

  b) If the result of the flip is heads, what is the probability that a second flip of the same coin will also be heads?

  c) If the selected coin is flipped twice and comes up heads twice in a row, what is the probability that the coin is two-headed?

2.38. Prove Lemma 2.3.2.

2.39. Let X be a random variable that takes on only values $1 - \frac{1}{n}$ for positive integers n. Furthermore, let $F(1 - \frac{1}{n}) = 1 - \frac{1}{n}$ be the cdf for X.

  a) What is $P(X = \frac{1}{2})$?

  b) What is $P(X = \frac{3}{4})$?

  c) Explain why the set $\{F(w) \mid w < 1\}$ has no maximum.

  d) What is $F(1)$?

  e) What is $\sup\{F(w) \mid w < 1\}$?

  f) What is $P(X = 1)$?

2.40. Early probability theory owes a debt to Chevalier de Mere, a rich Frenchman who liked gambling. More importantly for our story, he was an acquaintance of Blaise Pascal.

De Mere played a gambling game in which he bet that he could throw a six in four throws of a die. His empirical evidence suggested that this game was biased slightly in his favor, so that in the long run he won slightly more often than he lost, and hence made money.

Later de Mere became interested in a game where he rolled two dice and bet that he would roll double six at least once in 24 throws. He came up with 24 since there were 6 times as many possibilities for the outcome of two dice as there are of one, so he figured it would take 6 times as many throws to reach a game that was again slightly in his favor. But to check himself, he asked Pascal to do the computations for him.

Follow in Pascal's footsteps and determine the probability of winning each of these games. In fact, you should be able to do this at least two different ways (using two different distributions).

2.41. Amy is a college basketball player and a good free throw shooter. She claims that she makes 92% of her free throw attempts. Each week her coach has her shoot free throws until she makes 100 of them. One week this required 114 attempts, but she said that was an unusually large number for her.

Assuming Amy really is a 92% free throw shooter, how often will she require 114 or more attempts to make 100 free throws?

2.42. In your sock drawer are 5 pairs of socks. Two of them match what you are wearing, three of them do not. You continue blindly grabbing pairs of socks until a matching pair is selected, placing any non-matching socks on your bed to get them out of your way. Let X be the number of tries until you find a matching pair of socks. What is the pmf for X?

2.43. A random process flips a fair coin four times. Determine the pmf for each of the following random variables:

- X = number of heads before first tail (4 if no tails),
- Y = number of heads after first tail (0 if no tails),
- Z = absolute difference between number of heads and number of tails,
- W = product of number of heads and number of tails.

2.44. A certain type of Hamming code adds 3 bits to every 4-bit word. With this extra information, any single bit transmission error can be detected and corrected. Suppose a signal is sent over a poor channel where the probability is 0.05 that any given bit will be incorrectly received (a 0 sent but a 1 received or vice versa) and that errors are independent.

- **a)** What is the probability that an encoded 4-bit word will be correctly interpreted by the receiver?
- **b)** What is the probability that a Hamming-coded 4-bit word (7 total bits) will be correctly interpreted by the receiver?
- **c)** Make a graph that shows how the probability of correct interpretation depends on the bitwise error rate in each situation.

2.45. Free Throw Freddie is an 80% free throw shooter. What is more likely, that he makes at least 9 of 10 or that he makes at least 18 of 20?

2.46. Free Throw Freddie is an 80% free throw shooter. At the end of practice he shoots until he makes 10 shots.

- **a)** What is the probability that Freddie makes 10 in a row (without a miss, since he stops once 10 shots have been made)?
- **b)** What is the probability that Freddie needs 15 or more attempts?
- **c)** Free Throw Frank is only a 70% free throw shooter. He also takes shots until he makes 10. What is the probability that he needs 15 or more attempts?

2.47. Frank and Freddie (see previous problem) have decided to try a new free throw drill. Tonight they will shoot free throws until they make 5 in a row.

  a) What is the probability that Freddie will leave without missing a shot?

  b) What is the probability that Freddie will miss exactly one shot?

  c) What is the probability that Freddie will miss more than one shot before leaving?

  d) Let X be the number of free throws that Freddie misses. Make a plot of the pmf for X (over a reasonable range of values – in theory there is no limit to the number of misses he could have).

  e) Repeat the questions above for Frank.

2.48. A multiple choice test has 20 items with four possible answers given for each item. 12 is the minimum passing score.

  a) What is the probability of passing simply by guessing?

  b) What is the probability of passing if you are able to eliminate one choice each time but guess among the remaining three choices?

  c) Suppose the questions come from a large test bank and you know the correct answer to 50% of the questions in the test bank, can eliminate one choice for 40% of the questions, and have to guess blindly for the other 10% of the questions. If the questions are selected from the test bank independently at random, what is the probability that you pass the test?
     [Hint: What is your probability of getting the first question correct?]

2.49. You estimate the chances of your favorite sports team winning any game against its current play-off opponent to be $\pi = 0.6$. For the sake of this problem, assume that the outcome of each game is independent of the others.

  a) What is the probability that your team will win a best-of-three play-off (first team to win two games wins)?

  b) What is the probability that your team will win a best-of-5 play-off?

  c) What is the probability that your team will win a best-of-7 play-off?

2.50. Let X be a geometric random variable with parameter π.

  a) What is $P(X \geq k)$?   [Hint: geometric series.]

  b) Show that for all $x \geq k$,

$$P(X = x \mid X \geq k) = P(X = x - k) \,.$$

  c) Because of this result, we say that the geometric distribution is *memoryless*. Explain how this is an appropriate name for this property.

2.51. A child's game includes a spinner with four colors on it. Each color is one quarter of the circle. You want to test the spinner to see if it is fair, so you decide to spin the spinner 50 times and count the number of blues. You do this and record 8 blues. Carry out a hypothesis test, carefully showing the four steps. Do it "by hand" (using R but not `binom.test()`). Then check your work using `binom.test()`.

2.52. Fred wants to know whether his cat, Gus, has a preference for one paw or uses both paws equally. He dangles a ribbon in front of the cat and records which paw Gus uses to bat at it. He does this 10 times, and Gus bats at the ribbon with his right paw 8 times and his left paw 2 times. Then Gus gets bored with the experiment and leaves. Can Fred conclude that Gus is right-pawed, or could this result have occurred simply due to chance under the null hypothesis that Gus bats equally often with each paw?

2.53. Mendel [**Men65**] crossed pea plants that were heterozygotes for green pod/yellow pod. If this is inherited as a simple Mendelian trait, with green dominant over yellow, the expected ratio in the offspring is 3 green: 1 yellow. He observed 428 Green and 152 yellow.

What is the probability of being this *close* to what the model predicts? Do you have have evidence to suggest that Mendel's data are too good?

[Note: This is not quite the same thing as the question of a binomial test.]

2.54. Roptrocerus xylophagorum is a parasitoid of bark beetles. To determine what cues these wasps use to find their hosts, researchers placed female wasps in the base of a Y-shaped tube, with a different odor in each arm of the Y, and recorded the number of wasps that entered each arm [**SPSB00**].

a) In one experiment, one arm of the Y had the odor of bark being eaten by adult beetles, while the other arm of the Y had the odor of bark being eaten by larval beetles. Ten wasps entered the area with the adult beetles, while 17 entered the area with the larval beetles.

Calculate a p-value and draw a conclusion.

b) In another experiment that compared infested bark with a mixture of infested and uninfested bark, 36 wasps moved towards the infested bark, while only 7 moved towards the mixture.

Calculate a p-value and draw a conclusion.

2.55. Suppose you are going to test a coin to see if it is fair.

a) You decide to flip it 200 times and to conduct a binomial test with the data you collect. Suppose the coin is actually biased so that it comes up heads 55% of the time. What is the probability that your test will have a p-value less than 0.05?

b) How does your answer change if you flip the coin 400 times instead?

c) How many flips must you perform if you want a 90% chance of detecting a coin that comes up heads 55% of the time. (Assume here again that we will reject the null hypothesis if the p-value is less than 0.05.)

[Hint: How many heads would it take for you to get a p-value less than 0.05?]

2.56. Express the results in the previous problem using the terminology of statistical power.

2.57. Generally, power increases with sample size. But if you look carefully at the plots on page 64, you will see that although there is an overall increasing pattern to the plots, power is not quite monotone in the sample size. (You may want to recreate the plots at a larger size to see this more clearly.) That is, sometimes increasing the sample size actually *decreases* the power. This seems counterintuitive. Can you give a plausible explanation for why this is happening?

2.58. Find a random variable X and a transformation $t : \mathbb{R} \to \mathbb{R}$ such that

$$\mathrm{E}(t(X)) \neq t(\mathrm{E}(X)) \,.$$

2.59. Justify each equality in the proof of Theorem 2.5.4.

2.60. Provide a reason for each step in the proof of Theorem 2.5.7.

2.61. Prove Lemma 2.5.8.

[Hint: Use Lemma 2.5.3 and Theorem 2.5.7.]

2.62. Refer back to Exercise 2.42. What are the expected value and variance of X, the number of pairs of socks drawn from the sock drawer before a matching pair is found?

2.63. Let $X \sim \mathsf{NBinom}(3, 0.5)$, and let $Y \sim \mathsf{NBinom}(3, 0.2)$. Estimate $\mathrm{Var}(X)$ and $\mathrm{Var}(Y)$ numerically. Compare your results to the values given in Theorem 2.6.9.

2.64. A discrete uniform random variable has a pmf of the form

$$f(x) = \begin{cases} \frac{1}{n} & \text{if } x = 1, 2, \ldots, n, \\ 0 & \text{otherwise.} \end{cases}$$

Find the expected value and variance of a discrete uniform random variable. (Your answers will be functions of n.)

2.65. In computer science, it often occurs that items must be located within a list. How the items are stored in the list affects how long it takes to locate the items.

a) Suppose n items are stored in random order. To find a requested item, the items are searched sequentially until that item is found (assuming the requested item is on the list, of course). Let X be the number of items in the list that must be accessed (looked at) until the requested item is found. What is $\mathrm{E}(X)$?

b) Suppose that 255 items are stored in a list *in order* and that we can access any element in the list. Now a binary search can be used to locate the requested item. We can look in the middle location, and if the item is not there, we will know if it is to the left or the right, etc.

Now what is the expected number of items looked at? (Compare this with the result from part a) when $n = 255$.)

2.66. A roulette wheel has 38 slots numbered 1 through 36 plus 0 and 00. If you bet \$1 that an odd number will come up, you win \$1 if you are correct and lose \$1 otherwise. Let X denote your net winnings (negative for losings) in a single play of this game. Determine $\mathrm{E}(X)$ and $\mathrm{Var}(X)$.

2.67. Weird Willy offers you the following choice. You may have 1/3.5 dollars, or you may roll a fair die and he will give you $1/X$ dollars where X is the value of the roll. Which is the better deal? Compute $\mathrm{E}(1/X)$ to decide.

2.68. Is it possible to say in general which of $E(X^2)$ and $E(X)^2$ will be larger?

   a) Give an example of two random variables Y and Z such that $E(Y^2) > E(Y)^2$ and $E(Z^2) < E(Z)^2$ or explain why one of these two is impossible.

   b) Is it possible that $E(X^2) = E(X)^2$? If so, give an example. If not, explain why not.

2.69. Consider a simple gambling game where you place a bet of $\$n$, flip a fair coin, and win $\$n$ if it comes up heads and lose $\$n$ if it comes up tails.

   Now consider the following gambling strategy for playing this game.

   • Start by betting $\$1$.

   • Each time you lose, bet twice as much the next time.

   • As soon as you win, quit.

   a) If you win on the first try, how much money do you win?

   b) If you lose three times before winning, how much money do you win/lose (net over all the plays)?

   c) If you lose k times before winning, how much money do you win/lose (net over all the plays)?

   d) Let X be the amount of money bet in the last round (when you win). Determine a formula for $P(X = 2^k)$. (The only possible values for X are powers of two because you double the bet each round.)

   e) Use the definition of expected value to calculate $E(X)$. What is strange about this? What problem does it reveal with this betting scheme?

2.70. Find an example of three random variables that are pairwise independent but not independent.

2.71. There is yet another way for three random variables to be independent. What do you think it would mean to say that discrete random variables X and Y are conditionally independent given Z?

2.72. Derive a formula for $E(X^2)$ in terms of $E(X)$ and $Var(X)$.

2.73. Sums of independent random variables.

   a) Show that for any random variables X, Y, and Z,
$$E(X + Y + Z) = E(X) + E(Y) + E(Z) \,.$$

   b) Show that for independent random variables X, Y, and Z,
$$Var(X + Y + Z) = Var(X) + Var(Y) + Var(Z) \,.$$

   c) Generalize these results to sums of n random variables.

   d) Do we need full independence for the variance of a sum to be the sum of the variances, or is pairwise independence sufficient?

2.74. Show that if $X \sim \mathsf{NBinom}(s, \pi)$, then $E(X) = \dfrac{s}{\pi} - s$. This proves half of Theorem 2.6.9.

2.75. The following results generalize Lemma 2.6.11.

 a) Determine the formula for $\text{Var}(X + Y + Z)$ in the general case where the variables may not be independent. Show that your formula is correct.

 b) Generalize this to sums of k random variables $X = X_1 + X_2 + \cdots + X_k$.

2.76. Prove Lemma 2.6.12.

2.77. Describe a joint distribution of random variables X and Y such that $\text{Cov}(X, Y) = 0$, but X and Y are not independent.

2.78. Prove Lemma 2.6.14.

2.79. Alice: Let's take turns flipping a coin. The first person who gets heads wins. I'll go first.

Bob: But you might win on the very first throw. Let's compensate by giving me one extra throw on my first turn.

Alice: Fine.

Bob: And let's use my coin, which comes up heads with probability π.

Alice: Well, then I better know π before I agree.

What value of π makes the game fair to each player?

2.80. Customers arrive at a certain business at a rate of 6 per hour. Assuming a Poisson model for their arrival, answer the following questions.

 a) What is the probability that no customers come between 10:00 and 10:20?

 b) What is the probability that exactly two customers come between 10:00 and 10:20?

 c) What is the probability that more than 6 customers come between 10:00 and 11:00? Less than 6? Exactly 6?

 d) Sam is working a 4-hour shift. What is the probability that between 20 and 30 (inclusive) customers come on his shift?

 e) Why might a Poisson model not be a very good model for some businesses?

2.81. After a 2010 NHL play-off win in which Detroit Red Wings wingman Henrik Zetterberg scored two goals in a 3-0 win over the Phoenix Coyotes, Detroit coach Mike Babcock said, "He's been real good at playoff time each and every year. He seems to score at a higher rate."

Do the data support this claim? In 506 regular season games, Zetterberg scored 206 goals. In 89 postseason games, he scored 44 goals. Goal scoring can be modeled as a Poisson random process. Assuming a goal-scoring rate of $\frac{206}{506}$ goals per game, what is the probability of Zetterberg scoring 44 or more goals in 89 games? How does this probability relate to Coach Babcock's claim?

2.82. Repeat the analysis from Example 2.7.2 for the weeks of November 13 (`week2`) and November 20 (`week3`). Does a similar pattern hold all three weeks?

2.83. Repeat the analysis from Example 2.7.2 for the total number of fumbles for each team over the three weeks included in the `fumbles` data set.

2.84. Use `rpois(120,1.75)` to simulate fumbles for 120 teams. Compare the results of your simulation to a Poisson distribution with the same mean as your simulated data, just as in Example 2.7.2.

  a) How does your simulated data compare to the theoretical distribution?

  b) Repeat this for several simulated data sets. Comment on the fit of a Poisson distribution to the real football fumble data in light of these simulations?

2.85. Students were given one of two questions on a class survey:

  • Suppose that you have decided to see a play for which the admission charge is $20 per ticket. As you prepare to purchase the ticket, you discover that you have lost a $20 bill. Would you still pay $20 for a ticket to see the play?

  • Suppose that you have decided to see a play for which the admission charge is $20 per ticket. As you prepare to enter the theater, you discover that you have lost your ticket. Would you pay $20 to buy a new ticket to see the play?

Here is a tabulation of the responses from a recent class:

attend play?	lost cash	lost ticket
no	9	14
yes	13	9

The proportion of survey participants who would buy a new ticket differs depending on what was lost, but is this just random chance, or are people who lose cash really more likely to buy a ticket than those who lose a ticket? Do the appropriate analysis and draw a conclusion.

2.86. Refer to Exercise 2.85. Here is a tabulation of the responses from several recent classes:

attend play?	lost cash	lost ticket
no	61	69
yes	103	44

The proportion of survey participants who would buy a new ticket differs depending on what was lost, but is this just random chance, or are people who lose cash really more likely to buy a ticket than those who lose a ticket? Do the appropriate analysis and draw a conclusion.

Which data set gives the stronger evidence that people respond differently to losing cash and losing a ticket?

2.87. Compute the odds ratios for the tables given in Exercises 2.85 and 2.86. Note: Your results will differ slightly from the odds ratio reported in `fisher.test()`.

2.88. Repeat the analysis of Fisher's twin-conviction data set using a different cell as the reference cell. Do this both using the hypergeometric distribution directly and also using `fisher.test()`.

2.89. Show that Benford's Law is a valid probability mass function. That is, show that the sum of the probabilities is 1.

2.90. Read one of the articles that gives a theoretical explanation for the form of Benford's Law and write a brief summary.

2.91. Pick a large data set and see how well (or poorly) the frequency of first digits matches the predictions of Benford's Law. The following R function may be of use:

```
> firstDigit <- function(x) {
+     trunc(x / 10^(floor(log(abs(x),10))))
+ }
# lengths (mi) of 141 major North American rivers
> table(firstDigit(rivers))           # lengths in miles

 1  2  3  4  5  6  7  8  9
14 31 36 18 12 13  8  5  4
> table(firstDigit(1.61 * rivers))    # lengths in km

 1  2  3  4  5  6  7  8  9
29  7 13 26 21 17  9 10  9
> table(firstDigit(5280 * rivers))    # lengths in feet

 1  2  3  4  5  6  7  9
61 35 20 10  5  5  3  2
```

2.92. Five cards are dealt from a well-shuffled standard deck of cards (52 cards). Let K be the number of kings dealt and let Q be the number of queens dealt.

a) Are K and Q independent? Explain.

b) Compute $P(K = 2 \mid Q = 2)$.

2.93. Five cards are dealt from a well-shuffled standard deck of cards (52 cards). Let K be the number of kings dealt and let H be the number of hearts dealt.

a) Are K and H independent? Explain.

b) Compute $P(K = 2 \mid H = 2)$.

2.94. An urn contains 5 red balls, 3 blue balls, and 2 white balls. Three balls are drawn from the urn at random without replacement. Let R be the number of red balls; B, the number of blue balls; and W, the number of white balls.

a) What is $f_{B|W=w}$? Is this a familiar distribution?

b) What is $f_{R|W=w}$? Is this a familiar distribution?

2.95. A fair coin is tossed five times. Let Y be the number of heads in all five tosses. Let X be the number of heads in the first two tosses.

a) Are X and Y independent? Explain.

b) Determine the conditional pmf $f_{Y|X=x}$. Is it a familiar distribution?

c) Determine the conditional pmf $f_{X|Y=y}$. Is it a familiar distribution?

2.96. Let X and Y be independent $\mathsf{Binom}(n, \pi)$-random variables. Let $Z = X + Y$. Show the following:

a) $Z \sim \mathsf{Binom}(2n, \pi)$.

b) The conditional distributions of X on Z are hypergeometric. What are the parameters for a fixed value of z?

Hint: Let f be the joint pmf of X and Z. You must show that $f_{X|Z=z}$ is the pmf of a hypergeometric distribution.

## Chapter 3

Continuous Distributions

> The theory of probabilities is at bottom nothing but common sense reduced to calculus.
>
> P. Laplace [**Lap20**]

In Chapter 2 we studied discrete random variables and some important situations that can be modeled with them. Now we turn our attention to continuous random variables. Our goal is to model random variables that can take on any real value in some interval. The height of a human adult, for example, can take on any real value between 60 and 70 inches (and values beyond that range as well), and so it is not suited for the framework of discrete random variables.

Admittedly, most of the time when we actually measure height, there is a limit to the accuracy of the recorded measurements. Perhaps we measure to the nearest inch, or half-inch, or cm. One might argue that this means we are dealing with a discrete random variable with only a finite number of possible measured values. Nevertheless, in situations like this it is generally better to work with continuous random variables. There are two reasons for this. First, continuous random variables often approximate real world situations very well. When our measurements are discretized as a matter of practice (measurement limitations) but not as a matter of principle, we will nearly always work with continuous distributions. In fact, we will see that even some clearly discrete distributions have useful continuous approximations. Second, continuous mathematics (calculus) is often easier to work with than discrete mathematics.

3.1. pdfs and cdfs

So how do we define continuous random variables? We return to the key idea behind histograms, namely that

$$\text{area} = \text{probability}.$$

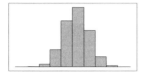 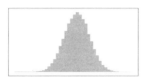 

Figure 3.1. Histograms with finer and finer bins.

Consider the histograms in Figure 3.1. Each one represents the same data, but the bin sizes are different. As the bin sizes get smaller, the resulting histogram gets smoother and smoother. These pictures, which should remind you of Riemann sums and integrals, suggest that we can define continuous random variables by replacing the step function that forms the upper boundary of the histogram with a smooth curve. We can then use integration to determine area and hence the probability that a random variable takes on a value in a particular interval.

Definition 3.1.1. A **probability density function** (pdf) is a function $f : \mathbb{R} \to \mathbb{R}$ such that

- $f(x) \geq 0$ for all $x \in \mathbb{R}$, and
- $\int_{-\infty}^{\infty} f(x)\, dx = 1$.

The **continuous random variable** X defined by the pdf f satisfies

$$P(a \leq X \leq b) = \int_{a}^{b} f(x)\, dx$$

for any real numbers $a \leq b$. □

The following simple lemma demonstrates one way in which continuous random variables are very different from discrete random variables.

Lemma 3.1.2. *Let X be a continuous random variable with pdf f. Then for any real number a:*

- $P(X = a) = 0$.
- $P(X < a) = P(X \leq a)$.
- $P(X > a) = P(X \geq a)$.

Proof. $P(X = a) = \int_{a}^{a} f(x)\, dx = 0$, $P(X \leq a) = P(X < a) + P(X = a) = P(X < a)$, and $P(X \geq a) = P(X > a) + P(X = a) = P(X > a)$. □

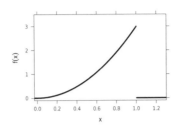

Figure 3.2. A plot of the pdf $f(x) = 3x^2$ on $[0, 1]$.

Example 3.1.1.

Q. Consider the function

$$f(x) = \begin{cases} 3x^2 & \text{if } x \in [0, 1], \\ 0 & \text{otherwise.} \end{cases}$$

Show that f is a pdf and calculate $\mathrm{P}(X \leq 1/2)$.

A. Figure 3.2 contains a graph of f. We can verify that f is a pdf by integration:

$$\int_{-\infty}^{\infty} f(x)\, dx = \int_{0}^{1} 3x^2\, dx = x^3\big|_{0}^{1} = 1\ ,$$

so f is a pdf and

$$\mathrm{P}(X \leq 1/2) = \int_{0}^{1/2} 3x^2\, dx = x^3\big|_{0}^{1/2} = 1/8\ .$$

R can calculate numerical approximations to these integrals with `integrate()`. The `fractions()` function in the MASS package uses a numerical algorithm to locate a nearby fraction. This can be useful if you expect that the exact value should be a rational number.

```
# define the pdf for X                                    pdf-example01
> f <- function(x,...) { 3 * x^2  * (0 <= x & x <= 1) }
# numerical integration gives approximation and tolerance
> integrate(f,0,1)
1 with absolute error < 1.1e-14
> integrate(f,0,0.5)
0.125 with absolute error < 1.4e-15
> integrate(f,0,0.5)$value              # just the approximation value
[1] 0.125
> require(MASS)                         # for the fractions() function
> fractions(integrate(f,0,0.5)$value)   # find nearby fraction
[1] 1/8                                                              ◁
```

The cumulative distribution function of a continuous random variable is defined just as it was for a discrete random variable, but we use an integral rather than a sum to get the cdf from the pdf in this case.

Definition 3.1.3. Let X be a continuous random variable with pdf f. Then the **cumulative distribution function** (cdf) for X is

$$F(x) = \mathrm{P}(X \le x) = \int_{-\infty}^{x} f(t)\, dt \ . \qquad\qquad \square$$

Example 3.1.2.

Q. Determine the cdf of the random variable from Example 3.1.1.

A. For any $x \in [0, 1]$,

$$F_X(x) = \mathrm{P}(X \le x) = \int_{0}^{x} 3t^2\, dt = t^3 \big|_0^x = x^3 \ .$$

So

$$F_X(x) = \begin{cases} 0 & \text{if } x \in [-\infty, 0), \\ x^3 & \text{if } x \in [0, 1], \\ 1 & \text{if } x \in (1, \infty). \end{cases} \qquad\qquad \triangleleft$$

Notice that the cdf F_X is an anti-derivative of the pdf f_X. This follows immediately from the definitions and leads to the following lemma.

Lemma 3.1.4. *Let F_X be the cdf of a continuous random variable X. Then the pdf f_X satisfies*

$$f_X(x) = \frac{d}{dx} F_X(x) \ . \qquad\qquad \square$$

Lemma 3.1.4 implies that we can define random variables by specifying a cdf instead of a pdf. In fact, for continuous distributions, the cdf is in many ways the more important of the two functions, since probabilities are completely determined by the cdf.

Definition 3.1.5. Two random variables X and Y are said to be **equal in distribution** if they have equal cdfs, that is, if

$$F_X(x) = F_Y(x)$$

for all $x \in \mathbb{R}$. $\qquad\qquad \square$

It is possible for two different pdfs to give rise to the same cdf. See Exercise 3.7.

3.1.1. Uniform Distributions

A discrete uniform distribution describes a random variable that takes on a number of possible values each with equal probability. (See Exercise 2.64.) Its pdf is a constant (where it is non-zero). A continuous uniform distribution has a pdf that is constant on some interval.

Definition 3.1.6. A **continuous uniform random variable** on the interval $[a, b]$ is the random variable with pdf given by

$$f(x; a, b) = \begin{cases} \frac{1}{b-a} & \text{if } x \in [a, b], \\ 0 & \text{otherwise.} \end{cases} \qquad\qquad \square$$

It is easy to confirm that this function is indeed a pdf. We could integrate, or we could simply use geometry. The region under the graph of the uniform pdf is a rectangle with width $b - a$ and height $\frac{1}{b-a}$, so the area is 1.

Example 3.1.3.

Q. Let X be uniform on $[0, 10]$. What is $P(X > 7)$? What is $P(3 \leq X < 7)$?

A. Again we argue geometrically. $P(X > 7)$ is represented by a rectangle with base from 7 to 10 along the x-axis and a height of 0.1, so $P(X > 7) = 3 \cdot 0.1 = 0.3$. Similarly $P(3 \leq X < 7) = 0.4$. In fact, for any interval of width w contained in $[0, 10]$, the probability that X falls in that particular interval is $w/10$.

We could also obtain these results using `integrate()`. We need to be a bit clever about defining the pdf, however, because `integrate()` requires a function that is *vectorized*. That is, when given a vector of input values, our function must produce a vector of output values. Most of the arithmetic operations in R are vectorized, but it takes some care to make sure that functions we define ourselves are vectorized. Here is an example showing how this can be done.

```
> x <- 5:15; x
 [1]  5  6  7  8  9 10 11 12 13 14 15
# this would be the natural way to define the pdf:
> tempf <- function(x) { 0.1 * (0 <= x && x <= 10) }
# but it only returns one value when given a vector:
> tempf(x)
[1] 0.1
# integrate() expects to see 0.1 0.1 0.1 0.1 0.1 0.1 0.0 0.0 0.0 0.0 0.0
# sapply() applies a function to each item in a vector:
> f <- function(x) { sapply(x,tempf) }
> f(x)
 [1] 0.1 0.1 0.1 0.1 0.1 0.1 0.0 0.0 0.0 0.0 0.0
# now we are ready to go
# numerical integration gives approximation and tolerance
> integrate(f,7,10)
0.3 with absolute error < 3.3e-15
> integrate(f,3,7)
0.4 with absolute error < 4.4e-15
> integrate(f,7,15)
0.3 with absolute error < 3.4e-15
```

cont-uniform

◁

Example 3.1.4.

Q. Let X be uniform on the interval $[0, 1]$ (which we denote $X \sim \mathsf{Unif}(0, 1)$). What is the cdf for X?

A. For $x \in [0, 1]$, $F_X(x) = \int_0^x 1 \, dx = x$, so

$$F_X(x) = \begin{cases} 0 & \text{if } x \in (\infty, 0), \\ x & \text{if } x \in [0, 1], \\ 1 & \text{if } x \in (1, \infty). \end{cases}$$

Graphs of the pdf and cdf appear in Figure 3.3.   ◁

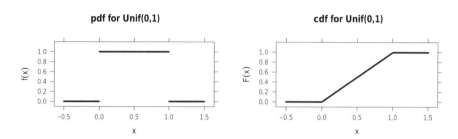

Figure 3.3. Graphs of the pdf and cdf for a $\mathsf{Unif}(0,1)$-random variable.

Although it has a very simple pdf and cdf, the $\mathsf{Unif}(0,1)$ random variable has several important uses. One such use is related to random number generation. Computers are not able to generate truly random numbers. Algorithms that attempt to simulate randomness are called pseudo-random number generators. $X \sim \mathsf{Unif}(0,1)$ is a model for an idealized random number generator. Computer scientists compare the behavior of a pseudo-random number generator with the behavior that would be expected for X to test the quality of the pseudo-random number generator.

We will see later that a good pseudo-random number generator for $X \sim \mathsf{Unif}(0,1)$ can be used to build good pseudo-random number generators for other distributions as well.

Example 3.1.5. The R functions `runif()`, `dunif()`, `punif()`, and `qunif()` work on the same principle as the functions we have been using for discrete distributions.

```
> runif(6,0,10)     # 6 random values on [0,10]          runif
[1] 5.4497 4.1245 3.0295 5.3842 7.7717 8.5714
> dunif(5,0,10)     # pdf is 1/10
[1] 0.1
> punif(5,0,10)     # half the distribution is below 5
[1] 0.5
> qunif(0.25,0,10)  # 1/4 of the distribution is below 2.5
[1] 2.5                                                    ◁
```

3.1.2. The cdf Method

Example 3.1.7 demonstrates a method that we will use repeatedly for deriving a pdf of a random variable. We will refer to it as the **cdf method** because we first derive the cdf of the new random variable and then use differentiation to obtain the pdf.

Example 3.1.6.

Q. Let $X \sim \mathsf{Unif}(0,1)$ and let $Y = X^2$. Drew and Josh want to find a pdf for Y. Drew thinks they should use $g(x) = [f(x)]^2$ and Josh thinks it should be $h(x) = f(x^2)$, where f is a pdf for X.

Show that neither of these is correct and find a correct pdf for Y.

A. Let f be a pdf for X. Recall that $f(x) = 1$ on $[0,1]$. It is tempting to guess (as did Drew and Josh) that the pdf for Y should be one of

- $g(x) = [f(x)]^2$ or
- $h(x) = f(x^2)$.

But neither of these is correct. Notice that for $x \in [0, 1]$,

$$g(x) = [f(x)]^2 = 1 = f(x) \text{ , and}$$
$$h(x) = f(x^2) = 1 = f(x) \text{ .}$$

But X and Y do not have the same distribution. X is uniformly distributed, but Y is more likely to take on smaller values than larger ones because squaring a number in $[0, 1]$ results in a smaller value. In particular,

$$X \in [0, \tfrac{1}{2}] \iff Y \in [0, \tfrac{1}{4}] \text{ ,}$$

so

$$P(Y \leq \tfrac{1}{4}) = \tfrac{1}{2} \text{ , but}$$
$$P(X \leq \tfrac{1}{4}) = \tfrac{1}{4} \text{ .}$$

Thus X and Y do not have the same distribution, and neither of these approaches is correct.

The reasoning of the calculations we just made holds the key to a solution, however. First we will determine the cdf of Y by relating it to the cdf for X. For $y \in [0, 1]$,

$$\begin{aligned}
F_Y(y) &= P(Y \leq y) \\
&= P(X^2 \leq y) \\
&= P(X \leq \sqrt{y}) \\
&= \sqrt{y} \text{ .}
\end{aligned}$$

Now we can determine a pdf for Y by differentiation:

$$\begin{aligned}
f_Y(y) &= \frac{d}{dy}(\sqrt{y}) \\
&= \frac{1}{2\sqrt{y}} \text{ .}
\end{aligned}$$

As an additional check, we can confirm that this is indeed a pdf.

```
> g <- function(y) { 1/(2*sqrt(y)) * (0 <= y & y <= 1) }
> integrate(g,0,1)
1 with absolute error < 2.9e-15
```
cdf-method01

◁

Example 3.1.7.

Q. Let $X \sim \mathsf{Unif}(0, 1)$ and let $Y = \max(X, 1 - X)$. What is the pdf for Y?

A. You may be able to guess the answer by thinking about the following situation. Suppose there are two species in a particular ecosystem. Let X represent the proportion of organisms belonging to one species in a randomly selected region. Then $1 - X$ is the proportion belonging to the other, and Y is the proportion belonging to the more prevalent species.

We can't derive the pdf for Y by trying to obtain a formula for

$$P(Y = y)$$

as we did when we derived pmfs for discrete random variables, because this value is zero for any continuous random variable. Once again we will start by deriving the cdf for Y. Note that Y only takes on values in the interval $[0.5, 1]$.

Suppose, for example, that $Y \leq 0.75$. What does this imply about X? Clearly X cannot be larger than 0.75. Neither can X be less than 0.25, because then $1 - X > 0.75$. So $X \in [0.25, 0.75]$. Generalizing, for $y \in [0.5, 1]$, we have

$$
\begin{aligned}
F_Y(y) = P(Y \leq y) &= P(1 - y \leq X \leq y) \\
&= \int_{1-y}^{y} 1 \, dy \\
&= y - (1 - y) = 2y - 1 .
\end{aligned}
$$

A pdf is now obtained by differentiation:

$$f_Y(y) = \frac{d}{dy}(2y - 1) = 2 \text{ on } [0.5, 1] .$$

We recognize this as the pdf for a uniform random variable on $[0.5, 1]$, so $Y \sim$ Unif$(0.5, 1)$.   ◁

3.1.3. Exponential Distributions

If $\int_{-\infty}^{\infty} f(x) \, dx = a \in (-\infty, \infty)$, then $\frac{1}{a} f(x)$ is a pdf. In this case, we refer to f as a **kernel** of the distribution, and $1/a$ is the **scaling constant**. It is often convenient to think about kernels rather than density functions.

Example 3.1.8.

Q. Let

$$
f(x; \lambda) = \begin{cases} e^{-\lambda x} & \text{if } x \in [0, \infty), \\ 0 & \text{if } x \in (-\infty, 0). \end{cases}
$$

For what values of λ is f the kernel of a distribution? What is the scaling constant associated with λ?

A. If $\lambda > 0$, then

$$
\int_0^{\infty} e^{-\lambda x} \, dx = \lim_{L \to \infty} -\frac{1}{\lambda}[e^{-\lambda L} - 1] = \frac{1}{\lambda} .
$$

(If $\lambda \leq 0$, then the integral diverges.) The scaling constant is $\frac{1}{1/\lambda} = \lambda$, so the pdf for the distribution is given by

$$
f(x; \lambda) = \begin{cases} \lambda e^{-\lambda x} & \text{if } x \in [0, \infty), \\ 0 & \text{if } x \in (-\infty, 0). \end{cases}
$$

                                                                                                            ◁

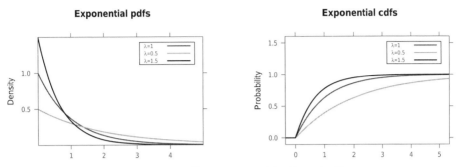

Figure 3.4. Some exponential pdfs and cdfs.

Definition 3.1.7. An **exponential random variable** X is a continuous random variable with pdf

$$f(x; \lambda) = \begin{cases} \lambda e^{-\lambda x} & \text{if } x \geq 0, \\ 0 & \text{if } x < 0. \end{cases}$$

We will denote this by $X \sim \mathsf{Exp}(\lambda)$. □

The exponential density functions are easily integrated, and the cdf of an $\mathsf{Exp}(\lambda)$-distribution is given by

$$F_X(x) = \int_0^x \lambda e^{-\lambda t} \, dt = 1 - e^{-\lambda x} \ .$$

The pdfs (left) and cdfs (right) of several exponential distributions are displayed in Figure 3.4.

3.1.4. Exponential and Poisson Distributions

There is an important connection between the exponential distributions and the Poisson distributions introduced in Section 2.7.1. This connection explains why exponential distributions are often used to model the lifetime or "time until failure" of objects like computer hardware or more generally to model the time until some event – like the arrival of a customer or an automobile accident at a particular location – occurs.

Suppose that failures can be modeled by a Poisson process with a rate of λ per unit time. As we saw in Chapter 2, this means that in 1 unit of time any number of failures could occur, but that average number of failures is λ. Now we can ask about a new random variable

$$Y = \text{ time until the next failure.}$$

Y is a continuous random variable (any non-negative real is a possible time until failure). We will begin by deriving the cdf for Y:

$$\mathrm{P}(Y \leq y) = 1 - \mathrm{P}(Y > y) = 1 - \mathrm{P}(0 \text{ occurrences in time } [0, y])$$
$$= 1 - \mathrm{P}(X = 0) = 1 - e^{-\lambda y} \ ,$$

where $X \sim \mathsf{Pois}(\lambda y)$ since a rate of λ occurrences per unit time implies a rate of λy per observation time y. If we don't already recognize the cdf, we can differentiate to get the pdf:

$$f_Y(y) = \frac{d}{dy}\left(1 - e^{-\lambda y}\right) = \lambda e^{-\lambda y} .$$

We recognize this as the pdf for an $\mathsf{Exp}(\lambda)$-random variable.

This duality between Poisson and exponential random variables is much like the duality between the binomial and geometric random variables. In the former case, time is measured continuously; in the latter, time is discrete. Otherwise, the relationship is the same.

Example 3.1.9.

Q. Let's consider an example at a bank. Suppose that customers arrive independently at an average rate of 10 per hour and that the Poisson (and hence exponential) distributions are good models for the number of arrivals in a fixed time and the time until the next arrival. What is the probability that no customers arrive in the next ten minutes?

A. If we measure time in hours, then since 10 minutes is $\frac{1}{6}$ hours, we are looking for $P(Y > \frac{1}{6})$, where $Y \sim \mathsf{Exp}(\lambda = 10)$.

Notice that $E(Y) = \frac{1}{\lambda} = \frac{1}{10}$ makes sense. If customers arrive at a rate of 10 per hour, we expect a customer about every tenth of an hour (every 6 minutes):

$$P(Y > \frac{1}{6}) = 1 - P(Y \le \frac{1}{6}) = 1 - (1 - e^{-\lambda \cdot \frac{1}{6}}) = e^{-\lambda \cdot \frac{1}{6}} = e^{-10 \cdot \frac{1}{6}} \approx 0.189 .$$

So nearly 20% of the time there will be a 10-minute gap between customers.

Of course, we could measure time in minutes instead of hours throughout. This would be modeled by a random variable Y_2 that is exponentially distributed with $\lambda = \frac{1}{6}$ and we would want to determine $P(Y_2 > 10)$. We get the same probability as before:

$$P(Y_2 > 10) = 1 - P(Y_2 \le 10) = 1 - (1 - e^{-\lambda \cdot 10}) = e^{-\lambda \cdot 10} = e^{-\frac{1}{6} \cdot 10} \approx 0.189 .$$

We could also use the related Poisson random variable with $\lambda = 10/6$ (customers per 10 minutes). All three methods are illustrated in the following R code.

```
> 1 - pexp(1/6,10)    #   10 customers per  1 hour          bank01
[1] 0.18888
> 1 - pexp(10,1/6)    #  1/6 customer  per  1 minute
[1] 0.18888
> dpois(0,10/6)       # 10/6 customers per 10 minutes
[1] 0.18888
```

It does not matter which method is used, but it is critically important to pay attention to the units in which time is expressed and to use them consistently. ◁

One interesting property of the exponential distributions is illustrated in the following example.

Example 3.1.10.

Q. Let $X \sim \text{Exp}(\lambda)$. What is $P(X > b \mid X > a)$? Assume $b > a > 0$.

A. Since $b > a$,

$$
\begin{aligned}
P(X > b \mid X > a) &= \frac{P(X > b)}{P(X > a)} \\
&= \frac{e^{-\lambda b}}{e^{-\lambda a}} \\
&= e^{-\lambda(b-a)} \\
&= P(X > b - a) \, .
\end{aligned}
$$

This shows that the probability depends only on the difference between a and b. In other words, for any t and k, $P(X > k) = P(X > t + k \mid X > t)$. A process that has this feature is said to be **memoryless**.   ◁

Example 3.1.10 has important consequences for deciding if an exponential distribution is a good model for a given situation. If, for example, we use an exponential distribution to model the time until failure, the model implies that the distribution of future time until failure does not depend on how old the part or system already is; such parts neither improve nor degrade with age. Presumably they fail because of things that happen to them, like being dropped or struck by lightening.

More general lifetime models exist which can take into account features like early failure (parts that have a high probability of failure but are likely to last a long time if they survive an initial burn-in phase) or wear (parts that degrade over time).

Example 3.1.11. It should not be thought that the assumptions of the Poisson model imply that Poisson events should occur at roughly regular intervals. Although the probability of an event occurring in a small interval is the same for all intervals of the same width, in any long enough run of a Poisson process, there will be clusters of events here and there. Similarly there may be long stretches without any events.

Using the connection between the Poisson and exponential distributions and the fact that the exponential distribution is memoryless, we can simulate the timing of Poisson events using the exponential distribution to model the time between consecutive events:

pois-sim

```
> runs <- 8; size <- 40
> time <- replicate(runs, cumsum(rexp(size)))
> df <- data.frame(time = as.vector(time), run = rep(1:runs,each=size) )
> stop <- min(apply(time,2,max))
> stop <- 5 * trunc(stop/5)
> df <- df[time <= stop,]
> myplot <- stripplot(run~time, df, pch=1, cex=.7, col='black',
+          panel=function(x,y,...){
+                  panel.abline(h=seq(1.5,7.5,by=1),col='gray60')
+                  panel.abline(v=seq(0,stop,by=5), col='gray60')
```

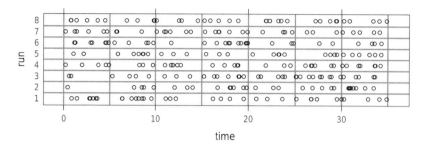

Figure 3.5. A simulation of Poisson events shows that although the location of event clusters and gaps cannot be predicted, their presence can.

```
+                           panel.stripplot(x,y,...)
+              })
```

The number of events depicted a boxed section in Figure 3.5 has a $\mathsf{Pois}(\lambda = 5)$-distribution. As we would expect, some of these sections are more densely populated with events than others. Viewed over time, we see that the events cluster in some places and are relatively widely separated in other places. This is all part of what the Poisson model predicts. The Poisson model does not, however, allow us to predict when the clusters and gaps will occur.   ◁

3.2. Mean and Variance

The mean and variance of a continuous random variable are computed much like they are for discrete random variables, except that we must replace summation with integration.

Definition 3.2.1. Let X be a continuous random variable with pdf f. Then:

- $\mathrm{E}(X) = \mu_X = \int_{-\infty}^{\infty} x f(x)\, dx$.
- $\mathrm{Var}(X) = \sigma_X^2 = \int_{-\infty}^{\infty} (x - \mu_X)^2 f(x)\, dx$.   □

Intuitively, our definition of variance looks like it could be restated as
$$\mathrm{Var}(X) = \mathrm{E}\left((X - \mu_X)^2\right) .$$
This is indeed true, as the following lemma shows.

Lemma 3.2.2. *Let X be a continuous random variable, and let $Y = t(X)$. Then*
$$\mathrm{E}(Y) = \int_{-\infty}^{\infty} t(x) f(x)\, dx .$$

Proof. Let f and F be the pdf and cdf for X, and let g and G be the pdf and cdf for Y. For notational ease, we'll handle the case where t is differentiable and increasing, and $\mathrm{P}(X \geq 0) = 1$.

In this case,
$$G(t(x)) = \mathrm{P}(Y \leq t(x)) = \mathrm{P}(X \leq x) = F(x) .$$

Differentiating, we see that

$$f(x) = F'(x) = G'(t(x))t'(x) = g(t(x))t'(x) \ .$$

Now using the substitution $u = t(x)$, we see that

$$\int_{-\infty}^{\infty} t(x)f(x)\,dx = \int_{-\infty}^{\infty} t(x)\,g(t(x))\,t'(x)\,dx = \int_{-\infty}^{\infty} u\,g(u)\,du = \mathrm{E}(Y) \ . \qquad \square$$

As with discrete distributions, the following theorem often simplifies determining the variance.

Theorem 3.2.3. *For any continuous or discrete random variable X,*

$$\mathrm{Var}(X) = \mathrm{E}(X^2) - \mathrm{E}(X)^2 \ .$$

Proof. The proof is similar to the proof of Theorem 2.5.7 but we replace summation with integration. $\qquad \square$

Lemma 3.2.4. *Let $X \sim \mathsf{Unif}(a,b)$. Then:*

- $\mathrm{E}(X) = \frac{a+b}{2}$.
- $\mathrm{Var}(X) = \frac{(b-a)^2}{12}$.

Proof.

$$\mathrm{E}(X) = \int_a^b x\frac{1}{b-a}\,dx = \frac{1}{b-a}\frac{x^2}{2}\Big|_a^b = \frac{b^2-a^2}{2(b-a)} = \frac{a+b}{2} \ ,$$

$$\mathrm{E}(X^2) = \int_a^b x^2\frac{1}{b-a}\,dx = \frac{1}{b-a}\frac{x^3}{3}\Big|_a^b = \frac{b^3-a^3}{3(b-a)} = \frac{a^2+ab+b^2}{3} \ ,$$

$$\mathrm{Var}(X) = \frac{a^2+ab+b^2}{3} - \frac{a^2+2ab+b^2}{4} = \frac{(b-a)^2}{12} \ . \qquad \square$$

Lemma 3.2.5. *Let $Y \sim \mathsf{Exp}(\lambda)$. Then:*

- $\mathrm{E}(X) = \dfrac{1}{\lambda}$.
- $\mathrm{Var}(X) = \dfrac{1}{\lambda^2}$.

Proof. Using integration by parts, we find that

$$\mathrm{E}(X) = \int_0^{\infty} \lambda x e^{-\lambda x}\,dx = \lim_{L\to\infty}\ (xe^{-\lambda x} - \frac{1}{\lambda}e^{-\lambda x})\Big|_0^L = \frac{1}{\lambda} \ .$$

The variance is found similarly, using integration by parts twice to evaluate $\int_0^{\infty} x^2\lambda e^{-\lambda x}$. $\qquad \square$

Example 3.2.1.

Q. Returning to the data in Example 2.7.2:

- How long should we expect to wait until the next fumble by a specified team?
- What is the probability that a team has no fumbles in the first half of a game?

A. A good model for the time until the next fumble is $Y \sim \mathsf{Exp}(\lambda = 1.75)$, so we should expect to wait about $1/1.75$ games. Football games are 60 minutes long, so this translates to approximately 34.28 minutes (football game time). We can calculate the probability of no fumbles in half of a football game using either an exponential or Poisson distribution.

```
> 1 - pexp(0.5, rate=1.75)
[1] 0.41686
> dpois(0,1.75/2)
[1] 0.41686
```
fumbles02

◁

Lemma 3.2.6. *Let X be a discrete or continuous random variable. Then:*

(1) $\mathrm{E}(aX + b) = a\,\mathrm{E}(X) + b$.

(2) $\mathrm{Var}(aX + b) = a^2\,\mathrm{Var}(X)$.

Proof. We have already proved these claims for discrete random variables. The proofs for continuous random variables are similar to the proofs given for their discrete analog.   □

3.3. Higher Moments

In statistics, higher moments are generalizations of the mean and variance, which are the first two moments.

Definition 3.3.1. Let X be a random variable.

- The kth **moment about the origin** is defined by

$$\mu_k = \mathrm{E}(X^k)\,,$$

  provided $\mathrm{E}(\mid X \mid^k) < \infty$. Note that $\mu_1 = \mu_X$.

- The kth **moment about the mean** is defined by

$$\mu'_k = \mathrm{E}((X - \mu)^k)\,,$$

  again provided $\mathrm{E}(\mid X \mid^k) < \infty$.   □

As the following lemma shows, we can recover moments about the mean from moments about the origin. This is a generalization of our method for calculating the variance as $\mathrm{E}(X^2) - \mathrm{E}(X)^2$.

Lemma 3.3.2. *Let X be a random variable for which the kth moment about the origin exists. Then:*

(1) μ_i *exists for all $i \leq k$.*

(2) $\mu'_k = \displaystyle\sum_{i=0}^{k} \binom{k}{i} (-1)^{k-i} \mu_i \mu^{k-i}$.

Proof. For the first claim we will handle the case that X is continuous. The discrete case is similar. Let f be the pdf for X and assume that

$$\int_{-\infty}^{\infty} \mid x \mid^k f(x) \, dx < \infty . \tag{3.1}$$

If $1 \leq i \leq k$ and $\mid x \mid > 1$, then $\mid x^k \mid > \mid x^i \mid$. On the other hand, if $\mid x \mid \leq 1$, then $\mid x^i \mid \leq 1$. Using these facts, we break up the integral in (3.1) into two parts:

$$
\begin{aligned}
\int_{-\infty}^{\infty} \mid x \mid^i f(x) \, dx &= \int_{|x| \geq 1} \mid x \mid^i f(x) \, dx && + \int_{|x| \leq 1} \mid x \mid^i f(x) \, dx \\
&\leq \int_{|x| \geq 1} \mid x \mid^k f(x) \, dx && + \int_{|x| \leq 1} 1 \cdot f(x) \, dx \\
&\leq \int_{|x| \geq 1} \mid x \mid^k f(x) \, dx && + 1 \\
&< \infty ,
\end{aligned}
$$

so μ_i exists.

The second claim follows directly from the binomial expansion of $(X - \mu)^k$ combined with Lemma 3.2.6. $\qquad\square$

Example 3.3.1.

$$
\begin{aligned}
\mu_3' &= \mathrm{E}\left[(X - \mu)^3\right] \\
&= \mathrm{E}\left[X^3 - 3X^2\mu + 3X\mu^2 - \mu^3\right] \\
&= \mu_3 - 3\mu_2\mu + 3\mu_1\mu^2 - \mu^3 \\
&= \mu_3 - 3\mu_2\mu + 2\mu^3 .
\end{aligned}
$$
$\qquad\triangleleft$

Higher moments are used to define additional measures of the shape of a distribution.

Definition 3.3.3. Let X be a random variable with $\mathrm{E}(X) = \mu$ and $\mathrm{Var}(X) = \sigma^2$.

- The **coefficient of skewness** is defined by

$$\gamma_1 = \frac{\mu_3'}{\sigma^3} = \frac{\mathrm{E}\left[(X - \mu)^3\right]}{\sigma^3} .$$

- The **coefficient of kurtosis** is defined by

$$\gamma_2 = \frac{\mu_4'}{\sigma^4} - 3 = \frac{\mathrm{E}\left[(X - \mu)^4\right]}{\sigma^4} - 3 . \qquad\square$$

Some comments on these definitions are in order.

(1) Dividing by σ^i makes these coefficients dimensionless and therefore insensitive to issues of units and scale.

(2) If X is symmetric, then $\gamma_1 = 0$. So lack of symmetry is measured by deviation from 0. The sign of γ_1 gives an indication of the direction in which the distribution is skewed.

(3) The coefficient of kurtosis is a measure of "peakedness" of a distribution. Relatively flat pdfs are said to be **platykurtic**; more peaked distributions are said to be **leptokurtic**. The point of reference is the normal distributions which we introduce in Section 3.4.1. Kurtosis is defined so that the normal distributions have $\gamma_2 = 0$. This is the reason for subtracting 3 in the definition.

3.3.1. Review of Power Series

Recall that a power series (about $t = 0$) is a series with the following form:

$$A(t) = \sum_{k=0}^{\infty} a_k t^k .$$

Before we discuss generating functions, we want to review some important properties of **power series**. These properties are stated largely without proof. Consult a calculus or analysis text for the justifications of these claims.

(1) A power series is a representation of a *function* since for different values of t the series might converge to different values (or fail to converge at all).

(2) Convergence: For any power series there is an extended real number $R \geq 0$ (possibly $R = \infty$) such that
  - $|t| < R \Rightarrow$ the power series converges;
  - $|t| > R \Rightarrow$ the power series diverges;

  R is called the **radius of convergence**.

(3) Derivatives: If a power series converges on $(-R, R)$, then it has derivatives of all orders on $(-R, R)$. The power series for the derivatives can be obtained by differentiating *term by term*. For example, the first derivative is

$$\frac{dA}{dt} = a_1 + 2a_2 t + 3a_3 t^2 + \cdots = \sum_{k=1}^{\infty} k a_k t^{k-1} = \sum_{k=0}^{\infty} (k+1) a_{k+1} t^k .$$

  Repeated differentiation shows that

$$A^{(k)}(0) = k! a_k , \quad \text{i.e., } a_k = \frac{A^{(k)}(0)}{k!} . \tag{3.2}$$

  - We can use this to obtain a power series representation of a function that has derivatives of all orders in an interval around 0 since the derivatives at 0 tell us the coefficients of the power series, although this doesn't guarantee convergence. Such series are called **Taylor series** and the first finitely many terms are called a **Taylor polynomial**.
    For example, consider $f(t) = e^t$. Then for all k,

$$f^{(k)}(t) = e^t \ \text{ and}$$

$$f^{(k)}(0) = e^0 = 1 .$$

  So the Taylor series for e^t is

$$e^t = \sum_{k=0}^{\infty} \frac{t^k}{k!} .$$

(We used this fact in our derivation of the pmf for the Poisson distributions.)

(4) Uniqueness: If two power series converge to the same sum on some interval $(-R, R)$, then all the coefficients are equal.

3.3.2. Generating Functions

In calculus, we used Taylor's Theorem to find a power series for a given function (property (3) above). Now we want to do the reverse. Instead of using the power series to learn about the function it converges to, we want to use the function a power series converges to, to learn about the *sequence of coefficients* of the series. In this case we refer to the power series function as the **generating function** of the sequence because we will see how to use it to generate the terms of the sequence.

Actually, there are two kinds of generating functions:

(1) the **ordinary generating function** for $\{a_k\}$ (as described above):

$$A(t) = \sum a_k t^k \; ;$$

(2) the **exponential generating function** for $\{a_k\}$ is the ordinary generating function for $\{\frac{a_k}{k!}\}$:

$$B(t) = \sum \frac{a_k}{k!} t^k \; .$$

The difference is whether or not we consider $\frac{1}{k!}$ in (3.2) to be part of the sequence we are interested in.

Obtaining generating functions from sequences

If we have a particular sequence in mind, it is a simple matter to define the (ordinary or exponential) generating function. For some series, we can express the generating function in closed form. In the examples below, our two favorite series appear: geometric series and the Taylor series for the exponential function.

Example 3.3.2.

Q. Let $\{a_k\} = 1, 1, 1, \dots$.   What are the ordinary and exponential generating functions for $\{a_k\}$?

A. Ignoring issues of convergence,

- ordinary: $A(t) = 1 + t + t^2 + t^3 + \cdots = \dfrac{1}{1-t}$,

- exponential: $B(t) = 1 + t + \dfrac{1}{2!}t^2 + \dfrac{1}{3!}t^3 + \cdots = e^t$.   ◁

Example 3.3.3.

Q. Let $\{a_k\} = \{2^k\} = 1, 2, 4, \dots$. What are the ordinary and exponential generating functions for $\{a_k\}$?

A. Ignoring issues of convergence,

- ordinary: $A(t) = 1 + 2t + 4t^2 + 8t^3 + \cdots = 1 + 2t + (2t)^2 + (2t)^3 + \cdots = \dfrac{1}{1 - 2t}$,

- exponential: $B(t) = 1 + 2t + \dfrac{1}{2!}(2t)^2 + \dfrac{1}{3!}(2t)^3 + \cdots = e^{2t}$.   ◁

Obtaining sequences from generating functions

Now suppose we know the generating function $A(t)$. How do we get the sequence? We use (3.2):

- ordinary: $a_k = \dfrac{A^{(k)}(0)}{k!}$,

- exponential: $\dfrac{a_k}{k!} = \dfrac{A^{(k)}(0)}{k!}$, so $a_k = A^{(k)}(0)$.

Example 3.3.4.

Q. $A(t) = \frac{1}{1-3t}$ is the generating function of a sequence. What is the sequence?

A. First we compute the derivatives of $A(t)$:

$$A^{(k)}(t) = k!\, 3^k (1 - 3t)^{-k-1} \ ,$$

which we evaluate at $t = 0$:

$$A^{(k)}(0) = k!\, 3^k \ .$$

If we are dealing with an ordinary generating function, then

$$a_k = 3^k \ .$$

If we consider A to be an exponential generating function for the sequence, then the sequence is defined by

$$a_k = k!\, 3^k \ .$$ ◁

3.3.3. Moment Generating Functions

We want to use generating functions to study the sequence of moments of a pdf. In this case, the exponential generating function is easier to work with than the ordinary generating function. So we define the moment generating function of a pdf of a random variable X to be the exponential generating function of the sequence of moments μ_k:

$$M_X(t) = \sum_{k=0}^{\infty} \frac{\mu_k}{k!} t^k \ , \tag{3.3}$$

where we define $\mu_0 = 1$ for the purposes of the series above.

Now, if we can just get our hands on $M_X(t)$, and if it is reasonably easy to differentiate, then we can get all of our moments via differentiation instead of summation or integration. This is potentially much easier. What we need is a function $A(t)$ such that

$$A^{(k)}(0) = \mu_k \ . \tag{3.4}$$

(See (3.2) again.) And the function is (drum roll please...):

Theorem 3.3.4. *Let* X *be a random variable with moment generating function* $M_X(t)$. *Then*

$$M_X(t) = \text{E}(e^{tX}) \,.$$

Proof. We need to show that (3.4) is satisfied for $A(t) = \text{E}(e^{tX})$. Note that A is a function of t, since we will get potentially different expected values depending on what t is.

If X is continuous, then we have (limits of integration omitted)

$$\text{E}(e^{tX}) = \int e^{tx} f_X(x) \, dx \,,$$

$$\frac{d}{dt} \text{E}(e^{tX}) = \frac{d}{dt} \int e^{tx} f_X(x) \, dx = \int \frac{d}{dt} e^{tx} f_X(x) \, dx$$

$$= \int x e^{tx} f_X(x) \, dx \,,$$

$$\frac{d}{dt} \text{E}(e^{tX}) \Big|_{t=0} = \int x e^{0 \cdot x} f_X(x) \, dx = \int x \cdot f_X(x) \, dx = \text{E}(X) = \mu_1 \,.$$

Similarly,

$$\frac{d^2}{dt^2} \text{E}(e^{tX}) \Big|_{t=0} = \int x^2 e^{0 \cdot x} f_X(x) \, dx \Big|_{t=0} = \text{E}(X^2) = \mu_2 \,,$$

$$\frac{d^i}{dt^i} \text{E}(e^{tX}) \Big|_{t=0} = \int x^i e^{0 \cdot x} f_X(x) \, dx \Big|_{t=0} = \text{E}(X^i) = \mu_i \,.$$

For discrete random variables, we replace the integrals above with sums.

$$\text{E}(e^{tX}) = \sum e^{tx} f_X(x) \,,$$

$$\frac{d}{dt} \text{E}(e^{tX}) = \frac{d}{dt} \sum e^{tx} f_X(x) = \sum \frac{d}{dt} e^{tx} f_X(x)$$

$$= \sum x e^{tx} f_X(x) \,,$$

so

$$\frac{d}{dt} \text{E}(e^{tX}) \Big|_{t=0} = \sum x e^{0 \cdot x} f_X(x) = \text{E}(X) = \mu_1 \,,$$

$$\frac{d^i}{dt^i} \text{E}(e^{tX}) \Big|_{t=0} = \sum x^i e^{tx} f_X(x) \Big|_{t=0} = \text{E}(X^i) = \mu_i \,. \qquad \square$$

Example 3.3.5.

Q. Let Y be an exponential random variable with pdf $f_Y(y) = \lambda e^{-\lambda y}$, $y > 0$. Use the moment generating function to determine $\mathrm{E}(Y)$ and $\mathrm{Var}(Y)$.

A.

$$M_Y(t) = \mathrm{E}(e^{tY}) = \int_0^\infty e^{ty}\lambda e^{-\lambda y}\, dy = \int_0^\infty \lambda e^{ty - \lambda y}\, dy = \int_0^\infty \lambda e^{(t - \lambda)y}\, dy\ .$$

Now suppose $|t| < \lambda$ and let $u = (t - \lambda)\, y$, so $du = (t - \lambda)\, dy$ and the integral becomes

$$M_Y(t) = \int_0^{-\infty} \frac{\lambda}{t - \lambda} e^u\, du = \frac{\lambda}{\lambda - t} = \frac{1}{1 - \frac{t}{\lambda}}\ .$$

($M_Y(y)$ doesn't exist for $|t| \geq \lambda$.) Now we can determine the moments of Y by differentiating:

- $M_Y'(t) = \lambda(-1)(\lambda - t)^{-2}(-1) = \lambda(\lambda - t)^{-2}$, so
- $\mathrm{E}(Y) = \mu_1 = M_Y'(0) = \lambda/\lambda^2 = 1/\lambda$.
- $M_Y''(t) = \lambda(-2)(\lambda - t)^{-3}(-1) = 2\dfrac{\lambda}{(\lambda - t)^3}$.
- $\mathrm{E}(Y^2) = \mu_2 = M_Y''(0) = \dfrac{2}{\lambda^2}$.
- $\mathrm{Var}(Y) = \mu_2 - \mu_1^2 = \dfrac{2}{\lambda^2} - \dfrac{1}{\lambda^2} = \dfrac{1}{\lambda^2}$.   ◁

Example 3.3.6.

Q. Let $X \sim \mathsf{Binom}(n, \pi)$. Determine the mgf for X.

A. The pdf for X is $f_X(x) = \binom{n}{x}\pi^x(1 - \pi)^{n-x}$, so

$$\mathrm{E}(e^{tX}) = \sum_x e^{tx} f_X(x) = \sum_x e^{tx}\binom{n}{x}\pi^x(1 - \pi)^{n-x}$$

$$= \sum_x \binom{n}{x}(e^t\pi)^x(1 - \pi)^{n-x}$$

$$= (\pi e^t + 1 - \pi)^n$$

for all values of t. Now we can find the moments by differentiating:

- $M_X'(t) = n(\pi e^t + 1 - \pi)^{n-1}(\pi e^t)$,
- $M_X'(0) = n(\pi + 1 - \pi)^{n-1}(\pi) = n\pi$,
- $M_X''(t) = n(n-1)(\pi e^t + 1 - \pi)^{n-2}(\pi e^t)^2 + n(\pi e^t + 1 - \pi)^{n-2}(\pi e^t)$,
- $M_X''(0) = n(n-1)(\pi + 1 - \pi)^{n-2}\pi^2 + n(\pi + 1 - \pi)^{n-2}\pi = n(n-1)\pi^2 + n\pi$,
- $\mathrm{Var}(X) = n(n-1)\pi^2 + n\pi - n^2\pi^2 = -n\pi^2 + n\pi = n(1 - \pi)\pi$.

Of course, each of these values matches what we found in Chapter 2.   ◁

We will not prove the following theorem here, but it is very important because it says that if we can recognize the mgf of a random variable, then we know what the random variable is.

Theorem 3.3.5. *Let X and Y be random variables that have moment generating functions M_X and M_Y. Then X and Y are **identically distributed** (i.e., have the same cdfs) if and only if*

$$M_X(t) = M_Y(t) \text{ for all } t \text{ in some interval containing } 0. \qquad \square$$

We will encounter some important applications of this theorem shortly.

3.3.4. Moment Generating Functions for Linear Transformations

Theorem 3.3.6. *If a and b are constants, then:*

(1) $M_{aX}(t) = M_X(at)$.

(2) $M_{X+b}(t) = e^{bt} M_X(t)$.

(3) $M_{aX+b}(t) = e^{bt} M_X(at)$.

Proof. The proof requires only algebraic manipulation and Lemma 3.2.6.

(1) $M_{aX}(t) = \mathrm{E}(e^{t \cdot aX}) = \mathrm{E}(e^{at \cdot X}) = M_X(at)$.

(2) $M_{X+b}(t) = \mathrm{E}(e^{t(X+b)}) = \mathrm{E}(e^{bt} \cdot e^{tX}) = e^{bt} \, \mathrm{E}(e^{tX}) = e^{bt} M_X(t)$.

(3) $M_{aX+b}(t) = e^{bt} M_X(at)$ by combining (1) and (2). $\qquad \square$

Example 3.3.7.

Q. Let $X \sim \mathsf{Binom}(n, \pi)$ and let $Y = 2X + 3$. What is the mgf for Y?

A. The mgf for Y is

$$e^{3t} M_X(2t) = e^{3t} \cdot (\pi e^{2t} + 1 - \pi)^n . \qquad \triangleleft$$

Example 3.3.8.

Q. Let $X \sim \mathsf{Exp}(\lambda)$. Let $Y = kX$ for some constant $k > 0$. Determine the mgf for Y.

A. $M_X(t) = \frac{\lambda}{\lambda - t}$, so

$$M_Y(t) = M_X(kt) = \frac{\lambda}{\lambda - kt} = \frac{\lambda/k}{\lambda/k - t} .$$

Notice that this is the mgf for an exponential function with parameter λ/k. By Theorem 3.3.5, this means that $Y \sim \mathsf{Exp}(\lambda/k)$. So all the exponential distributions are simply rescalings of each other. $\qquad \triangleleft$

3.4. Other Continuous Distributions

3.4.1. Normal Distributions

The normal distributions form arguably the most important family of distributions in all of probability and statistics. The normal distributions form a good model for many numerical populations. Heights, weights, and other physical measurements of organisms often follow an approximately normal distribution. So do measurement errors in many scientific experiments. Many discrete distributions are approximately normal, including the binomial and Poisson distributions (provided certain conditions are met by the parameters n, π, and λ). Even when the population in question is not well approximated by a normal distribution, sums and averages of samples from the population often approximate a normal distribution. This is a consequence of the Central Limit Theorem, which we will consider in Chapter 4.

Definition 3.4.1. A continuous random variable Z has the **standard normal distribution** (denoted $Z \sim \mathsf{Norm}(0,1)$) if its pdf is given by

$$\phi(z) = \frac{e^{-z^2/2}}{\sqrt{2\pi}} \; .$$

$\square$

The use of Z for a standard normal random variable and of ϕ for the pdf of the standard normal distribution is traditional in statistics. The cdf is denoted Φ.

We have not motivated this distribution with a particular model as we have done with the other distributions we have seen. In Chapter 4 we will learn why the standard normal distribution is so important for statistical inference. For now we will simply introduce some of its most important properties. We begin by showing that the function given in the definition is a legitimate pdf.

Theorem 3.4.2. $\phi(z) = \frac{e^{-z^2/2}}{\sqrt{2\pi}}$ *is a pdf. That is,*

$$\int_{-\infty}^{\infty} \phi(z) \, dz = 1 \; .$$

Proof. This proof requires a clever approach, since the kernel of the distribution is e^{-z^2}, which does not have an anti-derivative that can be expressed as an elementary function. The clever idea is to show that the *square* of the integral is 1. Let

$$T = \int_{-\infty}^{\infty} \frac{e^{-x^2/2}}{\sqrt{2\pi}} \, dx \; .$$

Then

$$T^2 = \int_{-\infty}^{\infty} \frac{e^{-x^2/2}}{\sqrt{2\pi}} \, dx \cdot \int_{-\infty}^{\infty} \frac{e^{-y^2/2}}{\sqrt{2\pi}} \, dy \; .$$

We want to show that $T^2 = 1$. We begin by observing that

$$T^2 = \int_{-\infty}^{\infty} \int_{-\infty}^{\infty} \frac{e^{-x^2/2}}{\sqrt{2\pi}} \frac{e^{-y^2/2}}{\sqrt{2\pi}} \, dx \, dy$$

$$= \int_{-\infty}^{\infty} \int_{-\infty}^{\infty} \frac{e^{-[x^2+y^2]/2}}{2\pi} \, dx \, dy \; .$$

The presence of the term $x^2 + y^2$ suggests a change of variables into polar coordinates, which we follow with the substitution $u = -r^2/2$:

$$T^2 = \int_0^{2\pi} \int_0^\infty \frac{e^{-r^2/2}}{2\pi} r \, dr \, d\theta$$

$$= \int_0^{-\infty} \int_0^{2\pi} -\frac{e^u}{2\pi} \, d\theta \, du$$

$$= \int_0^{-\infty} -2\pi \frac{e^u}{2\pi} \, du$$

$$= \lim_{L \to -\infty} - e^u \big|_{u=0}^{u=L}$$

$$= e^0 = 1 \, . \qquad \qquad \square$$

It is important to observe that although we have just shown that

$$\int_{-\infty}^\infty \phi(z) dz = 1 \, ,$$

the method we used does not help us evaluate

$$\Phi(a) = \int_{-\infty}^a \phi(z) dz \, ,$$

for $a \in \mathbb{R}$. The latter integral cannot be done by the usual means of calculus, so $\Phi(a)$ must be approximated using numerical methods.

Theorem 3.4.3. *Let $Z \sim \mathsf{Norm}(0,1)$. Then the moment generating function for Z is*

$$M_Z(t) = e^{t^2/2} \, .$$

Proof.

$$M_Z(t) = \mathrm{E}(e^{tZ}) = \int e^{tz} \frac{e^{-z^2/2}}{\sqrt{2\pi}} \, dz$$

$$= \frac{1}{\sqrt{2\pi}} \int e^{tz - z^2/2} \, dz \, .$$

The method of completing the square allows us to simplify the exponent:

$$z^2 - 2tz = z^2 - 2tz + t^2 - t^2$$

$$= (z - t)^2 - t^2 \, ,$$

so our integral becomes

$$M_Z(t) = \frac{e^{t^2/2}}{\sqrt{2\pi}} \int e^{-(z-t)^2/2} \, dz$$

$$= e^{t^2/2} \int \frac{e^{-u^2/2}}{\sqrt{2\pi}} \, du$$

$$= e^{t^2/2} \, .$$

The last step follows because $\frac{e^{-u^2}}{\sqrt{2\pi}} = \phi(u)$, so the integral must be 1. $\qquad \square$

Using the moment generating function, we can easily compute the mean and variance of the standard normal distribution.

Corollary 3.4.4. *Let* $Z \sim \mathsf{Norm}(0,1)$. *Then:*

- $\mathrm{E}(Z) = 0$.
- $\mathrm{Var}(Z) = 1$.

Proof. Exercise 3.21.   □

This explains the notation $\mathsf{Norm}(0,1)$; 0 and 1 are the mean and *standard deviation* of the standard normal distribution. We can obtain normal distributions with any mean and standard deviation we desire by using a linear transformation of the standard normal distribution.[1]

Definition 3.4.5. A continuous random variable X has a **normal distribution** with parameters μ and σ (denoted $X \sim \mathsf{Norm}(\mu, \sigma)$) if $-\infty < \mu < \infty$, $\sigma > 0$, and

$$X = \mu + \sigma Z$$

where $Z \sim \mathsf{Norm}(0,1)$.   □

Lemma 3.4.6. *Let* $X \sim \mathsf{Norm}(\mu, \sigma)$. *Then:*

(1) $\mathrm{E}(X) = \mu$.

(2) $\mathrm{Var}(X) = \sigma^2$.

(3) *The pdf for X is given by*

$$f(x; \mu, \sigma) = \frac{e^{-(x-\mu)^2/(2\sigma^2)}}{\sigma\sqrt{2\pi}} \ .$$

(4) *The moment generating function for X is*

$$M_X(t) = e^{\mu t + \sigma^2 t^2 / 2} \ .$$

Proof. The expected value and variance claims follow directly from Lemmas 2.5.3 and 2.5.8. Alternatively, we could obtain the mean and variance using the moment generating function (see Exercise 3.22).

The pdf can be obtained using the cdf method. Let $f(x)$ and $F(x)$ be the pdf and cdf of $X \sim \mathsf{Norm}(\mu, \sigma)$. Then

$$F(x) = \mathrm{P}(X \leq x) \tag{3.5}$$

$$= \mathrm{P}\left(\frac{X - \mu}{\sigma} \leq \frac{x - \mu}{\sigma}\right) \tag{3.6}$$

$$= \Phi\left(\frac{x - \mu}{\sigma}\right) \ . \tag{3.7}$$

---

[1]Since R uses the mean and standard deviation for its parameterization of the normal distributions, we will follow that convention. Others use the mean and *variance*. For the standard normal distribution it makes no difference since the standard deviation and variance are equal. But for other normal distributions it will matter which convention has been adopted.

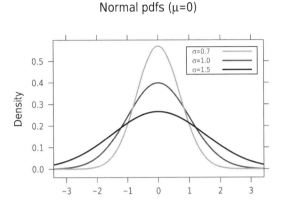

Figure 3.6. Some example normal pdfs.

So

$$f(x) = \frac{d}{dx}F(x) = \frac{d}{dx}\Phi(\frac{x-\mu}{\sigma})$$

$$= \phi(\frac{x-\mu}{\sigma}) \cdot \frac{1}{\sigma}$$

$$= \frac{e^{-(x-\mu)^2/2\sigma^2}}{\sigma\sqrt{2\pi}} .$$

Finally, the moment generating function follows directly from Theorem 3.3.6. □

The normal distributions are an important family of distributions with many applications, so it will be important to become familiar with their properties. The normal distributions have a symmetric, "bell-shaped" pdf (see Figure 3.6). The inflection point of this graph (where the concavity switches direction) is always 1 standard deviation away from the mean. (See Exercise 3.28.)

As usual, R supplies the d-, p-, q-, and r-operators `dnorm()`, `pnorm()`, `qnorm()`, and `rnorm()`. But it is useful to memorize some benchmark values for the $\mathsf{Norm}(0,1)$-distribution.

If $X \sim \mathsf{Norm}(\mu, \sigma)$, then $X = \mu + \sigma Z$ for $Z \sim \mathsf{Norm}(0,1)$. Read backwards, this says that $Z = \frac{X-\mu}{\sigma}$. The following lemma shows how to relate probabilities for any normal distribution to probabilities for the standard normal distribution.

Lemma 3.4.7. *Let $X \sim \mathsf{Norm}(\mu, \sigma)$ and let $Z \sim \mathsf{Norm}(0,1)$. Then*

$$\mathrm{P}(X \le x) = \mathrm{P}(Z \le \frac{x-\mu}{\sigma}) .$$ □

Lemma 3.4.7 has some important practical consequences. For any value x, the expression $\frac{x-\mu}{\sigma}$ is called the **standardized score** (or the z-**score**) for x. The z-score of x tells how many standard deviations above or below the mean x is.

By Lemma 3.4.7, for the purposes of computing probabilities, the z-scores suffice. In the past, statistics books would include tables of probabilities for the standard normal distribution. To get the probability of an event involving some other normal distribution, one would convert to z-scores and look things up in the table. All modern statistical packages (and many calculators) are capable of returning probabilities (values of the cdf), so such tables are no longer needed. In R, we are not even required to translate to z-scores.

```
# these two should return the same value:
> pnorm(5, mean=3, sd=2)    # 5 is 1 st dev above the mean of 3
[1] 0.84134
> pnorm(1)
[1] 0.84134
```

Nevertheless, understanding the relationship of normal distributions to the standard normal distribution allows us to have a universal calibration scheme for knowing if some value of x is unusually large or small. It is useful to learn a few benchmark values. In particular, for *any* normal distribution,

- approximately 68% of the distribution lies within 1 standard deviation of the mean,

- approximately 95% of the distribution lies within 2 standard deviations of the mean,

- approximately 99.7% of the distribution lies within 3 standard deviations of the mean.

These benchmarks are often referred to as the **68-95-99.7 rule**, also known as **the empirical rule**.

```
# these two should return the same value:
> pnorm(5, mean=3, sd=2)    # 5 is 1 st dev above the mean of 3
[1] 0.84134
> pnorm(1)
[1] 0.84134
```

Example 3.4.1.

Q. Originally, the mathematics and verbal portions of the Scholastic Aptitude Test (SAT-M and SAT-V) were normalized based on a sample of 10,000 students who took the test in 1941. The normalization was designed to give a distribution of scores that was approximately normal with a mean of 500 and a standard deviation of 100. Scores were not allowed to exceed 800 or fall below 200, however.

(1) Approximately what percentage of students taking this exam scored between 400 and 700?

(2) Approximately what percentage of students taking this exam scored 800?

A. We can answer these questions using the 68-95-99.7 rule. Let $X \sim \mathsf{Norm}(500, 100)$. Then:

(1) $P(400 \le X \le 600) \approx 68\%$, so $P(400 \le X \le 500) \approx 34\%$.
Similarly, $P(300 \le X \le 700) \approx 95\%$, so $P(500 \le X \le 700) \approx 47.5\%$.
Combining, we get $P(400 \le X \le 700) \approx 34\% + 47.5\% = 81.5\%$.

(2) Any scores that "should be" over 800 are truncated to 800, so the proportion of students receiving a score of 800 is approximately

$$P(X \geq 800) \approx \frac{1}{2} P(X \leq 200 \,\text{or}\, X \geq 800) = \frac{1}{2}(1 - 0.997) = \frac{1}{2}(0.003) = 0.15\% \,.$$

We can check these results using R.

```
> pnorm(700,500,100) - pnorm(400,500,100)      normal-sat
[1] 0.8186
> 1- pnorm(800,500,100)
[1] 0.0013499                                          ◁
```

Example 3.4.2. Over time, the scores on the SAT drifted so that by 1990, among 1,052,000 students used as the *1990 Reference Group* [**Dor02**], the mean SAT-M score was 475 and the mean SAT-V score was only 422.

Q. Assuming the scores remained normal and that the standard deviation remained 100, approximately what portion of the students scored below 500 on each portion of the exam?

A. Our benchmarks now only serve to give very crude approximations. For example, since the z-score for 500 on the SAT-M test is $\frac{500-422}{100} = 0.78$, we know that the proportion scoring below 500 on this exam will be a bit less than $50\% + 34\% = 84\%$. Similarly, for the SAT-M scores, the percentage must be a somewhat above 50%. R can calculate more exact answers.

```
> pnorm(500,422,100)                              sat-rescale
[1] 0.7823
> pnorm(500,475,100)
[1] 0.59871
```

The actual percentages reported in [**Dor02**] were 75% and 57%.   ◁

Example 3.4.3.

Q. Using both the old ($\mathsf{Norm}(500, 100)$) and new ($\mathsf{Norm}(500, 110)$) normalization schemes, determine the 80th percentile on an SAT exam.

A. The function `qnorm()` answers these questions easily.

```
> qnorm(0.80,500,100)                             sat-percentile
[1] 584.16
> qnorm(0.80,500,110)
[1] 592.58                                                ◁
```

3.4.2. Gamma Distributions

The gamma distributions can be thought of as a generalization of the exponential distributions. Recall from calculus that for each $n \geq 0$,

$$\int_0^\infty x^n e^{-x} \, dx$$

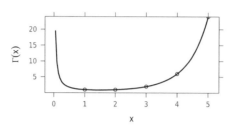

Figure 3.7. The gamma function.

converges and, therefore, is the kernel of a distribution. When $n = 0$, this is an exponential distribution. For $n \geq 0$, this gives rise to a **gamma distribution**.

Before introducing the gamma distributions formally, we first define a generalization of the factorial function that will allow us to express the scaling constants of the gamma distributions.

Definition 3.4.8. The **gamma function** is defined for all $\alpha \in [0, \infty)$ by

$$\Gamma(\alpha) = \int_0^\infty x^{\alpha-1} e^{-x} \, dx \, . \qquad \square$$

Lemma 3.4.9. *The gamma function satisfies the following:*

(1) $\Gamma(1) = 1$.

(2) $\Gamma(\alpha) = (\alpha - 1)\Gamma(\alpha - 1)$.

(3) *If α is a positive integer, then $\Gamma(\alpha) = (\alpha - 1)!$.*

Proof. For (1) observe that

$$\Gamma(1) = \int_0^\infty e^{-x} \, dx = \lim_{L \to \infty} -e^{-x}\big|_0^L = -0 + e^0 = 1 \, .$$

Using integration by parts, we see that

$$\begin{aligned}
\Gamma(\alpha) &= \int_0^\infty x^{\alpha-1} e^{-x} \, dx \\
&= \lim_{L \to \infty} x^{\alpha-1}(-e^{-x})\big|_{x=0}^{x=L} + \int_0^\infty (\alpha - 1) x^{\alpha-2} e^{-x} \, dx \\
&= 0 + (\alpha - 1) \int_0^\infty x^{\alpha-2} e^{-x} \, dx \\
&= (\alpha - 1)\Gamma(\alpha - 1) \, .
\end{aligned}$$

So (2) holds.

(3) follows from (1) and (2) by induction. Note, for example, that

$$\Gamma(4) = 3 \cdot \Gamma(3) = 3 \cdot 2 \cdot \Gamma(2) = 3 \cdot 2 \cdot 1 \cdot \Gamma(1) = 3! \, . \qquad \square$$

The gamma function provides us with continuous interpolation of the factorial function shifted by 1.[2] (See Figure 3.7.)

---

[2] $\Gamma(\alpha)$ can also be defined for complex-valued α provided the real part of α is non-negative.

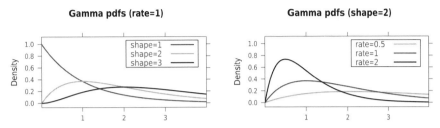

Figure 3.8. Some example gamma pdfs.

Definition 3.4.10. A **gamma random variable** X with shape parameter α and rate parameter λ (denoted $X \sim \mathsf{Gamma}(\alpha, \lambda)$) has the pdf

$$f(x; \alpha, \lambda) = \frac{\lambda^\alpha x^{\alpha-1} e^{-\lambda x}}{\Gamma(\alpha)} \ .$$

The scale parameter β of a gamma distribution is $\beta = 1/\lambda$. An equivalent form of the pdf using the scale parameter β is

$$f(x; \alpha, \beta) = \frac{x^{\alpha-1} e^{-x/\beta}}{\beta^\alpha \Gamma(\alpha)} \ . \qquad \square$$

Notice that when $\alpha = 1$, then the gamma distribution is an exponential distribution.

Once again, R supplies the d-, p-, q-, and r-operators `dgamma()`, `pgamma()`, `qgamma()`, and `rgamma()` for the gamma distributions. The parameters can be specified either as `shape=`α and `rate=`λ (the default) or as `shape=`α and `scale=`β. We will follow R's convention and use λ as our default parameter for the gamma distributions and denote a gamma distribution with scale parameter β as $\mathsf{Gamma}(\alpha, \frac{1}{\beta})$ or $\mathsf{Gamma}(\alpha, \beta = \frac{1}{\lambda})$.

Theorem 3.4.11. *Let* $X \sim \mathsf{Gamma}(\alpha, \lambda)$. *Then:*

(1) $M_X(t) = \left[\frac{\lambda}{\lambda - t} \right]^\alpha$.

(2) $\mathrm{E}(X) = \alpha/\lambda$.

(3) $\mathrm{Var}(X) = \alpha/\lambda^2$.

Proof. The moment generating function for a gamma distribution is given by

$$M_X(t) = \int_0^\infty e^{tx} \frac{\lambda^\alpha x^{\alpha-1} e^{-\lambda x}}{\Gamma(\alpha)} \ dx$$

$$= \frac{\lambda^\alpha}{\Gamma(\alpha)} \int_0^\infty x^{\alpha-1} e^{(t-\lambda)x} \ dx \ .$$

Provided $t < \lambda$, the last integral above is a kernel of the pdf of $Y \sim \mathsf{Gamma}(\alpha, \lambda - t)$. Thus the integral is the reciprocal of the scaling constant, i.e.,

$$M_X(t) = \frac{\lambda^\alpha}{\Gamma(\alpha)} \frac{\Gamma(\alpha)}{(\lambda - t)^\alpha}$$

$$= \frac{\lambda^\alpha}{(\lambda - t)^\alpha} \ .$$

The mean and variance can now be obtained from the moment generating function:

$$M'(0) = \frac{\alpha}{\lambda} \, ,$$

$$M''(0) = \frac{\alpha(\alpha+1)}{\lambda^2} \, ,$$

$$\text{Var}(X) = \frac{\alpha(\alpha+1)}{\lambda^2} - \frac{\alpha^2}{\lambda^2} = \frac{\alpha}{\lambda^2} \, . \qquad \square$$

The reason that $\beta = 1/\lambda$ is called the scale parameter of the gamma and exponential distributions comes from the following lemma which says that if X is a gamma random variable with scale parameter β, then kX is a gamma random variable with scale parameter $k\beta$. In other words, scaling X scales the scale parameter β.

Lemma 3.4.12. *Let $X \sim \mathsf{Gamma}(\alpha, \lambda)$, and let $Y = kX$. Then $Y \sim \mathsf{Gamma}(\alpha, \lambda/k)$.*

Proof. The mgf for Y is

$$M_Y(t) = M_X(kt) = \left[\frac{\lambda}{\lambda - kt} \right]^\alpha = \left[\frac{\lambda/k}{\lambda/k - t} \right]^\alpha .$$

Since this is the mgf for a $\mathsf{Gamma}(\alpha, \frac{\lambda}{k})$-distribution, the lemma is proved. $\qquad \square$

3.4.3. Weibull Distributions

The Weibull distributions are another family of distributions that generalize the exponential distributions. The pdf of the Weibull distributions is based on the observation that by substituting $u = x^\alpha$, we can evaluate

$$\int x^{\alpha-1} e^{x^\alpha} \, dx \, .$$

Definition 3.4.13. A random variable X is said to have the **Weibull distribution** with shape parameter $\alpha > 0$ and scale parameter $\beta > 0$ if the pdf for X is

$$f(x; \alpha, \beta) = \begin{cases} \frac{\alpha}{\beta^\alpha} x^{\alpha-1} e^{-(x/\beta)^\alpha} & \text{if } x \in [0, \infty), \\ 0 & \text{otherwise.} \end{cases} \qquad \square$$

The parameters of the Weibull distributions are called `shape` and `scale` in the R functions `rweibull()`, `dweibull()`, `pweibull()`, and `qweibull()`.

There are some situations in which a theoretical justification can be given for selecting a Weibull distribution as a model, but often Weibull distributions are used simply because they provide a good fit to data. When $\alpha = 1$, the Weibull random variable is an exponential random variable with $\lambda = 1/\beta$, so the exponential distributions are a special case both of the Weibull distributions and of the beta distributions. The relationship between the exponential and Weibull distributions is easier to see by comparing the cdfs of these distributions:

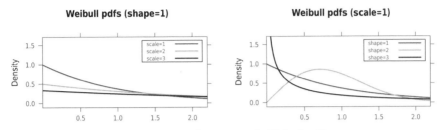

Figure 3.9. Some example Weibull pdfs.

$$\frac{\text{exponential cdf}}{1 - e^{-\frac{x}{\beta}}} \quad \frac{\text{Weibull cdf}}{1 - e^{-\left(\frac{x}{\beta}\right)^{\alpha}}}$$

The Weibull distributions provide models for lifetime that are not memoryless and so can model wear (expected remaining lifetime decreases over time) or "infant mortality" (expected remaining lifetime increases as early failures are removed).

The following lemma is stated without proof.

Lemma 3.4.14. *Let $X \sim \text{Weibull}(\alpha, \beta)$. Then:*

- $E(X) = \beta\Gamma(1 + \frac{1}{\alpha})$.
- $\text{Var}(X) = \beta^2 \left[\Gamma(1 + \frac{2}{\alpha}) - \left[\Gamma(1 + \frac{1}{\alpha})\right]^2\right]$. $\qquad\qquad\qquad\square$

3.4.4. Beta Distributions

Like the uniform distribution, the beta distributions are defined on a finite interval.

Definition 3.4.15. A random variable X is said to have the **beta distribution** with shape parameters $\alpha > 0$ and $\beta > 0$ if the pdf for X is

$$f(x; \alpha, \beta) = \begin{cases} \frac{\Gamma(\alpha+\beta)}{\Gamma(\alpha)\Gamma(\beta)} x^{\alpha-1}(1-x)^{\beta-1} & \text{if } x \in [0,1], \\ 0 & \text{otherwise.} \end{cases} \qquad \square$$

Notice that when $\alpha = \beta = 1$, then the beta distribution is the uniform distribution on $[0,1]$. Beta distributions are often used to model proportions that may vary from sample to sample. A linear transformation can be used to get beta distributions on different intervals. The two shape parameters (called `shape1` and `shape2` in R) allow for quite a variety of distributions. (See Figure 3.10 for examples.) One way to get a feel for the shape of a beta pdf is to ignore the scaling constant and focus on the kernel

$$x^{\alpha-1}(1-x)^{\beta-1} .$$

- When $\alpha = \beta$, then the pdf will be symmetric about $1/2$.
- When $\alpha < 1$ and $\beta < 1$, then the pdf will be bowl-shaped
- When $\alpha < 1$ and $\beta \geq 1$, or $\alpha = 1$ and $\beta > 1$, then the pdf is strictly decreasing.
- When $\alpha = 1$ and $\beta < 1$ or $\alpha > 1$ and $\beta \leq 1$, then the pdf is strictly increasing.
- When $\alpha > 1$ and $\beta > 1$, then the pdf is unimodal.

Beta pdfs

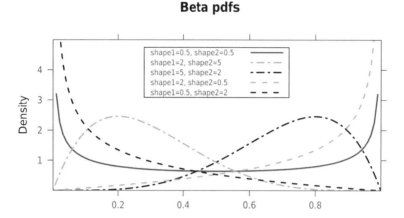

Figure 3.10. Some example beta pdfs.

As usual, R provides `rbeta()`, `dbeta()`, `pbeta()`, and `qbeta()` for working with beta distributions. We state the following important properties of the beta distributions without proof.

Lemma 3.4.16. *Let* $X \sim \mathsf{Beta}(\alpha, \beta)$. *Then:*

- $\mathrm{E}(X) = \frac{\alpha}{\alpha+\beta}$.

- $\mathrm{Var}(X) = \frac{\alpha\beta}{(\alpha+\beta)^2(\alpha+\beta+1)}$.

- $M_X(t) = 1 + \sum_{k=1}^{\infty}\left(\prod_{r=0}^{k-1}\frac{\alpha+r}{\alpha+\beta+r}\right)\frac{t^k}{k!}$. $\qquad\qquad\square$

Now that we have introduced a number of distributions, both discrete and continuous, a reasonable question is how we determine which distributions are good models for data in particular situations. This is an important and very general question that has many sorts of answers, and we will only begin to provide answers here.

Often it is possible to select a good family of distributions for a given situation because the conditions that motivated the definition of a particular family seem to be (at least approximately) satisfied the phenomenon at hand. This still leaves the question of selecting an appropriate member of the family – that is, estimating the parameters. In Chapters 4, 5, and 6 we will develop three general methods for estimating the parameters of a distribution that are used repeatedly in a wide variety of situations.

For now we focus on two graphical methods. In Section 3.5 we introduce a numerical method for estimating the pdf of a distribution from data. For the moment, our primary use for these estimated pdfs will be for plotting. In Section 3.6 we introduce a graphical tool for comparing sampled data to a theoretical distribution to see how well they "fit". In subsequent chapters we will return to the question of "goodness of fit" and develop methods for measuring goodness of fit quantitatively.

3.5. Kernel Density Estimation

Kernel density estimation is a method for extrapolating from data to an estimated pdf for the population distribution. Kernel density estimation can be thought of as a generalization of a histogram. Notice that the outline of a histogram (on the density scale) is a legitimate pdf since the area of the histogram is 1. But histograms are artificially jagged and often we expect the true population distribution to be smoother than our histograms.

Our goal is to arrive at a smoother estimated pdf. Figure 3.11 shows a histogram and a kernel density estimate based on the `faithful` data of Old Faithful eruption times. The idea behind kernel density estimation is to replace each data value with the kernel of some distribution. The resulting estimate is the sum of all these kernels, appropriately scaled to be a pdf.

More formally, let $K(x)$ be a kernel function centered at 0. Then $K(x - x_i)$ is a shifting of this kernel function that is centered at x_i. The kernel density estimate from a data set $x_1, x_2, \ldots, x_n$ is then

$$\hat{f}_K(x) = \sum_{i=1}^{n} \frac{1}{na} K(x - x_i) \, ,$$

where $1/a$ is the scaling constant for our kernel K. Before returning to the Old Faithful eruption data, let's consider a smaller data set.

Example 3.5.1.

Q. Consider the following small data set:

$$2.2 \quad 3.3 \quad 5.1 \quad 5.5 \quad 5.7 \quad 5.7 \quad 6.9 \quad 7.8 \quad 8.4 \quad 9.6$$

Compute kernel density estimates from this data using various kernels.

A. We begin by entering the data.

```
> x <- c(2.2,3.3,5.1,5.5,5.7,5.7,6.9,7.8,8.4,9.6)
```

kde01-data

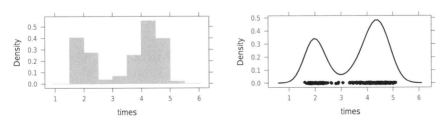

Figure 3.11. A histogram and a density plot of Old Faithful eruption times.

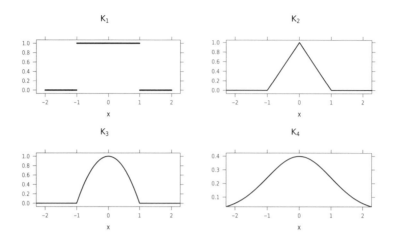

Figure 3.12. Four kernel functions for kernel density estimation.

Now consider four kernel functions:

$$K_1(x) = 1 \cdot [\![x \in (-1, 1)]\!] = \begin{cases} 1 & \text{if } x \in (-1, 1), \\ 0 & \text{otherwise,} \end{cases}$$

$$K_2(x) = (1 - |x|) \, [\![x \in (-1, 1)]\!] = \begin{cases} 1 - |x| & \text{if } x \in (-1, 1), \\ 0 & \text{otherwise,} \end{cases}$$

$$K_3(x) = (1 - x^2) [\![x \in (-1, 1)]\!] = \begin{cases} 1 - x^2 & \text{if } x \in (-1, 1), \\ 0 & \text{otherwise,} \end{cases}$$

$$K_4(x) = \phi(x), \text{ the standard normal pdf.}$$

The four kernel functions are displayed in Figure 3.12 and defined in R as follows.

```
> K1 <- function(x) { # rectangular                              kde01
+    return( as.numeric( -1 < x & x < 1 ) )
+ }
> K2 <- function(x) { # triangular
+    return( (1 - abs(x)) * as.numeric(abs(x) < 1) )
+ }
> K3 <- function(x) {      # parabola / Epanechnikov
+    return( (1 - x^2) * as.numeric(abs(x) < 1) )
+ }
> K4 <- dnorm          # Gaussian
```

The following R function computes a kernel density estimate from a kernel function and data. The kernel density estimate is returned as a function which can then be evaluated or plotted.

```
> kde <- function(data,kernel=K1,...) {                          kde02
+    n <- length(data)
+    scalingConstant=integrate(function(x){kernel(x,...)},-Inf,Inf)$value
+    f <- function(x) {
```

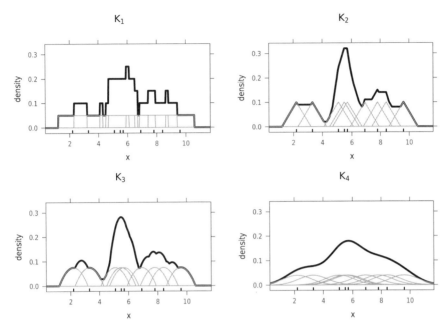

Figure 3.13. Kernel density estimation using different kernel functions with the data from Example 3.5.1.

```
+              mat <- outer(x,data, FUN=function(x,data) {kernel(x-data,...)} )
+              val <- apply(mat,1,sum)
+              val <- val/(n*scalingConstant)
+              return(val)
+      }
+      return(f)
+ }
```

The resulting kernel density estimates are displayed graphically in Figure 3.13. The lighter lines show $\frac{1}{na}K(x-x_i)$ for each x_i. The darker line is the sum of these, i.e., $\hat{f}_K$. $\lhd$

3.5.1. The Effect of Bandwidth

As our kernel density estimates in Example 3.5.1 indicate, the resulting density estimate depends greatly on the kernel functions used. Both the shape and the variance of the kernel K affect $\hat{f}_K$. Comparing $\hat{f}_{K_1}$ with $\hat{f}_{K_2}$, we readily see the flatness of K_1 and the peakedness of K_2 reflected in the resulting density estimates. As the following example shows, however, the differences between $\hat{f}_{K_2}$ and $\hat{f}_{K_4}$ have more to do with the different variances of K_2 and K_4 than with the smoother shape of K_4.

Example 3.5.2.

Q. Compare triangular and normal kernel density estimates using kernels that are pdfs with the same standard deviation.

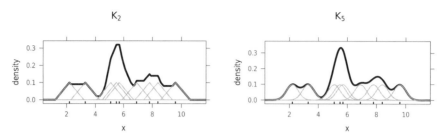

Figure 3.14. Kernel density estimation using different kernel functions with the data from Example 3.5.1.

A. The triangular kernel we used is a pdf. Since the mean is 0, the variance of this distribution is

$$\mu_{K_2} = \int_{-1}^{1} x^2 (1 - |x|) \, dx = 1/6 .$$

A comparison of kernel density estimates using K_2 and the pdf of a normal distribution with mean 0 and variance $1/6$ is shown in Figure 3.14. Although the density estimate using normal kernels is still a bit smoother, the two density estimates are now quite similar.   ◁

As the preceding example shows, the amount of smoothing provided by a kernel density estimate depends directly on the spread (i.e., the variance) of the kernels used. This is similar to the effect that bin sizes have on histograms. (Indeed, histograms *are* kernel density estimates; see Exercise 3.36.) In the context of kernel density estimates, this is referred to as **bandwidth**. Just as with histograms, the bandwidth must be chosen with care. If the bandwidth is too small, our density estimate may be too jagged to reveal any patterns in the underlying distribution. If the bandwidth is too large, we may smooth away interesting patterns altogether.

Various algorithms have been proposed to select a suitable kernel and bandwidth for density estimation. Typically these involve an attempt to minimize the mean integrated squared error (MISE) of $\hat{f}$

$$\text{MISE} = \text{E} \left(\int (f(x) - \hat{f}_K(x))^2 \, dx \right) .$$

The expectation is taken over random samples from the underlying distribution. (The value of the integral is a random variable since $\hat{f}_K(x)$ depends on the random data.) One of the challenges, of course, is that f is unknown, so MISE must be estimated.

3.5.2. Density Estimation in R

Kernel density estimates can be calculated in R using the `density()` function and can be plotted using the `densityplot()` function from the `lattice` package. The (slightly shortened) function prototypes below reveal that `densityplot()` passes along to `density()` all of the arguments necessary to select the desired kernel.

```
density(x, bw = "nrd0", adjust = 1,
        kernel = c("gaussian", "epanechnikov", "rectangular",
```

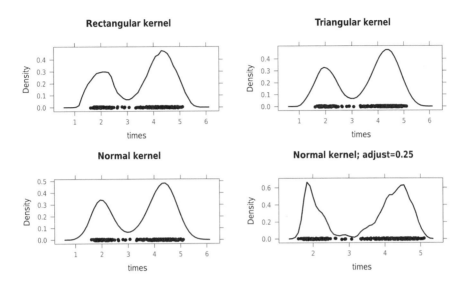

Figure 3.15. Kernel density estimates for Old Faithful data.

```
                  "triangular", "biweight",
                  "cosine", "optcosine"),
       weights = NULL, window = kernel, width,
       give.Rkern = FALSE,
       n = 512, from, to, cut = 3, na.rm = FALSE, ...)

densityplot(x, data,
            allow.multiple = is.null(groups) || outer,
            bw, adjust, kernel, window, width, give.Rkern,
            n = 50, from, to, cut, na.rm, ...)
```

In R the bandwidth argument is the standard deviation of the smoothing kernels used. (Some authors and texts define the bandwidth differently.) The argument `bw` can supply either a bandwidth or an algorithm for computing a bandwidth. A number of algorithms are available (see `?bw.nrd` for details). The bandwidth actually used is `adjust*bandwidth`. This makes it easy to select bandwidths like 'half the default bandwidth' without knowing what the default bandwidth is. A number of different kernel shapes (including the four we have already investigated) are available. Gaussian (i.e., normal) is the default.

The value returned from `density()` is not a function but (along with some additional information) a list of x- and y-values at which the kernel density estimate has been (approximately) calculated. Additional details are available in the help for `density()` and `densityplot()`.

Example 3.5.3. We will demonstrate the use of `densityplot()` with the Old Faithful eruption times. Plots of the resulting kernel density estimates appear in Figure 3.15.

faithful-kde

```
> times <- faithful$eruptions
> kdeFaithfulRect <- densityplot(~times,kernel="rectangular",
```

```
+      main="Rectangular kernel")
> kdeFaithfulTri <- densityplot(~times,kernel="triangular",
+      main="Triangular kernel")
> kdeFaithfulNormal <- densityplot(~times,
+      main="Normal kernel")
> kdeFaithfulNormal2 <- densityplot(~times,adjust=0.25,
+      main="Normal kernel; adjust=0.25")
> density(times)        # display some information about the kde

Call:
        density.default(x = times)

Data: times (272 obs.);        Bandwidth 'bw' = 0.3348

      x                  y
 Min.   :0.596    Min.   :0.000226
 1st Qu.:1.973    1st Qu.:0.051417
 Median :3.350    Median :0.144701
 Mean   :3.350    Mean   :0.181346
 3rd Qu.:4.727    3rd Qu.:0.308607
 Max.   :6.104    Max.   :0.484209
```

◁

3.6. Quantile-Quantile Plots

Suppose we have some data, and we believe (or would like) that the data come from a distribution that is normal or approximately normal. Or perhaps we think some other distribution would be a good model. How can we tell if there is a good fit between the theoretical model and the empirical data? We could produce a histogram and overlay the pdf or pmf of the distribution (as we did in Example 2.7.2). This is a rather crude method, however, because it is difficult to judge the fit of a histogram to a pdf or pmf and because the histogram itself depends on our choice of bins.

A **quantile-quantile plot** makes a more direct comparison of the data (empirical distribution) to a model (theoretical distribution). We will motivate quantile-quantile plots in the context of checking for normality. Such plots are usually called **normal-quantile plots**. The same principles can be used with any proposed theoretical distribution, as the examples later in this section will show.

3.6.1. Normal-Quantile Plots

The idea of a quantile-quantile plot is straightforward: We want to form a scatterplot that relates our data values to the ideal values of the theoretical distribution. The only subtle detail to work out is how we should define those "ideal values". As the name suggests, we do this using quantiles. We begin by arranging our n data values in increasing order:

$$x_1 < x_2 < \cdots < x_n \, .$$

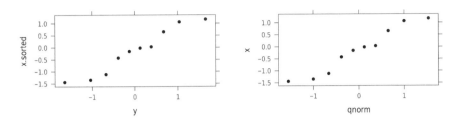

Figure 3.16. Normal-quantile plots from scratch (left) and using qqmath().

For each data value x_i, we use the data to estimate the probability q_i that a random value in the distribution we are sampling from is less than x_i:

$$q_1 < q_2 < \cdots < q_n .$$

(There are several methods that can be used to generate the q_i's. We present one simple method in the example below.) Finally, the ideal values (called theoretical quantiles)

$$y_1 < y_2 < \cdots < y_n$$

are chosen so that $P(Y \le y_i) = q_i$ if Y is sampled from our comparison distribution. That is, x_i is the same quantile in the data as y_i is in the comparison distribution.

Example 3.6.1.

Q. Construct a normal-quantile plot for the following data:

$$-0.16 \quad 1.17 \quad -0.43 \quad -0.02 \quad 1.06 \quad -1.35 \quad 0.65 \quad -1.12 \quad 0.03 \quad -1.44$$

A. The vector q below is chosen using the ruler method (see Section 1.2.4). Once those values have been selected, obtaining the normal quantiles can be done using qnorm.

normal-quantile-01

```
> x <- c(-0.16,1.17,-0.43,-0.02,1.06,-1.35,0.65,-1.12,0.03,-1.44)
# sort the data
> x.sorted <- sort(x); x.sorted
 [1] -1.44 -1.35 -1.12 -0.43 -0.16 -0.02  0.03  0.65  1.06  1.17
> q <- seq(0.05,0.95, by=0.1); q
 [1] 0.05 0.15 0.25 0.35 0.45 0.55 0.65 0.75 0.85 0.95
> y <- qnorm(q); y
 [1] -1.64485 -1.03643 -0.67449 -0.38532 -0.12566  0.12566  0.38532
 [8]  0.67449  1.03643  1.64485
```

Following the convention of the R functions that construct quantile-quantile plots, the scatterplot in Figure 3.16 has our original data along the vertical axis and the theoretical quantiles along the horizontal axis.

normal-quantile-01p

```
> qqplot <- xyplot(x.sorted~y)
```

R provides two functions that construct normal-quantile plots: the base graphics function qnorm() and the lattice graphics function qqmath(). A slightly

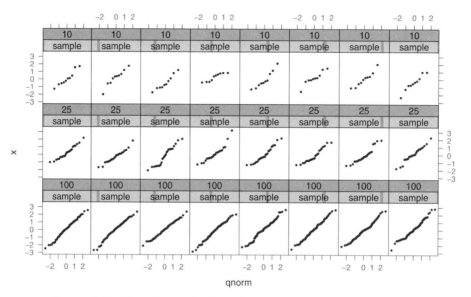

Figure 3.17. Normal-quantile plots of simulated samples from a standard normal distribution. There are 8 samples each of size 10, 25, and 100.

modified version of the ruler method is used to obtain the theoretical quantiles. The percentages used are calculated by the function `ppoints()`.

```
> ppoints(10)           # percentages for 10 data values    normal-quantile-02
 [1] 0.060976 0.158537 0.256098 0.353659 0.451220 0.548780 0.646341
 [8] 0.743902 0.841463 0.939024
> myplot <- qqmath(x)    # generate the normal-quantile plot        ◁
```

The points in the normal-quantile plots in Example 3.6.1 fall roughly, but not perfectly, along a line. If our data were ideal, they would fall directly on the line with equation $y = x$, since the empirical and theoretical quantiles would be the same. Of course, there will be some random variation about the line even if the underlying process generating our data has a true standard normal distribution. The plots in Figure 3.17 serve to calibrate our eyes to the amount of variation that can be expected when the underlying distribution is standard normal. These plots were generated with the following code:

```
> dat10 <- data.frame(                                      normal-quantile-03
+           x = rnorm(8*10),              # 8 samples of size 10
+           size=rep(10,8*10),           # record sample size
+           sample=rep(1:8,each=10)      # record sample number
+           )
> dat25 <- data.frame(
+           x = rnorm(8*25),              # 8 samples of size 25
+           size=rep(25,8*25),           # record sample size
+           sample=rep(1:8,each=25)      # record sample number
+           )
> dat100 <- data.frame(
```

```
+                x = rnorm(8*100),              # 8 samples of size 100
+                size=rep(100,8*100),           # record sample size
+                sample=rep(1:8,each=100)       # record sample number
+                )
> simdata <- rbind(dat10,dat25,dat100)
>
# generate the normal-quantile plots for each of the 30 samples
> myplot <- qqmath(~x|sample*factor(size),data=simdata,
+       layout=c(8,3), as.table=TRUE,cex=0.5)
```

If data are selected from a normal distribution with mean μ and standard deviation σ, then the normal-quantile plot will still approximate a line, but the intercept and slope of the line will be the mean and standard deviation of the distribution, since non-standard normal distributions are linear transformations of the standard normal distribution. In particular, this means that we can use normal-quantile plots to compare data to the family of normal distributions without having to specify a particular normal distribution.

Many discrete distributions are approximately normal. The normal-quantile plots look a bit different when applied to discrete data because often there are many observations with the same value.

Example 3.6.2. When $n\pi$ and $n(1-\pi)$ are both large enough, the binomial random variables are approximately normal.

normal-quantile-04

```
> x <- rbinom(40,50,0.4); x          # sample of size 40 from Binom(50,0.4)
 [1] 18 23 19 24 25 14 20 24 20 20 26 20 22 21 16 24 18 14 18 26 24 22 21
[24] 29 21 22 20 21 18 16 26 25 22 23 13 20 22 17 18 17
> myplot <- xqqmath(~x,fitline=TRUE)
```

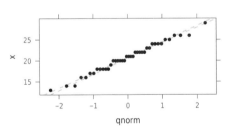

The function xqqmath() in the fastR package makes it simple to add reference lines to a qqmath() plot. Three types of lines can be added:

- $y = x$ (idline=TRUE).

- $y = a + bx$ (fitline=TRUE).
    The default values of the intercept and slope are the mean and standard deviation of the data, but the defaults can be supplied using the arguments slope and intercept.

- A line that connects the first and third quantiles (mathline=TRUE).   ◁

3.6.2. Quantile-Quantile Plots for Other Distributions

Quantile-quantile plots can be made comparing data to any distribution; we just need to change the way the quantiles are calculated. To make quantile-quantile plots using `qqmath()`, the user must provide the quantile function. Typically this will require the user to define a wrapper function around one of the built-in quantile functions as in the following example.

Example 3.6.3.

Q. Ten units were all tested until failure [**Kec94**]. The times-to-failure were 16, 34, 53, 75, 93, 120, 150, 191, 240, and 339 hours. A parameter estimation method indicates that the best fitting Weibull distribution has parameters $\alpha = 1.2$ and $\beta = 146$.

How well is this data modeled by a Weibull($\alpha = 1.2, \beta = 146$)-distribution? When $\alpha = 1$, the Weibull distribution is exponential. How well is this data modeled by an exponential distribution?

A. The Weibull fit looks very good. (See Figure 3.18.)

```
> life01 <- c(16, 34, 53, 75, 93, 120, 150, 191, 240, 339)        qq-weibull-01
> qweib <- function(x) { qweibull(x,1.2,146) }
> myplot <- xqqmath(~life01,distribution=qweib,idline=TRUE,qqmathline=FALSE)
```

Although we have not yet talked in general about parameter estimation, in the case of an exponential distribution there is a natural method for estimating a parameter λ that fits the data well. Since $E(X) = 1/\lambda$ for $X \sim \mathsf{Exp}(\lambda)$, a good choice for $\hat{\lambda}$ (our estimate of λ) is $1/\bar{x}$.

```
> life01 <- c(16, 34, 53, 75, 93, 120, 150, 191, 240, 339)        qq-exp-01
> mean(life01); 1/mean(life01)
[1] 131.1
[1] 0.0076278
> qf <- function(x) { qexp(x,1/mean(life01)) }
> myplot02 <- xqqmath(~life01,distribution=qf,idline=TRUE,qqmathline=FALSE)
```

The fit for the exponential model (see Figure 3.19) is not as good as for the Weibull model, and the reference lines show that there is a clear curvilinear trend to the quantile-quantile plot. Comparing the exponential quantile-quantile plot to the reference line, we see that the units that lasted longest did not last as long as

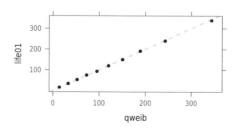

Figure 3.18. A quantile-quantile plot for lifetime data.

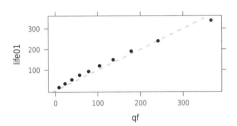

Figure 3.19. jn exponential quantile-quantile plot for lifetime data.

an exponential model would predict. Similarly, the early-failing units did not fail as quickly as would have been expected. This may be an indication that there is a "wear effect". The units are better when new and worse when old. The exponential model, because it is memoryless, does not model wear. An increasing failure rate over time is also indicated by the fact that $\alpha > 1$ in the Weibull fit.   ◁

Recall that by Example 3.3.8 all exponential distributions are rescalings of $\mathsf{Exp}(\lambda = 1)$. This means that (just as for the normal distributions) it isn't necessary to estimate λ when making a quantile-quantile plot for an exponential distribution.

3.7. Joint Distributions

3.7.1. Probability Density Functions

In this section will we make use of the abbreviation for a vector of values introduced in Appendix B:

$$\boldsymbol{x} = \langle x_1, x_2, \ldots, x_k \rangle \; .$$

This notation hides the number of values in the vector – this must be clear from context (or irrelevant) for this notation to be used. One advantage of this notation is that dealing with multiple variables looks notationally just like dealing with a single variable. It is also reminiscent of the way R deals with vectors.

We are especially interested in **random vectors**, i.e., jointly distributed random variables.

Definition 3.7.1. A **joint probability density function** (pdf) is a function $f : \mathbb{R}^k \to \mathbb{R}$ such that

- $f(x_1, x_2, \ldots, x_k) = f(\boldsymbol{x}) \geq 0$ for all $\boldsymbol{x} \in \mathbb{R}^k$, and

- $\displaystyle\int_{-\infty}^{\infty} \int_{-\infty}^{\infty} \cdots \int_{-\infty}^{\infty} f(x_1, x_2, \ldots, x_k) \, dx_1 \, dx_2 \cdots dx_k = \int_{\boldsymbol{x} \in \mathbb{R}^k} f(\boldsymbol{x}) \, d\boldsymbol{x} = 1.$

The **joint distribution** of $X_1, X_2, \ldots, X_k$ defined by the pdf f satisfies

$$\mathrm{P}(\boldsymbol{X} \in A) = \int_{\boldsymbol{x} \in A} f(\boldsymbol{x}) \, d\boldsymbol{x} \; .$$ □

For simplicity, we will often deal with the case where there are two jointly distributed random variables, but the same principles can be extended to any number of random variables. For notational ease, we will drop the limits of integration when the integral is over all of $\mathbb{R}^k$.

Example 3.7.1. Let $f(x, y) = 1$ on $[0, 1]^2$. It is easy to show that

$$\int_0^1 \int_0^1 f(x, y) \, dx \, dy = 1 \ ,$$

so f is a joint pdf. We can think of the point $\langle X, Y \rangle$ as a random point on a unit square.

Probabilities of events are determined by integrating over the region representing the event. For example,

$$\mathrm{P}(X < Y) = \int_0^1 \int_x^1 1 \, dy \, dx = \frac{1}{2} \ .$$

In this case we can evaluate the integral geometrically as well as algebraically because the integral gives the volume of a solid with triangular base and constant height of 1, which has volume $\frac{1}{2} \cdot 1 = \frac{1}{2}$.   ◁

Here is a more interesting example.

Example 3.7.2.

Q. Let $g(x, y) = 6xy^2$ on $[0, 1]^2$ be the joint pdf for X and Y. What is $\mathrm{P}(X \leq Y)$?

A. First, let's check that we have a legitimate pdf:

$$\int_0^1 \int_0^1 g(x, y) \, dx \, dy = \int_0^1 \int_0^1 6xy^2 \, dx \, dy = \int_0^1 \left(3x^2 y^2\right)\big|_{x=0}^{x=1} \, dy = \int_0^1 3y^2 \, dy = 1 \ .$$

To obtain our desired probability, we need to choose limits of integration so that we are integrating over the region $A = \{\langle x, y \rangle \mid 0 \leq x \leq y \leq 1\}$. Here is one way to do this:

$$\int_0^1 \int_x^1 6xy^2 \, dy \, dx = 3/5 \ .$$

The integral is easily done by hand. The R packages `cubature` and `R2Cuba` contain functions that will perform multivariate numerical integration. We will illustrate using `adaptIntegrate()` from the `cubature` package. This function can only handle rectangular regions of integration, but we can fix that by a clever definition of the integrand.

<code>joint02</code>

```
> require(cubature)
> f <- function(x) { 6 * x[1] * x[2]^2 }
> adaptIntegrate(f,c(0,0),c(1,1))
$integral
[1] 1

$error
[1] 3.3307e-16

$functionEvaluations
```

```
[1] 17

$returnCode
[1] 0

> g <- function(x) {
+     if (x[1] > x[2]) {return(0)}    # set value to 0 if X > Y
+     return(f(x))                    # else return joint pdf
+     }
> adaptIntegrate(g,c(0,0),c(1,1),tol=0.01)   # get less accuracy
$integral
[1] 0.60192

$error
[1] 0.0058386

$functionEvaluations
[1] 1547

$returnCode
[1] 0                                                              ◁
```

The definition of a joint cdf is similar to the definition of the cdf of a single random variable.

Definition 3.7.2. The **joint cdf** of a random vector $\boldsymbol{X} = \langle X_1, \ldots, X_n \rangle$ is the function

$$F(\boldsymbol{x}) = \mathrm{P}(\boldsymbol{X} \leq \boldsymbol{x}) = \mathrm{P}(X_i \leq x_i \text{ for all } i) = \int_{-\infty}^{x_1} \cdots \int_{-\infty}^{x_n} f(\boldsymbol{t}) \, d\boldsymbol{t} \,. \qquad \square$$

As with distributions of a single random variable, we can recover the pdf from the cdf.

Lemma 3.7.3. *Let $F(\boldsymbol{x})$ be a cdf. The pdf f for this distribution satisfies*

$$\frac{\partial}{\partial x_1} \frac{\partial}{\partial x_2} \cdots \frac{\partial}{\partial x_n} F(\boldsymbol{x}) = f(\boldsymbol{x}) \,. \qquad \square$$

3.7.2. Marginal and Conditional Distributions

Our work with discrete distributions motivates the following definitions. As usual, we replace sums with integrals.

Definition 3.7.4. Let f be the pdf of jointly distributed random variables X and Y. The **marginal distribution** of X is given by the pdf

$$f_X(x) = \int f(x, y) \, dy \,. \qquad \square$$

Definition 3.7.5. Let f be the pdf of jointly distributed random variables X and Y. The **conditional distribution** of X given $Y = y$ is given by the pdf

$$f_{X|Y=y}(x) = \frac{\text{joint}}{\text{marginal}} = \frac{f(x, y)}{f_Y(y)} \,.$$

An alternative notation for the conditional pdf is

$$f_{X|Y}(x \mid y) = f_{X|Y=y}(x) \ . \qquad \square$$

We will also the notation $\mathrm{P}(X \leq x | Y = y)$ to mean $\mathrm{P}(W \leq x)$ where W is the random variable with pdf given by $f_{X|Y=y}$. Note that this use of conditional probability notation is not covered by Definition 2.2.9 since $\mathrm{P}(Y = y) = 0$.

Example 3.7.3.

Q. Let $f(x, y) = 1$ on $[0, 1]^2$. Determine the marginal and conditional pdfs.

A. The marginal pdf for X is given by

$$f_X(x) = \int_0^1 1 \, dy = 1 \ .$$

So $X \sim \mathsf{Unif}(0, 1)$. By symmetry $Y \sim \mathsf{Unif}(0, 1)$, too.

$$f(x \mid y) = \frac{1}{1} = 1 \ .$$

So the conditional distribution is also uniform. $\triangleleft$

Example 3.7.4. Here is a more interesting example. Let $g(x, y) = 6xy^2$ on $[0, 1]^2$. The marginal distributions of X and Y are given by

$$f_X(x) = \int_0^1 6xy^2 \, dy = 2xy^3 \big|_{y=0}^{y=1} = 2x \ , \text{ and}$$

$$f_Y(y) = \int_0^1 6xy^2 \, dx = 3x^2y^2 \big|_{x=0}^{x=1} = 3y^2 \ .$$

The conditional distributions are given by

$$f_{X|Y}(x \mid y) = \frac{\text{joint}}{\text{marginal}} = \frac{6xy^2}{3y^2} = 2x \ , \text{ and}$$

$$f_{Y|X}(y \mid x) = \frac{\text{joint}}{\text{marginal}} = \frac{6xy^2}{2x} = 3y^2 \ . \qquad \triangleleft$$

3.7.3. Independence

Notice that in the previous example $f_{X|Y}(x \mid y)$ does not depend on y; in fact, $f_X(x) = f_{X|Y}(x \mid y)$ for all y. Similarly, $f_{Y|X}(y) = f_Y(y)$. Also, notice that $f(x, y) = f_X(x)f_Y(y)$. As our work above shows, whenever the conditional distribution is the same as the marginal distribution, the joint distribution will be the product of the marginals. This becomes our definition of independence.

Definition 3.7.6. Two continuous random variables X and Y with joint pdf f are **independent** if

$$f(x, y) = f_X(x)f_Y(y)$$

for every x and y. $\square$

Lemma 3.7.7. *If X and Y are independent continuous random variables, then for any x and y,*

- $f_X(x) = f_{X|Y}(x \mid y)$, *and*
- $f_Y(y) = f_{Y|X}(y \mid x)$.

Proof. Suppose $f(x, y) = f_X(x)f_Y(y)$. Then

$$f_{X|Y}(x \mid y) = \frac{\text{joint}}{\text{marginal}} = \frac{f_X(x)f_Y(y)}{\int f_X(x)f_Y(y)\, dx} = \frac{f_X(x)f_Y(y)}{f_Y(y)\int f_X(x)\, dx} = \frac{f_X(x)}{1} = f_X(x)\,.$$

Reversing the roles of X and Y shows that $f_{Y|X}(y \mid x) = f_Y(y)$. $\qquad\square$

Example 3.7.5.

Q. Let $f(x, y) = x + y$ on $[0, 1]^2$ be the joint pdf for X and Y. Check that f is a legitimate pdf and determine whether X and Y are independent.

A. First we check that the total probability is 1:

$$\int\int f(x, y)\, dx\, dy = \int_0^1 \left. \left(xy + \frac{y^2}{2} \right) \right|_{y=0}^{y=1} dx$$

$$= \int_0^1 (x + 1/2)\, dx$$

$$= \left. \frac{x^2}{2} + \frac{x}{2} \right|_{x=0}^{x=1}$$

$$= 1/2 + 1/2 = 1\,.$$

It does not appear that we can factor f into $f_X(x) \cdot f_Y(y)$. We can confirm that X and Y are not independent by comparing marginal and conditional distributions.

$$f_X(x) = \int f(x, y)\, dy = \int_0^1 x + y\, dy$$

$$= \left. xy + \frac{y^2}{2} \right|_{y=0}^{y=1}$$

$$= x + \frac{1}{2}\,,$$

and by symmetry,

$$f_Y(y) = y + \frac{1}{2}\,.$$

But

$$f_{X|Y}(x \mid y) = \frac{f(x, y)}{f_Y(y)}$$

$$= \frac{x + y}{y + \frac{1}{2}}\,.$$

In particular,

$$P(X \le 1/2 \mid Y = 0) = \int_0^{1/2} f_{X|Y}(x \mid 0) \, dx$$

$$= \int_0^{1/2} \frac{x + 0}{1/2} \, dx$$

$$= x^2 \big|_0^{1/2} = 1/4 \,,$$

but

$$P(X \le 1/2 \mid Y = 1) = \int_0^{1/2} \frac{x + 1}{3/2} \, dx$$

$$= \frac{4}{3}x^2 + \frac{2}{3}x \bigg|_0^{1/2} = 1/3 + 1/3 = 2/3 \,.$$

So X is more likely to be small ($\le 1/2$) when Y is large (1) than when Y is small (0). ◁

We state the following important and intuitive property of independence without proof. You are asked to prove the case where X and Y are both discrete in Exercise 3.43.

Lemma 3.7.8. *Let X and Y be independent random variables, and let f and g be transformations. Then $f(X)$ is independent of $g(Y)$.* ☐

3.7.4. Distributions of Sums and Means

We start by considering just two random variables.

Theorem 3.7.9. *Let X and Y be random variables. Then:*

(1) $E(X + Y) = E(X) + E(Y)$.

(2) $E(X \cdot Y) = E(X) \cdot E(Y)$, *provided X and Y are independent.*

(3) $\mathrm{Var}(X + Y) = \mathrm{Var}(X) + \mathrm{Var}(Y)$, *provided X and Y are independent.*

(4) $\mathrm{Var}(X + Y) = \mathrm{Var}(X) + \mathrm{Var}(Y) + 2\,\mathrm{Cov}(X, Y)$, *whether or not X and Y are independent.*

Proof. The proof is nearly identical to the corresponding proof of Theorem 2.6.7. Note that the definition of covariance was in terms of expected values and applies to continuous random variables without modification. ☐

Theorem 3.7.9 can be extended to joint distributions of more random variables by induction.

Example 3.7.6.

Q. Suppose X, Y, and Z are independent and $X \sim \mathsf{Norm}(10, 2)$, $Y \sim \mathsf{Norm}(12, 3)$, and $Z \sim \mathsf{Norm}(11, 1)$. What are the mean and variance of $X + Y + Z$?

A. The mean is $10 + 12 + 11$, and the variance is $2^2 + 3^2 + 1^2 = 14$. ◁

$s = 0.5$

$s = 1.25$

Figure 3.20. Calculating $F_S(s)$ geometrically.

In several important situations we can determine more than just the mean and variance of a sum of independent random variables; we can determine the *distribution* of the sum. We begin with a simple example that can be done by the cdf method.

Example 3.7.7.

Q. Let $X \sim \mathsf{Unif}(0,1)$ and $Y \sim \mathsf{Unif}(0,1)$ be independent. What is the distribution of $X + Y$?

A. Let $S = X + Y$. We have already seen that the joint pdf is $f(x,y) = 1$. $F_S(s) = \mathrm{P}(S \le s)$ can be calculated geometrically (see Figure 3.20):

$$F_S(x) = \begin{cases} \frac{s^2}{2} & \text{if } s \in [0,1], \\ 1 - \frac{(2-s)^2}{2} & \text{if } s \in [1,2]. \end{cases}$$

So

$$f_S(s) = \begin{cases} s & \text{if } s \in [0,1], \\ 2 - s & \text{if } s \in [1,2]. \end{cases} \qquad \lhd$$

For other examples, it is easier to work with moment generating functions.

Theorem 3.7.10. *Let X and Y be independent random variables with moment generating functions M_X and M_Y defined on an interval containing 0. Let $S = X + Y$. Then*

$$M_S(t) = M_X(t) \cdot M_Y(t)$$

on the intersection of the intervals where M_X and M_Y are defined.

Proof.

$$M_S(t) = \mathrm{E}(e^{tS}) = \mathrm{E}(e^{tX+tY}) = \mathrm{E}(e^{tX} \cdot e^{tY}) = \mathrm{E}(e^{tX}) \cdot \mathrm{E}(e^{tY}) = M_X(t) \cdot M_Y(t) . \quad \square$$

Example 3.7.8. Suppose $X \sim \mathsf{Binom}(n, \pi)$, $Y \sim \mathsf{Binom}(m, \pi)$, and X and Y are independent. Then

$$M_{X+Y}(t) = M_X(t) \cdot M_Y(t) = (\pi e^t + 1 - \pi)^n \cdot (\pi e^t + 1 - \pi)^m = (\pi e^t + 1 - \pi)^{n+m} .$$

Notice that this is the mgf for a $\mathsf{Binom}(m + n, \pi)$-random variable, so $X + Y \sim \mathsf{Binom}(m + n, \pi)$. $\qquad \lhd$

Example 3.7.9. Let X and Y be independent normal random variables with $X \sim \mathsf{Norm}(\mu_1, \sigma_1)$ and $Y \sim \mathsf{Norm}(\mu_2, \sigma_2)$. Then

$$M_{X+Y}(t) = e^{\mu_1 t + \sigma_1^2 t^2/2} e^{\mu_2 t + \sigma_2^2 t^2/2} = e^{(\mu_1 + \mu_2)t + (\sigma_2^2 + \sigma_2^2)t^2/2} ,$$

which we recognize as the mgf of a normal random variable with mean $\mu_1 + \mu_2$ and variance $\sigma_1^2 + \sigma_2^2$. So $X + Y \sim \mathsf{Norm}(\mu_1 + \mu_2, \sqrt{\sigma_1^2 + \sigma_2^2})$.   ◁

It is important to note that it is rather unusual for independent sums to have a distribution that is in the same family as the original variables, as is the case with the normal and binomial random variables. Additional examples are presented in the exercises.

A very important special case of jointly distributed random variables is given in the following definition.

Definition 3.7.11. Jointly distributed random variables $X_1, X_2, \ldots, X_n$ are said to be **independent and identically distributed** (abbreviated iid) if the variables are independent and all the marginal distributions are the same. That is, if f is the common pdf for each X_i, then

$$f_{\boldsymbol{X}}(\boldsymbol{x}) = \prod_{i=1}^{n} f(x_i) \, .$$

We will denote this situation as

$$\boldsymbol{X} \overset{\text{iid}}{\sim} \mathsf{Dist}$$

when each $X_i \sim \mathsf{Dist}$.   □

Independent identically distributed random variables are a model for random sampling from a population and are the theoretical basis for much of the inferential statistics we will encounter.

Lemma 3.7.12. *Suppose* $X_1, X_2, \ldots, X_n$ *are iid and that* $\mathrm{E}(X_i) = \mu$ *and* $\mathrm{Var}(X_i) = \sigma^2$. *Let* $S = X_1 + X_2 + \cdots + X_n$, *and let* $\overline{X} = \frac{1}{n}S$. *Then:*

(1) $\mathrm{E}(S) = n\mu$, *and* $\mathrm{Var}(S) = n\sigma^2$.

(2) $\mathrm{E}(\overline{X}) = \mu$, *and* $\mathrm{Var}(\overline{X}) = \frac{\sigma^2}{n}$.

Proof. This is an immediate consequence of Theorem 3.7.9.   □

If the population distribution is normal, we can say even more about $\overline{X}$.

Lemma 3.7.13. *Suppose* $\boldsymbol{X} \overset{\text{iid}}{\sim} \mathsf{Norm}(\mu, \sigma)$. *Then:*

(1) $S = X_1 + X_2 + \cdots + X_n \sim \mathsf{Norm}(n\mu, \sqrt{n}\sigma)$.

(2) $\overline{X} = \dfrac{X_1 + X_2 + \cdots + X_n}{n} \sim \mathsf{Norm}(\mu, \dfrac{\sigma}{\sqrt{n}})$.

Proof.

$$M_S(t) = \left(e^{\mu t + \sigma^2 t^2/2} \right)^n$$
$$= e^{n\mu t + n\sigma^2 t^2/2} \, .$$

This is the moment generating function for a normal random variable with mean $n\mu$ and variance $n\sigma^2$. The result for the mean follows by dividing by n.   □

3.8. Summary

3.8.1. Continuous Distributions

Continuous random variables are defined in terms of a **probability density function** (pdf) $f : \mathbb{R} \to \mathbb{R}$ such that

- $f(x) \geq 0$ for all $x \in \mathbb{R}$, and
- $\int_{-\infty}^{\infty} f(x)\, dx = 1$.

The continuous random variable X defined by the pdf f satisfies

$$P(a \leq X \leq b) = \int_a^b f(x)\, dx$$

for any real numbers $a \leq b$. Much of what is true about discrete random variables carries over to continuous random variables by replacing sums with integrals. So, for example, the expected value and variance of a continuous random variable with pdf f are defined by

$$E(X) = \int_{-\infty}^{\infty} x f(x)\, dx \,,$$

$$\text{Var}(X) = \int_{-\infty}^{\infty} (x - E(X))^2 f(x)\, dx \,.$$

By the Fundamental Theorem of Calculus, the cdf of a continuous random variable is an antiderivative of the pdf. Because $P(X = a) = 0$ for any continuous random variable X and any $a \in \mathbb{R}$, the pdf of a continuous random variable is typically derived by the **cdf method**, that is, by deriving the cdf first and differentiating to obtain the pdf.

Quantile-quantile plots provide a visualization of how well data match a theoretical distribution. They are especially useful for distributions such as the normal distribution that are closed under linear transformations. In that case, the quantile-quantile plot will look the same (up to the labeling of the axes) for any member of the family of theoretical distributions.

Kernel density estimation builds an empirically derived distribution by replacing each value in a data set with the kernel of a distribution (usually centered on the data value). Appropriately scaled, the sum of these kernels is a pdf. Kernel density estimates are generalizations of a histogram, which can be considered as a kernel density estimate formed by replacing each data value with the kernel of a uniform distribution (in this case, not centered on the data value but on its bin).

Joint distributions of continuous random variables are defined via multivariate pdfs. The marginal distributions of X and Y can be recovered from their joint pdf f by integration:

$$f_X(x) = \int_{-\infty}^{\infty} f(x, y)\, dy \,, \qquad f_Y(y) = \int_{-\infty}^{\infty} f(x, y)\, dx \,.$$

Conditional pdfs are defined using the same conditional $= \dfrac{\text{joint}}{\text{marginal}}$ pattern that was used for discrete distributions. For example,

$$f_{X|Y=y}(x) = \frac{f(x,y)}{f_Y(y)} \, .$$

Two continuous random variables X and Y with joint pdf f are **independent** if the joint pdf factors into a product of marginals:

$$f(x,y) = f_X(x) \cdot f_Y(y) \, .$$

Variables $X_1, X_2, \ldots, X_n$ are said to be **independent and identically distributed** (iid) if their joint distribution is the product of the marginal distributions, which are all the same:

$$f_{\boldsymbol{X}}(\boldsymbol{x}) = \prod_{i=1}^{n} f(x_i) \, .$$

We denote this situation as $\boldsymbol{X} \overset{\text{iid}}{\sim} \mathsf{Dist}$ where Dist is the common marginal distribution of the variables. This is an important model for statistical sampling.

Table B (see the inside back cover) summarizes important properties of the continuous distributions we have encountered thus far.

3.8.2. Moment Generating Functions

The moment generating function $M_X(t)$ for a random variable X is the exponential generating function of the sequence of moments of the pdf for X. The following list summarizes the key properties of moment generating functions that make them a powerful tool for working with both continuous and discrete random variables.

(1) $M_X(t) = \mathrm{E}(e^{tX})$. This gives us a way to compute the moment generating function in many important cases.

(2) The moment generating function may not be defined (because the moments don't exist or because the power series doesn't converge).

(3) If two random variables have the same moment generating functions on an interval containing 0, then they are equal in distribution (i.e., have the same cdf).

(4) If $M_X(t)$ is defined on an interval containing 0, then the moments of X can be recovered from the $M_X(t)$ using the identity

$$M^{(k)}(0) = \mu_k = \mathrm{E}(X^k) \, .$$

(5) We can easily compute the moment generating function of a linear transformation:

$$M_{aX+b}(t) = e^{bt} \cdot M_X at \, .$$

(6) If X and Y are independent random variables with moment generating functions that are defined on $(-r, r)$, then

$$M_{X+Y}(t) = M_X(t) \cdot M_Y(t) \text{ on } (-r, r).$$

Moment generating functions for particular distributions are included in the summary of distributions in Tables A and B (see the inside back cover and facing page).

Properties (5) and (6) allowed us to prove several important closure properties of families of distributions. Among these, the most important is that if X and Y are independent normal random variables, then $aX + b$, $X + Y$, and $X - Y$ are also normally distributed with means and standard deviations that are easily derived. From this it follows that if $\boldsymbol{X}$ is an iid random sample of size n from a $\mathsf{Norm}(\mu, \sigma)$ population, then

$$\overline{X} \sim \mathsf{Norm}\left(\mu, \frac{\sigma}{\sqrt{n}}\right).$$

In Chapter 4, we will use moment generating functions to derive a form of the Central Limit Theorem, which will (approximately) extend this result to many other populations.

3.8.3. R Commands

Here is a table of important R commands introduced in this chapter. Usage details can be found in the examples and using the R help.

`dnorm(x,mean,sd)`	pdf for $X \sim \mathsf{Norm}$(mean,sd).
`pnorm(q,mean,sd)`	$P(X \leq q)$ for $X \sim \mathsf{Norm}$(mean,sd).
`qnorm(p,mean,sd)`	x such that $P(X \leq x) = $ p for $X \sim$ Norm(mean,sd).
`rnorm(n,mean,sd)`	Simulate n random draws from a Norm(mean, sd)-distribution.
`dunif(...); punif(...);` `qunif(...); runif(...);` `dexp(...); pexp(...);` `qexp(...); rexp(...);` `dgamma(...); pgamma(...);` `qgamma(...); rgamma(...);` `dbeta(...); pbeta(...);` `qbeta(...); rbeta(...);` `dweibull(...); pweibull(...);` `qweibull(...); rweibull(...)`	Similar to the functions above but for uniform, exponential, gamma, beta, and Weibull distributions.
`f <- function(...) { }`	Define a function.
`integrate(f,lower,upper,...)`	Numerically approximate $\int_{\texttt{lower}}^{\texttt{upper}} \texttt{f}(x)\, dx$.
`adaptIntegrate(f,lowerLimit,` `    upperLimit,tol,...)`	Numerically approximate multivariate integrals [`cubature`].
`fractions(x,...)`	Find a rational number near x [MASS].

`sapply(X,FUN)`	Apply the function `FUN` to each element of the vector `X`.
`gamma(x)`	$\Gamma(x)$
`density(x,bw,adjust,kernel,...)`	Kernel density estimate.
`densityplot(x,data,allow.multiple,` `  bw,adjust,kernel,...)`	Kernel density plot.
`qnorm(x,...); qqmath(x,...);` `xqqmath(x,...)`	Normal-quantile plot for x. (Other distributions are also possible.)
`data.frame(...);`	Construct a new data frame.

Exercises

3.1. Let $f(x) = \begin{cases} k(x-2)(x+2) & \text{if } -2 \le x \le 2, \\ 0 & \text{otherwise.} \end{cases}$

a) Determine the value of k that makes f a pdf. Let X be the corresponding random variable.

b) Calculate P$(X \ge 0)$.

c) Calculate P$(X \ge 1)$.

d) Calculate P$(-1 \le X \le 1)$.

3.2. Let $g(x) = \begin{cases} kx(x-3) & \text{if } 0 \le x \le 3, \\ 0 & \text{otherwise.} \end{cases}$

a) Determine the value of k that makes g a pdf. Let X be the corresponding random variable.

b) Calculate P$(X \le 1)$.

c) Calculate P$(X \le 2)$.

d) Calculate P$(1 \le X \le 2)$.

3.3. Describe a random variable that is neither continuous nor discrete. Does your random variable have a pmf? A pdf? A cdf?

3.4. Show that if f and g are pdfs and $\alpha \in [0, 1]$, then $\alpha f + (1 - \alpha)g$ is also a pdf.

3.5. Determine the median and first and third quartiles of an exponential distribution.

3.6. Let $F(x) = \frac{1}{2} + \frac{1}{\pi} \arctan(x)$.

   a) Show that F is a cdf.

   b) Find the corresponding pdf.

   c) Let X be the random variable with cdf F. Find x such that $P(X > x) = 0.1$.

This random variable is called the **Cauchy random variable**.

3.7. Define two pdfs f and g such that f and g are not identical but give rise to the same cdfs. The corresponding random variables are equal in distribution.

3.8. Prove Theorem 3.2.3.

3.9. Use integration by parts to derive the variance of an exponential random variable. (See Lemma 3.2.5.)

3.10. The cdf of the random variable X is $F(x) = x^2/4$ on $[0,2]$. Compute the following:

   a) $P(X \le 1)$,

   b) $P(0.5 \le X \le 1)$,

   c) $P(X > 1.5)$,

   d) the median of X,

   e) the pdf of X,

   f) $E(X)$,

   g) $Var(X)$.

3.11. The cdf of the random variable X is $G(x) = x^3/8$ on $[0,2]$. Compute the following:

   a) $P(X \le 1)$,

   b) $P(0.5 \le X \le 1)$,

   c) $P(X > 1.5)$,

   d) the median of X,

   e) the pdf of X,

   f) $E(X)$,

   g) $Var(X)$.

3.12. The time X between two randomly selected consecutive cars in a traffic flow model is modeled with the pdf $f(x) = k/x^4$ on $[1, \infty)$.

   a) Determine the value of k.

   b) Obtain the cdf of X.

   c) What is $P(2 \le X \le 3)$?

   d) What is $E(X)$, and what does it tell you about traffic in this model?

   e) What is the median of X, and what does the median tell you about traffic in this model?

   f) What is the standard deviation of X?

3.13. Prove Lemma 3.2.6.

3.14. The discrete uniform distribution has the following pmf:

$$f(x) = \begin{cases} 1/n & \text{if } x \in \{1, 2, \ldots, n\}, \\ 0 & \text{otherwise.} \end{cases}$$

Determine the moment generating function for this random variable.

[Hint: Recall that $e^{ab} = (e^a)^b$ and look for a geometric sum.]

3.15. Find the moment generating function for $X \sim \mathsf{Geom}(\pi)$.

[Hint: Recall that $e^{ab} = (e^a)^b$ and look for a geometric series.]

3.16. Let Y be a continuous random variable with pdf

$$f(y) = \begin{cases} y & \text{if } y \in [0, 1], \\ 2 - y & \text{if } y \in [1, 2], \\ 0 & \text{otherwise.} \end{cases}$$

Find $M_Y(t)$.

3.17. Derive the moment generating function for $X \sim \mathsf{Pois}(\lambda)$.

[Hint: Recall that $\sum_{x=0}^{\infty} \frac{r^x}{x!} = e^r$ for any $r \in \mathbb{R}$.]

3.18. Let Y be a continuous random variable with pdf

$$f(y) = ye^{-y} \text{ on } y \in [0, \infty) .$$

Find $M_Y(t)$.

3.19. The moment generating function for a random variable X is

$$M_X(t) = (1 - \pi_1 - \pi_2) + \pi_1 e^t + \pi_2 e^{2t} .$$

Find the mean and variance of X.

3.20. The moment generating function for a random variable X is

$$M_X(t) = e^{-\lambda + \lambda e^t} .$$

Find the mean and variance of X.

3.21. The moment generating function for a random variable X is

$$M_X(t) = e^{t^2/2} .$$

Find the mean and variance of X.

3.22. The moment generating function for a random variable X is

$$M_X(t) = e^{\alpha t + \beta t^2/2} .$$

Find the mean and variance of X.

3.23. The moment generating function for a random variable X is

$$M_X(t) = (1 - \alpha t)^{-k} .$$

Find the mean and variance of X.

3.24. The moment generating function for a random variable X is

$$M_X(t) = \frac{e^{2t}}{1 - t^2} \ .$$

Find the mean and variance of X.

3.25. The moment generating function for a random variable X is

$$M_X(t) = \frac{2}{2 - t} \ .$$

Find the mean and variance of X.

3.26. The moment generating function for a random variable X is

$$M_X(t) = \left[\frac{3}{3 - t} \right]^2 \ .$$

Find the mean and variance of X.

3.27.

a) Derive the mgf for $X \sim \mathsf{Geom}(\pi)$.

b) Derive the mgf for $Y \sim \mathsf{NBinom}(r, \pi)$.

c) Use these results to calculate $\mathrm{E}(X)$ and $\mathrm{Var}(X)$. This proves Theorem 2.6.9.

3.28. Show that the pdf of a $\mathsf{Norm}(\mu, \sigma)$-distribution has inflection points at $\mu - \sigma$ and $\mu + \sigma$.

3.29. Beginning in 1995, SAT-V and SAT-M scores were normalized using new formulas designed so that the reference group had a mean of 500 and a standard deviation of 110 on each exam [**Dor02**]. As before, scores are truncated to the range 200–800.

Given the new system, approximately what percentage of test takers receive a score of 800?

3.30. Identify the random variables (family and parameters) with the following moment generating functions:

a) $M_W(t) = (\frac{e^t + 1}{2})^{10}$.

c) $M_Y(t) = \frac{1}{1 - 2t}$.

b) $M_X(t) = e^{t + t^2/2}$.

d) $M_Z(t) = \left[\frac{1}{1 - 2t} \right]^3$.

3.31. Suppose $X \sim \mathsf{Gamma}(\alpha, \lambda)$. Does $3X$ have a gamma distribution? If so, what are the parameters?

3.32. Approximately 68% of a normal distribution is within 1 standard deviation of the mean. For each of the following distributions, determine the percentage that is within 1 standard deviation of the mean:

a) $\mathsf{Exp}(\lambda = 1)$,

b) $\mathsf{Exp}(\lambda = 2)$,

c) $\mathsf{Unif}(0, 1)$,

d) $\mathsf{Beta}(\alpha = 2, \beta = 4)$.

3.33. The lifetime of a certain type of vacuum tube is modeled by $X \sim$ Weibull($\alpha = 2, \beta = 3$). Compute the following:

    a) $E(X)$ and $\text{Var}(X)$,

    b) the median of X,

    c) $P(X \leq E(X))$,

    d) $P(1.5 \leq X \leq 6)$,

    e) the probability that X is within 1 standard deviation of the mean.

3.34. A scientist suggests that $X \sim$ Beta($\alpha = 5, \beta = 2$) is a good model for the proportion of surface area in a randomly selected quadrat that is covered by a certain plant that she is studying. Assuming this model is (approximately) correct, compute the following:

    a) $E(X)$ and $\text{Var}(X)$,

    b) the median of X,

    c) $P(X \leq E(X))$,

    d) $P(0.2 \leq X \leq 0.4)$,

    e) the probability that X is within 1 standard deviation of the mean.

3.35. Show that for any kernel K and any data set, $\hat{f}_K$ computed from the data is a legitimate pdf.

3.36. A histogram *is* a kernel density estimate. Define a kernel function so that the resulting kernel density estimate will be equivalent to a histogram of the data.

3.37. In a simulation where the underlying distribution is known, MISE can be estimated by averaging the integrated squared error over a number of random samples. Do this for all combinations of

    • distribution: Norm($0, 1$), Exp(1), Beta($\frac{1}{2}, \frac{1}{2}$);

    • kernel: `'gaussian'`, `'triangular'`, `'rectangular'`, `'epanechnikov'`;

    • sample size: $10, 30, 100$;

    • adjust: $1, 3, \frac{1}{3}$.

Display your results in a suitable table and comment on any patterns you observe. (You may want to try additional examples to confirm.)

Hints:

    • `density()` does not return a density function but a list of x and y values for the density function. How can you use this information to estimate integrated squared error? Write a function `ise()` to do this.

    • Write a function `mise(size,reps,rdist,args,...)` to compute (estimated) MISE based on `reps` samples of size `size` from a distribution randomly sampled using the function `rdist` with arguments `args`. The "dots argument" ($\ldots$) can be used to pass arguments to `density()`.

        The following code can be used to call the function `rdist` with the appropriate arguments:

```
| sampleData <- do.call(rdist,args=c(list(n=size),args));
```
Note that `args` must be a list. The default value should be an empty list: `args=list()`.

3.38. Each normal-quantile plot below shows significant departure from normality. Describe how the distribution sampled differs from a normal distribution.

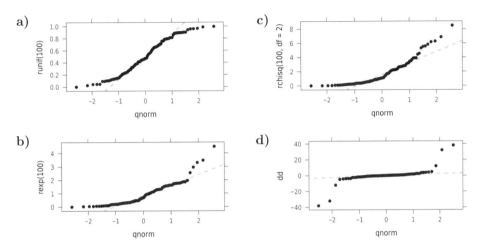

3.39. The `fastR` data set `Jordan8687` contains Michael Jordan's point totals for every game of his 1986–1987 regular season. Is a normal distribution a good model for Michael Jordan's point production that year?

3.40. The `pheno` data set contains phenotype information for 2457 subjects from the Finland-United States Investigation of NIDDM Genetics (FUSION) study of type 2 diabetes [**SMS**+**04**]. Among the physical measurements recorded is the height (in cm) of the subjects.

   a) Make a normal-quantile plot of the heights of all subjects in this data set. Do the heights appear to be normally distributed?

   b) Now make conditional normal-quantile plots, conditioned on `sex`. Do the heights of the men and women in this study appear to be approximately normally distributed? If not, how do the distributions differ from a normal distribution?

3.41. The `pheno` data set contains phenotype information for 2457 subjects from the Finland-United States Investigation of NIDDM Genetics (FUSION) study of type 2 diabetes [**SMS**+**04**]. Among the physical measurements recorded is the weight (in kg) of the subjects.

   a) Make a normal-quantile plot of the weights of all subjects in this data set. Do the weights appear to be normally distributed?

   b) Now make conditional normal-quantile plots, conditioned on `sex`. Do the weights of the men and women in this study appear to be approximately normally distributed? If not, how do the distributions differ from a normal distribution?

3.42. Express the binomial coefficient $\binom{n}{k}$ using the gamma function. Now look at the documentation for the negative binomial distribution in R. What do you notice?

3.43. Prove Lemma 3.7.8 in the case where X and Y are both discrete.

3.44. The kernel of a joint continuous distribution is $x^2 y^3$ on $[0,1]^2$.

  a) Determine the pdf.
  b) Determine $P(X \leq Y)$.
  c) Are X and Y independent?

3.45. Alice and Bob agree to meet at the dining hall for supper. Unfortunately, neither of them is very organized or very patient. Suppose that the random variables

$$X = \text{the time Alice arrives at the dining hall, and}$$
$$Y = \text{the time Bob arrives at the dining hall}$$

are independent uniform random variables on the interval $[5,6]$. (The units are hours after noon.)

  a) What is the joint pdf for X and Y?
  b) What is the probability that both of them arrive before 5:30?
  c) If neither of them is willing to wait more than 10 minutes for the other to show up, what is the probability that they eat together? That is, what is the probability that they arrive within 10 minutes of each other?

3.46. Suppose you have 10 light bulbs and that the lifetimes of the bulbs are independent and exponentially distributed with a mean lifetime of 100 hours. Let T be the time when the last light bulb burns out. What is the distribution of T?

  [Hint: What is the probability that all 10 light bulbs fail before time t?]

3.47. Suppose you want to choose a point within a circle of radius R "uniformly at random". One way to interpret this is that the joint pdf for X and Y (the two coordinates of point in a circle of radius R centered at the origin) should be a constant function within the circle and 0 elsewhere.

  a) Determine the joint pdf of X and Y.
  b) What is the probability that the point $\langle X, Y \rangle$ is in the "middle half" of the circle, i.e., that $\sqrt{X^2 + Y^2} \leq R/2$?
  c) What is the probability that $|X - Y| \leq R$?
  d) Determine the marginal pdf of X.
  e) Are X and Y independent?

[Hint: Because the pdf is so simple, you shouldn't need to do any integration. Just use geometry.]

3.48. Another way to pick a point in a circle of radius R "at random" is to independently pick an angle $A \sim \mathsf{Unif}(0, 2\pi)$ and a distance from the center $D \sim \mathsf{Unif}(0, R)$. Does this give the same joint distribution as in the preceding problem or a different one?

3.49. Where in the proof of Theorem 3.7.10 do we use the assumption that X and Y are independent?

3.50. Ralph and Claudia are playing a game in which the higher score wins. Ralph's scores in this game are (approximately) normally distributed with a mean of 100 and a standard deviation of 20. Claudia's scores in this game are (approximately) normally distributed with a mean of 110 and a standard deviation of 15.

a) Who is more likely to score above 150?

b) Assuming their scores are independent (which is not true for many games but is at least approximately true for some other games), what is the (approximate) probability that Ralph beats Claudia?

c) Now suppose they play three games and declare the winner to be the one who gets the highest total for the three games together. What is the (approximate) probability that Ralph beats Claudia in this format?

d) This is a pretty silly problem. But why is the mathematics involved so important? (Think of some examples where we have used this same sort of reasoning to do more useful things.)

e) One more change. This time they play three games but the winner is the one who wins at least two of the three. What is the (approximate) probability that Ralph beats Claudia in this format?

3.51. Suppose X and Y are independent random variables with $X \sim \mathsf{Pois}(\lambda_1)$ and $Y \sim \mathsf{Pois}(\lambda_2)$. Is $X + Y$ a Poisson random variable? If so, what is the rate parameter of the distribution?

3.52. Suppose X and Y are independent random variables with $X \sim \mathsf{Binom}(n, \pi_1)$ and $Y \sim \mathsf{Binom}(n, \pi_2)$. Is $X + Y$ a binomial random variable? If so, what are the parameters of the distribution?

3.53. Suppose X and Y are independent random variables with $X \sim \mathsf{Gamma}(\alpha_1, \lambda)$ and $Y \sim \mathsf{Gamma}(\alpha_2, \lambda)$. Is $X + Y$ a gamma random variable? If so, what are the parameters of the distribution?

3.54. Suppose X and Y are independent random variables with $X \sim \mathsf{Gamma}(\alpha, \lambda_1)$ and $Y \sim \mathsf{Gamma}(\alpha, \lambda_2)$. Is $X + Y$ a gamma random variable? If so, what are the parameters of the distribution?

## Chapter 4

Parameter Estimation and Testing

All models are wrong, but some are useful.

George Box [**BD87**]

In this chapter we begin the work of statistical inference in earnest. The typical situation is the following: We are interested in some population or process and obtain information about it by collecting data on a sample. The big question is, *What does our particular sample allow us to infer about the population or process of interest?* In order to answer this question, we first need a **model** for that population or process.

4.1. Statistical Models

The use of models is important in many areas of mathematics, the physical sciences, engineering, and statistics. Each discipline has its own set of well-established models. In many cases, the giants of the field are those who first developed a new model or category of models. Additional progress in the field is made by improving the models, or techniques for understanding the models, or methods for applying the models to "real world" situations.

Clearly there is a wide range of what is considered a model. A biologist studying mice to learn about human response to a drug refers to the mouse as a model organism. A mathematician may write down an equation and call that a model. Engineers build prototypes and test systems. Computer scientists write programs and call them computational models. All these models share a common feature: They are *abstract.* That is, they distill from a (usually complex) situation certain

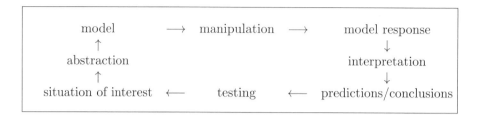

Figure 4.1. A schematic illustrating the modeling process.

features that are of particular importance. By manipulating and studying the (often simpler) model, one hopes to learn about the original situation.

This process is illustrated schematically in Figure 4.1. A model is an abstraction of the situation of interest. Once we have a model, we can manipulate it to see how it responds. In the case of a physical model, this manipulation may involve intervening to control the environment of the model to see how the model behaves. In a computational model, we may run the software with different inputs and record the varying output. In the case of a mathematical model, the manipulation is conceptual – we can ask what our mathematical formulation implies about various situations. By interpreting how the model responds to manipulation, we can make predictions about the original situation. When possible, we test these predictions against the original situation of interest. If the predictions are not good enough, we may start the entire process over with a model that we hope is an improvement over the previous attempt.

The utility of a model is judged by how well it helps us understand, predict, or manipulate the original situation of interest. Good models share some key features.

(1) Focus.

Since no model can represent everything, a good model must focus on those aspects of the situation that matter most for the purposes at hand.

(2) Simplicity.

The simpler the model, the easier it is to manipulate, to understand, and to communicate.

(3) Comprehensiveness.

In opposition to our desire for simplicity is the need for the model to include all the important aspects of the situation. In words attributed to Einstein, we want our models to be "as simple as possible, but no simpler".

(4) Falsifiability.

All models are wrong, but we should be able to tell when our model is "good enough" and when it is so wrong that it is no longer useful. Typically we do this by comparing the model's predictions with actual outcomes.

One of the challenges of statistical models is that they are by nature stochastic. That is, they have a random component and so are not completely deterministic. This makes falsifying a model, for example, more challenging. Usually we can't conclude that some result is *incompatible* with our model but merely that it is *unlikely* given our model. When the probability is low enough – assuming we believe our data are correct – we are forced to reject the model. When the probability is very low, such conclusions are relatively easy to make, but there are always gray areas.

Many of the models we will study can be expressed using **model formulas**. At a high level, many of these have the general form

$$\text{data} = \text{smooth} + \text{rough}$$

or

$$\text{data} = \text{pattern} + \text{error}.$$

For example, we could model a sample from a population with a normal distribution as

$$X_i = \mu + \varepsilon_i, \quad \varepsilon_i \sim \text{Norm}(0, \sigma), \tag{4.1}$$

or using vector notation as

$$\boldsymbol{X} = \mu + \sigma\boldsymbol{\varepsilon}, \quad \varepsilon \overset{\text{iid}}{\sim} \text{Norm}(0, 1). \tag{4.2}$$

This model could also be expressed as

$$\boldsymbol{X} \overset{\text{iid}}{\sim} \text{Norm}(\mu, \sigma). \tag{4.3}$$

Each of these formulations say that observation i consists of a mean level (μ) plus some amount that varies from individual to individual. The model also makes some assumptions (normality and independence) about the individual-level variation. Clearly there are many situations where one or both of these assumptions is inappropriate.

This type of statistical model is really a set or **family** of distributions (one distribution for each combination of the parameters μ and σ). An important part of statistical modeling is *fitting* the model, by which we mean deciding what values of the parameters correspond to "reasonable" distributions. It may also be that we decide the model is not a good fit for any values of the parameters. This latter conclusion may lead us to try a different model.

4.2. Fitting Models by the Method of Moments

So how do we estimate the parameters of a model from data? As it turns out, there are several general methods that provide estimates for broad classes of models. In Chapters 5 and 6 we will learn about two of the most important methods: the maximum likelihood method and the method of least squares. Here we will introduce a third method: the method of moments.

4.2.1. Fitting One-Parameter Models

The idea behind the method of moments is quite natural. Suppose, for example, we want to estimate the mean μ of a population from sample data. A natural estimate for μ is the sample mean $\bar{x}$. This is probably the first idea you would have come up with if someone asked you for an estimate. Let's examine the reasoning that makes this seem like a good method.

(1) Our sample comes from a population with unknown mean μ.

(2) It's natural to think that the mean of the sample should be close to the mean of the population.

(3) Therefore, we define our estimate to be the sample mean:
$$\hat{\mu} = \overline{x} \ .$$

The "hat notation" used here indicates that $\hat{\mu}$ is an estimated value for μ.

The method of moments extends this line of reasoning to estimate parameters in a wide range of situations. We will demonstrate the method first for one-parameter models and then discuss how the method can estimate multiple parameters simultaneously.

Example 4.2.1.

Q. Suppose we sample from a uniform distribution $\mathsf{Unif}(0,\theta)$ with unknown θ. Our model formula for this is
$$X \overset{\text{iid}}{\sim} \mathsf{Unif}(0,\theta) \ .$$
What is the method of moments estimator for θ?

Apply this to estimate θ if the (sorted) sample values are

$$1.6 \quad 2.8 \quad 6.2 \quad 8.2 \quad 8.5 \quad 8.7$$

A. The steps are the same as above.

(1) First we determine the population mean for $X \sim \mathsf{Unif}(0,\theta)$ as a function of θ. In this case
$$\mu_X = \mathrm{E}(X) = \theta/2 \ .$$

(2) Next we determine the sample mean $\overline{x}$ from our (randomly sampled) data.

(3) Finally, we let $\hat{\theta}$ be the value of θ that makes these two equal:
$$\hat{\theta}/2 = \overline{x} \ .$$

A bit of simple algebra gives us a formula for our estimate:
$$\hat{\theta} = 2\overline{x} \ .$$

Applying this method to our data, our estimate is $\hat{\theta} = 2\overline{x} = 2 \cdot 6 = 12$.

```
> x <- c(1.6,2.8,6.2,8.2,8.5,8.7);  mean(x)
[1] 6
```
◁

Example 4.2.2.

Q. As in the previous example, assume we are sampling from a $\mathsf{Unif}(0,\theta)$-distribution. Determine the method of moments estimate for θ if the (sorted) sample data are

$$0.2 \quad 0.9 \quad 1.9 \quad 2.2 \quad 4.7 \quad 5.1$$

A. The method of moments estimate is $\hat{\theta} = 2 \cdot 2.5 = 5$.

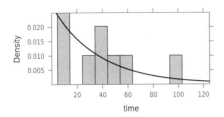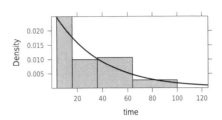

Figure 4.2. Comparing intervals between repairs with an exponential distribution fit using the method of moments.

```
> x <- c(0.2,0.9,1.9,2.2,4.7,5.1); mean(x);
[1] 2.5
```

This example points out a potential issue with the method of moments. Since one of our sample values was 5.1, we *know* that $\theta \geq 5.1$, so we *know* that our estimate is wrong. Exercise 4.3 asks you to explore how often this situation occurs.

◁

Example 4.2.3.

Q. The time between repairs of a device can be reasonably modeled with an exponential distribution if we believe that the events that cause the device to fail occur according to the Poisson model. Use the method of moments to fit an exponential distribution to the following repair intervals (measured in hours of service time between repairs):

$$49.0 \quad 60.4 \quad 8.9 \quad 43.4 \quad 34.8 \quad 8.2 \quad 13.6 \quad 11.5 \quad 99.4 \quad 31.9$$

A. The mean of an exponential distribution is $1/\lambda$. We obtain the method of moments estimate by solving

$$\overline{x} = 1/\hat{\lambda} \,,$$

so

$$\hat{\lambda} = 1/\overline{x} \,.$$

We can use R to perform the calculations.

```
> time <- c(49.0,60.4,8.9,43.4,34.8,8.2,13.6,11.5,99.4,31.9)     mom-exp
> mean(time)
[1] 36.11
> lambda.hat = 1/mean(time); lambda.hat
[1] 0.027693
```

It can be difficult to assess the shape of a distribution based on a histogram of just 10 values, but Figure 4.2 overlays the exponential distribution on a histogram of our data. The histogram on the right uses unequal bin sizes, which works better for such a skewed distribution. Alternatively, we could look at a quantile-quantile plot.

◁

This same method can be used to estimate an unknown parameter in any situation where our model has one unknown parameter and we have a formula for the mean of the model distribution as a function of the parameter. Additional examples are given in the exercises.

4.2.2. Estimating More Than One Parameter

The method of moments generalizes to fit models with more than one parameter by matching multiple sample moments to the population moments. Here is an outline of the method.

(1) Determine m moments of the population distribution (the model) as functions of the (unknown) parameters.

(2) Determine m sample moments from the data.

(3) Solve a system of equations setting the sample moments equal to the population moments. A solution to this system of equations gives the method of moments estimates for the parameters.

That is, the method of moments estimates are obtained by solving the following system of "equations":

$$\text{first model moment} \quad = \quad \text{first data moment,}$$

$$\text{second model moment} \quad = \quad \text{second data moment,}$$

$$\vdots \qquad\qquad = \qquad\qquad \vdots$$

The model moments (left sides) are to be expressed in terms of the model parameters. The data moments (right side) are defined by

Definition 4.2.1. The kth **sample moment (about** 0**)** is given by

$$\hat{\mu}_k = \frac{1}{n} \sum_{i=1}^{n} (x_i)^k .$$

The kth **sample moment about the (sample) mean** is given by

$$\hat{\mu}'_k = \frac{1}{n} \sum_{i=1}^{n} (x_i - \bar{x})^k . \qquad\qquad \square$$

For the higher moments, we can use either moments about 0 or moments about the mean when we use the method of moments. The estimates are the same using either approach. (See Exercise 4.6.) Take note that $\hat{\mu}'_2 = \sum \frac{(x_i - \bar{x})^2}{n} \neq s^2 = \sum \frac{(x_i - \bar{x})^2}{n-1}$. The correct formula is

$$\hat{\mu}'_2 = s^2 \frac{n-1}{n} .$$

Example 4.2.4.

Q. Suppose we sample from a population with mean μ and variance σ^2.

(1) Derive the method of moments estimates for μ and σ^2.

(2) Use the method of moments to estimate μ and σ from the following (sorted) sample of size 10:

$$57.9 \quad 70.8 \quad 86.3 \quad 92.3 \quad 94.2 \quad 117.0 \quad 118.4 \quad 122.4 \quad 125.8 \quad 134.4$$

A. The population moments are $\mu_1 = \mu$ and $\mu_2' = \sigma^2$. We obtain the method of moments estimates $\hat{\mu}$ and $\hat{\sigma}^2$ by solving the equations

$$\hat{\mu} = \overline{x},$$

$$\hat{\sigma}^2 = \hat{\mu}_2' = \frac{1}{n} \sum (x_i - \overline{x})^2.$$

In this case, we see that the sample mean is the method of moments estimate for μ, but the sample variance is *not* the method of moments estimate for the variance. Instead, the method of moments estimate for σ^2 is the very natural estimate that uses n in the denominator instead of $n - 1$. In Section 4.6 we will learn why it is advantageous to use $n - 1$.

Expressed in terms of the sample variance, our estimates are

$$\hat{\mu} = \overline{x},$$

$$\hat{\sigma}^2 = \frac{n-1}{n} s^2.$$

Applying this to our data set, we see that $\hat{\mu} = 101.95$ and $\hat{\sigma} = 24.18$ ($\hat{\sigma}^2 = 584.6$):

> mom-norm

```
###sink:mom-norm
> x<-c(57.9, 70.8, 86.3, 92.3, 94.2, 117.0, 118.4, 122.4, 125.8, 134.4);
> mean(x);
[1] 101.95
> sd(x);
[1] 25.486
> sqrt(9/10 * var(x));
[1] 24.178
```
◁

Example 4.2.5.

Q. Suppose we sample from a uniform population on an unknown interval $[\alpha, \beta]$. Derive the method of moments estimates for these parameters.

A. The mean and variance of the uniform distribution are $\frac{\alpha+\beta}{2}$ and $\frac{(\beta-\alpha)^2}{12}$. So our estimates are the solutions to

$$\frac{\hat{\alpha} + \hat{\beta}}{2} = \overline{x},$$

$$\frac{(\hat{\beta} - \hat{\alpha})^2}{12} = \hat{\mu}_2' = \frac{n-1}{n} s^2.$$

If we reparameterize using $\theta = (\alpha + \beta)/2$ and $\delta = (\beta - \alpha)/2$, then the population distribution is $\mathsf{Unif}(\theta - \delta, \theta + \delta)$, and

$$\hat{\theta} = \overline{x},$$

$$2\hat{\delta}/\sqrt{12} = \sqrt{\hat{\mu}_2'},$$

$$\hat{\delta} = \sqrt{3}\sqrt{\hat{\mu}_2'}.$$

From this we can easily derive $\hat{\alpha}$ and $\hat{\beta}$:

$$\hat{\alpha} = \overline{x} - \sqrt{3}\sqrt{\hat{\mu}_2'} = \overline{x} - s\sqrt{\frac{3(n-1)}{n}},$$

$$\hat{\beta} = \overline{x} + \sqrt{3}\sqrt{\hat{\mu}_2'} = \overline{x} + s\sqrt{\frac{3(n-1)}{n}}. \qquad \triangleleft$$

Example 4.2.6.

Q. Derive the method of moments estimates for the parameters of a $\mathsf{Beta}(\alpha, \beta)$-distribution.

A. We begin by setting up our equations:

$$\frac{\hat{\alpha}}{\hat{\alpha} + \hat{\beta}} = \overline{x}, \qquad (4.4)$$

$$\frac{\hat{\alpha}\hat{\beta}}{(\hat{\alpha} + \hat{\beta})^2(\hat{\alpha} + \hat{\beta} + 1)} = v = s^2 \frac{n-1}{n}. \qquad (4.5)$$

Notice that the (4.4) implies that

$$\frac{\hat{\alpha} + \hat{\beta}}{\hat{\alpha}} = 1 + \frac{\hat{\beta}}{\hat{\alpha}} = \frac{1}{\overline{x}}.$$

and that

$$\hat{\alpha} + \hat{\beta} = \frac{\hat{\alpha}}{\overline{x}}$$

Letting $R = \frac{1}{\overline{x}} - 1 = \frac{\hat{\beta}}{\hat{\alpha}}$, we see that the (4.5) reduces to

$$v = \frac{R\hat{\alpha}^2}{\left(\frac{\hat{\alpha}}{\overline{x}}\right)^2 \left(\frac{\hat{\alpha}}{\overline{x}} + 1\right)}. \qquad (4.6)$$

This can be solved algebraically, but first we demonstrate how to obtain a numerical solution using `uniroot()` in R.

```
> beta.mom <- function(x,lower=0.01,upper=100) {          mom-beta01
+     x.bar <- mean (x)
+     n <- length(x)
+     v <- var(x) * (n-1) / n
+     R <- 1/x.bar - 1
+
+     f <- function(a){            # note: undefined when a=0
+         R * a^2 / ( (a/x.bar)^2 * (a/x.bar + 1) ) - v
+     }
```

```
+
+      u <- uniroot(f,c(lower,upper))
+
+      return( c(shape1=u$root, shape2=u$root * R) )
+ }
> x <- rbeta(50,2,5); beta.mom(x)
shape1 shape2
1.6088 5.9083
```

Now for the algebraic solution. Equation (4.6) reduces to

$$v = \frac{R}{\left(\frac{1}{\bar{x}}\right)^2 \left(\frac{\hat{\alpha}}{\bar{x}} + 1\right)}$$

which only has one occurrence of $\hat{\alpha}$. Algebraic unraveling yields

$$\hat{\alpha} = \bar{x} \left(\frac{\bar{x}(1 - \bar{x})}{v} - 1\right),$$

$$\hat{\beta} = (1 - \bar{x}) \left(\frac{\bar{x}(1 - \bar{x})}{v} - 1\right).$$

```
# algebraic solutions                                    mom-beta01a
> x.bar <- mean(x); x.bar
[1] 0.21402
> v <- var(x) * (length(x) - 1) / length(x); v
[1] 0.019750
> x.bar*( x.bar*(1-x.bar)/v - 1 );          # alpha=shape1
[1] 1.6088
> (1-x.bar) * ( x.bar*(1-x.bar)/v - 1);     # beta=shape2
[1] 5.9083                                                     ◁
```

4.3. Estimators and Sampling Distributions

4.3.1. Estimates and Estimators

The estimates we have been computing provide a "best guess" for the value of a parameter, but they don't provide any indication of how good the estimate is. For this reason, they are sometimes called **point estimates** (in contrast to **interval estimates**, which we will learn about shortly).

If we obtain data by random sampling, then any number (statistic) computed from that data is a random variable since its value depends on a random process (the sampling). The distribution of this random variable is called a **sampling distribution**. It is important to distinguish the sampling distribution from two other distributions involved in statistical inference. The population distribution is what we really want to know about, but it is typically unknown. What we do know is our sample, which also has a distribution (called the sample distribution). The sampl*ing* distribution is a theoretical tool that allows us to ask and answer questions about what would happen if we repeatedly obtained random samples.

If we are interested in estimating a parameter, we typically do this by calculating some number (a statistic) from our data. The parameter to be estimated is called the **estimand**. The value associated with a particular data set is called an **estimate**. If we consider this as a random variable (with a sampling distribution), we will call it an **estimator**. That is, an estimator is a function that maps random data to real numbers.

Example 4.3.1. Suppose we want to estimate the mean μ of a population. We decide to do it in a very natural way: We collect a sample of size 100 and use the sample mean as our *estimate*. If the mean of our data is 23.5, we denote this as

$$\hat{\mu} = \overline{x} = 23.5 \ .$$

This is our *estimate* of μ. If we consider what would happen over many samples of size 100, we are considering the *estimator* (a random variable)

$$\hat{\mu} = \overline{X} = \frac{1}{n}\sum_{i=1}^{100} X_i \ . \qquad \triangleleft$$

Example 4.3.1 makes use of some important notation conventions. If θ is a parameter, then $\hat{\theta}$ will be used to denote either an estimate or an estimator for θ. This notation provides a way of succinctly describing both what parameter we wish to estimate and what method we are using.[1]   We will sometimes use a subscript to indicate the sample size involved. Thus

$$\hat{\mu}_n = \frac{X_1 + \cdots + X_n}{n}$$

indicates that for a sample of size n we are estimating μ by summing the values in the data set and dividing by the sample size.

If the population distribution is normal (under certain conditions about how the sampling is performed), we already know the sampling distribution of $\overline{X}_n$, by Lemma 3.7.13:

$$\overline{X}_n \sim \mathsf{Norm}(\mu, \sigma/\sqrt{n}) \ .$$

Example 4.3.2. We can inspect this result via simulations. Suppose we obtain a sample of size 16 from a normal distribution with mean 100 and standard deviation 12. The sampling distribution for $\overline{X}$ should be $\overline{X} \sim \mathsf{Norm}(100,3)$. Figure 4.3 displays a histogram with the $\mathsf{Norm}(100,3)$ pdf overlayed and a vertical line drawn at 100. The `xhistogram()` function in `fastR` simplifies adding these additional features to the histogram.

```
# 1000 sample means of samples of size 16 from N(100,12):    sample-means
> sampleMeans <- replicate(5000,mean(rnorm(16,100,12)))
> mean(sampleMeans)
[1] 99.969
> sd(sampleMeans)
[1] 3.0112
> myplot<-xhistogram(~sampleMeans,n=20,v=100,
+     density=TRUE,args=list(mean=100,sd=3))
```

[1] Despite the importance of this distinction between estimators and estimates, it is traditional to rely on context to determine whether $\hat{\theta}$ is an estimate or an estimator. This typically causes no difficulty since an estimate is a fixed number (e.g., $\hat{\theta} = 2.4$) and an estimator is a random variable (e.g., $\mathrm{E}(\hat{\theta}) = 2.4$).

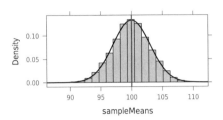

Figure 4.3. 5000 simulated sample means.

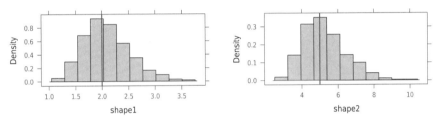

Figure 4.4. Histograms of method of moments estimators for the parameters of a beta distribution (simulated data).

The `replicate()` function is very handy for doing multiple simulations like this. In our example we use `replicate()` to repeat the command `mean(rnorm(16,100,12))` 5000 times. The 5000 resulting numbers are stored in a vector (`sampleMeans`) which can then be manipulated however we like. We will use `replicate()` on several other occasions in this chapter. ◁

If the population is not normal or if we are interested in a different parameter, the sampling distribution may not be a normal distribution. Furthermore, if we estimate multiple parameters, the sampling distributions of the various parameters may or may not be independent.

Example 4.3.3.

Q. Use simulation to investigate the sampling distributions for the two shape parameters in Example 4.2.6.

A. In practice, we will not know the true values of the parameters of the model nor that the model is correct. But when we do simulations, we can be sure that the model is correct and we can compare the estimated parameter values with the actual values. The code below does this and produces a histograms of the two shape parameters.

mom-beta01-sim

```
> results<-as.data.frame(t(replicate(1000,beta.mom(rbeta(50,2,5)))))
> plot1 <- xhistogram(~shape1, results, type='density',v=2)
> plot2 <- xhistogram(~shape2, results, type='density',v=5)
```

From the histograms in Figure 4.4, we see that the estimates do cluster around the correct values but that there is also quite a bit of variation from sample to sample. Furthermore, the sampling distributions are skewed, so the sampling distributions in this case are not normal distributions. Understanding the properties of this sort of distribution will be an important part of statistical inference.

It is also interesting to look at the joint distribution of the estimated shape parameters, which are clearly not independent.

```
mom-beta02-sim
> plot3 <- xyplot(shape2~shape1, results, panel=function(x,y,...){
+              panel.abline(a=0,b=5/2)
+              panel.xyplot(x,y,...)
+              })
> plot4 <- xhistogram(~shape2/shape1, results, type='density', v=2.5)
```

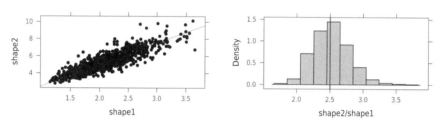

Figure 4.5. The joint distribution of the method of moments estimators for the two shape parameters of a beta distribution.

Although there is a high degree of variability in the individual estimators, the ratio of the two estimators has much less variability. This suggests that it would be easier to estimate this ratio than to estimate either of the two shape parameters individually.   ◁

4.3.2. Sampling Methods

The sampling distribution of an estimator clearly depends on the distribution of the population being sampled. But it also depends on the sampling method, a fact that we have largely ignored in the discussion so far. In many situations obtaining a random sample is quite challenging, so complicated, multi-stage sampling schemes are employed. For now will focus our attention on two relatively simple sampling methods. In both cases we will denote a random sample of size n as $\boldsymbol{X} = \langle X_1, X_2, \ldots, X_n \rangle$ and a particular sample as $\boldsymbol{x} = \langle x_1, x_2, \ldots, x_n \rangle$. That is, X_i (or x_i) is the ith value in our sample. If we are interested in a repeatable process and can repeat it in such a way that the outcome of each replicate comes from the same distribution and is independent of the others, then a sample of size n consists of n independent and identically distributed random variables. We will refer to this method of sampling as **iid random sampling**. Mathematically, this is the easiest situation to handle. This is the sampling method we were simulating in the preceding section.

Although an iid sampling method could be used when sampling from a finite population as well (and would, in fact, be simpler to describe and analyze mathematically), there is another sampling method that is more common in practice.

Definition 4.3.1. Simple random sampling is a process that selects a sample of size n from a population of size N in such a way that each of the $\binom{N}{n}$ different subsets of size n from the population is equally likely to be selected. $\square$

Example 4.3.4. A simple random sample can be obtained if we have a complete list of the members of the population and a way of obtaining measurements from any members of the population we select. The `sample()` function can simulate both simple random sampling and iid random sampling (by setting `replace=TRUE`). The example below uses the `VonBort` data set, which contains records from the Prussian army about the number of deaths by horse kick in each of 14 army corps for each year from 1874 until 1894.

```
> sample(1:1000,25)              # 25 random numbers in 1-1000          sample-srs
 [1] 288 788 409 881 937  46 525 887 548 453 948 449 670 566 102 993 243
[18]  42 323 996 872 679 627 972 640
> sample(1:10,8,replace=TRUE)    # iid random sample
[1]  8  6  6  3  2 10 10  7
> require(vcd)
> sample(VonBort$deaths,10)                # show a sample of size 10
 [1] 2 0 0 1 1 2 0 1 0 0
> mean(sample(VonBort$deaths,10))          # mean of a (different) sample
[1] 1
> replicate(10,mean(sample(VonBort$deaths,10))) # do it 10 times
 [1] 0.6 0.4 0.3 0.8 0.5 0.9 0.9 0.5 0.7 0.7
> mean(VonBort$deaths)                      # mean of entire data set
[1] 0.7                                                                    ◁
```

When applied to a finite population, the difference between an iid sampling scheme and simple random sampling is that in the former we sample *with replacement* and in the latter *without replacement*. Because of this, the X_i's are not independent when X is a simple random sample. That's the bad news. The good news contained in the following lemma and corollary is that the marginal distribution of X_i is the same as the population distribution in each case and that the degree of dependence among the X_i's in a simple random sample is small if the population is large.

Lemma 4.3.2. *Let* $X = \langle X_1, X_2, \ldots, X_n \rangle$ *be a simple random sample from a population of size* N. *Suppose the mean and variance of the population are* μ *and* σ^2. *Then*

$$\mathrm{Cov}(X_i, X_j) = \begin{cases} \sigma^2 & i = j, \\ -\sigma^2/(N-1) & i \neq j. \end{cases}$$

Proof. $\mathrm{Cov}(X_i, X_i) = \mathrm{Var}(X_i) = \sigma^2$.

If $i \neq j$, let the distinct population values be $v_1, v_2, \ldots, v_m$ and let n_k be the number of occurrences of v_k in the population. It is possible that $m < N$ since there may be multiple individuals with identical values. (We will have $m = N$ only if $n_k = 1$ for all k.) Then

$$\mathrm{E}(X_i) = \sum_{k=1}^{m} v_k \frac{n_k}{N} = \mu \ ,$$

$$\mathrm{E}(X_i^2) = \sum_{k=1}^{m} v_k^2 \frac{n_k}{N} = \sigma^2 + \mu^2 \ ,$$

$$\mathrm{E}(X_i)\,\mathrm{E}(X_j) = \mu^2 \ , \ \text{ and}$$

$$\mathrm{Cov}(X_i, X_j) = \mathrm{E}(X_i X_j) - \mathrm{E}(X_i)\,\mathrm{E}(X_j) = \mathrm{E}(X_i X_j) - \mu^2 \ .$$

All that remains is to determine the value of $\mathrm{E}(X_i X_j)$:

$$\mathrm{E}(X_i X_j) = \sum_{k=1}^{m} \sum_{l=1}^{m} v_k v_l \, \mathrm{P}(X_i = v_k \text{ and } X_j = v_l)$$

$$= \sum_{k=1}^{m} \sum_{l=1}^{m} v_k v_l \, \mathrm{P}(X_i = v_k)\, \mathrm{P}(X_j = v_l \mid X_i = v_k)$$

$$= \sum_{k=1}^{m} v_k \, \mathrm{P}(X_i = v_k) \sum_{l=1}^{m} v_l \, \mathrm{P}(X_j = v_l \mid X_i = v_k)$$

$$= \sum_{k=1}^{m} v_k \frac{n_k}{N} \left(v_k \frac{n_k - 1}{N - 1} + \sum_{l \neq k} v_l \frac{n_l}{N - 1} \right)$$

$$= \sum_{k=1}^{m} v_k \frac{n_k}{N} \left(\frac{-v_k}{N - 1} + \sum_{l=1}^{m} v_l \frac{n_l}{N - 1} \right)$$

$$= \frac{1}{N - 1} \sum_{k=1}^{m} v_k n_k \left(\frac{-v_k}{N} + \sum_{l=1}^{m} v_l \frac{n_l}{N} \right)$$

$$= \frac{1}{N - 1} \sum_{k=1}^{m} v_k n_k \left(\frac{-v_k}{N} + \mu \right)$$

$$= \frac{1}{N - 1} \left(-\sum_{k=1}^{m} v_k^2 \frac{n_k}{N} + N\mu \sum_{k=1}^{m} v_k \frac{n_k}{N} \right)$$

$$= \frac{1}{N - 1} \left(-(\mu^2 + \sigma^2) + N\mu^2 \right)$$

$$= \mu^2 - \frac{\sigma^2}{N - 1} \ .$$

Putting this all together, we see that

$$\mathrm{Cov}(X_i, X_j) = \left[\mu^2 - \frac{\sigma^2}{N - 1} \right] - \mu^2 = -\frac{\sigma^2}{N - 1} \ . \qquad \square$$

Corollary 4.3.3. *Let $\boldsymbol{X}$ be a simple random sample of size n from a population of size N with mean μ and variance σ^2. Then:*

- $\mathrm{E}(\overline{X}) = \mu$.
- $\mathrm{Var}(\overline{X}) = \dfrac{\sigma^2}{n} \dfrac{N-n}{N-1}$.

Proof. We have already shown that $\mathrm{E}(\overline{X}) = \mu$.

$$\mathrm{Var}(\overline{X}) = \frac{1}{n^2} \mathrm{Var}(X_1 + X_2 + \cdots + X_n)$$

$$= \frac{1}{n^2} \sum_{i,j} \mathrm{Cov}(X_i, X_j)$$

$$= \frac{1}{n^2} \left[\sum_i \mathrm{Var}(X_i) + \sum_{i \neq j} \mathrm{Cov}(X_i, X_j) \right]$$

$$= \frac{1}{n^2} \left[n\sigma^2 - n(n-1)\frac{\sigma^2}{N-1} \right]$$

$$= \frac{\sigma^2}{n} - \frac{n-1}{n}\frac{\sigma^2}{N-1}$$

$$= \frac{\sigma^2}{n} \left(1 - \frac{n-1}{N-1} \right) . \qquad \square$$

By the results above, we see that whether we use an iid sample or a simple random sample, as an estimator, the sample mean has two desirable properties.

Definition 4.3.4. An estimator $\hat{\theta}$ for a parameter θ is an **unbiased estimator** if

$$\mathrm{E}(\hat{\theta}) = \theta . \qquad \square$$

Lemma 4.3.5. *Let $\boldsymbol{X}$ be an iid sample or a simple random sample from a population. Then the sample mean is an unbiased estimator of the population mean.*

Proof. Independence is not an issue for expected value, so $\mathrm{E}(\overline{X}) = \mu$ follows directly from Lemma 3.7.13. $\qquad \square$

Definition 4.3.6. A sequence of estimators $\hat{\theta}_n$ for a parameter θ is **consistent** if for any $\varepsilon > 0$,

$$\lim_{n \to \infty} \mathrm{P}(|\hat{\theta}_n - \theta| < \varepsilon) = 1 . \qquad \square$$

In the sequence of estimators, n typically refers to sample size. The idea of a consistent estimator then is that we can make the probability of obtaining an estimate that is close to the estimand large by choosing a large enough sample size. As the following sequence of results shows, to demonstrate that an unbiased estimator is consistent, it suffices to show that the variance decreases to zero as the sample size increases.

Lemma 4.3.7 (Chebyshev's Inequality). *For any random variable X with mean μ and variance σ^2,*

$$\mathrm{P}(|X - \mu| \geq t) \leq \frac{\sigma^2}{t^2} .$$

Proof. Let $\varepsilon = \mathrm{P}(|X - \mu| \geq t)$. Of all distributions satisfying this equation, the one with the smallest variance is the discrete distribution described by

value	$\mu - t$	μ	$\mu + t$
probability	$\varepsilon/2$	$1 - \varepsilon$	$\varepsilon/2$

It is easy to check that the variance of this distribution is εt^2. The variance of any other distribution must be at least this much, so

$$\varepsilon t^2 \leq \sigma^2 \;,$$

from which it follows that

$$\varepsilon \leq \frac{\sigma^2}{t^2} \;. \qquad \square$$

Corollary 4.3.8. *For any random variable X with mean μ and variance σ^2,*

$$\mathrm{P}(|X - \mu| \geq k\sigma) \leq \frac{1}{k^2} \;.$$

Proof. Let $t = k\sigma$. Then $\frac{\sigma^2}{t^2} = \frac{\sigma^2}{k^2\sigma^2} = \frac{1}{k^2}$. The result follows. $\qquad \square$

Example 4.3.5. Let X be a random variable. Then at least $3/4$ of the distribution must lie within 2 standard deviations of the mean since Chebyshev's Inequality implies that at most $1/2^2 = 1/4$ of the distribution is outside that range. $\qquad \triangleleft$

The estimates given by Chebyshev's Inequality are often very crude. For example, for a normal distribution we know that approximately 95% of the distribution lies within 2 standard deviations of the mean. Nevertheless, even a crude approximation is sufficient to obtain the result we want.

Lemma 4.3.9. *If $\hat{\theta}_n$ is a sequence of unbiased estimators for θ and $\lim\limits_{n\to\infty} \mathrm{Var}(\hat{\theta}_n) = 0$, then $\hat{\theta}_n$ is also consistent.*

Proof. Since $\hat{\theta}_n$ is unbiased, $\mathrm{E}(\hat{\theta}_n - \theta) = 0$. Applying Chebyshev's Inequality to $\hat{\theta}_n - \theta$ gives

$$\mathrm{P}(|\hat{\theta}_n - \theta| < \varepsilon) \geq 1 - \frac{(\sigma^2/n)}{\varepsilon^2} = 1 - \frac{\sigma^2}{n\varepsilon^2} \to 1 \text{ as } n \to \infty. \qquad \square$$

Corollary 4.3.10. *Using either iid sampling or simple random sampling, the sample mean is a consistent estimator of the population mean.* $\qquad \square$

Corollary 4.3.10 is often referred to as the **Weak Law of Large Numbers** because it tells us that the mean of ever larger samples approaches the population mean. There is also a Strong Law of Large Numbers which guarantees an even stronger form of convergence, namely that the probability that an infinite random sequence will converge to the mean of the distribution it is sampled from is 1.

Example 4.3.6. We can see the Law of Large Numbers at work in the following simulation. Data are sampled from an exponential distribution with $\lambda = 1$. As the sample size grows, the running mean appears to be converging to 1.

```
> x <- rexp(1000)
> runningMean <- cumsum(x) / 1:length(x)
> expPlot <-
+     xyplot(runningMean~1:1000,ylab="running mean", xlab="n",type="l",
+         panel=function(...){ panel.abline(h=1,col='gray70');
+             panel.xyplot(...); });
```

law-large-numbers

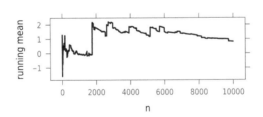

Figure 4.6. Running mean from an exponential distribution.

◁

Example 4.3.7. It is worth noting that the Law of Large Numbers does not hold for all distributions. The Cauchy distribution does not have an expected value (see Exercise 3.6) and its running mean does not appear to converge even if we simulate a sample of size 10,000. Furthermore, if we generate multiple random samples and plots of the running means for each, they may look quite different one from another.

Figure 4.7. Running mean from a Cauchy distribution.

```
> x <- rcauchy(10000)
> runningMean <- cumsum(x) / 1:length(x)
> cauchyPlot <- xyplot(runningMean~1:10000,
+     ylab="running mean",xlab="n", type="l");
```

lln-cauchy

◁

4.4. Limit Theorems

4.4.1. The Central Limit Theorem

The results of the previous section tell us that the mean of an iid sample or a simple random sample is an unbiased and consistent estimator of the population mean.

Furthermore, if the population is normal, then the distribution of $\overline{X}$ is also normal (at least for the case of iid sampling). The **Central Limit Theorem** tells us that in many situations, even if the population is not normal, the sampling distribution of $\overline{X}$ is still approximately normal.

Our proof of the Central Limit Theorem relies on a result about moment generating functions that we will not prove.

Lemma 4.4.1. *Let $W_1, W_2, W_3, \ldots$ be an infinite sequence of random variables with cdfs $F_1, F_2, F_3, \ldots$ and moment generating functions $M_1, M_2, M_3, \ldots$. If there is a random variable W and an interval containing 0 such that for all t in the interval*

$$\lim_{n \to \infty} M_n(t) = M_W(t) ,$$

then

$$\lim_{n \to \infty} F_n(w) = F_W(w)$$

for all $w \in (-\infty, \infty)$. □

There are several types of convergence of sequences of random variables that are used in statistics. The convergence of cdfs as in the preceding lemma is called **convergence in distribution**.

Definition 4.4.2. Let $X_1, X_2, \ldots$ be a sequence of random variables with $F_1, F_2, \ldots$ the corresponding sequence of cdfs. We say that $\{X_n\}$ **converges in distribution** to X if

$$\lim_{n \to \infty} F_n(x) = F_X(x)$$

for all x where F_X is continuous. This is denoted $X_n \overset{D}{\to} X$. □

Theorem 4.4.3 (Central Limit Theorem). *Let $X_1, X_2, X_3, \ldots$ be a sequence of iid random variables from a distribution that has a moment generating function defined on an interval containing 0. Then*

$$\lim_{n \to \infty} \mathrm{P} \left(\frac{\overline{X}_n - \mu}{\sigma / \sqrt{n}} \le z \right) = \Phi(z),$$

where $\overline{X}_n = \frac{X_1 + \cdots + X_n}{n}$. In other words,

$$\frac{\overline{X}_n - \mu}{\sigma / \sqrt{n}} \overset{D}{\to} \mathsf{Norm}(0, 1) .$$

Proof. We begin by defining some random variables:

$$Z_i = \frac{X_i - \mu}{\sigma} ,$$

$$W_n = \frac{\overline{X}_n - \mu}{\sigma / \sqrt{n}} = \frac{Z_1 + \cdots + Z_n}{\sqrt{n}} = \frac{Z_1}{\sqrt{n}} + \cdots + \frac{Z_n}{\sqrt{n}} .$$

By Lemmas 2.5.3 and 2.5.8,

$$\mathrm{E}(\overline{X}_n) = \mu , \qquad\qquad \mathrm{Var}(\overline{X}_n) = \sigma / \sqrt{n} ,$$
$$\mathrm{E}(Z_i) = 0 , \qquad\qquad \mathrm{Var}(Z_i) = 1 ,$$
$$\mathrm{E}(W_n) = 0 , \qquad\qquad \mathrm{Var}(W_n) = 1 .$$

Let $M(t)$ be the moment generating function of the Z_i's. Then

$$M_{W_n}(t) = \left[M\left(\frac{t}{\sqrt{n}}\right) \right]^n .$$

Since $E(Z_i) = 0$ and $\text{Var}(Z_i) = 1$, we know that

$$M(0) = 1 , \qquad\qquad M'(0) = 0 , \quad\text{and}\qquad\qquad M''(0) = 1 .$$

By Taylor's Theorem, for each value of t there is a real number $r \in (-t, t)$ such that

$$M(t) = M(0) + M'(0)t + \frac{1}{2}M''(r)t^2$$

$$= 1 + \frac{t^2}{2}M''(r) .$$

If we apply this to $\frac{t}{\sqrt{n}}$ in place of t, we see that for some $s \in (-\frac{t}{\sqrt{n}}, \frac{t}{\sqrt{n}})$

$$M\left(\frac{t}{\sqrt{n}}\right) = 1 + \frac{t^2}{2n}M''(s) ,$$

and

$$M_{W_n}(t) = \left[1 + \frac{t^2}{2n}M''(s) \right]^n = \left[1 + \frac{\frac{t^2}{2}M''(s)}{n} \right]^n .$$

Now as $n \to \infty$, $s \to 0$ since $s \in \left(-\frac{t}{\sqrt{n}}, \frac{t}{\sqrt{n}}\right)$, and so $M''(s) \to M''(0) = 1$ since M'' is continuous. (M has derivatives of all orders, so they are all continuous.) Also,

$$\lim_{n\to\infty} \left[1 + \frac{a_n}{n} \right]^n = e^a$$

if $a_n \to a$ as $n \to \infty$. So

$$\lim_{n\to\infty} M_{W_n}(t) = e^{t^2/2} .$$

But this is the moment generating function for the standard normal distribution, so

$$\lim_{n\to\infty} F_{W_n}(w) = \Phi(w) . \qquad\qquad \square$$

Notice that Theorem 4.4.3 implies that

$$\overline{X}_n \approx \text{Norm}(\mu, \sigma/\sqrt{n}) , \tag{4.7}$$

with the approximation improving as $n \to \infty$. The standard deviation of a sampling distribution is called the **standard error**.[2] We will denote the standard error as SE or $\sigma_{\hat\theta}$. In the case of iid random sampling for the sample mean, we already knew from Lemma 3.7.12 that

$$E(\overline{X}) = \mu_{\overline{X}} = \mu , \quad\text{and}$$

$$SE = \sigma_{\overline{X}} = \frac{\sigma}{\sqrt{n}} .$$

The additional information from the Central Limit Theorem is that we now know the distribution, at least approximately, provided the sample size is large enough.

---

[2]Some authors reserve the term standard error only for an *estimate* of the standard deviation of the sampling distribution.

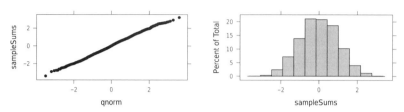

Figure 4.8. Approximating a $\mathsf{Norm}(0,1)$ distribution with the sum of 12 independent $\mathsf{Unif}(-1/2, 1/2)$-random variables.

There are many Central Limit Theorems. The version of the Central Limit Theorem that we prove here is weak in that it relies on moment generating functions, and not all distributions have moment generating functions. Stronger versions of the Central Limit Theorem have been proved that avoid this restriction by using characteristic functions rather than moment generating functions. (The characteristic function is defined by $\mathrm{E}(e^{itX})$ and requires some complex analysis to evaluate.) Other versions of the Central Limit Theorem relax the conditions that all the X_i's have the same distribution or that they are independent (but the dependence must not be too great).

The Central Limit Theorem explains why so many naturally occurring distributions are approximately normal. If the value of a measurement (like the height of a person) depends on a large number of small and nearly independent influences (nutrition, various genetic factors, etc.) and the effects are additive (getting a boost from two different genes makes you especially tall), then the Central Limit Theorem suggests that the normal distribution may be a good model.

The importance of the Central Limit Theorem is not as a limit result, but as a result applied with particular finite values of n. The Central Limit Theorem says that when n is large enough, the approximation $\overline{X} \approx \mathsf{Norm}(\mu, \sigma/\sqrt{n})$ becomes as accurate as we like. The good news is that the convergence is typically quite fast, so "large enough" is practically manageable. We won't prove theorems about the rate of convergence here. Instead, we will explore this issue via simulations.

If the population distribution is normal, then the approximation is exact for all values of n. So we would expect the approximation to be quite good, even for small samples, when the population distribution is close to normally distributed. This is in fact the case.

If the distribution we are interested in is less like a normal distribution (strongly skewed, bimodal, etc.), then the sample must be larger for the distribution of sample means to be usefully approximated by the normal distribution, but very often samples of size 30 (or even smaller) suffice even for population distributions that are quite skewed.

Example 4.4.1. Let $X \sim \mathsf{Unif}(-\frac{1}{2}, \frac{1}{2})$. Then $\mathrm{E}(X) = 0$ and $\mathrm{Var}(X) = 1/12$. The sum of 12 independent variables with this distribution will have mean 0 and variance 1. The normal approximation is already quite good for a sample of this size as the following simulation shows.

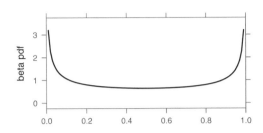

Figure 4.9. The pdf of the Beta$(0.5, 0.5)$-distribution.

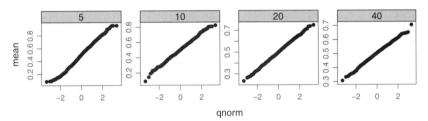

Figure 4.10. Convergence to a normal distribution of the sampling distribution for the sample mean sampled from a bimodal distribution.

```
> sampleSums <- replicate(2000,sum(runif(12,-0.5,0.5)))    unif12
> myplot1<-qqmath(~sampleSums)
> myplot2<-histogram(~sampleSums)
```

The fit is so good (see Figure 4.8), that this used to be a common way to generate pseudo-random normally distributed values from a standard good pseudo-random number generator for the uniform distribution.   ◁

Example 4.4.2. A Beta$(0.5, 0.5)$-distribution (see Figure 4.9) seems like the sort of distribution that would pose a challenge to the convergence of the Central Limit Theorem. But as the following simulation demonstrates, even for this distribution convergence is quite rapid.

```
> sampleMeans05 <- replicate(1000,mean(rbeta(5,0.5,0.5)))    betaCLT
> sampleMeans10 <- replicate(1000,mean(rbeta(10,0.5,0.5)))
> sampleMeans20 <- replicate(1000,mean(rbeta(20,0.5,0.5)))
> sampleMeans40 <- replicate(1000,mean(rbeta(40,0.5,0.5)))
> betaSim <- data.frame(
+     mean=c(sampleMeans05,sampleMeans10,sampleMeans20,sampleMeans40),
+     size=rep(c(5,10,20,40),each=1000))
> myplot <- qqmath(~mean|factor(size),betaSim,scales=list(relation='free'))
```

The plots are shown in Figure 4.10.   ◁

4.4.2. Other Limit Theorems

The Central Limit Theorem (and its proof) can be used to obtain other limit results as well. An especially important example is that under many conditions, binomial random variables are approximately normal.

Example 4.4.3. Let $X_n \sim \mathsf{Binom}(n, \pi)$ for some $\pi \in (0, 1)$. By Example 3.7.8, we can express X_n as

$$X_n = \sum_{i=1}^{n} Y_i \ , \ \text{where}$$

$$\boldsymbol{Y} \overset{\text{iid}}{\sim} \mathsf{Binom}(1, \pi) \ .$$

Recall that $\mathrm{E}(Y_i) = \pi$ and $\mathrm{Var}(Y_i) = \pi(1-\pi)$. But X_n is also $n\overline{Y}$, so by the Central Limit Theorem, we know that when n is large enough,

- X_n is approximately normal,
- $\mathrm{E}(X_n) = n\,\mathrm{E}(Y_i) = n\pi$, and
- $\mathrm{Var}(X_n) = n^2 \cdot \mathrm{Var}(\frac{X_n}{n}) = n^2\pi(1-\pi)/n = n\pi(1-\pi)$.

More technically, a straightforward application of the Central Limit Theorem yields

$$\frac{X_n - n\pi}{\sqrt{n\pi(1-\pi)}} \overset{D}{\to} \mathsf{Norm}\,(0, 1) \ .$$

The plots in Figure 4.12 compare the quantiles of binomial distributions with their approximating normal distributions. As the plots indicate, the approximation is better when n is larger and also when π is near $1/2$ than when π is near 0 or 1. A general rule of thumb is that the approximation is good enough to use for estimating probabilities provided $n\pi \geq 10$ and $n(1-\pi) \geq 10$. In other words, we should expect to see 10 successes and 10 failures if we want to use this approximation. When $n\pi$ is small, then the binomial distribution is skewed and a sizable proportion of the approximating normal distribution is below 0 (see Figure 4.11). A similar statement holds when $n(1 - \pi)$ is small. Exercise 4.19 relates this to the rule of thumb above.   ◁

Example 4.4.4. This approximation in Example 4.4.3 can be improved if we use the so-called **continuity correction**. Recalling that the binomial distributions are

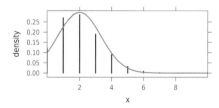

Figure 4.11. The normal distribution is a poor approximation to the binomial distribution when $n\pi$ or $n(1 - \pi)$ are small.

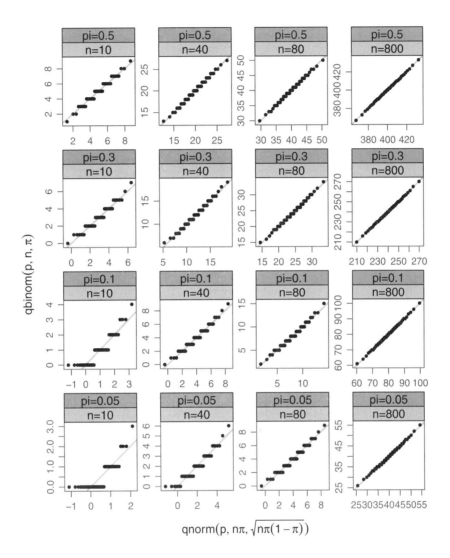

Figure 4.12. A comparison of binomial and normal quantiles.

discrete and integer-valued while the normal distributions are continuous, it makes sense to use the equivalence

$$P(a \leq X \leq b) = P(a - 0.5 \leq X \leq b + 0.5)$$

when X is binomial and a and b are integers.

Suppose, for example, that $X \sim \mathsf{Binom}(100, 0.6)$. Then we can approximate $P(55 \leq X \leq 65) = P(54.5 \leq X \leq 65.5)$ using the corresponding $\mathsf{Norm}(60, \sqrt{24})$-distribution with and without the continuity correction. As the following examples show, significant improvements to the probability estimates can result from using the continuity correction.

```
# P(55 <= X <= 65)                                    continuity-correction
> pbinom(65,100,0.6) - pbinom(54,100,0.6)
[1] 0.73857
# without continuity correction:
> pnorm(65,60,sqrt(100*0.6*0.4)) -  pnorm(55,60,sqrt(100*0.6*0.4))
[1] 0.69257
# with continuity correction:
> pnorm(65.5,60,sqrt(100*0.6*0.4)) -  pnorm(54.5,60,sqrt(100*0.6*0.4))
[1] 0.73843                                                        ◁
```

The fact that the binomial distribution is approximately normal (given a large enough value of n) provides us with an alternative to the binomial test. Since $X \sim \mathsf{Binom}(n, \pi_0)$ implies that $X \approx \mathsf{Norm}\left(n\pi_0, \sqrt{n\pi_0(1 - \pi_0)}\right)$, an alternative test statistic for conducting a test of the hypotheses

- H_0: $\pi = \pi_0$,
- H_a: $\pi \neq \pi_0$

is the statistic

$$Z = \frac{\hat{\pi} - \pi_0}{\sqrt{\pi_0(1 - \pi_0)/n}},$$

where

$$\hat{\pi} = \frac{X}{n}.$$

If the null hypothesis is true and n is sufficiently large, then $Z \approx \mathsf{Norm}(0, 1)$.

Example 4.4.5.

Q. Suppose that we flip a coin 100 times and obtain 60 heads. Is this sufficient evidence to conclude that the coin is biased?

A. We can compare the results from `binom.test()` with those from the z-test outlined above. As we see below, the p-value of the test is quite close to 0.05 so we are in the gray area. We have some evidence that the coin may be biased, but not overwhelming evidence. Perhaps we should obtain a larger sample by flipping the coin some more.

```
# "exact" p-value                                       binomial-ztest
> binom.test(60,100);

        Exact binomial test

data:  60 and 100
number of successes = 60, number of trials = 100, p-value =
0.05689
alternative hypothesis: true probability of success is not equal to 0.5
95 percent confidence interval:
 0.49721 0.69671
sample estimates:
probability of success
                0.6
```

```
# approximate p-value
> z <- ( 0.6 - 0.5 ) / sqrt(0.5 * 0.5/100); z;
[1] 2
> 2 * (1 - pnorm(z) )
[1] 0.0455
>
# approximate p-value with continuity correction
> z <- ( 0.595 - 0.5 ) / sqrt(0.5 * 0.5 / 100); z;    #0.595 = 59.5/100
[1] 1.9
> 2 * (1 - pnorm(z) )
[1] 0.057433
>
# R can automate the approximate version too:
> prop.test(60,100)            # uses continuity correction by default

        1-sample proportions test with continuity correction

data:  60 out of 100, null probability 0.5
X-squared = 3.61, df = 1, p-value = 0.05743
alternative hypothesis: true p is not equal to 0.5
95 percent confidence interval:
 0.49700 0.69522
sample estimates:
  p
0.6

> prop.test(60,100,correct=FALSE)     # turn off continuity correction

        1-sample proportions test without continuity correction

data:  60 out of 100, null probability 0.5
X-squared = 4, df = 1, p-value = 0.0455
alternative hypothesis: true p is not equal to 0.5
95 percent confidence interval:
 0.5020 0.6906
sample estimates:
  p
0.6
```

We make two observations about this example:

(1) `prop.test()` uses a test statistic called X^2 (what looks like an X is really a capital χ; the statistic is pronounced "chi square" or "chi squared"). In this particular case, $X^2 = z^2$, the square of the z-statistic.

(2) Without the continuity correction, the estimated p-values are not only less accurate, but they are too small. That is, the uncorrected z-test is **anti-conservative.**   ◁

Essentially the same argument that shows that a binomial random variable is approximately normal (under certain conditions) can be used for other families of distributions that are closed under addition. See the exercises for some examples.

4.5. Inference for the Mean (Variance Known)

Recall the general framework for statistical inference.

(1) There is a population or process of interest.

(2) Typically our only access to information about the population or process is via sampling.

That is, we select and measure some members of the population or repeat the process some number of times and record measurements each time.

(3) Because of this, the parameters of (a model for) the population or process distribution are *unknown*.

(4) What we do know is our sample data. We can calculate numerical summaries (statistics) from this data.

(5) Statistical inference seeks to obtain information about the unknown parameters from the known statistics (in the context of an assumed model).

We will focus our attention on two types of statistical inference:

- Hypothesis Testing.

  We have already seen some examples of hypothesis testing. Recall that hypothesis testing attempts to determine whether the evidence represented by the data speak for or against a particular statement about the parameter(s) of a distribution.

  Example. We want to test the hypothesis that the average fill weight of a box of cereal is at least as much as the amount stated on the packaging.

- Parameter Estimation via Confidence Intervals.

  On the other hand, we may not have a particular statement we wish to test. We might just want to give a reasonable estimate for the unknown parameter(s) of interest. Confidence intervals provide one way of doing this that takes into account the degree of precision we can expect from our sample.

  Example. We want to estimate the mean birth weight of children born in a certain country.

Inference for means is based on the fact that in most situations of interest the Central Limit Theorem guarantees that if the population has mean μ and standard deviation σ, then

$$\overline{X} \approx \mathsf{Norm}(\mu, \frac{\sigma}{\sqrt{n}}), \text{ so}$$

$$Z = \frac{\overline{X} - \mu}{\sigma/\sqrt{n}} \approx \mathsf{Norm}(0, 1) \ .$$

As we will see, the fact that we don't know μ is not a difficulty. The situation with σ, on the other hand, is different. The more common situation is that we don't know σ either, and we will eventually learn a method for dealing with this situation. For the moment we will handle only the simpler and less frequently used inference procedures that assume σ is known.

4.5.1. Hypothesis Tests

Now that we know an approximate sampling distribution for sample means, the hypothesis test procedure (sometimes called the z-test) is a straightforward application of our general four-step outline for hypothesis tests (see Section 2.4.1).

(1) State the null and alternative hypotheses.
- H_0: $\mu = \mu_0$,
- H_a: $\mu \neq \mu_0$ (or $\mu > \mu_0$ or $\mu < \mu_0$).

(2) Calculate the test statistic.

Our test statistic is $z = \dfrac{\overline{x} - \mu_0}{\sigma/\sqrt{n}}$.

(3) Determine the p-value.

Since $Z = \dfrac{\overline{X} - \mu}{\sigma/\sqrt{n}} \approx \mathsf{Norm}(0, 1)$, determining the (approximate) p-value for either one-sided or two-sided tests is a straightforward application of normal distributions.

(4) Draw a conclusion.

Example 4.5.1.

Q. A manufacturer claims that its 5-pound free weights have a mean weight of 5 pounds with a standard deviation of 0.05 pounds. You purchase 10 of these weights and weigh them on an accurate scale in a chemistry laboratory and find that the mean weight is only 4.96 pounds. What should you conclude? What assumptions are you making about the population of free weights?

A. We can use R to calculate the p-value.

```
> ( 4.96 - 5.0 ) / (0.05 / sqrt(10) ) -> z; z    # test statistic   free-weights
[1] -2.5298
> 2 * (pnorm(z))                                 # 2-sided p-value
[1] 0.011412
> 2 * (1 - pnorm(abs(z)))                         # 2-sided p-value again
[1] 0.011412
```

The p-value is quite small, suggesting (at least) one of the following:

- The mean weight is not 5.
- The standard deviation is larger than stated (perhaps quality control is not as good as the manufacturers think).
- Our sample was not sufficiently random (that is, not sufficiently like an iid or simple random sample).
- The distribution of weights produced by this manufacturer is different enough from a normal distribution that our approximations are too crude to be used reliably.
- We just got "unlucky" with our particular sample.

The test does not give us any indication about which of these is the cause of the low p-value. ◁

4.5.2. Confidence Intervals

Confidence intervals take a different approach to inference. Instead of gathering evidence to evaluate the truth of a particular statement about μ, we simply want to give some sense of the accuracy of our estimate ($\bar{x}$). That is, we want to give some indication of how close we believe $\bar{x}$ is to μ.

Recall that approximately 95% of a normal distribution lies within 2 (more accurately 1.96) standard deviations of the mean. This means that

$$\mathrm{P}(|\overline{X} - \mu| \leq 1.96SE) \approx 0.95$$

where $SE = \frac{\sigma}{\sqrt{n}}$ is the standard error. But this means that

$$\mathrm{P}(\overline{X} - 1.96SE \leq \mu \leq \overline{X} + 1.96SE) \approx 0.95 \ .$$

So if we have a particular random sample with sample mean $\bar{x}$, we will call the interval

$$(\bar{x} - 1.96SE, \bar{x} + 1.96SE)$$

the (approximate) 95% **confidence interval** for the mean. Often this sort of interval is represented using "$\pm$" notation:

$$\bar{x} \pm 1.96SE \ .$$

The expression $1.96SE$ is known as the **margin of error**.

Example 4.5.2.

Q. A sample of 25 oranges weighed an average (mean) of 10 ounces per orange. The standard deviation of the population of weights of oranges is 2 ounces. Find a 95% confidence interval (CI) for the population mean.

A. The (approximate) 95% confidence interval is

$$\bar{x} \pm 1.96SE = 10 \pm 1.96\frac{2}{\sqrt{25}} = 10 \pm 0.0784 = (9.9226, 10.0784) \ . \qquad \triangleleft$$

It is important to understand what the 95% refers to in the example above. To understand this, we must consider what is (and what is not) random in this situation.

(1) The process by which the interval was constructed is a random process.

Sampling was a random process, so an interval produced by random sampling gives rise to a random interval (A, B), where $A = \overline{X} - z_*SE$ and $B = \overline{X} + z_*SE$. Usually (for approximately 95% of these intervals) this random interval will contain μ. Sometimes (for approximately 5% of the intervals), it will not.

(2) The parameter μ is not random.

The parameter is unknown, but fixed. A particular observed sample and its particular confidence interval are also not random. So it is *incorrect* to say in the previous example that there is an approximately 95% probability that $\mu \in (9.9226, 10.0784)$. There is not anything random in that statement, so we can't quantify anything with probability.

If we generated many samples, we would expect approximately 95% *of the samples* to produce confidence intervals that correctly contain the population mean. The other 5% would not. The percentage of intervals that correctly contain the parameter being estimated is called the **coverage rate**. Of course, we don't ever know for a particular interval if it contains the mean we are trying to estimate or not. So the probability is a measure of the *method* used, not of the particular result obtained.

We can use simulation to see how well the nominal coverage rate (95%) matches the actual coverage rate.

Example 4.5.3. We will sample from a Norm(500, 100)-population to investigate coverage rates of 95% confidence intervals. We begin by defining a function that will compute a confidence interval and return both the confidence interval and the point estimate, much the way `prop.test()` and `binom.test()` do.

z-ci

```
> ci <- function(x, sd=100, conf.level=0.95) {
+     alpha = 1 - conf.level
+     n = length(x)
+     zstar <- - qnorm(alpha/2)
+     interval <- mean(x)  + c(-1,1) * zstar * sd / sqrt(n)
+     return(list(conf.int=interval, estimate=mean(x)))
+     }
```

The `CIsim()` function in the `fastR` package computes confidence intervals from simulated samples from a specified distribution, reporting the coverage rate and returning a data frame that can be used to display the intervals graphically (see Figure 4.13).

CI-vis

```
# simulate 100 intervals and plot them.
> results <- CIsim(n=20, samples=100, estimand=500,
+         rdist=rnorm, args=list(mean=500,sd=100),
+         method=ci, method.args=list(sd=100))
Did the interval cover?
  No  Yes
0.04 0.96
> coverPlot <- xYplot(Cbind(estimate,lower,upper) ~ sample, results,
+         groups=cover, col=c('black','gray40'),cap=0,lwd=2,pch=16)
```

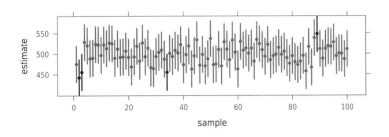

Figure 4.13. One hundred random 95% confidence intervals.

The simulations below compute the coverage rate for confidence intervals computed from samples of various sizes.

```
# an example CI from a sample of size 20                          simulateCI
> ci(rnorm(20,500,100))$conf.int
[1] 426.09 513.74
# 10,000 simulated samples of size 20
> CIsim(n=20, samples=10000, rdist=rnorm, args=list(mean=500,sd=100),
+        estimand=500, method=ci, method.args=list(sd=100))
Did the interval cover?
    No    Yes
0.0451 0.9549
#
# an example CI from a sample of size 5
> ci(rnorm(5,500,100))$conf.int
[1] 425.0 600.3
# 10,000 simulated samples of size 5
> CIsim(n=5,samples=10000, rdist=rnorm, args=list(mean=500,sd=100),
+        estimand=500, method=ci, method.args=list(sd=100))
Did the interval cover?
    No    Yes
0.0471 0.9529
#
# an example CI from a sample of size 2
> ci(rnorm(2,500,100))$conf.int
[1] 302.12 579.30
# 10,000 simulated samples of size 2
> CIsim(n=2,samples=10000, rdist=rnorm, args=list(mean=500,sd=100),
+        estimand=500, method=ci, method.args=list(sd=100))
Did the interval cover?
    No    Yes
0.0503 0.9497
```

As we see, our empirical coverage rates are all very close to 95% – as they should be, since when the population is normal, the sampling distribution for the sample mean is exactly normal for any sample size. Of course, the confidence intervals are much wider when based on smaller samples.   ◁

Example 4.5.4. We can repeat the simulation above with a different population distribution to see how well the approximations of the Central Limit Theorem are working. In this case our samples are drawn from a uniform distribution.

```
> mu = 1/2; v = 1/12            # mean and variance          simulateCI-unif
#
# 10,000 simulated samples of size 20
> CIsim(n=20, samples=10000, rdist=runif, estimand=mu,
+        method=ci, method.args=list(sd=sqrt(v)))
Did the interval cover?
    No    Yes
0.0539 0.9461
#
# 10,000 simulated samples of size 5
```

```
> CIsim(n=5, samples=10000, rdist=runif, estimand=mu,
+         method=ci, method.args=list(sd=sqrt(v)))
Did the interval cover?
    No    Yes
0.0471 0.9529
#
# 10,000 simulated samples of size 2
> CIsim(n=2, samples=10000, rdist=runif, estimand=mu,
+         method=ci, method.args=list(sd=sqrt(v)))
Did the interval cover?
    No    Yes
0.0428 0.9572
```

The empirical coverage rates are not as close to 95% as they were in the previous example, but they are still quite good. Exercise 4.25 asks you to investigate whether these differences are large enough to cause concern. ◁

Example 4.5.5. The simulation below uses the beta distribution with shape parameters 0.4 and 0.6.

simulateCI-beta

```
> mu <- 0.4 / (0.4 + 0.6); mu                     # mean for beta dist
[1] 0.4
> v <- (0.4*0.6) / ((0.4 + 0.6)^2 * (0.4+0.6+1)); v    # var for beta dist
[1] 0.12
#
# 10,000 simulated samples of size 20
> CIsim(n=20, samples=10000, rdist=rbeta, args=list(shape1=0.4,shape2=0.6),
+         estimand=mu, method=ci, method.args=list(sd=sqrt(v)))
Did the interval cover?
    No    Yes
0.0478 0.9522
#
# 10,000 simulated samples of size 5
> CIsim(n=5, samples=10000, rdist=rbeta, args=list(shape1=0.4,shape2=0.6),
+         estimand=mu, method=ci, method.args=list(sd=sqrt(v)))
Did the interval cover?
    No    Yes
0.0487 0.9513
#
# 10,000 simulated samples of size 2
> CIsim(n=2, samples=10000, rdist=rbeta, args=list(shape1=0.4,shape2=0.6),
+         estimand=mu, method=ci, method.args=list(sd=sqrt(v)))
Did the interval cover?
    No    Yes
0.0356 0.9644
```
◁

Example 4.5.6. The simulation below uses an exponential distribution with $\lambda = 1/10$.

simulateCI-exp

```
> rate = 1/10
> v = (1/rate)^2                                  # var of exponential
> mu = 10                                         # mean of exponential
```

```
> ci(rexp(20,rate),sd=sqrt(v))$conf.int          # an example CI
[1]   3.7291 12.4943
#
# 10,000 simulated samples of size 20
> CIsim(n=20, samples=10000, rdist=rexp, args=list(rate=rate),
+        estimand=mu, method=ci, method.args=list(sd=sqrt(v)))
Did the interval cover?
    No    Yes
0.0445 0.9555
#
# 10,000 simulated samples of size 5
> CIsim(n=5, samples=10000, rdist=rexp, args=list(rate=rate),
+        estimand=mu, method=ci, method.args=list(sd=sqrt(v)))
Did the interval cover?
    No    Yes
0.0435 0.9565
#
# 10,000 simulated samples of size 2
> CIsim(n=2, samples=10000, rdist=rexp, args=list(rate=rate),
+        estimand=mu, method=ci, method.args=list(sd=sqrt(v)))
Did the interval cover?
    No    Yes
0.0456 0.9544                                                    ◁
```

Taken together, these examples suggest that the coverage rate is quite close to the nominal rate of 95% for a wide range of situations. See Exercise 4.25 for a formal comparison of these simulated results to the nominal coverage rate of 95%.

Other confidence levels

There is, of course, no reason to restrict ourselves to a confidence level of 95%. Before describing a general confidence interval, we introduce a bit of notation. For any $\alpha \in [0, 1]$, let z_α be the number satisfying

$$\mathrm{P}(Z \geq z_\alpha) = \mathrm{P}(Z \leq -z_\alpha) = \alpha \;,$$

where $Z \sim \mathsf{Norm}(0, 1)$. In R,

$$z_\alpha = \mathtt{qnorm}(1 - \alpha) = \mathtt{-qnorm}(\alpha) \;.$$

When α is clear from context, we will often use an abbreviated notation:

$$z_* = z_\alpha \;.$$

Definition 4.5.1. Let $\boldsymbol{X} = X_1, X_2, \ldots, X_n$ be a random sample of size n from a population with unknown mean μ and known variance σ. The (approximate) $100C\%$ confidence interval for μ is

$$\overline{x} \pm z_{\alpha/2} SE \;,$$

where $\alpha = 1 - C$.   □

Example 4.5.7.

Q. A sample of size 25 has mean 10. Assuming the population has a standard deviation of 2, compute a 99% confidence interval from this data.

A. A 99% confidence interval requires that we determine $z_{0.005}$, which equals 2.58. The standard error is $SE = \frac{2}{\sqrt{25}} = 0.4$, so the resulting interval is

$$10 \pm (2.58)(0.4) = 10 \pm 1.03 = (8.97, 11.03) .$$

99CI

```
> zstar <- - qnorm(0.005); zstar
[1] 2.5758
> se <- 2 / sqrt(25); se
[1] 0.4
> zstar * se;
[1] 1.0303
> 10 + c(-1,1) * zstar * se              # confidence interval
[1]  8.9697 11.0303
```
◁

4.5.3. A Connection Between Hypothesis Tests and Confidence Intervals

The following example illustrates an important connection between confidence intervals and hypothesis tests.

Example 4.5.8.

Q. Suppose we perform a hypothesis test with hypotheses

- H_0: $\mu = \mu_0$,
- H_a: $\mu \neq \mu_0$

using a sample of size n with sample mean $\bar{x}$. For what values of μ_0 will we reject H_0 with $\alpha = 0.05$?

A. We reject when $|z| = \frac{|\bar{x} - \mu_0|}{SE} \geq 1.96$. Solving for μ_0, we see that we reject when

$$\mu_0 \notin (\bar{x} - 1.96SE, \bar{x} + 1.96SE) . \qquad \lhd$$

As the example above shows, the values of μ_0 that lie in the 95% confidence interval for the mean are precisely the values that are not rejected. So we can interpret this confidence interval as a set of "plausible values for μ" in the sense that we would not reject them based on our sample.

A similar situation arises for any confidence level:

$$\mu_0 \text{ is in the } (1 - \alpha) \text{ confidence interval}$$
$$\Updownarrow$$
the two-sided p-value of the corresponding hypothesis test is greater than α.

This duality between hypothesis testing and confidence intervals holds much more generally. See Section 5.3.

4.6. Estimating Variance

The previous section handled situations in which the population variance σ^2 can be considered known. This is not typically the case. Usually we need to estimate σ from sample data.

Lemma 4.6.1. *Given an iid random sample, the sample variance is an unbiased estimator for the population variance. That is,*

$$\mathrm{E}\left(\frac{\sum(X_i - \overline{X})^2}{n-1}\right) = \sigma^2 .$$

Proof. We begin by deriving an algebraically equivalent form of the numerator of the variance:

$$\begin{aligned}
\sum(X_i - \overline{X})^2 &= \sum(X_i^2 - 2X_i\overline{X} + \overline{X}^2) \\
&= \sum X_i^2 - 2\sum X_i\overline{X} + \sum \overline{X}^2 \\
&= \sum X_i^2 - 2\overline{X}\sum X_i + n\overline{X}^2 \\
&= \sum X_i^2 - 2n\overline{X}^2 + n\overline{X}^2 \\
&= \sum X_i^2 - n\overline{X}^2 .
\end{aligned} \tag{4.8}$$

Now let μ and σ^2 be the common mean and variance of the X_i's. Recall that for any random variable W, $\mathrm{Var}(W) = \mathrm{E}(W^2) - \mathrm{E}(W)^2$, so $\mathrm{E}(W^2) = \mathrm{E}(W)^2 + \mathrm{Var}(W)$. Applying this to the X_i's and to $\overline{X}$, we have

$$\mathrm{E}(X_i^2) = \mu^2 + \sigma^2 ,$$

and

$$\mathrm{E}(\overline{X}^2) = \mu^2 + \frac{\sigma^2}{n} .$$

Using this, we see that

$$\begin{aligned}
\mathrm{E}\left(\sum(X_i - \overline{X})^2\right) &= \mathrm{E}\left(\sum X_i^2 - n\overline{X}^2\right) \\
&= \sum \mathrm{E}(X_i^2) - n\,\mathrm{E}(\overline{X}^2) \\
&= n(\mu^2 + \sigma^2) - n(\mu^2 + \frac{\sigma^2}{n}) \\
&= n\sigma^2 - \sigma^2 \\
&= (n-1)\sigma^2 .
\end{aligned}$$

From this it easily follows that $\mathrm{E}\left(\frac{\sum(X_i-\overline{X})^2}{n-1}\right) = \sigma^2$. $\qquad\square$

Lemma 4.6.1 justifies our definition of the sample variance. A more natural definition would have been

$$\hat{\sigma}^2 = \frac{\sum(X_i - \overline{X})^2}{n} ,$$

but this estimator is biased:

$$E\left(\hat{\sigma}^2\right) = \frac{n-1}{n}\sigma^2 \ .$$

It is worth noting that $S = \sqrt{\frac{(\sum X_i - \overline{X})^2}{n-1}}$ is not an unbiased estimator of the standard deviation. (See Exercise 4.30.)

The characterization of the sample variance given in (4.8) is important enough to separate out as its own corollary (to the proof).

Corollary 4.6.2. *Let* $\boldsymbol{X} = \langle X_1, X_2, \ldots, X_n \rangle$ *be any sample. Then*

$$\sum (X_i - \overline{X})^2 = \sum X_i^2 - n\overline{X}^2 \ . \qquad \square$$

An important interpretation of Corollary 4.6.2 via linear algebra motivates the remainder of this section. $\sum X_i^2$ is the squared length of the vector

$$\boldsymbol{X} = \langle X_1, X_2, \ldots, X_n \rangle \ ,$$

and $n\overline{X}^2$ is the squared length of the vector

$$\overline{\boldsymbol{X}} = \langle \overline{X}, \overline{X}, \ldots, \overline{X} \rangle \ .$$

Let $\boldsymbol{V} = \boldsymbol{X} - \overline{\boldsymbol{X}}$ be the difference between these two vectors. Then

$$|\boldsymbol{V}|^2 = \sum (X_i - \overline{X})^2$$

is the lefthand side of (4.8), so

$$(n-1)S^2 = |\boldsymbol{V}|^2 \ .$$

Our goal is to determine the distribution of $|\boldsymbol{V}|^2$. By Corollary 4.6.2 we know that

$$|\boldsymbol{V}|^2 = |\boldsymbol{X}|^2 - |\overline{\boldsymbol{X}}|^2 \ .$$

But this implies that

$$|\boldsymbol{V}|^2 + |\overline{\boldsymbol{X}}|^2 = |\boldsymbol{X}|^2 = |\boldsymbol{V} + \overline{\boldsymbol{X}}|^2 \ ,$$

so by the converse to the Pythagorean Theorem, we know that $\boldsymbol{V} \perp \overline{\boldsymbol{X}}$. We could also have proved this using the dot product as follows:

$$\boldsymbol{V} \cdot \overline{\boldsymbol{X}} = \sum (X_i - \overline{X})\overline{X} = \sum (\overline{X}X_i - \overline{X}^2) = \overline{X}\left(\sum X_i\right) - n\overline{X}^2 = n\overline{X}^2 - n\overline{X}^2 = 0 \ .$$

Since $\boldsymbol{V} \perp \overline{\boldsymbol{X}}$, $\boldsymbol{V}$ can be expressed as

$$\boldsymbol{V} = \sum_{i=2}^{n} \boldsymbol{V}_i$$

with $\boldsymbol{V}_i \perp \boldsymbol{V}_j$ whenever $i \neq j$ and $\boldsymbol{V}_i \perp \overline{\boldsymbol{X}}$ for all $i \geq 2$. (Choose any orthogonal unit vectors $\boldsymbol{u}_i$ with $\boldsymbol{u}_1 = \boldsymbol{1}$, and project $\boldsymbol{V}$ along each $\boldsymbol{u}_i$ to obtain $\boldsymbol{V}_i$.) So

$$(n-1)S^2 = |\boldsymbol{V}|^2 = \sum_{i=2}^{n} |\boldsymbol{V}_i|^2 \ .$$

Our new goal is to determine the joint distribution of $|\boldsymbol{V}_2|^2, |\boldsymbol{V}_3|^2, \ldots, |\boldsymbol{V}_n|^2$.

Since $\boldsymbol{V}_i = (\boldsymbol{V} \cdot \boldsymbol{u}_i)\boldsymbol{u}_i$ and $|\boldsymbol{V}_i|^2 = (\boldsymbol{V} \cdot \boldsymbol{u}_i)^2$, we would like to know the distribution of the dot product of random vector with a (fixed) unit vector.

Lemma 4.6.3. *Let $\boldsymbol{X} = \langle X_1, X_2, \ldots, X_n \rangle$ be an iid random sample with each $X_i \sim \mathsf{Norm}(\mu, \sigma)$, and let $\boldsymbol{u} = \langle u_1, u_2, \ldots, u_n \rangle$ be a unit vector in $\mathbb{R}^n$. Then*

$$\boldsymbol{X} \cdot \boldsymbol{u} \sim \mathsf{Norm}(\mu \sum u_i, \sigma) \ .$$

Proof. Since $\boldsymbol{u}$ is a unit vector, $|\boldsymbol{u}|^2 = \sum u_i^2 = 1$, and

$$\boldsymbol{X} \cdot \boldsymbol{u} = \sum u_i X_i \sim \mathsf{Norm}(\sum u_i \mu, \sum u_i^2 \sigma^2) = \mathsf{Norm}(\mu \sum u_i, \sigma^2) \ . \qquad \square$$

Lemma 4.6.4. *If $\boldsymbol{X} = \langle X_1, X_2, \ldots, X_n \rangle$ is a vector of independent random variables, and $\boldsymbol{u}$ and $\boldsymbol{w}$ are (fixed) unit vectors in $\mathbb{R}^n$, then*

$$\boldsymbol{X} \cdot \boldsymbol{u} \text{ and } \boldsymbol{X} \cdot \boldsymbol{w} \text{ are independent if and only if } \boldsymbol{u} \perp \boldsymbol{w}.$$

Proof. We will prove something slightly weaker, namely that $\mathrm{Cov}(\boldsymbol{X} \cdot \boldsymbol{u}, \boldsymbol{X} \cdot \boldsymbol{w}) = 0$ if and only if $\boldsymbol{u} \perp \boldsymbol{w}$:

$$
\begin{aligned}
\mathrm{Cov}(\boldsymbol{X} \cdot \boldsymbol{u}, \boldsymbol{X} \cdot \boldsymbol{w}) &= \sum_{i,j} \mathrm{Cov}(X_i \cdot u_i, X_j \cdot w_j) \\
&= \sum_{i,j} u_i w_j \, \mathrm{Cov}(X_i, X_j) \\
&= \sum_{i} u_i w_i \, \mathrm{Cov}(X_i, X_i) \\
&= \sum_{i} u_i w_i \, \mathrm{Var}(X_i) \\
&= \sigma^2 \sum_{i} u_i w_i \\
&= \sigma^2 \, \boldsymbol{u} \cdot \boldsymbol{w} \\
&= 0 \text{ if and only if } \boldsymbol{u} \perp \boldsymbol{w} \ .
\end{aligned}
$$

This proves the lemma in one direction and strongly hints at the truth in the other direction. A proof of independence would require that we look at the joint and marginal distributions to see that we really have independence. We'll omit that part of the proof. $\qquad \square$

Now let's apply all this to an iid random sample from a normal population. Let $\boldsymbol{u}_i$ be an orthonormal basis for $\mathbb{R}^n$ such that $\boldsymbol{u}_1$ is in the direction of $\overline{\boldsymbol{X}}$, i.e., in the direction of $\boldsymbol{1} = \langle 1, 1, \ldots, 1 \rangle$. (Exercise 4.34 provides an explicit definition of the $\boldsymbol{u}_i$'s.) Then

$$\boldsymbol{V} = \boldsymbol{X} - \overline{\boldsymbol{X}} = \sum_{i=2}^{n} (\boldsymbol{V} \cdot \boldsymbol{u}_i) \boldsymbol{u}_i = \sum_{i=2}^{n} (\boldsymbol{X} \cdot \boldsymbol{u}_i) \boldsymbol{u}_i \ .$$

Since $\boldsymbol{u}_1 \parallel \langle 1, 1, \ldots 1 \rangle$, the sum of the coefficients of each of the other $\boldsymbol{u}_i$'s must be 0. Thus by Lemma 4.6.3

$$\boldsymbol{V} \cdot \boldsymbol{u}_i \sim \mathsf{Norm}(0, \sigma) \ ,$$

and by Lemma 4.6.4, the $|\boldsymbol{V}_i|$ are independent. This implies the following theorem.

Theorem 4.6.5. *Let $\boldsymbol{X} = \langle X_1, X_2, \ldots, X_n \rangle$ be an iid random sample. Then $\overline{X}$ and S^2 are independent random variables.*

Proof. $V_1 = \overline{X}$, so $\overline{X}$ is a function of $|V_1| = \sqrt{n} \cdot \overline{X}$. Since $|V_1|$ is independent of the other $|V_i|$ and S^2 is a function of $|V_2|, |V_3|, \ldots, |V_n|$, it follows that $\overline{X}$ and S^2 are independent. $\qquad\square$

This also means that $(n-1)S^2$ can be expressed as the sum of $n-1$ independent random variables, each of which is the square of a normal random variable with mean 0 and standard deviation σ. That is,

$$(n-1)S^2 = W_1^2 + W_2^2 + \cdots + W_{n-1}^2$$

with each $W_i \sim \mathsf{Norm}(0, \sigma)$ (iid).

The distribution of a sum of squares of iid normal random variables is important enough to warrant its own name:

Definition 4.6.6. Let $\{Z_i\}_{i=1}^n$ be a collection of independent standard normal random variables. Then

$$Z_1^2 + Z_2^2 + \cdots + Z_n^2$$

has the **chi-squared distribution** with n **degrees of freedom**. $\qquad\square$

Using the chi-squared distributions, we can succinctly express the result we have been working toward in this section.

Lemma 4.6.7. *Let $X = \langle X_1, X_2, \ldots, X_n \rangle$ be an iid random sample from a normal population with variance σ^2. Then*

$$\frac{(n-1)S^2}{\sigma^2} = \frac{\sum (X_i - \overline{X})^2}{\sigma^2} \sim \mathsf{Chisq}(n-1) \;.$$

Proof. We have already shown that

$$(n-1)S^2 = W_1^2 + W_2^2 + \cdots + W_{n-1}^2$$

with each $W_i \sim \mathsf{Norm}(0, \sigma)$. So

$$\frac{(n-1)S^2}{\sigma^2} = \left(\frac{W_1}{\sigma}\right)^2 + \left(\frac{W_2}{\sigma}\right)^2 + \cdots + \left(\frac{W_{n-1}}{\sigma}\right)^2$$

with each $\frac{W_i}{\sigma} \sim \mathsf{Norm}(0, 1)$. $\qquad\square$

The cdf method can be used to show that chi-squared random variables are a special case of gamma random variables.

Lemma 4.6.8. *If $X \sim \mathsf{Chisq}(n)$. Then $X \sim \mathsf{Gamma}(\alpha = \frac{n}{2}, \lambda = \frac{1}{2})$.*

Proof. We first handle the case when $n = 1$. Let $Z \sim \mathsf{Norm}(0, 1)$ and let $x \geq 0$. Then

$$\begin{aligned} F_X(x) &= \mathrm{P}(Z^2 \leq x) \\ &= \mathrm{P}(-\sqrt{x} \leq Z \leq \sqrt{x}) \\ &= \Phi(\sqrt{x}) - \Phi(-\sqrt{x}) \;. \end{aligned}$$

So the pdf for X is

$$\frac{d}{dx} F_X(x) = \frac{1}{2\sqrt{x}} \left[\phi(\sqrt{x}) + \phi(-\sqrt{x}) \right]$$

$$= \frac{1}{\sqrt{x}} \phi(\sqrt{x})$$

$$= \frac{x^{-1/2} e^{-x/2}}{\sqrt{2\pi}}$$

on $[0, \infty)$. The kernel of this density function is the kernel of the $\Gamma(\frac{1}{2}, \beta = 2)$ distribution, so $X \sim \Gamma(\frac{1}{2}, \lambda = \frac{1}{2})$. Note that this also implies that $\Gamma(\frac{1}{2}) = \sqrt{\pi}$.

For the full result, we simply note that by Exercise 3.53, independent sums of such random variables are again gamma random variables with the stated parameter values. $\qquad \square$

Corollary 4.6.9. *Let* $X \sim \mathsf{Chisq}(n)$. *Then*

- $\mathrm{E}(X) = n$.
- $\mathrm{Var}(X) = 2n$.

Proof. This follows immediately from what we already know about gamma distributions. $\qquad \square$

4.7. Inference for the Mean (Variance Unknown)

When the variance of the population was known, our inference procedures for the mean were based on the distribution of

$$Z = \frac{\overline{X} - \mu}{\sigma/\sqrt{n}} = \frac{\overline{X} - \mu}{SE} .$$

It seems natural then that when the population variance is not known we should use

$$T = \frac{\overline{X} - \mu}{S/\sqrt{n}}$$

instead. We will refer to $S/\sqrt{n}$ (or $s/\sqrt{n}$) as the estimated standard error (of the sample mean) because it is an estimate of the standard deviation of the sampling distribution. The estimated standard error of an estimator θ can be denoted as $\hat{\sigma}_{\hat{\theta}}$ (an estimation of $\sigma_{\hat{\theta}}$) or as $s_{\hat{\theta}}$.

For large enough n, T should be approximately normal since S will usually be very close to σ so there is little difference between values of T and values of Z. When sample sizes are smaller, however, this approximation is not good. The variation in sample variances from sample to sample causes T to be more variable than Z. This is especially bad since it could cause us to reject null hypotheses too readily and to form confidence intervals that are narrower than our data warrant.

So what is the correct distribution of

$$T = \frac{\overline{X} - \mu}{S/\sqrt{n}} ?$$

The first person to ask (and answer) this question was William S. Gosset, who worked at the time for a brewery in Dublin. Unable to publish using his own name (as a condition of employment at the Guinness Brewery), Gosset published his work on small sample inference under the name of Student, and the distribution above has come to be called **Student's *t*-distribution**.

Notice that

$$T = \frac{\overline{X} - \mu}{S/\sqrt{n}}$$

$$= \frac{\left(\frac{\overline{X} - \mu}{\sigma/\sqrt{n}}\right)}{S/\sigma}.$$

The numerator is a standard normal random variable. The denominator is independent and its square is

$$\frac{S^2}{\sigma^2} = \frac{\left(\frac{(n-1)S^2}{\sigma^2}\right)}{n-1},$$

which is a chi-squared random variable divided by its degrees of freedom. This leads us to define two new families of random variables.

Definition 4.7.1. Let $Z \sim \mathsf{Norm}(0,1)$ and let $V \sim \mathsf{Chisq}(n)$ be independent random variables. The distribution of

$$\frac{Z}{\sqrt{V/n}}$$

is called the *t*-distribution with n degrees of freedom. $\qquad\square$

Definition 4.7.2. Let $U \sim \mathsf{Chisq}(m)$ and let $V \sim \mathsf{Chisq}(n)$ be independent random variables. The distribution of

$$\frac{U/m}{V/n}$$

is called the *F*-distribution with m and n degrees of freedom. $\qquad\square$

Lemma 4.7.3. *If* $T \sim \mathsf{t}(n)$, *then* $T^2 \sim \mathsf{F}(1, n)$.

Proof. Let $T = \frac{Z}{\sqrt{V/n}}$ where $Z \sim \mathsf{Norm}(0,1)$ and $V \sim \mathsf{Chisq}(n)$. Then

$$T^2 = \frac{Z^2/1}{V/n} \sim \mathsf{F}(1, n). \qquad\square$$

Importantly, the *t*-distributions depend on the sample size but not on the mean and variance of the population. The pdfs for the *t* and *F* distributions can be obtained using the cdf method since they are quotients of independent random variables with familiar pdfs. We omit these derivations but give the pdfs in the following lemmas.

Lemma 4.7.4. *Let* $T \sim \mathsf{t}(n)$. *Then the pdf for* T *is given by*

$$f_T(t; n) = \frac{\Gamma(\frac{n+1}{2})}{\Gamma(n/2)\sqrt{n\pi}} \left(1 + \frac{t^2}{n}\right)^{-(n+1)/2}. \qquad\square$$

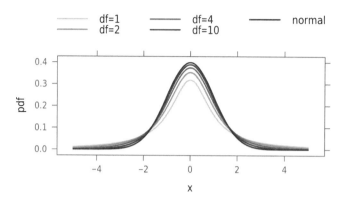

Figure 4.14. A comparison of t-distributions with the standard normal distribution. As the degrees of freedom increase, the t-distributions become more and more like the standard normal distribution.

Lemma 4.7.5. *Let* $X \sim \mathsf{F}(m, n)$. *Then the pdf for* X *is given by*

$$f_X(x; m, n) = \frac{\Gamma(\frac{m+n}{2})}{\Gamma(m/2)\Gamma(n/2)} \left(\frac{m}{n}\right)^{m/2} x^{\frac{m}{2}-1} \left(1 + \frac{m}{n}x\right)^{-(m+n)/2} \quad on \ [0, \infty). \quad \square$$

As with the other important distributions we have encountered, numerical approximations can be obtained in R using `dt()`, `df()`, `pt()`, `pf()`, etc. The t-distributions are much like the standard normal distribution. They are symmetric about 0 and unimodal, but the standard deviation of a t-distribution is larger and more of the probability is in the tails of a t-distribution than in the tails of a standard normal distribution.

Hypothesis tests and confidence intervals using the t-distributions are no more difficult than their z-based counterparts. We simply replace z with t (with the appropriate degrees of freedom).

Example 4.7.1.

Q. A sample of size $n = 10$ has mean $\bar{x} = 10.3$ and standard deviation $s = 0.4$. Compute a 95% confidence interval for the population mean.

A. The confidence interval has the form

$$\bar{x} \pm t_{0.025}\frac{s}{\sqrt{n}} \ .$$

We can use R to do the calculations as follows:

```
> tstar <- qt(0.975,df=9); tstar;
[1] 2.2622
> 10.3 + c(-1,1) * tstar * 0.4 / sqrt(10);
[1] 10.014 10.586
```

tCI-01

◁

Example 4.7.2.

Q. Use the data from the previous example to do a hypothesis test with a null hypothesis that $\mu = 10.0$.

A. Since the 95% confidence interval does not include 10, we know that the p-value will be less than 0.05.

t-test-01

```
> t <- ( 10.3 - 10 )/ (0.4 / sqrt(10)); t;       # test statistic
[1] 2.3717
> 2 * pt(-abs(t),df=9);              # p-value using t-distribution
[1] 0.041792
> 2 * pnorm(-abs(t));               # "p-value" using normal distribution
[1] 0.017706
```

Notice how incorrectly using the normal distribution makes the "p-value" quite a bit smaller. ◁

Example 4.7.3. If we have data in R then the function t.test() can do all of the computations for us. The following example computes a confidence interval and performs a hypothesis test using the sepal widths of one species of iris from the iris data set.

t-test-iris

```
# for CI; p-value not interesting here
> t.test(iris$Sepal.Width[iris$Species=="virginica"])


        One Sample t-test

data:  iris$Sepal.Width[iris$Species == "virginica"]
t = 65.208, df = 49, p-value < 2.2e-16
alternative hypothesis: true mean is not equal to 0
95 percent confidence interval:
 2.8823 3.0657
sample estimates:
mean of x
    2.974


# this gives a more interesting p-value
> t.test(iris$Sepal.Width[iris$Species=="virginica"],mu=3)


        One Sample t-test

data:  iris$Sepal.Width[iris$Species == "virginica"]
t = -0.5701, df = 49, p-value = 0.5712
alternative hypothesis: true mean is not equal to 3
95 percent confidence interval:
 2.8823 3.0657
sample estimates:
mean of x
    2.974
```

The use of the t-distribution is based on the assumption that the population is normal. As we will soon see, the t-procedures are remarkably robust – that is, they are still very good approximations even when the underlying assumptions do not hold. Since the sample is reasonably large ($n = 50$) and unimodal, we can be quite confident that the t-procedures are appropriate for this data. ◁

4.7.1. Robustness

The t-procedures were developed under the assumption that we have an indepen-dent random sample from a normally distributed population. In practice these assumptions are rarely met. The first assumption, independence, depends on the design of the study. In some situations obtaining an independent (or nearly in-dependent) random sample is relatively easy. In other situations it can be quite challenging. In many situations, researchers do their best to approximate indepen-dent random sampling and then proceed using the methods we have just described. In other situations other sampling designs are used intentionally, and this requires the use of different analysis tools as well.

The issue of population distribution is largely beyond the control of the re-searchers, and in many cases, the population will not be normal, perhaps not even approximately normal. Thus we would like to know how well our procedures per-form even when the population is not normally distributed. Procedures that per-form well even when some of the assumptions on which they are based do not hold are said to be **robust**.

One way to explore the robustness of a procedure is via simulations. We can simulate a large number of samples drawn from various types of distributions to see how well they perform. For example, we could compute confidence intervals and see how well the empirical coverage rate compares to the nominal coverage rate. Or we could compute p-values and either compare empirical and nominal type I error rates or compare the distribution of p-values with the $\mathsf{Unif}(0,1)$-distribution that p-values should have when the null hypothesis is true.

Example 4.7.4. Let's begin by seeing how well the t-procedures work with a normal population. We will once again use `CIsim()` to tally the results of simulated confidence intervals. The default `method` for `CIsim()` is `t.test()`.

```
# an example CI from a sample of size 20                    t-simulations
> t.test(rnorm(20,500,100))$conf.int
[1] 468.62 546.68
attr(,"conf.level")
[1] 0.95
# 10,000 simulated samples of size 20
> CIsim(n=20, samples=10000, estimand=500,
+       rdist=rnorm, args=list(mean=500,sd=100))
Did the interval cover?
    No    Yes
0.0475 0.9525
#
# an example CI from a sample of size 5
> t.test(rnorm(5,500,100))$conf.int
[1] 329.14 604.96
attr(,"conf.level")
[1] 0.95
# 10,000 simulated samples of size 5
> CIsim(n=5, samples=10000, estimand=500,
+       rdist=rnorm, args=list(mean=500,sd=100))
Did the interval cover?
```

```
     No    Yes
0.0501 0.9499
#
# an example CI from a sample of size 2
> t.test(rnorm(2,500,100))$conf.int
[1] -345.53 1313.28
attr(,"conf.level")
[1] 0.95
# 10,000 simulated samples of size 2
> CIsim(n=2, samples=10000, estimand=500,
+        rdist=rnorm, args=list(mean=500,sd=100))
Did the interval cover?
     No    Yes
0.0469 0.9531
```

As we would expect, the coverage rates are all near 95% – even if the samples are small. Notice that we cannot even apply the procedure on a sample of size 1 since we are unable to make any estimate of variance based on one value. Also notice that the intervals formed from small samples are quite wide. In practice they may be too wide to be of any use.   ◁

Example 4.7.5. A t-distribution has heavier tails than a normal distribution, even if we adjust for the differences in variance. For $n \geq 3$, the mean and variance of the $t(n)$-distributions are 0 and $\frac{n}{n-2}$. (The $t(2)$-distribution has no standard deviation, and the $t(1)$-distribution is the Cauchy distribution, which has no mean.) Comparing the $t(3)$-distribution with the $\mathsf{Norm}(0, \sqrt{3})$-distribution, we see that $t(3)$ has slightly heavier tails. It is interesting to note that the $t(3)$-distribution also has more density near 0 than the $\mathsf{Norm}(0, \sqrt{3})$-distribution.

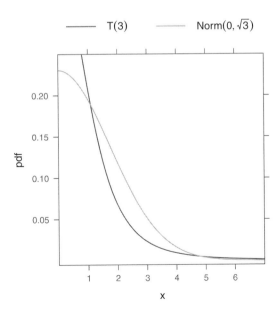

Figure 4.15. Comparing the tails of $\mathsf{Norm}(0, \sqrt{3})$- and $t(3)$-distributions.

The simulations below indicate perhaps a slight overcoverage of the confidence intervals when we sample from a t(3)-distribution. (See Exercise 4.25 for a more formal treatment of whether this really is evidence of overcoverage.)

```
                                                                    t-robust-heavytails
# an example CI from a sample of size 20
> t.test(rt(20,3))$conf.int
[1] -0.25047  0.92753
attr(,"conf.level")
[1] 0.95
# 10,000 simulated samples of size 20
> CIsim(n=20, samples=10000, estimand=0, rdist=rt, args=list(df=3))
Did the interval cover?
    No    Yes
0.0429 0.9571
#
# an example CI from a sample of size 5
> t.test(rt(5,3))$conf.int
[1] -0.22976  1.00851
attr(,"conf.level")
[1] 0.95
# 10,000 simulated samples of size 5
> CIsim(n=5, samples=10000, estimand=0, rdist=rt, args=list(df=3))
Did the interval cover?
    No    Yes
0.0382 0.9618
#
# an example CI from a sample of size 2
> t.test(rt(2,3))$conf.int
[1] -21.624  20.836
attr(,"conf.level")
[1] 0.95
# 10,000 simulated samples of size 2
> CIsim(n=2, samples=10000, estimand=0, rdist=rt, args=list(df=3))
Did the interval cover?
    No    Yes
0.0392 0.9608                                                             ◁
```

Example 4.7.6. Now let's see how the t-procedures fare when the population distribution is skewed. Here are some simulation results using an exponential distribution for the population.

```
                                                                       t-robust-exp
# an example CI from a sample of size 20
> t.test(rexp(20,1/10))$conf.int
[1]   5.773 15.664
attr(,"conf.level")
[1] 0.95
# 10,000 simulated samples of size 20
> CIsim(n=20, samples=10000, estimand=10, rdist=rexp, args=list(rate=1/10))
Did the interval cover?
    No    Yes
0.0755 0.9245
```

```
#
# an example CI from a sample of size 5
> t.test(rexp(5,1/10))$conf.int
[1]  2.4761 17.5586
attr(,"conf.level")
[1] 0.95
# 10,000 simulated samples of size 5
> CIsim(n=5, samples=10000, estimand=10, rdist=rexp, args=list(rate=1/10))
Did the interval cover?
    No    Yes
0.1204 0.8796
#
# an example CI from a sample of size 2
> t.test(rexp(2,1/10))$conf.int
[1] -13.458  19.048
attr(,"conf.level")
[1] 0.95
# 10,000 simulated samples of size 2
> CIsim(n=2, samples=10000, estimand=10, rdist=rexp, args=list(rate=1/10))
Did the interval cover?
    No    Yes
0.0918 0.9082
```

Notice that there is marked undercoverage here for small samples.   ◁

Example 4.7.7. Another way to check robustness is to compute p-values from simulated samples under the null hypothesis and to compare the distribution of these p-values to the uniform distribution that p-values should have. Figure 4.16 contains quantile-quantile plots for samples of various sizes from t(3)- and Exp(1)-distributions. (The R-code used to generate this plot appears at the end of this example.)

Notice how when sampling from the exponential distribution – especially with small sample sizes – there are too many small p-values. This is precisely the part of the distribution that we are most interested in. The lower panel of Figure 4.16 zooms in so we can take a closer look. We see from this that using a t-test carelessly can cause us to over-interpret the significance of our results. (The p-values we compute will be smaller than the data really warrant.) This problem decreases as the sample size increases and is of little concern for samples of size 40 or more.

The distribution of p-values computed from a population with the t(3)-distribution, on the other hand, is much closer to the uniform distribution. If anything, it is slightly conservative.

```
> tpval <- function(x,mu=0) {                                    t-robust-qq
+       xbar <- mean(x)
+       n <- length(x)
+       s <- sd(x)
+       t <- (xbar - mu) / (s / sqrt(n))
+       return ( 2 * pt(-abs(t), df=n-1) )
+ }
> ddd <- data.frame(
+       pval = c(
```

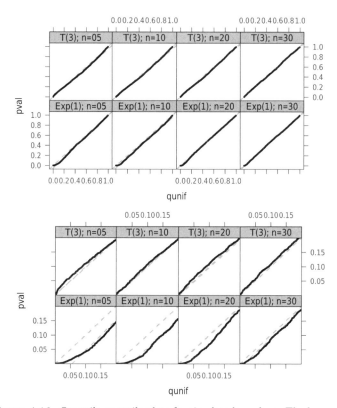

Figure 4.16. Quantile-quantile plots for simulated p-values. The lower panel zooms in on the lower tail of the distributions.

```
+           replicate( 2000, tpval(rt(5,df=3)) ),
+           replicate( 2000, tpval(rt(10,df=3)) ),
+           replicate( 2000, tpval(rt(20,df=3)) ),
+           replicate( 2000, tpval(rt(30,df=3)) ),
+           replicate( 2000, tpval(rexp(5,1), mu=1) ),
+           replicate( 2000, tpval(rexp(10,1), mu=1) ),
+           replicate( 2000, tpval(rexp(20,1), mu=1) ),
+           replicate( 2000, tpval(rexp(30,1), mu=1) )
+           ),
+      dist = c(
+        rep("T(3); n=05",2000), rep("T(3); n=10",2000),
+        rep("T(3); n=20",2000), rep("T(3); n=30",2000),
+        rep("Exp(1); n=05",2000), rep("Exp(1); n=10",2000),
+        rep("Exp(1); n=20",2000), rep("Exp(1); n=30",2000)
+        ))
> myplot1 <- xqqmath(~pval|dist,ddd,dist=qunif,idline=TRUE,cex=0.4)
> myplot2 <- xqqmath(~pval|dist,ddd,dist=qunif,idline=TRUE,cex=0.4,
+      xlim=c(0,0.2), ylim=c(0,0.2))
```

◁

4.7.2. Rules of Thumb

There are, of course, many other possible population distributions than the few that we have simulated. Experience, simulations, and theoretical work have led statisticians to formulate "rules of thumb" to help guide practitioners as to when various statistical procedures are warranted. (See, for example, [**vB02**].) These are somewhat complicated by the fact that the most important assumptions are about the population, and if the researcher has little or no information about the population, then it can be difficult or impossible to justify these assumptions. Nevertheless, the following rules of thumb are offered as a useful starting point.

These rules of thumb all suppose that we have a sufficiently good sample. That is, our sample should be an iid random sample, a simple random sample, or something closely approximating one of these. (Other methods exist for handling more complicated sampling methods.)

(1) **Small Samples.** If the population is normal, then the t-procedures are exact for all sample sizes.

This isn't really a rule of thumb; it is a theorem. But it is our starting point. Of course, no population is *exactly* normal. But then again, we don't really need our p-values and confidence intervals to be *exactly* correct either. If we are quite sure that our population is very closely approximated by a normal distribution, we will proceed with t-procedures even with small sample sizes.

The important question, however, is this: How much are the t-procedures affected by various violations of this normality assumption?

(2) **Medium Samples.** The t-procedures work well for most populations that are roughly symmetric and unimodal as long as the sample size is at least 10–15. There is no magic number here. The larger the sample, the better. The more symmetric, the better. So there is a balance between lack of symmetry and sample size.

(3) **Large Samples.** The t-distributions can be used even when the population is clearly skewed, provided the the sample size is at least 30–40.

Again, there are no magic numbers. But recall that as n increases, several things happen. First, by the Central Limit Theorem the sampling distribution for $\overline{X}$ becomes more and more like a normal distribution. Second, the t-distribution also becomes more and more like a normal distribution. Third, the sample variance is more and more likely to be quite close to the population variance (because the sample variance is an unbiased and consistent estimator).

Some textbooks suggest simply using the normal distribution (z-procedures) for large samples. This is primarily because this was simpler before the easy availability of computers that can calculate p-values and critical values for t-distributions with arbitrary degrees of freedom. We advocate using the t-distribution whenever the population variance is unknown and estimated by the sample variance. The difference between these two approaches will be small when n is large, but using the t-distributions is more conservative (produces slightly wider confidence intervals and slightly larger p-values) and makes intuitive sense.

These rules of thumb beg an important question: How do we know if the population is approximately normal, roughly symmetric and unimodal, moderately or strongly skewed? Sometimes we have *a priori* information (or at least belief) about the population. This may be based on data from other situations believed to be very similar or it may be based on some theoretical considerations. For example, the Central Limit Theorem (especially in more general forms than the theorem we have proved) suggests that distributions that result from adding or averaging many small contributions should be at least approximately normally distributed. This is one reason why the normal distribution arises so often in nature.

In other situations, we have to base our judgment primarily on our sample data. In this case we look at our sample data for evidence of normality, symmetry, unimodality, and outliers.

(1) **Small Samples: You Gotta Have Faith.**
It is very difficult to judge the normality of a population from a small sample. So for small samples, we must have *a priori* justification for using the *t*-procedures. Barring this, we need to look for ways to increase the sample size or choose a procedure that does not make a normality assumption.

(2) **Medium and Large Samples: Look Before You Leap.**
Always look at your sample data. As sample sizes increase, the sample will give a better and better idea about the population distribution while at the same time it will matter less and less just what the population distribution is like.

With a sample of modest size, evidence of skew or the presence of outliers should be cause for concern. Outliers are problematic for two reasons. First, small samples from distributions with approximately normal distributions rarely have outliers, so an outlier is some evidence that the population may be skewed or have heavy tails. Second, the presence of an outlier has a large effect on the sample mean and sample variance which are used in our calculations. It may be that a single value from our sample is driving the p-value or estimate.

Outliers may also be an indication of errors in the data. If it can be established that the outlier is due to some error (data entry or measurement error, for example), then it can be removed from the data set (or corrected, if possible). Of course, we should also look at the non-outliers to see if there are such problems and removing them as well if we detect errors. The removal of such values from a data set is called **data cleaning**. In some situations, very elaborate methods are used for data cleaning in an attempt to begin the analysis with data that are as accurate as possible.

Sometimes outliers are the interesting story in the data. Their presence in situations where they were not expected may be an indication that there is some part of the situation that is not yet fully understood. Perhaps there is a small portion of the population that is very different from the majority. It may be interesting to study this subpopulation on its own. In any case, *it is not acceptable to remove outliers simply because they are outliers.* Outliers should be investigated, not ignored.

4.8. Confidence Intervals for a Proportion

4.8.1. A Simple But Imprecise Method

We can use the fact that the binomial distribution is approximately normal to produce the following simple confidence interval for a proportion π:

$$\underbrace{\hat{\pi}}_{\text{estimate}} \quad \pm \quad \underbrace{z_*}_{\text{crit. val.}} \quad \underbrace{\sqrt{\frac{\hat{\pi}(1 - \hat{\pi})}{n}}}_{\text{est. } SE} . \tag{4.9}$$

This confidence interval is called the **Wald interval**. As it turns out, confidence intervals formed this way are not very good. The nominal and actual coverage rates (the percentage of intervals that contain the parameter being estimated) of these confidence intervals may not be as close as we would like, especially for small and moderate sample sizes. That is, it is not necessarily the case that 95% of the 95% confidence intervals formed this way contain π. The problem is that the estimated standard error used is a poor estimator of the actual standard error unless n is large.

4.8.2. A Better Approach

We could derive a $1 - \alpha$ confidence interval in a different way by asking the question:

> What values of π_0 would we not reject if we tested the null hypothesis $\pi = \pi_0$ (with a two-sided alternative)?

These values are plausible values for π, since we would not reject them based on our sample. A value of π_0 is plausible if

$$z = \frac{|\hat{\pi} - \pi_0|}{\sqrt{\frac{\pi_0(1 - \pi_0)}{n}}} \le z_{\alpha/2} . \tag{4.10}$$

The equation corresponding to (4.10) is quadratic in π_0, so we can solve for π_0 using the quadratic formula. The resulting solutions are

$$\pi_0 = \frac{\hat{\pi} + \frac{z_*^2}{2n} \pm z_* \sqrt{\frac{\hat{\pi}(1 - \hat{\pi})}{n} + \frac{z_*^2}{4n^2}}}{1 + z_*^2/n} . \tag{4.11}$$

This interval between these two solutions is known as the **score interval**. Notice that when n is very large, some terms have negligible magnitude, and this is very close to our easier formula (4.9). Furthermore, for a 95% confidence interval, $z_* \approx 2$, and (4.11) is very nearly

$$\tilde{\pi} \quad \pm \quad z_* \sqrt{\frac{\tilde{\pi}(1 - \tilde{\pi})}{n + 4}} , \tag{4.12}$$

where $\tilde{\pi} = \frac{x+2}{n+4}$. (See Exercise 4.46.)

This means we would get nearly the same result if we added two successes and two failures to our data and then calculated the Wald interval from the modified

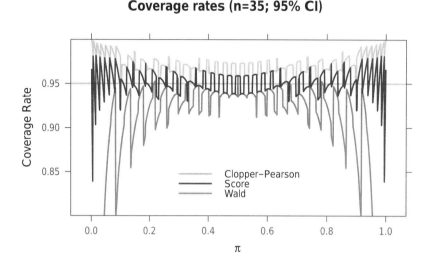

Figure 4.17. A comparison of coverage rates for three confidence interval methods.

data. This is sometimes called the **plus four confidence interval**, or the **Wilson confidence interval**, since Wilson observed the potential usefulness of this adjustment to the Wald interval in the 1920s.

Relatively recently, the plus four interval has gained popularity, and it now appears in a number of elementary statistics books. It has been shown [**AC98**] that the coverage rates for this confidence interval are very close to 95% even for quite small sample sizes, and since it is no more difficult to use than (4.9), it is recommended that (4.12) or (4.11) be used rather than (4.9) for all sample sizes when the confidence level is near 95% (say between 90% and 99%). For other confidence levels we must make adjustments to (4.12) or use (4.11).

Figure 4.17 shows a graphical comparison of coverage rates of three confidence intervals for π. In addition to the Wald and score intervals, it also displays the coverage rate of **Clopper-Pearson intervals**. These intervals are designed to ensure that the actual coverage rate is always at least the nominal confidence level. Clopper-Pearson intervals are computed by inverting the p-values from the binomial test and ensuring that the 1-sided p-values are at least 0.025 (for a 95% confidence interval) in each direction.

The take-away message from Figure 4.17 is that the score intervals seem to be a nice compromise. Note that it is not possible to give a deterministic confidence interval with exactly a 95% confidence rate in all situations. This is because the underlying sampling distribution is binomial, so for many values of π and n it is not possible to select tails of the distribution that sum to exactly 5%. Exercise 4.47 asks you to explain why all three methods lead to plots that are very jagged.

A more detailed comparison of various methods of calculating confidence intervals for a proportion can be found in [**AC98**].

4.9. Paired Tests

The paired t-test

One common use of the t-procedures is in a situation where we have two quantitative measurements for each unit of investigation and we want to compare the two measurements. These might be before and after measurements, or measurements under different experimental conditions, or measurements of spouses (each couple is a unit with one measurement for each spouse), etc.

This situation can be handled by first forming the difference (or some other combination of the two measurements) and then using a t-test or confidence interval to analyze the new variable. Aside from the initial transformation of the data, there is nothing new to the analysis. Paired tests are important from the perspective of the design and power of the study, however.

Example 4.9.1. Some researchers at the Ohio State University did a study to see if a football filled with helium travels farther than one filled with air when punted. One novice punter kicked 39 pairs of balls. One of each pair was filled with helium, the other with air. The punter was not aware of which balled was filled with which gas. The order was determined randomly with an equal probability of each type of football being punted first. Pairing was used to avoid the possible effects of learning (punts become better over time) or fatigue (punts become worse over time). Had the punter been given all of one preparation first followed by 39 of the other footballs, then any observed effects would have been confounded with the time (early or late) of punting.

```
                                                          helium-football
> football <- heliumFootballs      # give it a shorter name
> head(football,3)
  Trial Air Helium
1     1  25     25
2     2  23     16
3     3  18     25
>
> football$diff <- football$Helium - football$Air
> t.test(football$diff)

        One Sample t-test

data:  football$diff
t = 0.4198, df = 38, p-value = 0.677
alternative hypothesis: true mean is not equal to 0
95 percent confidence interval:
 -1.7643  2.6874
sample estimates:
mean of x
  0.46154
```

We do not have evidence suggesting that the gas used to fill the football makes any difference in the distance punted.   ◁

The sign test

In the previous example we did not discuss whether using a t-test is appropriate. Since our sample is reasonably large, the t-test should work for all but a very highly skewed distribution, so we are safe in using the t-distribution here. It is important to note that when we do a paired test, it is the distribution *of differences* in the population that is our primary concern. Thus, for small samples, we must be convinced that these differences are distributed approximately normally. The distributions of the two measurements separately do not matter.

Should we be concerned about the use of a t-distribution, we have another option. Instead of computing quantitative differences, we could consider only which of the two values is larger. If there is really no difference in the distribution of the two variables, then which is larger should be the same thing as the toss of a fair coin. That is, under the null hypothesis, either type of ball should be equally likely to travel farther. So the sampling distribution of the number of helium-filled balls traveling farther is $\mathsf{Binom}(n, 0.5)$, where n is the sample size (number of pairs of footballs kicked). In other words, we are conducting a test of

$$H_0 : \pi = 0.5$$

using a binomial sampling distribution. We can do this in R using `binom.test()` (or `prop.test()`). This idea leads to a test known as the **sign test**, because it looks only at the sign of the difference.

Example 4.9.2. Returning to our helium-filled footballs, we can do a sign test by comparing the distances for air- and helium-filled footballs to see which is larger.

> helium-football-sign

```
> n <- nrow(football)
> x <- sum(football$Helium > football$Air)
> binom.test(x,n)

        Exact binomial test

data:  x and n
number of successes = 20, number of trials = 39, p-value = 1
alternative hypothesis: true probability of success is not equal to 0.5
95 percent confidence interval:
 0.34780 0.67582
sample estimates:
probability of success
               0.51282
```

In this particular case, we see that there is nearly a 50-50 split between pairs of kicks where the helium-filled ball traveled farther and pairs of kicks where the air-filled ball traveled farther. ◁

4.10. Developing New Tests

In this chapter we have focused on methods of estimation and testing for two parameters: a population mean (especially of a normal distribution) and a population proportion. These methods relied upon knowing two things:

- a good point estimator for the parameter in question and
- the (approximate) sampling distribution for that estimator.

The methods we have used can be generalized for estimation and testing of other parameters, and we will see many examples of this in the chapters to come.

Along the way, we will encounter two important alternatives to the method of moments. **Maximum likelihood estimators** are the most frequently used estimators in statistical practice, primarily because there is an accompanying theory that enables us to derive (approximate) sampling distributions for maximum likelihood estimators in many situations. We will discuss maximum likelihood estimators in Chapter 5. For a special class of models called linear models, **least squares estimators** are often used. Linear models and least squares estimators will be introduced in Chapter 6.

We end this chapter with some methods that rely heavily upon computers. Although these methods were known to early statisticians like Fisher, they were largely unused until computers with sufficient power to perform the analyses became readily available. Now they form an important category of statistical analysis.

4.10.1. Empirical p-Values

In the hypothesis tests we have done so far, computing the p-value has required that we know the sampling distribution of our test statistic. But what if we don't know the sampling distribution? In some situations, we can use simulation to compute an empirical p-value.

Example 4.10.1. The owners of a residence located along a golf course collected the first 500 golf balls that landed on their property. Most golf balls are labeled with the make of the golf ball and a number, for example "Nike 1" or "Titleist 3". The numbers are typically between 1 and 4, and the owners of the residence wondered if these numbers are equally likely (at least among golf balls used by golfers of poor enough quality that they lose them in the yards of the residences along the fairway).

Of the 500 golf balls collected, 486 of them bore a number between 1 and 4. The results are tabulated below:

number on ball	1	2	3	4
count	137	138	107	104

Q. What should the owners conclude?

A. We will follow our usual four-step hypothesis test procedure to answer this question.

(1) State the null and alternative hypotheses.

- The null hypothesis is that all of the numbers are equally likely. We could express this as

$$H_0 : \pi_1 = \pi_2 = \pi_3 = \pi_4$$

  where π_i is the true proportion of golf balls (in our population of interest) bearing the number i. Since these must sum to 1, this is equivalent to

$$H_0 : \pi_1 = \tfrac{1}{4}, \pi_2 = \tfrac{1}{4}, \pi_3 = \tfrac{1}{4}, \text{ and } \pi_4 = \tfrac{1}{4} \ .$$

- The alternative is that there is some other distribution (in the population).

(2) Calculate a test statistic.

We need to devise a test statistic that will measure how well the data match the distribution of the null hypothesis. We will use better test statistics in Chapter 5, but for now consider the statistics $R = \text{maxcount} - \text{mincount}$. When R is small, all of the counts are nearly the same, as we would expect under the null hypothesis. A large value of R gives us evidence against the null hypothesis.

(3) Determine the p-value.

In our data, the value of r is $138 - 104 = 34$, so we are looking for

$$\text{p-value} = P(R \geq 34)$$

under the assumption that the null hypothesis is true. But to answer this question we need to know something about the sampling distribution of R (when the null hypothesis is true).

We will not develop a theoretical distribution for R to answer this question. Instead we will calculate an empirical p-value estimate by simulation. We will simulate many samples of 486 golf balls under the assumption that each number is equally likely and determine the proportion of these that have a range statistic at least as large as 34.

To sample the golf balls under the assumption that each number is equally likely, we use the R function `rmultinom()` which we use for the moment without comment. (See Section 5.1.3 for further information.) The `statTally()` function in the `fastR` package tabulates the test statistic on our random data and produces a graphical and numerical summary.

golfballs-range

```
> stat <- function(x) { diff(range(x)) }
> plot <- statTally(golfballs,rgolfballs,stat,
+          xlab="test statistic (range)")
Test Stat function:

function (x)
{
    diff(range(x))
}


Test Stat applied to sample data = 34


Test Stat applied to random data:
```

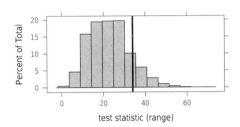

Figure 4.18. Empirical distribution of the range statistic for testing goodness of fit in the golf ball data set.

```
50% 90% 95% 99%
 22  36  40  49
        Of the random samples
                8615 ( 86.15 % ) had test stats < 34
                 203 (  2.03 % ) had test stats = 34
                1182 ( 11.82 % ) had test stats > 34
```

From this it appears that

$$\text{p-value} = P(R \geq 34) > 0.10 .$$

(See also Figure 4.18.)

(4) Interpret the p-value and draw a conclusion.

From the output above we see that our test statistic ($r = 34$) is on the large side but not wildly unusual. This analysis is not enough to cause us to reject the null hypothesis.   ◁

The empirical approach used in the preceding example could be used in many situations. All we need is

- some reasonable test statistic and

- a means for simulating data under the assumption that the null hypothesis is true.

There are advantages, however, to taking a more systematic approach (as we will do in subsequent chapters).

(1) Although any reasonable test statistic can give us our desired type I error rates and legitimate p-values, some test statistics have much greater power (i.e., lower type II error rates). The most powerful test may depend on the types of alternatives that we are most interested in detecting.

The code in Example 4.10.1 can be used to explore other possible test statistics for the golf ball situation described there and can be modified for use in other situations where it is easy to sample from the null distribution. Exercise 4.56 asks you to repeat the i simulations above with other test statistics.

(2) If we use a test statistic that has a known (approximate) sampling distribution, we can avoid using simulation to calculate p-values, and it can be easier to

understand the power of the test. Fortunately, there are general methods for producing test statistics that in many situations have desirable theoretical properties, including known asymptotic sampling distributions.

(3) In Example 4.10.1 it is clear how to do the simulation. In other situations, it may be less clear how to generate random data under the null hypothesis. One solution to this problem that works in a wide variety of situations is a **permutation test**. See Example 4.10.2.

4.10.2. Permutation Tests

Example 4.10.2. Recall from Example 2.7.4 Fisher's data from a twin study. The data are repeated here in tabular form:

	Not Convicted	Convicted
Dizygotic	15	2
Monozygotic	3	10

Suppose we were unaware of the hypergeometric distribution but wanted to test the null hypothesis that there is no association between conviction of the second twin and whether the twins are identical (monozygotic) or not (dizygotic). In other words the probability of conviction should be the same for both types of twins.

Now consider the raw data, which look something like the following:

Twin Type	Conviction
Di	No
$\vdots$	$\vdots$
Di	Yes
$\vdots$	$\vdots$
Mono	No
$\vdots$	$\vdots$
Mono	Yes

The key idea to permutation testing is this:

> *If the null hypothesis is true (i.e., there is no association between the two columns above), then if we shuffle the values in one column (conviction, say), we will be sampling from the same (null) distribution.*

We can then compute a p-value in one of two ways:

- We could generate *all* possible reshufflings of the conviction column and compute a test statistic for each.

- We could simulate many random shufflings and compute an empirical p-value instead.

It is easy to get R to perform the second of these. (There are ways to systematically generate all of the permutations, too, but remember there are 30! of them in this case. More on this in a moment.)

fisher-twins-perm

```
>
> numSims <- 20000
> ft <- data.frame(
+        twin=rep(c("Di","Mono"),times=c(17,13)),
+        conviction=rep(c("No","Yes","No","Yes"),times=c(15,2,3,10))
+ )
# check to see that table matches
> xtabs(~twin+conviction,ft);
      conviction
twin   No Yes
  Di   15   2
  Mono  3  10
# test statistic is value in top left cell
> xtabs(~twin+conviction,ft)[1,1];
[1] 15
>
# simulated data sets
> testStats <- replicate(numSims, {
+      sft <- ft;
+      sft$conviction <- sample(ft$conviction);
+      xtabs(~twin+conviction,sft)[1,1];
+      });
>
# for p-value
> table(testStats);
testStats
    5    6    7    8    9   10   11   12   13   14   15
    1   29  368 1649 3970 5789 4989 2432  659  106    8
# tail probabilities
> sum(testStats >= 15)/ numSims;
[1] 4e-04
> sum(testStats <= 5)/ numSims;
[1] 5e-05
# 2-sided p-value
> sum(testStats >= 15 | testStats <= 5)/ numSims ;
[1] 0.00045
```

Now let's compare the results above to the results from `fisher.test()`.

fisher-twins-exact

```
> fisher.test(xtabs(~twin+conviction,ft));

        Fisher's Exact Test for Count Data

data:  xtabs(~twin + conviction, ft)
p-value = 0.0005367
alternative hypothesis: true odds ratio is not equal to 1
95 percent confidence interval:
   2.7534 300.6828
```

```
sample estimates:
odds ratio
   21.305
```

The p-values don't match exactly, but since we used only 20,000 random permutations, we shouldn't consider that to be an exact value. Let's calculate a confidence interval for our permutation p-value.

```
> binom.test( sum(testStats >= 15 | testStats <= 5), numSims ) ;    fisher-twins-ci

        Exact binomial test

data:  sum(testStats >= 15 | testStats <= 5) and numSims
number of successes = 9, number of trials = 20000, p-value <
2.2e-16
alternative hypothesis: true probability of success is not equal to 0.5
95 percent confidence interval:
 0.00020579 0.00085407
sample estimates:
probability of success
              0.00045
```

From this we see that the p-value from `fisher.test()` is within the 95% confidence interval that results from our 20,000 permutations.

In fact, Fisher's exact test is equivalent to determining the probabilities based on all permutations being equally likely. Each shuffling of the `convicted` column corresponds to a 2 × 2 table, and the hypergeometric distribution helps us count how permutations give rise to each table. In this sense, Fisher's exact test *is* a permutation test – but one where the distribution of permutation p-values can be calculated exactly (without generating the permutations) rather than estimated empirically by generating some permutations at random.   ◁

4.10.3. Using Permutations to Test for Independence

How could we use permutations to test for independence? Suppose we have a sample with the values of two quantitative variables recorded for each observational unit. We would like to know if these values are independent.

Example 4.10.3.

Q. Within a given species of iris, one could imagine that sepal length and petal length are positively correlated (larger plants have larger sepal lengths and larger petal lengths). Do Fisher's iris data support this conjecture?

A. First we need a test statistic. One reasonable choice is the sum of the pairwise products. Recall from our discussion of covariance that

$$\sum_i (x_i - \bar{x})(y_i - \bar{y}) = \sum_i x_i y_i - \sum_i x_i \bar{y} - \sum_i \bar{x} y_i + \sum_i \bar{x}\,\bar{y} = \left(\sum_i x_i y_i\right) - n\bar{x}\,\bar{y}\,.$$

From this we learn several things:

- Either of the test statistics

$$S = \sum x_i y_i \quad \text{or}$$
$$\tilde{S} = \sum (x_i - \overline{x})(y_i - \overline{y}) = S - n\overline{x}\,\overline{y}$$

  would make reasonable test statistics.

- If we are sampling from a population where X and Y are independent, $E(X_i) = \mu_X$ and $E(Y_i) = \mu_Y$, then $E(X_i Y_i) = E(X_i)E(Y_i) = \mu_X \mu_Y$ and $E(\overline{X}\,\overline{Y}) = E(\overline{X})E(\overline{Y}) = \mu_X \mu_Y$, so

$$E(S) = n\mu_X \mu_Y \ , \quad \text{and}$$
$$E(\tilde{S}) = 0 \ .$$

- Larger values of the test statistics indicate positively correlated variables; smaller values indicate negatively correlated variables.

- This is really a test that the variables are uncorrelated rather than that they are independent. It does not have power to detect uncorrelated variables that are nevertheless not independent.

Now we can use R to calculate a permutation p-value for us.

```
                                                              iris-perm
> data(iris); numSims <-10000;
> setosa <- iris[iris$Species == "setosa",];
> corStat <- function(x,y) { sum( x * y ) - length(x) * mean(x) * mean(y) };
> testStat <- with(setosa, corStat(Sepal.Length,Petal.Length)); testStat;
[1] 0.8014
> simResults <- with(setosa, replicate(numSims,
+               corStat(Sepal.Length,sample(Petal.Length))));
# 1-sided p-value
> sum( simResults >= testStat ) / numSims;
[1] 0.0315
# a reasonable 2-sided p-value
> sum( abs(simResults) >= abs(testStat) ) / numSims;
[1] 0.0628
```

The p-value is near the edge of significance, but we don't have enough data to conclude confidently that there is a correlation between sepal and petal lengths. The evidence is much stronger for the other two varieties, however. (What sort of plot might have let us suspect this before we even ran the tests?)

```
                                                          iris-perm-versi
> numSims <-10000;
> versi <- iris[iris$Species == "versicolor",];
> corStat <- function(x,y) { sum( x * y ) - length(x) * mean(x) * mean(y) };
> testStat <- with(versi, corStat(Sepal.Length,Petal.Length)); testStat;
[1] 8.962
> simResults <- with(versi, replicate(numSims,
+               corStat(Sepal.Length,sample(Petal.Length))));
# 1-sided p-value
> sum( simResults >= testStat ) / numSims;
[1] 0
# a reasonable 2-sided p-value
```

```
> sum( abs(simResults) >= abs(testStat) ) / numSims;
[1] 0
```

Note that the p-value of 0 means that none of our 10,000 permutations produced a test statistic as extreme as the one in the unpermuted data. We would need to do more permutations to determine more accurately just how small the p-value is. ◁

4.10.4. A Permutation Test for One Variable

What if we only have one variable in our data set? Can permutations help us here? The answer is yes, and once again, we have already seen this test. The sign test can be thought of as a permutation test where we permute positive and negative labels.

To test a hypothesis of the form

$$H_0 : \mu = \mu_0 \,,$$

we first compute all of the differences $x - \mu_0$. Assuming the population distribution is symmetric about its mean,[3] if μ_0 is the true mean, then the number of positive differences should be $\mathsf{Binom}(n, 0.5)$. There is no need to actually permute the + and − signs since we can use the binomial distribution to determine the p-value exactly, just as we did in the example of helium- and air-filled footballs.

4.11. Summary

4.11.1. Estimating Parameters

Using data to estimate the parameters of a statistical model is called **fitting** the model. A parameter that is estimated in this way is called an **estimand**. An **estimate** (or **point estimate**) is a number computed for this purpose from a particular data set. If our data come from random sampling, then each random sample will yield an estimate, so these estimates form a random variable called an **estimator**. The distribution of an estimator is called the **sampling distribution** to distinguish it from two other distributions: the population distribution and the distribution of an individual sample.

The sampling distribution depends upon the estimator, the population sampled from, and the sampling method used. The **method of moments** is a general method for deriving estimators by solving a system of equations equating the moments of a distribution with the moments of the sample data. In the simplest case – when there is only one parameter in the model – method of moments estimates are obtained by equating the mean of the population with the sample mean. In subsequent chapters we will will study other estimation methods, including **likelihood estimators** and **least squares estimators**.

Sampling distributions are the foundation for **statistical inference**. Inference for the mean of a distribution is greatly simplified by the **Central Limit Theorem**

---

[3]Alternatively, we can think of this as a test for the median, in which case the symmetry assumption is no longer needed.

which says that no matter what the population distribution is, for large enough samples the sampling distribution for $\overline{X}$ is approximately normal. In fact,

$$\overline{X} \approx \mathsf{Norm}(\mu, \frac{\sigma}{\sqrt{n}}) ,$$

where μ and σ are the mean and standard deviation of the population. (The approximation is exact for **iid random sampling** from a normal population.) This implies that

$$Z = \frac{\overline{X} - \mu}{\sigma/\sqrt{n}} \approx \mathsf{Norm}(0, 1) ,$$

and from this we can derive hypothesis test and confidence interval procedures, provided σ is known.

Unfortunately, we do not usually know σ. A parameter that is not of primary interest but is required in this manner is called a **nuisance parameter**. William Gosset was the first to discover that the earlier practice of simply replacing σ with the sample standard deviation s is not accurate for small sample sizes. He was able to derive the correct sampling distribution for this statistic in the case of iid random sampling from a normal population:

$$T = \frac{\overline{X} - \mu}{\sigma/\sqrt{n}} \sim \mathsf{t}(n - 1) ,$$

where n is the sample size. This distribution is called Student's t-distribution and $n - 1$ is referred to as the **degrees of freedom**. The resulting t-procedures are quite robust and yield good approximations even for modest sample sizes provided the population is unimodal and not severely skewed.

Hypothesis tests are conducted by comparing a sample t-statistic to the $\mathsf{t}(n-1)$-distribution in the usual manner. An alternative to testing a particular hypothesis is to form a **confidence interval** for the parameter of interest. The confidence interval for the mean is

$$\overline{x} \pm t_* \frac{s}{\sqrt{n}} ,$$

where t_* is the appropriate **critical value** for the desired level of confidence. If the true parameter value of the population lies within a confidence interval, we say that the interval **covers** the parameter. Confidence intervals formed from random samples are random. Some of these random intervals will cover the parameter; others will not. The **coverage rate** is the probability that a random confidence interval covers the parameter. A procedure is called **conservative** if the coverage rate is higher than the **confidence level** (also called the **nominal coverage rate**) and **anti-conservative** if the coverage rate is lower than the confidence level.

A **paired t-test** is simply a t-test performed on a variable calculated from two other variables, most commonly by taking the difference between two observations for the same subject or the difference between observations of two **matched** subjects. Paired tests can be used, for example, to compare pre- and post-treatment conditions to see if there is a **treatment effect**. Using a paired test helps reduce the effect of the variability from subject to subject prior to treatment.

The mean is not the only parameter that can be estimated. The method of moments estimator for the variance is

$$V = \frac{\sum (X - \overline{X})^2}{n} \, .$$

Unfortunately, this estimator is **biased**: $E(V) \neq \sigma^2$. The bias is easily calculated, however, and leads us to the following **unbiased estimator** for variance (in the case of an iid random sample from a normal population):

$$S^2 = \frac{\sum (X - \overline{X})^2}{n - 1} \, .$$

This explains the otherwise mysterious denominator in the definition of sample variance. For an iid random sample from a normal population,

$$\frac{(n-1)S^2}{\sigma^2} \sim \mathsf{Chisq}(n - 1) \, .$$

From this we can derive hypothesis tests and confidence intervals for σ^2, but these procedures are less robust than those based on the t-distribution of the sample mean.

When sampling distributions are not known, it is possible to estimate them using computer simulations to generate an empirical distribution of test statistics. Although the idea of **permutation testing** has been around for a long time, it has gained new popularity recently due to the availability of computers fast enough to simulate large numbers of sample statistics. Some permutation tests are equivalent to **exact tests** (such as Fisher's exact test) based on known distributions. In this case, we can compute the p-value without the need for many permutations.

4.11.2. Distributions Based on the Standard Normal Distribution

In this chapter we introduced four new families of random variables. Each of them is derived from the standard normal distribution. Here we summarize their most important properties.

(1) Z^2 where $Z \sim \mathsf{Norm}(0, 1)$
- $Z^2 \sim \mathsf{Gamma}(\alpha = \frac{1}{2}, \lambda = \frac{1}{2})$.
- $E(Z^2) = \mathrm{Var}(Z) + E(Z)^2 = 1 + 0 = 1$.

  This also follows from gamma distribution facts.
- $\mathrm{Var}(Z^2) = 2$.

(2) $\mathsf{Chisq}(n)$: sum of squares of independent standard normal variables
- $X^2 = \displaystyle\sum_{i=1}^{n} Z_i^2 \sim \mathsf{Chisq}(n) = \mathsf{Gamma}(\alpha = \frac{n}{2}, \lambda = \frac{1}{2})$.
- $E(X^2) = n$ and $\mathrm{Var}(X^2) = 2n$.
- $\dfrac{(n-1)S^2}{\sigma^2} \sim \mathsf{Chisq}(n-1)$ for a sample $\boldsymbol{X} \stackrel{\text{iid}}{\sim} \mathsf{Norm}(\mu, \sigma)$ of size n.

(3) $\mathsf{F}(m,n) = \dfrac{\mathsf{Chisq}(m)/m}{\mathsf{Chisq}(n)/n}$ (numerator and denominator independent)

- $\mathrm{E}(F) = \frac{n}{n-2}$ when $F \sim \mathsf{F}(m,n)$ and $n > 2$.

(4) $\mathsf{t}(n) = \dfrac{\mathsf{Norm}(0,1)}{\sqrt{\mathsf{Chisq}(n)/n}}$ (numerator and denominator independent)

- $T^2 \sim \mathsf{F}(1,n)$ if $T \sim \mathsf{t}(n)$.

- If $T \sim \mathsf{t}(n)$, then $\mathrm{E}(T) = 0$ for $n > 1$ and $\mathrm{Var}(T) = \frac{n}{n-2}$ if $n > 2$.

- $\dfrac{\overline{X} - \mu}{S/\sqrt{n}} \sim \mathsf{t}(n-1)$ for an iid random sample $\boldsymbol{X} \overset{\text{iid}}{\sim} \mathsf{Norm}(\mu, \sigma)$ of size n.

4.11.3. R Commands

Here is a table of important R commands introduced in this chapter. Usage details can be found in the examples and using the R help.

`uniroot(f,interval,...)`	Numerically approximate a solution to $f(x) = 0$ with x within the interval specified by `interval`.
`sample(x,size,replace=FALSE)`	Select a sample of size `size` from x.
`binom.test(x,n,p=0.50,...)`	Use binomial distributions to conduct a hypothesis test or construct a confidence interval for a proportion.
`prop.test(x,n,p=0.50,...)`	Use normal approximations to the binomial distributions to conduct a hypothesis test or construct a confidence interval for a proportion.
`t.test(x,...)`	t-tests and confidence intervals.
`replicate(n,expr,...)`	Evaluate expression `expr` n times.
`dt(x,df)`	Evaluate pdf for $\mathsf{t}(\texttt{df})$-distribution
`pt(q,df)`	Evaluate cdf for $\mathsf{t}(\texttt{df})$-distribution
`qt(p,df)`	Compute quantiles for $\mathsf{t}(\texttt{df})$-distribution
`rt(,df)`	Simulate n random draws from a $\mathsf{t}(\texttt{df})$-distribution.
`dchisq(x,df); pchisq(q,df); qchisq(p,df); rchisq(n,df)`	Similar to the functions above but for $\mathsf{Chisq}(\texttt{df})$-distributions.
`df(x,df1,df2); pf(q,df1,df2); qf(p,df1,df2); rf(n,df1,df2)`	Similar to the functions above but for $\mathsf{F}(\texttt{df1},\texttt{df2})$-distributions.

Exercises

4.1. Let $X = X_1, X_2, \ldots, X_n$ be an iid random sample with $X_i \sim \text{Binom}(1, \pi)$. (This is a model for sampling for a population proportion of some categorical variable.) Derive the method of moments estimate for π.

4.2. Suppose $X \overset{\text{iid}}{\sim} \text{Unif}(-\theta, \theta)$. Explain why the method of moments is not a useful method to estimate θ.

4.3. Let $X \sim \text{Unif}(0, \theta)$ and let $\hat{\theta}$ denote the method of moments estimator for θ. When $\hat{\theta} < \max(X)$, then we know that $\hat{\theta}$ is incorrect. We want to know the probability that $\hat{\theta} < \max X$. Estimate this probability using simulations for $n = 6$, 12, and 24.

4.4. Derive the method of moments estimator for π for a sample of size n from a $\text{NBinom}(s, \pi)$-distribution. (Treat s as known, as it would be in a typical situation where you would be collecting data by repeating the Bernoulli trials.)

4.5. Write an R function `moment(k,x,centered=(k>1))` that takes a numeric vector and returns the kth sample moment (either $\hat{\mu}_k$ or $\hat{\mu}'_k$ depending on the value of `centered`). Test your function by comparing its output to the output of `mean()` and `var()`.

How does your function handle missing data (`NA` values)?

4.6. This problem justifies our use of moments about the mean rather than moments about 0 when applying the method of moments.

a) Show that for any distribution, the moments satisfy

$$\mu_2 = \mu'_2 + \mu^2 .$$

b) Show that for any sample, the sample moments satisfy

$$\hat{\mu}_2 = \hat{\mu}'_2 + \hat{\mu}^2 .$$

c) Suppose we have a model with two parameters $\boldsymbol{\theta} = \langle \theta_1, \theta_2 \rangle$. Show that the method of moments using $\hat{\mu}_1$ and $\hat{\mu}_2$ yields the same estimates of $\boldsymbol{\theta}$ as the method of moments using $\hat{\mu}_1$ and $\hat{\mu}'_2$.

d) Generalize these results to higher moments.

4.7. The `miaa05` data set contains statistics on each of the 134 players in the MIAA 2005 Men's Basketball season. How well does a beta distribution fit these players' free throw shooting percentage?

a) Use the method of moments to estimate the two parameters of the beta distribution.

b) Use a quantile-quantile plot to assess the goodness of fit.

c) Are there any players you should remove from the data before performing this analogy? Decide on an elimination rule, apply it, and repeat the analysis. Do you like the results better?

4.8. The `miaa05` data set contains statistics on each of the 134 players in the MIAA 2005 Men's Basketball season. How well does a beta distribution fit these players' field goal shooting percentage?

a) Use the method of moments to estimate the two parameters of the beta distribution.

b) Use a quantile-quantile plot to assess the goodness of fit.

c) Are there any players you should remove from the data before performing this analogy? Decide on an elimination rule, apply it, and repeat the analysis. Do you like the results better?

4.9. A sample of 10 units were all tested until failure [**Kec94**]. The times-to-failure were 16, 34, 53, 75, 93, 120, 150, 191, 240, and 339 hours. Use the method of moments to fit an exponential and a gamma distribution and compare the results.

4.10. In this problem you are asked to find your own data set. There are numerous online sites that have data in various formats. Alternatively, you can collect your own data (or use data you have already collected in a different class).

a) Use the method of moments to fit the model to the data.

b) Make a quantile-quantile plot and report on the quality of the fit.

4.11. Is the estimator in Examples 4.2.1 and 4.2.2 unbiased?

4.12. You can regenerate an identical set of "random" data by using the `set.seed()` function in R. This is primarily a tool for debugging purposes, but it can be used whenever we want to repeat a randomized simulation exactly. For example, the plot in Figure 4.7 can be regenerated using

> lln-cauchy-seed

```
> set.seed(123)
> x <- rcauchy(10000)
> runningMean <- cumsum(x) / 1:length(x)
> cauchyPlot <- xyplot(runningMean~1:10000,
+     ylab="running mean",xlab="n", type="l");
```

Use `set.seed()` to generate several of these plots and record the seeds you used to produce three or four of the most unusual plots you find.

4.13. Let $\boldsymbol{w} = \langle w_1, w_2, \ldots, w_n \rangle$ be a vector of fixed numbers (weights). For a sample $\boldsymbol{X} = \langle X_1, X_2, \ldots X_n \rangle$, let the **weighted sum** be defined by

$$\overline{X}_{\boldsymbol{w}} = \sum_{i=1}^{n} w_i X_i \ .$$

Answer the following assuming $\boldsymbol{X}$ is obtained by iid sampling.

a) Find conditions on $\boldsymbol{w}$ that make $\overline{X}_{\boldsymbol{w}}$ an unbiased estimator of μ, the population mean. Under these conditions we will call $\overline{X}_{\boldsymbol{w}}$ a **weighted mean** or **weighted average**.

b) Determine the variance of a weighted average (in terms of the weights $\boldsymbol{w}$).

c) Show that the variance of a weighted average is smallest when the weights are all equal. [Hint: First handle the case where $n = 2$. Then use that result to show the general case.]

4.14. Suppose we use iid sampling to obtain a sample of size 16 from a normal population with standard deviation $\sigma = 10$. What is the probability that our sample mean will be within 2 of the population mean?

4.15. SENSITIVE QUESTIONS. It can be difficult to rely on human responses to certain types of questions like "Have you ever used illegal drugs?" or "Have you ever cheated on an exam?" because the subjects may be concerned about the surveyor learning their response.

The method of **randomized response** was developed to deal with such questions. In this method, randomness (spinning a spinner, drawing a ball out of an urn, etc.) is used to decide which of two questions the subject responds to. The surveyor does not know which question it is. For example, the two questions might be

(A) True or False: I have never cheated on an exam.

(B) True or False: I have cheated on an exam.

The subject simply replies with true or false; the surveyor records the response without knowing which version ((A) or (B)) the subject is responding to. The hope is that this will make the subjects more truthful. Of course, this only helps if the information is still usable.

Let θ be the true proportion of people who would answer "true" to version (A) (and hence "false" to version (B)). The parameter θ is what we would really like to know. Let π be the probability (determined in advance by the design of the study) that the subject receives version (A). The probability π is known to the researchers (in fact, they choose it). Finally, let x and n be the number of "true" responses and the total number of subjects in the sample.

a) Let ρ be the probability that a randomly selected person will answer "true" (to whichever version they happen to get). Show that $\rho = (2\pi - 1)\theta + (1 - \pi)$.

b) If ρ and π are known, how can we determine the value of θ?

c) Let $\hat{\rho} = \frac{X}{n}$ be the random variable representing the proportion of subjects who say "true" in a random sample of size n. Show that $\hat{\rho}$ is an unbiased estimator for ρ. Does it matter whether the sampling is done with (iid) or without (SRS) replacement?

d) Find an unbiased estimator $\hat{\theta}$ for θ and prove that it is unbiased.

e) Assuming an iid sample, show that

$$\mathrm{Var}(\hat{\rho}) = \frac{\rho(1 - \rho)}{n} .$$

f) Assuming an iid sample, find an expression for $\mathrm{Var}(\hat{\theta})$. This will be approximately correct for simple random samples as well.

g) Are these estimators consistent?

4.16. Consider a very small population with only five individuals. The values of a variable for this population are $1, 2, 4, 4, 9$.

  a) Determine the mean and variance of the sampling distribution for the sample mean of samples of size 2 by calculating the means of all $\binom{5}{2}$ possible samples of a simple random sample.

  b) Compare your results with Corollary 4.3.3.

  c) Repeat this for an iid sample.

4.17. Suppose $X_n \overset{D}{\to} X$. Show that $aX_n + b \overset{D}{\to} aX + b$.

  [Hint: Start by expressing the cdf for $aY + b$ in terms of the cdf for Y.]

4.18. Formalize and prove statement (4.7). Why is it not quite right to say that $\overline{X}_n \overset{D}{\to} \mathsf{Norm}(\mu, \sigma/\sqrt{n})$ as $n \to \infty$?

4.19. Let $X \sim \mathsf{Binom}(n, \pi)$. Then by the Central Limit Theorem, if n is large enough, $X \approx \mathsf{Norm}\left(n\pi, \sqrt{n\pi(1-\pi)}\right)$. Show that if $\pi < 0.5$ and $n\pi \geq 10$, then $3\sqrt{\pi(1-\pi)/n} < \pi$.

  This explains our rule of thumb for the normal approximation to the binomial distributions. The rule of thumb says the approximation is good if the central 99.7% of the normal distribution stays in the interval $[0, n]$.

4.20. A local newspaper conducts a phone survey to judge public opinion about a referendum. The referendum passes if a majority of voters vote yes. Of the 950 respondents to the survey, 450 say they plan to vote yes. Is this sufficient information for the newspaper to predict the outcome of the referendum vote, or should they say it is "too close to call"?

  Conduct your test twice, once using the binomial distribution directly and again using the normal approximation to the binomial distribution. How do the results compare?

4.21. You are using a sensitive scale to weigh a rock sample. Measurements with this scale are known to be normal with a standard deviation of $\sigma = 0.02$ grams. You decide to weigh the sample 3 times and record the mean of the three weighings. What is the probability that the weight you record is within 0.02 grams of the actual mass of the rock? How much would the probability change if you increased the number of weighings to 4?

4.22. Write a function `z.test()` that performs the z test of the hypothesis $H_0 : \mu = \mu_0$. Begin your function with

```
z.test <- function (x, alternative = c("two.sided", "less", "greater"),
                                  mu = 0, sigma=1, conf.level = 0.95)
{
    DNAME <- deparse(substitute(x))     # record name of data coming in
    alternative <- match.arg(alternative)    # fancy argument matching

    # your code goes here
}
```

If you want to be especially fancy, end your function with a modification of

```
Z <- ??? ; names(Z) <- "z"
SIGMA <- sigma; names(SIGMA) <- "sigma"
MU <- mu; names(MU) <- "mean"
ESTIMATE <- ??? ; names(ESTIMATE) <- "sample mean"
CINT <- ???; attr(CINT, "conf.level") <- conf.level
PVAL <- ???;

structure( list( statistic = Z, parameter = SIGMA, p.value = PVAL,
    conf.int = CINT, estimate = ESTIMATE, null.value = MU,
    alternative = alternative, method = "Z test for a mean",
    data.name = DNAME),
    class = "htest")
```

filling in each ??? with the appropriate information for your test. This will return an object of class htest, and R will automatically know how to print a summary of your results that looks very much like the summary from binom.test(), fisher.test(), or t.test().

4.23. State and prove a limit theorem for Poisson random variables.

4.24. State and prove a limit theorem for gamma random variables.

4.25. Landon is doing some simulations to see how well the simulated coverage rate compares to the nominal 95% coverage rate in various situations. For each situation Landon simulates 10,000 samples and computes the coverage rate.

a) How high or low must the simulated coverage rate be for Landon to suspect that the true coverage rate is not 95%? Assume our usual threshold of significance, $\alpha = 0.05$.

b) Do we have evidence that any of the 95% confidence interval simulations performed in this chapter do not have coverage rates of 95%?

c) Landon realizes that his methods are only approximations. How high or low must the simulated coverage rate be for Landon to suspect that the true coverage rate is not between 94% and 96%?

d) Do we have evidence that any of the 95% confidence interval simulations performed in this chapter do not have coverage rates between 94% and 96%?

[Hint: What is Landon's null hypothesis?]

4.26. A 95% confidence interval for the mean weight gain of adult mice on a certain diet is $(11.2, 54.7)$ grams.

a) What was the mean weight gain of the mice in this study?

b) What can you say about the p-value of a hypothesis test with a null hypothesis that there is no weight gain (on average) for mice on this diet?

4.27. A statistical study reports a 95% confidence interval for the mean weight of a certain species of bird as 1545 g $\pm$ 182 g. Your biologist friend says, "Oh, I get it, 95% of these birds weigh between 1363 g and 1727 g."

How should you respond? (In particular your response should make it clear whether you think the biologist's answer is completely correct, mostly correct, partly correct, or completely wrong.)

4.28. Just as we had one-sided and two-sided p-values, we can define one-sided as well as two-sided confidence intervals. A one-sided confidence interval has the form (L, ∞) or $(-\infty, L)$ for some value L computed from the sample data. Just as for two-sided confidence intervals, we want the coverage rate to match the confidence level.

Suppose we have a sample of size $n = 25$ with mean $\bar{x} = 10$ drawn from a population with $\sigma = 3$. Calculate the two one-sided 95% confidence intervals.

4.29. Show that the sample variance of a simple random sample is a biased estimator of the population variance but that the bias is small in typical applications.

4.30. Show that in all interesting cases, if $\hat{\theta}^2$ is an unbiased estimator of θ^2, then $\hat{\theta}$ is *not* an unbiased estimator of θ. What are the uninteresting cases?

Note that this implies that the sample standard deviation is *not* an unbiased estimator of the population standard deviation.

4.31. Show that if μ is known, then for iid random sampling

$$\hat{\sigma}^2 = \frac{\sum(X_i - \mu)^2}{n}$$

is an unbiased estimator of σ^2.

4.32. Suppose you will gather two random samples of size n from the same normal population with unknown mean and known variance σ^2. If you use the first sample to form a 95% confidence interval of the form

$$\text{sample mean} \pm z_* \frac{\sigma}{\sqrt{n}} \ ,$$

what is the probability that the sample mean from the second sample lies within the confidence interval formed from the first sample?

Hint: Consider the distribution of $\overline{X} - \overline{Y}$, the difference between the two sample means.

4.33. Show that the variance of a $\mathsf{t}(n)$-random variable is equal to the mean of a $\mathsf{F}(1, n)$-random variable. (They are, in fact, both equal to $\frac{n}{n-2}$, but you do not have to prove that.)

4.34. This exercise provides an explicit basis for the decomposition of a sample vector.

a) Show that the following vectors are orthonormal (orthogonal and unit length):
- $\boldsymbol{u}_1 = \frac{1}{\sqrt{n}}\langle 1, 1, \ldots, 1 \rangle$,
- $\boldsymbol{u}_2 = \frac{1}{\sqrt{2}}\langle 1, -1, 0, \ldots, 0 \rangle$,
- $\boldsymbol{u}_3 = \frac{1}{\sqrt{6}}\langle 1, 1, -2, 0, \ldots, 0 \rangle$,
- $\boldsymbol{u}_i = \frac{1}{\sqrt{i(i-1)}} \langle \underbrace{1, 1, \ldots, 1}_{(i-1) \text{ 1's}}, 1-i, 0, \ldots, 0 \rangle$ for $i > 2$,

- $u_n = \frac{1}{\sqrt{n(n-1)}} \langle 1, 1, \ldots, 1, 1 - n \rangle.$

b) Show that for any x, $x \cdot u_1 = \bar{x}\sqrt{n}$.

c) Show that for any x, $(x \cdot u_1)u_1 = \bar{x} = \langle \bar{x}, \bar{x}, \ldots, \bar{x} \rangle$.

d) Let $x = \langle 3, 4, 4, 7, 7 \rangle$ and let $v = x - \bar{x}$. For $1 \le i \le 5$, compute

$$x \cdot u_i = |\operatorname{proj}(x \to u_i)| \qquad \text{and} \qquad v \cdot u_i = |\operatorname{proj}(v \to u_i)| .$$

What do you notice? Prove that this is true for any x.

4.35. Let $x = \langle 3, 4, 5, 8 \rangle$, let u_i be defined as in Exercise 4.34, and let

$$p_i = (x \cdot u_i)u_i = \operatorname{proj}(x \to u_i) .$$

a) Calculate the sample mean and sample variance for this data set.

b) Determine the projection vectors p_i and show that $x = \sum_i p_i$.

c) Determine the projection coefficients $l_i = |p_i|$ for each i.

d) Show that $\sum_{i=2}^{4} l_i^2 = 3s^2$.

4.36. Define the vectors w_i by

- $w_1 = u_1 = \frac{1}{2}\langle 1, 1, 1, 1 \rangle$,
- $w_2 = \frac{1}{2}\langle 1, 1, -1, -1 \rangle$,
- $w_3 = \frac{1}{2}\langle 1, -1, -1, 1 \rangle$,
- $w_4 = \frac{1}{2}\langle 1, -1, 1, -1 \rangle$.

a) Show that these vectors form an orthonormal basis of $\mathbb{R}^4$. That is, show that they each have unit length and are pairwise orthogonal.

b) Repeat parts b)–d) of the previous question using w_i in place of u_i. What do you notice?

4.37. Repeat the previous problem using

- $w_1 = u_1 = \frac{1}{2}\langle 1, 1, 1, 1 \rangle$,
- $w_2 = \frac{1}{2}\langle 1, 1, -1, -1 \rangle$,
- $w_3 = \frac{1}{\sqrt{2}}\langle 1, -1, 0, 0 \rangle$,
- $w_4 = \frac{1}{\sqrt{2}}\langle 0, 0, 1, -1 \rangle$.

4.38. Notice that in the preceding two problems, the sums of the coefficients of w_2, w_3, and w_4 are 0 in each case.

a) Why will this be the case no matter how we define $w_2, \ldots, w_n$?

b) What does this tell us about the distributions of the corresponding projection coefficients?

4.39. Compute 95% confidence intervals for sepal width for each of the species in the `iris` data set. Intuitively, what do these confidence intervals indicate about the different species? Explain.

4.40. Compute 95% confidence intervals for sepal length for each of the species in the `iris` data set. Intuitively, what do these confidence intervals indicate about the different species? Explain.

4.41. Compute 95% confidence intervals for the ratio of sepal length to sepal width for each of the species in the `iris` data set. Intuitively, what do these confidence intervals indicate about the different species? Explain.

4.42. Exercise 4.28 introduced one-sided confidence intervals for the population mean under the assumption that σ is known.

  a) Explain how to compute a one-sided confidence interval when σ is unknown.

  b) Demonstrate your method using the sepal lengths of versicolor irises using the `iris` data set.

  c) Check your work using the `alternative` argument to `t.test()`.

4.43. Suppose you want to estimate the mean shoe size of adults in Big City. You would like to have a confidence interval that is no wider than 0.5 shoe sizes. (The margin of error would be at most 0.25.) How large a sample must you get?

  a) This calculation will require that you make a guess about approximately what the standard deviation will be. What are the implications of guessing too high or too low? Should you guess on the high side or the low side?

  b) Should you include both men and women in your sample or just one or the other? Why?

  c) Suppose you estimate that the standard deviation of the population will be approximately 2. How large must your sample be to get the desired confidence interval?

4.44. The `chickwts` data set presents the results of an experiment in which chickens are fed six different feeds. If we assume that the chickens were assigned to the feed groups at random, then we can assume that the chickens can be thought of as coming from one population. For each feed, we can assume that the chickens fed that feed are a random sample of the (theoretical) population that would result from feeding all chickens that feed.

  a) For each of the six feeds, compute 95% confidence intervals for the mean weight of chickens fed that feed.

  b) From an examination of the six resulting confidence intervals, is there convincing evidence that some diets are better than others?

  c) Since you no doubt used the t-distribution to generate the confidence intervals in a), you might wonder whether that is appropriate. Are there any features in the data that suggest that this might not be appropriate?

4.45. The `miaa05` data set contains statistics on each of the 134 players in the MIAA 2005 Men's Basketball season. Choose 20 different random samples of size 15 from this data set. (The `sample()` function in R will automate the selection of samples; `replicate()` can be used to automate multiple samples.)

a) From each sample compute a 90% confidence interval for the mean PTSG (points per game) of MIAA players.

b) Of the 20 confidence intervals you computed in part a), how many actually did contain the true mean? (You can compute this since you know the population.)

c) How many of the 20 confidence intervals in part a) would you have expected (before you actually generated them) to contain the true mean?

d) In light of your answer in c), are you surprised by your answer in b)?

4.46. Show that (4.11) and (4.12) are nearly the same for 95% confidence intervals.

4.47. One immediately noticeable feature of Figure 4.17 is that the plots are very jagged. Explain why *any* method that constructs confidence intervals deterministically from the data will exhibit this effect. (There are methods that avoid this problem by inserting some randomness into the calculation of the confidence interval from the data. But this has the side effect that if you analyze the same data twice, you obtain different confidence intervals.)

4.48. In this problem, you will develop a hypothesis test for a new situation. Suppose that you believe the waiting time until you are served at the drive-thru window at the Big Burger Barn is a random variable with an exponential distribution but with unknown λ. The sign at the drive-thru window says that the average wait time (after placing the order until it is ready at the window) is 1 minute. You actually wait 2 minutes and 12 seconds. A friend in the car with you says that this is outrageous since it is more than twice as long as the average is supposed to be. He is sure that the claimed average must be incorrect.

a) What are the null and alternative hypotheses?

b) What is your test statistic and what distribution does it have if the null hypothesis is true?

c) Determine the p-value for your particular sample.

d) Write a sentence that explains clearly to your friend the meaning of that p-value. Your friend has not yet been fortunate enough to take a statistics course, so don't use statistical jargon.

e) What is the shortest wait time that would cause you to be suspicious of Big Burger Barn's claim? Explain.

f) How would things change if you decided to return 5 times and record the wait time each time? Do we have appropriate methods to deal with this situation?

4.49. The data set `cats` in the `MASS` package contains data on 144 cats. The variable `Bwt` gives the body weight of each cat, and `Sex` gives the sex of the cat. Compute 95% confidence intervals for the mean body weight of male and female cats.

4.50. The `endurance` data set contains data from a 1983 paper [**KM83**] testing the effects of vitamin C on grip strength and muscular endurance. This data set is also available from OzDASL (`http://www.statsci.org/data`), where the following description is given.

The effect of a single 600 mg dose of ascorbic acid versus a sugar placebo on the muscular endurance (as measured by repetitive grip strength trials) of fifteen male volunteers (19–23 years old) was evaluated. The study was conducted in a double-blind manner with crossover.

Three initial maximal contractions were performed for each subject, with the greatest value indicating maximal grip strength. Muscular endurance was measured by having the subjects squeeze the dynamometer, hold the contraction for three seconds, and repeat continuously until a value of 50% maximum grip strength was achieved for three consecutive contractions. Endurance time was defined as the number of repetitions required to go from maximum grip strength to the initial 50% value. Subjects were given frequent positive verbal encouragement in an effort to have them complete as many repetitions as possible.

a) Use a paired t-test to compare the two treatments. Is there a significant difference?

b) Now repeat the analysis, but first take the logarithm of each endurance measurement before doing the analysis

c) Since the difference of logarithms is the logarithm of the quotient, repeat the analysis using the quotient of the two measurements in place of the difference.

d) Let's try one more transformation. This time use the reciprocal of each endurance measurement.

e) Now analyze these data using the sign test.

f) We have several different analyses, each with a different p-value. How do we decide which is the "right" analysis?

4.51. Joe flips a coin 200 times and records 115 heads.

a) Give a 95% confidence interval for the true proportion of heads that this coin yields using Wald, score, and Wilson methods.

b) Does the confidence interval produced by `prop.test()` match any of these? If so, which one? (Try `prop.test()` with `correct` set to both `TRUE` and `FALSE`.)

Note that `?prop.test` does not specify just what method is being used but does reference two 1998 papers that compare multiple methods (seven in one paper, eleven in the other) for computing confidence intervals for a single proportion.

If you use `uniroot()`, be sure to consider the estimated precision when making your comparisons.

4.52. In a blind taste test, 45 coffee drinkers sampled fresh brewed coffee versus a gourmet instant coffee. When stating their preferences, 14 chose the instant, 26 chose the fresh brewed, and 5 gave no preference.

Note that in the helium vs. air football example there were two trials where no difference was measured. We ignored that detail there (and it doesn't really affect the overall conclusion in that case.)

a) In what situations will the way we deal with 'no difference' or 'no response' matter most? Explain.

b) What options can you think of for how to deal with this?

c) What should we do with those who had no preference here?

d) Test the claim that coffee drinkers tend to prefer fresh brewed coffee.

4.53. Can pleasant aromas help a student learn better? Hirsch and Johnston, of the Smell & Taste Treatment and Research Foundation, believe that the presence of a floral scent can improve a person's learning ability in certain situations. In their experiment, 22 people worked through a set of two pencil and paper mazes six times, three times while wearing a floral-scented mask and three times wearing an unscented mask. Individuals were randomly assigned to wear the floral mask on either their first three tries or their last three tries. Participants put on their masks one minute before starting the first trial in each group to minimize any distracting effect. Subjects recorded whether they found the scent inherently positive, inherently negative, or if they were indifferent to it. Testers measured the length of time it took subjects to complete each of the six trials.

The data are available in the `scent` data set.

a) Make a list of questions one might like to ask about this situation and for which the data collected provide some information.

b) For which of these questions do we know a statistical method?

c) Perform an analysis to answer one of the questions identified in b).

4.54. William Gosset (a.k.a. Student) did an experiment in which regular and kiln dried seeds were planted in adjacent plots in 11 fields. The yield (lbs/acre) for each plot was recorded at harvest. The data are available in the `gosset` data set.

a) Why was it important to plant seeds in adjacent plots in a variety of fields?

b) Do the data suggest that one method of preparing the seeds is better than the other?

4.55. When obtaining empirical p-values by computer simulations, the precision of our estimate increases with the number of replications. Typically, we only need a high degree of precision when the p-values are small, and the smaller the p-values, the more precision is desired. The quantity $Q = m/\hat{\pi}$ where m is the margin of error for a 95% confidence interval and $\hat{\pi} = x/n$ is a useful way to quantify this. For example, if $Q = 1$, then our confidence interval is

$$\hat{\pi} \pm m = \hat{\pi} \pm Q\hat{\pi} = \hat{\pi} \pm \hat{\pi} \;,$$

so we can be quite confident that the true p-value is no more than twice our estimated value.

a) Show that if $x \geq 4$, then $Q \leq 1$. (You may use Wald confidence intervals since they make the algebra simplest.)

b) How large must x be to guarantee that $Q \leq \frac{1}{2}$?

c) How large must x be to guarantee that $Q \leq 0.10$?

Note: This can be used to design an adaptive algorithm for computing empirical p-values. One starts by determining the desired value of Q. From this one can solve for a threshold x_Q, the required number of extreme test statistic values. Running the simulation until we find x extreme test statistics will suffice for our desired degree of precision. So, for example, simulating until we see 4 extreme statistics gives us 95% confidence that our empirical p-value is at worst half of the correct value, but it could be much better. By increasing Q to 0.10, we can be quite confident that our empirical p-value is correct up to an order of magnitude either way.

4.56. Repeat Example 4.10.1 with other test statistics. Be sure to find examples that give both higher and lower p-values than the p-value in Example 4.10.1.

4.57. We can also use simulations to estimate power.

   a) Use simulations to estimate the power of the range statistic from Example 4.10.1 if the true proportions are 30% 1's, 30% 2's, 20% 3's, and 20% 4's?

   b) Use simulations to estimate power for the estimators you used in Exercise 4.56.

4.58. Make a plot like the one in Figure 4.17 for the Wilson confidence interval.

## Chapter 5

Likelihood-Based Statistics

> Although this may seem a paradox, all exact science is dominated by the
> idea of approximation.
>
> Bertrand Russell [**AK66**]

5.1. Maximum Likelihood Estimators

In Chapter 4 we introduced method of moments estimation which was a formalization of the following idea:

> The moments of the sample data should be close to the moments of the
> model.

In particular, for models with one parameter, the method of moments estimate is
obtained by determining which parameter value causes the model to have a mean
equal to the mean of the sample data.

Maximum likelihood is a different, and more widely used, method for estimating
parameters. The maximum likelihood method is based on formalizing the following
intuitive question:

> Which values of the model parameters would make my data most likely?

As we did with the method of moments, we will begin by considering one-
parameter models and then move on to the more general case.

5.1.1. Estimating a Single Parameter

We introduce the maximum likelihood method with the following example.

Example 5.1.1. Michael has three dice in his pocket. One is a standard die with
six sides, another has four sides, and the third has ten sides. He challenges you to a
game. Without showing you which die he is using, Michael is going to roll a die 10

times and report to you how many times the resulting number is a 1 or a 2. Your challenge is to guess which die he is using.

Q. Michael reports that 3 of the 10 rolls resulted in a 1 or a 2. Which die do you think he was using?

A. The probability π of obtaining a 1 or a 2 is one of $\frac{1}{2}$, $\frac{1}{3}$, or $\frac{1}{5}$, depending on which die is being used. Our data are possible with any of the three dice, but let's see how likely they are in each case.

- If $\pi = \frac{1}{2}$, then the probability of obtaining this result is

$$\texttt{dbinom(3, 10, 1/2)} = \binom{10}{3}\left(\frac{1}{2}\right)^3\left(\frac{1}{2}\right)^7 = 0.1172 .$$

- If $\pi = \frac{1}{3}$, then the probability of obtaining this result is

$$\texttt{dbinom(3, 10, 1/3)} = \binom{10}{3}\left(\frac{1}{3}\right)^3\left(\frac{2}{3}\right)^7 = 0.2601 .$$

- If $\pi = \frac{1}{3}$, then the probability of obtaining this result is

$$\texttt{dbinom(3, 10, 1/5)} = \binom{10}{3}\left(\frac{1}{5}\right)^3\left(\frac{4}{5}\right)^7 = 0.2013 .$$

Of these, the largest likelihood is for the case that $\pi = \frac{1}{3}$, i.e., for the standard, six-sided die. Our data would be more likely to occur with that die than with either of the other two – it is the maximum likelihood die.   ◁

The next example extends this reasoning to a situation where we are not limited to only three possible parameter values.

Example 5.1.2.

Q. Suppose we want to estimate a population proportion π. We obtain an iid random sample of size 40. Our data contains 14 successes and 26 failures. For any value of π other than 0 or 1, it is possible to get 14 successes and 26 failures, but the probability of doing so depends on π. What value of π makes this most likely?

A. Our approach, in outline, has two steps.

(1) Define a **likelihood function** $L(\pi)$ that tells us how likely our data are for a given value of π.

(2) Determine what value of π makes the likelihood function largest. That value is the **maximum likelihood estimate** (MLE).

Suppose for a moment that our data is $\boldsymbol{x} = \langle 0, 1, 1, \ldots, 0, 0\rangle$ (using 0 for failure and 1 for success). Then

$$L(\pi; \boldsymbol{x}) = (1 - \pi) \cdot \pi \cdot \pi \cdots (1 - \pi) \cdot (1 - \pi) = \pi^{14}(1 - \pi)^{26}$$

is our likelihood function. Note that the likelihood function doesn't depend on the *order* of the successes and failures; it only depends on the *number* of successes. This is an important observation to which we will return later.

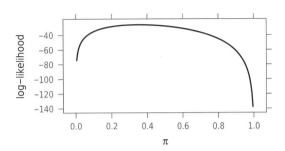

Figure 5.1. The log-likelihood function for Example 5.1.2.

Now that we have the likelihood function, we need to maximize it. The methods used to maximize a function depend on the form of the function. One frequently used approach relies on differentiation. For a differentiable function on a closed interval, the maximum must occur at an endpoint or where the derivative is 0.

Often we will begin by making some simplifications to the formula. Remember that we are primarily interested in the value of π that makes the likelihood function largest and not in the value of the likelihood function itself. Because of this, we may apply any monotone transformation to the likelihood function and work with the transformed likelihood function instead since monotone functions preserve the locations of maxima and minima. Constant multiple, power, root, and especially logarithmic transformations are often very useful here. The advantage of the logarithmic transformation is that it transforms powers and products (which occur frequently in likelihood functions) into products and sums, which are typically easier to work with if we need to differentiate.

In our example, it is sufficient to find the value of π that maximizes

$$l(\pi) = \log(L(\pi)) = 14\log(\pi) + 26\log(1 - \pi) \, .$$

Now we are ready to maximize l. Remember that the variable of interest here is π – we want to know what value of π will make l (and hence L) largest. We can find extreme values by determining when $\frac{dl}{d\pi} = 0$. Since

$$\frac{dl}{d\pi} = 14\frac{1}{\pi} + 26\frac{-1}{1-\pi}$$
$$= \frac{14}{\pi} - \frac{26}{1-\pi} \, ,$$

$\frac{dl}{d\pi} = 0$ when

$$\frac{14}{\pi} = \frac{26}{1-\pi} \, ,$$
$$26\pi = 14(1 - \pi) \, ,$$
$$40\pi = 14 \, ,$$
$$\pi = \frac{14}{40} \, .$$

It is easy to see that this value maximizes l, so our maximum likelihood estimate is $\hat{\pi} = \frac{14}{40}$, the sample proportion.

Figure 5.1 displays the graph of $l(\pi)$ (for $\boldsymbol{x}$ as in our sample). Notice that it is quite flat in the region surrounding $\hat{\pi} = 14/40 = 0.35$. This means that for other values of π near $\hat{\pi}$, our data are not much less likely than for our MLE. In general, the flatness of the curve near the MLE gives some indication of the variability of the estimator from one sample to another. We will return to this important idea later in this chapter.   ◁

Example 5.1.3.

Q. Generalize the previous example. That is, suppose now we have a sample of size n with t successes. What is the maximum likelihood estimator for π?

A. Since the likelihood function only depends on the number of successes (t) and not on any other features of the data, we can express our likelihood function as

$$L(\pi; t) = \pi^t (1 - \pi)^{n-t} .$$

From here we proceed just as before, beginning with a logarithmic transformation:

$$l(\pi) = t \log(\pi) + (n - t) \log(1 - \pi) .$$

We want to find the value of π that makes l largest and our method from before generalizes easily:

$$\frac{dl}{d\pi} = t \frac{1}{\pi} + (n - t) \frac{-1}{1 - \pi}$$
$$= \frac{t}{\pi} - \frac{n - t}{1 - \pi} .$$

So $\frac{dl}{d\pi} = 0$ when

$$\frac{t}{\pi} = \frac{(n - t)}{1 - \pi} ,$$
$$(n - t)\pi = t(1 - \pi) ,$$
$$n\pi = t ,$$
$$\pi = \frac{t}{n} .$$

Thus our maximum likelihood estimator for π is $\hat{\pi} = \frac{t}{n}$, the sample proportion.   ◁

The previous examples are illustrative in several ways. The first thing you may notice is that although the idea sounds simple enough, carrying out the method can get quite involved notationally and computationally. In fact, we will often use numerical rather than analytical methods to obtain maximum likelihood estimates. Fortunately R has tools to help us with this task. We'll learn about them shortly.

It is also important to notice how the likelihood function is related to the pmf of the model distribution. The difference between them is a matter of perspective. In Example 5.1.3, the pmf and likelihood function are

$$f(\boldsymbol{x}; \pi) = L(\pi; \boldsymbol{x}) = \pi^t (1 - \pi)^{n-t} , \tag{5.1}$$

where t is the number of successes in $\boldsymbol{x}$. When we consider the pmf f, we consider π fixed and use (5.1) to determine probabilities for different values of $\boldsymbol{x}$. When we consider the likelihood function L, we consider the *data* to be fixed and determine likelihood values for different parameters π.

Finally, we note that in this case the likelihood function only depends on $\boldsymbol{x}$ through the statistic t (the number of successes). That is, if $\boldsymbol{x}$ and $\boldsymbol{y}$ each represent samples with t successes, then $L(\pi; \boldsymbol{x}) = L(\pi; \boldsymbol{y})$. In particular, this means that the MLE for π with data $\boldsymbol{x}$ will be the same as the MLE for π with data $\boldsymbol{y}$ (since *all* the likelihoods are the same). Since *all* the information that the data $\boldsymbol{x}$ provide about π is contained in the statistic

$$T(\boldsymbol{x}) = \text{number of successes} \,,$$

we call T a **sufficient statistic** for π. The notion of sufficiency is important for theoretical reasons and is closely related to maximum likelihood methods.

The remainder of this section and the exercises provide several additional examples of this method. Many of the situations are the same as the ones we addressed with the method of moments in Chapter 4. Sometimes the two methods lead to the same estimators, but often they do not.

Example 5.1.4.

Q. Suppose $\boldsymbol{X} \overset{\text{iid}}{\sim} \text{Unif}(0, \theta)$. What is the maximum likelihood estimator for θ? Apply this to estimate θ if the (sorted) sample values are

$$1.6 \quad 2.8 \quad 6.2 \quad 8.2 \quad 8.5 \quad 8.7$$

A. We will begin by handling the special case where $\boldsymbol{x} = \langle 1.6, 2.8, 6.2, 8.2, 8.5, 8.7 \rangle$. It no longer makes sense to calculate the *probability* that $\boldsymbol{X} = \boldsymbol{x}$, but we can still use the maximum likelihood method by replacing pmfs with pdfs.

Since the pdf of the uniform distribution (for a single variate) is $\frac{1}{\theta}$ on $[0, \theta]$, the likelihood function is

$$L(\theta; \boldsymbol{x}) = \begin{cases} 0 & \text{if } \theta < 8.7, \\ \left(\frac{1}{\theta}\right)^6 & \text{if } \theta \geq 8.7. \end{cases}$$

We might be tempted to use calculus to maximize L as in our previous examples, but there is a much easier method this time. For positive values of θ, $\left(\frac{1}{\theta}\right)^6$ is a monotone decreasing function, so L is maximized at the smallest possible value of θ for which $L(\theta) \neq 0$. Thus the maximum likelihood estimate is $\hat{\theta} = 8.7$.

The general case is quite similar. Suppose the data are $\boldsymbol{X} = \langle x_1, \ldots, x_n \rangle$. Let M be the maximum value of the x_i's. Then

$$L(\theta; \boldsymbol{x}) = \begin{cases} 0 & \text{if } \theta < M, \\ \left(\frac{1}{\theta}\right)^n & \text{if } \theta \geq M. \end{cases}$$

By the same reasoning as above, the MLE is $\hat{\theta} = M = \max(\boldsymbol{X})$. Notice that in this situation the method of moments estimator and the maximum likelihood estimator are different. ◁

Example 5.1.5.

Q. An instructor believes that the number of students who arrive late for class should follow a Poisson distribution. The table below indicates the number of such students in 10 consecutive lectures:

count	0	1	2	3	4
frequency	3	3	2	1	1

Derive a general formula for the MLE for λ if $\boldsymbol{X} \overset{\text{iid}}{\sim} \mathsf{Pois}(\lambda)$ and apply the result to this data.

A. The likelihood function is

$$L(\lambda; \boldsymbol{x}) = \prod_x \frac{e^{-\lambda}\lambda^x}{x!} \ .$$

To find the MLE, it suffices to maximize $\tilde{l}(\lambda; \boldsymbol{x}) = \log(L(\lambda; \boldsymbol{x}) \cdot (x!)^n)$, since $(x!)^n$ is constant for all values of λ:

$$\tilde{l}(\lambda; \boldsymbol{x}) = \sum_x \left[\log(e^{-\lambda}\lambda^x) \right] = -n\lambda + \sum_x x \log(\lambda) = -n\lambda + n\bar{x}\log(\lambda) \ .$$

Differentiating, we obtain

$$\frac{\partial}{\partial \lambda}\tilde{l}(\lambda; \boldsymbol{x}) = -n + \frac{n\bar{x}}{\lambda} \ ,$$

so $\dfrac{\partial \tilde{l}}{\partial \lambda} = 0$ when $\lambda = \bar{x}$, and

$$\hat{\lambda} = \bar{x} \ . \tag{5.2}$$

This is the same estimate we found using the method of moments. Using this with the current data yields $\hat{\lambda} = 1.4$. Assuming a Poisson model, this is the parameter value that makes our data more likely than any other parameter value does. ◁

Example 5.1.6. A field biologist wants to estimate the density of a species of plant in a certain region. She uses a Poisson model that says the number of plants observed in a region of area A should be $\mathsf{Pois}(\lambda A)$, where the parameter λ is the plant density (plants per unit area) to be estimated.

One sampling method for this would be to pick n positions at random and then measure the distance from that position to the nearest plant of the species of interest. The sample $\boldsymbol{Y}$ consists of the n measurements.

Q. What is the maximum likelihood estimator for λ using this kind of data?

A. First we need to derive a likelihood function $L(\lambda; \boldsymbol{y}) = f_{\boldsymbol{Y}}(\boldsymbol{y}; \lambda)$ for this situation. Assuming the locations are far enough apart, we can assume each measurement is independent of the others.

First we derive the pdf for a single measurement Y using the cdf method. For $y \geq 0$,

$$P(Y \leq y) = 1 - P(Y > y) = 1 - P(\text{no plants within radius } y)$$
$$= 1 - \texttt{dpois}(0, \lambda \pi y^2)$$
$$= 1 - \frac{e^{-\lambda \pi y^2}(\lambda \pi y^2)^0}{0!}$$
$$= 1 - e^{-\lambda \pi y^2} .$$

So on $[0, \infty)$,

$$f_Y(y; \lambda) = \frac{\partial}{\partial y}\left(1 - e^{-\lambda \pi y^2}\right)$$
$$= e^{-\lambda \pi y^2} \cdot 2\pi \lambda y .$$

Thus $Y \sim \mathsf{Weibull}(2, 1/\sqrt{\lambda \pi})$ and the log-likelihood function is

$$l(\lambda; \boldsymbol{y}) = \sum_i \left(-\lambda \pi y_i^2 + \log(2\pi y_i) + \log(\lambda)\right) .$$

Differentiating, we find that

$$\frac{\partial}{\partial \lambda} l(\lambda; \boldsymbol{y}) = \sum_i \left(-\pi y_i^2 + \frac{1}{\lambda}\right) ,$$

which is 0 when

$$\pi \sum_i y_i^2 = \frac{n}{\lambda} ,$$

so

$$\hat{\lambda} = \frac{n}{\pi \sum_i y_i^2} .$$

This estimator has a natural interpretation. The numerator is the number of plants observed, and the denominator is the total area inspected to find them. So this estimator is the sample average plant density. ◁

Example 5.1.7.

Q. In genetics, if allele frequencies are in **Hardy-Weinberg equilibrium**, then the genotypes AA, Aa, and aa occur in the population with frequencies given by

AA	Aa	aa
θ^2	$2\theta(1-\theta)$	$(1-\theta)^2$

where θ is the population frequency of the A allele. Assuming Hardy-Weinberg equilibrium, estimate θ if a sample of genotypes gives the following data:

AA	Aa	aa
83	447	470

A. We begin by deriving the likelihood function for the general case where there are x_i genotypes in cell i of the table. (Implicitly we are using the fact that the three cell counts form a sufficient statistic for θ.)

$$L(\theta; x_1, x_2, x_3) = \binom{1000}{x_1} \binom{1000 - x_1}{x_2} (\theta^2)^{x_1} (2\theta(1 - \theta))^{x_2} ((1 - \theta)^2)^{x_3}$$

$$= \left[\binom{1000}{x_1} \binom{1000 - x_1}{x_2} 2^{x_2} \right] (\theta^2)^{x_1} \theta^{x_2} (1 - \theta)^{x_2} ((1 - \theta)^2)^{x_3}.$$

Ignoring the bracketed term, which is constant, and taking logarithms, we get

$$\tilde{l}(\theta; \boldsymbol{x}) = 2x_1 \log(\theta) + x_2 \log(\theta) + x_2 \log(1 - \theta) + 2x_3 \log(1 - \theta)$$

$$= (2x_1 + x_2) \log(\theta) + (2x_3 + x_2) \log(1 - \theta).$$

This we must maximize over $\theta \in [0, 1]$. Taking the derivative, we obtain

$$\frac{\partial}{\partial \theta} \tilde{l}(\theta; \boldsymbol{x}) = \frac{2x_1 + x_2}{\theta} - \frac{2x_3 + x_2}{1 - \theta},$$

which is 0 when

$$\frac{1 - \theta}{\theta} = \frac{1}{\theta} - 1 = \frac{2x_3 + x_2}{2x_1 + x_2},$$

i.e., when

$$\frac{1}{\theta} = 1 + \frac{2x_3 + x_2}{2x_1 + x_2}$$

$$= \frac{2x_1 + x_2 + 2x_1 + x_2}{2x_1 + x_2}.$$

So

$$\hat{\theta} = \frac{2x_1 + x_2}{2x_1 + 2x_2 + 2x_3}$$

$$= \frac{2x_1 + x_2}{2n}.$$

Our estimator has a natural interpretation: It is the sample allele frequency.

Our data yield the estimate

$$\hat{\theta} = \frac{2 \cdot 83 + 447}{2000} = 0.3065.$$

Of course, we don't know yet how accurate this estimate is (how many decimal places should we trust?) nor how to test how well the data agree with the assumption of Hardy-Weinberg equilibrium. We will address these issues later in this chapter.   ◁

5.1.2. Models with More Than One Parameter

The maximum likelihood method works equally well when there are multiple parameters in the model – provided we can perform the maximization step.

Example 5.1.8.

Q. Suppose we sample from a uniform population on an unknown interval $[\alpha, \beta]$. Derive the maximum likelihood estimators for these parameters.

A. This is very similar to Example 5.1.4.

$$L(\alpha, \beta; \boldsymbol{x}) = \begin{cases} 0 & \text{if } \min(\boldsymbol{x}) < \alpha \text{ or } \max(\boldsymbol{x}) > \beta, \\ \left(\frac{1}{\beta - \alpha}\right)^n & \text{otherwise.} \end{cases}$$

The maximum value of L occurs when $\beta - \alpha$ is as small as possible, that is, when $\alpha = \min(\boldsymbol{x})$ and $\beta = \max(\boldsymbol{x})$. So

$$\hat{\alpha} = \min(\boldsymbol{x}) \qquad \text{and} \qquad \hat{\beta} = \max(\boldsymbol{x}) . \qquad \triangleleft$$

Example 5.1.9.

Q. Derive the MLE for $\langle \mu, \sigma \rangle$ if $\boldsymbol{X} \overset{\text{iid}}{\sim} \mathsf{Norm}(\mu, \sigma)$.

A. The log-likelihood function in this case is

$$l(\mu, \sigma; \boldsymbol{x}) = \sum_i -\frac{1}{2} \log(2\pi\sigma^2) - \frac{(x_i - \mu)^2}{2\sigma^2} = -\frac{n}{2} \log(2\pi\sigma^2) - \frac{1}{2\sigma^2} \sum_i (x_i - \mu)^2 .$$

From this we obtain the partial derivatives:

$$\frac{\partial l}{\partial \mu} = \frac{1}{2\sigma^2} \sum_i (x_i - \mu) , \text{ and}$$

$$\frac{\partial l}{\partial \sigma} = -\frac{n}{2} \frac{4\pi\sigma}{2\pi\sigma^2} + \sigma^{-3} \sum_i (x_i - \mu)^2$$

$$= -\frac{1}{\sigma^3} \left(\sigma^2 n - \sum_i (x_i - \mu)^2 \right) ,$$

from which it follows that

$$\hat{\mu} = \overline{x} ,$$

$$\hat{\sigma}^2 = \frac{\sum_i (x_i - \overline{x})^2}{n} = \frac{n-1}{n} s^2 . \qquad \triangleleft$$

5.1.3. Multinomial Distributions

The **multinomial distributions** are a generalization of the binomial distributions. Recall that in a binomial random process, each trial had two possible outcomes (success and failure) and the binomial random variable counted the number of successes. For a multinomial random process, each trial has two or more possible outcomes, and we record how many trials resulted in each possible outcome. Thus a multinomial distribution is a joint distribution of these several counts.

Definition 5.1.1 (Multinomial Distributions). If

(1) a random process is repeated n times,

(2) each trial has k possible outcomes,

(3) the probability of outcome i is π_i for each trial, and

(4) the outcome of each trial is independent of the others,

then the vector $\boldsymbol{X} = \langle X_1, X_2, \ldots, X_k \rangle \sim \mathsf{Multinom}(n, \boldsymbol{\pi})$, where X_i is the number of trials with outcome i, and $\boldsymbol{\pi} = \langle \pi_1, \pi_2, \ldots \pi_k \rangle$. $\square$

The parameterization given here is a bit redundant since $\pi_1 + \pi_2 + \cdots + \pi_k = 1$. Similarly, $X_1 + X_2 + \cdots + X_k = n$. Thus given any $k - 1$ values of $\boldsymbol{\pi}$ or $\boldsymbol{X}$, the remaining value is determined. The binomial distribution is essentially the same as a multinomial distribution with $k = 2$ but expressed without this redundancy.

The joint multinomial pmf is given by

$$\mathrm{P}(\boldsymbol{X} = \boldsymbol{x}) = f(\boldsymbol{x}; \boldsymbol{\pi}) = \binom{n}{x_1 x_2 \cdots x_k} \pi_1^{x_1} \pi_2^{x_2} \cdots \pi_k^{x_k} ,$$

where the **multinomial coefficient** is defined by

$$\binom{n}{x_1 \cdots x_k} = \binom{n}{x_1}\binom{n - x_1}{x_2}\binom{n - x_1 - x_2}{x_3} \cdots \binom{x_k}{x_k}$$

$$= \frac{n!}{x_1! x_2! \cdots x_k!} . \tag{5.3}$$

As with binomial random variables, multinomial random variables represent a tabulation of counts of the results from the individual trials. The multinomial distributions have a wide variety of uses, especially as models for categorical responses with more than 2 levels.

The marginal distribution of each X_i is $\mathsf{Binom}(n, \pi_1)$, since the probability of outcome i is π_i and the probability of obtaining some other outcome is $1 - \pi_i$. The X_i's are not independent, however. $\mathrm{Cov}(X_i, X_j) < 0$ since when there are more outcomes of type i, there are fewer remaining trials that might have outcome j. The conditional distributions of X_i given a fixed value of X_j are discussed in Exercise 5.5.

As with other distributions we have encountered, R provides the functions `rmultinom()` and `dmultinom()`. Since the multinomial distributions are joint distributions of several random variables, there are no multinomial analogs to `pbinom()` and `qbinom()`.

For the moment we are primarily interested in determining the maximum likelihood estimator for $\boldsymbol{\pi} = \langle \pi_1, \pi_2, \ldots, \pi_k \rangle$. We will see some important applications of this result later in this chapter.

Example 5.1.10.

Q. Suppose $\boldsymbol{X} \sim \mathsf{Multinom}(\boldsymbol{\pi})$. What are the maximum likelihood estimators for $\boldsymbol{\pi}$?

A. Let $\boldsymbol{x}$ be the vector of k cell counts from a sample of size $n = \sum x_i$. Then

$$l(\boldsymbol{\pi}; \boldsymbol{x}) = \log L(\boldsymbol{\pi}; \boldsymbol{x}) = \sum_i \log(\mathtt{dmultinom}(x_i, \boldsymbol{\pi}))$$

$$\propto x_1 \log(\pi_1) + x_2 \log(\pi_2) + \cdots + x_k \log(\pi_k) = \tilde{l}(\boldsymbol{\pi}; \boldsymbol{x}) .$$

Unfortunately, we cannot find the MLE by solving the following system of equations:

$$0 = \frac{\partial}{\partial \pi_1} \tilde{l} = \frac{x_1}{\pi_1},$$

$$0 = \frac{\partial}{\partial \pi_2} \tilde{l} = \frac{x_2}{\pi_2},$$

$$\vdots = \vdots$$

$$0 = \frac{\partial}{\partial \pi_r} \tilde{l} = \frac{x_k}{\pi_k}.$$

This system has no solutions since the x_i's are not all 0.

In fact, $\tilde{l}$ has no maximum value if $\boldsymbol{\pi}$ is unconstrained. We can make each term in the sum larger by increasing any of the π_i's. But $\boldsymbol{\pi}$ is only an acceptable parameter if $\sum_i \pi_i = 1$, so we really want the maximum value of $\tilde{l}$ subject to that constraint. The method of Lagrange multipliers is a standard tool for solving constrained optimization problems of this sort. Using this method (see Exercise 5.7), one can show that the MLE satisfies

$$\hat{\pi}_i = \frac{x_i}{n}. \tag{5.4}$$

$\triangleleft$

5.1.4. Numerical Maximum Likelihood Methods

In the preceding section we were able to derive formulas for the maximum likelihood estimators. There are many situations, however, where closed form expressions for the maximum likelihood estimators are not so easy to derive. In these situations, a variety of numerical methods can be employed to obtain the maximum likelihood estimates. The fact that a computer can often be used to obtain (good approximations to) the maximum likelihood estimates is one of the reasons that maximum likelihood estimation is so important in modern statistical applications.

In this section we present a number of examples where the estimates are derived numerically rather than analytically. These numerical methods rely on a numerical optimization algorithm. There are several such algorithms that are established tools for this. In the examples below, we use `nlmax()` from the `fastR` package. This function is based on the non-linear minimizer `nlm()`, modified to find a maximum rather than a minimum and to produce somewhat different output. Other options in R include `optim()`, `constrOptim()`, and a variety of optimizers in specialty packages.

Example 5.1.11.

Q. Recall that the beta distributions are a useful model for distributions of proportions. Below are the batting averages of 15 randomly selected major league baseball players (from the 2000 season, minimum 200 at bats):

$$\begin{array}{cccccccc} 0.320 & 0.297 & 0.264 & 0.306 & 0.291 & 0.290 & 0.316 & 0.324 \\ 0.295 & 0.294 & 0.235 & 0.283 & 0.273 & 0.295 & 0.327 \end{array}$$

(1) Use maximum likelihood to fit a beta distribution to this data.

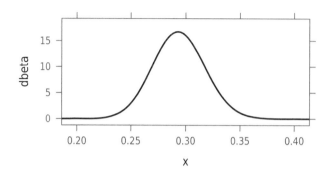

Figure 5.2. Fitting a beta distribution to a sample of batting averages using maximum likelihood.

(2) In 2000, 314 major league players had at least 200 at bats. Using the maximum likelihood estimates, estimate the highest and lowest batting averages in the major leagues in 2000. (The actual values are 0.372 and 0.197.)

A. We will use R heavily in this solution. First, we enter our data into a vector

```
> ba <- c(0.320,0.297,0.264,0.306,0.291,0.290,0.316,0.324,
+          0.295,0.294,0.235,0.283,0.273,0.295,0.327)
```
`baseballBA1`

Next we need to derive the likelihood function, which we can express as

$$L(\alpha, \beta; \boldsymbol{x}) = \prod_{i=1}^{15} \mathtt{dbeta}(x_i, \alpha, \beta).$$

It is more efficient to work with the log-likelihood:

$$l(\alpha, \beta; \boldsymbol{x}) = \sum_{i=1}^{15} \mathtt{dbeta}(x_i, \alpha, \beta, \mathtt{log=TRUE}).$$

We will use `nlmax()` to find the maximum likelihood estimates. This function is a wrapper provided by `fastR` around the `nlm()` function, which is a minimizer. The wrapper allows us to work with the maximum directly and also provides some modified output. We must put our likelihood function into a particular format to use it with `nlmax()`. The parameters must be presented as a vector in the first argument. We will call this vector `theta`. Any other arguments to the function (in this case, the data vector) may be given after `theta`.

```
# log likelihood function
> loglik <- function(theta,x) { sum(dbeta(x,theta[1],theta[2],log=T)) }
```
`baseballBA2`

Now we are ready to use `nlmax()`. We need to seed the algorithm with a starting "point" p. Here we choose `p=c(1,1)` which corresponds to $\mathsf{Beta}(1,1) = \mathsf{Unif}(0,1)$. This (unreasonable) model would predict that batting averages are equally likely to take any value between 0 and 1. Starting from this, `nlmax()` will search numerically for the maximum likelihood estimates. Even given this initial value, convergence

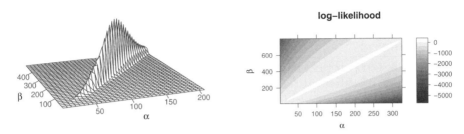

Figure 5.3. Two representations of the likelihood function for fitting batting averages with a beta distribution (Example 5.1.11).

to the MLE is quick. In more complicated situations, it is common to use method of moments estimates to seed the search for the maximum likelihood estimates.

```
                                                              baseballBA3
# suppress warnings from log(0)
> oldopt <- options(warn=-1)
> summary(nlmax(loglik,p=c(1,1),x=ba))

      Maximum: 34.7864
      Estimate:107.2568 257.5854
      Gradient: 7.4209e-08 -3.0807e-08
    Iterations: 28

Relative gradient is close to zero, current iterate is probably an
approximate solution.[Code=1]
# get just the mle
> nlmax(loglik,p=c(1,1),x=ba)$estimate
[1] 107.26 257.59
> options(oldopt)                              # reset options
```

We see that our MLE is $\hat{\theta} = \langle 107.3, 257.6 \rangle$. We can check that the MLE is at least reasonable by plotting the pdf for $\mathsf{Beta}(107.3, 257.6)$ (Figure 5.2). According to this fit, nearly all batting averages should be between 0.200 and 0.400. Using our ruler method, we can estimate the lowest and highest batting averages predicted by the model.

```
                                                              baseballBA4
# using the ruler method
> qbeta(313.5/314,107.3,257.6)
[1] 0.36692
> qbeta(0.5/314,107.3,257.6)
[1] 0.22695
```

We see that the lowest batting average achieved was a bit lower than our method estimates but that the maximum batting average is actually quite close. That's pretty good for using only 15 batting averages as input.

It is instructive to look at a plot of the likelihood function in this example (Figure 5.3). Notice that the likelihood falls off dramatically as the relationship between α to β departs from the apparently linear relationship along the "ridge" of the likelihood function but that the surface is much flatter along that ridge in the plot. This is an informal indication that our data provide a much better estimate for this linear relationship between α and β than for the values of α and β themselves.   ◁

Example 5.1.12.

Q. The data set `faithful` contains information on eruption times of the Old Faithful geyser in Yellowstone National Park. The distribution is clearly bimodal, leading one to believe that there may be two distinct types or classes of eruption. Use maximum likelihood to fit a mixture of normals model to this data.

A. By a mixture of normals we mean a model with 5 parameters:

- α: the proportion of the distribution in the group with the smaller mean,

- μ_1, σ_1: the mean and standard deviation of the group with the smaller mean, and

- μ_2, σ_2: the mean and standard deviation of the group with the larger mean.

We will denote such a mixture as $\alpha\mathsf{Norm}(\mu_1, \sigma_1) : (1-\alpha)\mathsf{Norm}(\mu_2, \sigma_2)$. The density function (see Exercise 5.12) is given by

$$\alpha\varphi(x; \mu_1, \sigma_1) + (1 - \alpha)\varphi(x; \mu_2, \sigma_2) \,,$$

where $\varphi(x; \mu, \sigma)$ is the pdf for a $\mathsf{Norm}(\mu, \sigma)$-distribution.

We can now use `nlm()` to estimate the five parameters:

```
# density function for mixture of normals                          mle-faithful
> dmix <- function(x, alpha,mu1,mu2,sigma1,sigma2) {
+     if (alpha < 0) return (dnorm(x,mu2,sigma2))
+     if (alpha > 1) return (dnorm(x,mu1,sigma1))
+
+     alpha * dnorm(x,mu1,sigma1) + (1-alpha) * dnorm(x,mu2,sigma2)
+     }
>
# log-likelihood
> loglik <- function(theta, x) {
+     alpha <- theta[1]
+     mu1 <- theta[2]
+     mu2 <- theta[3]
+     sigma1 <- theta[4]
+     sigma2 <- theta[5]
+     density <- function (x) {
+         if (alpha < 0) return (Inf)
+         if (alpha > 1) return (Inf)
+         if (sigma1<= 0) return (Inf)
+         if (sigma2<= 0) return (Inf)
+         dmix(x,alpha,mu1,mu2,sigma1,sigma2)
+     }
+     sum( log ( sapply( x, density) ) )
+ }
>
# seed the algorithm
> m <- mean(faithful$eruptions)
```

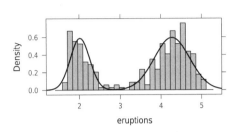

Figure 5.4. Fitting a mixture of normals to Old Faithful eruption times.

```
> s <- sd(faithful$eruptions)
>
> oldopt <- options(warn=-1)              # suppress warnings from log(0)
> mle <- nlmax(loglik,p=c(0.5,m-1,m+1,s,s),x=faithful$eruptions)$estimate
> mle
[1] 0.34840 2.01861 4.27334 0.23562 0.43706
> options(oldopt)
>
> d <- function(x) {
+     dmix(x,mle[1],mle[2],mle[3],mle[4],mle[5])
+ }
>
> myplot <- xhistogram(~eruptions,data=faithful,
+     n=30,
+     density=T,
+     dmath=dmix,
+     args=list(
+         alpha=mle[1],
+         mu1=mle[2],
+         mu2=mle[3],
+         sigma1=mle[4],
+         sigma2=mle[5])
+     )
```

Overlaying the density function determined by the maximum likelihood estimates on a histogram of eruption times (Figure 5.4) reveals a reasonably good fit, although it appears that there is an excess of especially short and especially long eruptions, an indication that the two groups might each be slightly skewed (and in opposite directions).   ◁

When using numerical optimizers – especially when one doesn't know or understand the details completely – it is important to realize that the results may not always optimize the objective function. It is essentially impossible for a numerical algorithm to distinguish between a local minimum and a global minimum, for example, so one common way that such an algorithm can fail is by converging to a local minimum or even a saddle point. Optimization algorithms typically place a bound on the number of iterations, so another cause for incorrect results is convergence that is too slow.

The following example shows how an apparently harmless choice of initial point can cause `nlm()` (and hence `nlmax()`) to fail to find the MLE.

Example 5.1.13. The choice of an initial point for `nlmax()` in the preceding example deserves some comment. Observe how `nlmax()` fails to find the MLE if we choose an unfortunate initial point:

```
# seed the algorithm                                    mle-faithful2
> m <- mean(faithful$eruptions)
> s <- sd(faithful$eruptions)
>
> mle <- nlmax(loglik,p=c(0.5,m,m,s,s),x=faithful$eruptions)$estimate
> mle
[1] 0.5000 3.4878 3.4878 1.1393 1.1393
```

We have started the algorithm at a saddle point. While it would be better to move one mean up and the other down, `nlmax()` is unable to decide which to move in which direction, so it "converges" to the best single normal distribution rather than to the best mixture of two normals.   ◁

We conclude with a numerical example that supports the result from Example 5.1.10.

Example 5.1.14.

Q. Fit a multinomial distribution to data summarized in the following table:

value	1	2	3	4
count	10	20	30	40

A. Before using `nlm()`, we remove the redundancy in the parameterization of the multinomial distribution.

```
> oldopt <- options(warn=-1)     # suppress warnings from log(0)   mle-multinom
> loglik <- function(theta,x) {
+     probs <- c(theta, 1-sum(theta))
+     if (any (probs < 0)) {return(Inf)}
+     return( dmultinom(x,size=100,prob=probs, log=T) )
+     }
> nlmax(loglik,p=rep(0.25,3),x=c(10,20,30,40))$estimate -> mle; mle
[1] 0.1 0.2 0.3
> options(oldopt)                        # restore options
> round(mle,6)
[1] 0.1 0.2 0.3
```

Notice that this agrees (to 6 decimal places) with the results from Example 5.1.10.
                                                                        ◁

5.2. Likelihood Ratio Tests

The likelihood ratio statistic provides a very general and powerful method for using the likelihood function to develop hypothesis tests about the parameters of a model.

It is the most used hypothesis test method in all of statistics. This is due in large part to three features of the method:

(1) It suggests a test statistic.

(2) The asymptotic distribution of the likelihood ratio statistic is often known.

(3) There are theoretical reasons to believe that likelihood ratio tests are often more powerful than other competing tests for the same hypotheses.

This means that in a wide range of applications, if we can state the null and alternative hypotheses within the likelihood ratio framework, likelihood methods will provide a testing procedure that is at least approximately correct when sample sizes are large enough. Furthermore, likelihood methods typically have good power as well.

5.2.1. The Basic Likelihood Ratio Test

Before introducing the general method, we begin with the simpler situation where our model has only one parameter θ and the hypotheses of interest are

- H_0: $\theta = \theta_0$,
- H_a: $\theta \neq \theta_0$.

The likelihood ratio statistic in this case is

$$\lambda = \frac{L(\theta_0; \boldsymbol{x})}{L(\hat{\theta}; \boldsymbol{x})} , \tag{5.5}$$

where $\hat{\theta}$ is the MLE and L is the likelihood function.

The numerator of λ is the likelihood if the null hypothesis is true. The denominator is the maximum value the likelihood can have for any value of θ. So this ratio is at most 1. Intuitively, when the ratio is much less than 1, then the data are much more likely when the parameter is equal to the MLE than when the parameter is as specified in the null hypothesis, so we should reject H_0. In order to calculate a p-value, of course, we will need to know something about the distribution of Λ, the value of the likelihood ratio for a random sample when H_0 is true. In the following example, the distribution of Λ is easily obtained by direct calculation.

Example 5.2.1.

Q. Let $\boldsymbol{X} \overset{\text{iid}}{\sim} \text{Unif}(0, \theta)$ be an iid random sample. Use the likelihood ratio to test the null hypothesis that $\theta = 10$ if the maximum value in a sample of size $n = 5$ is 8.

A. We will follow our usual four-step method (see Section 2.4.1).

(1) The null and alternative hypotheses are
- H_0: $\theta = 10$,
- H_a: $\theta \neq 10$.

(2) We calculate the likelihood ratio test statistic as follows.

Recall from Example 5.1.4 that $\hat{\theta} = \max(\boldsymbol{x}) = 8$ and that

$$L\left(\theta; \boldsymbol{x}\right) = \left(\frac{1}{\theta}\right)^n,$$

provided $\theta \geq \max(\boldsymbol{x})$. Thus

$$\Lambda = \frac{L(\theta_0; \boldsymbol{X})}{L(\hat{\theta}; \boldsymbol{X})} = \frac{(1/\theta_0)^n}{(1/\hat{\theta})^n} = \left(\frac{\max(\boldsymbol{X})}{\theta_0}\right)^n,$$

and

$$\lambda = \frac{L(\theta_0; \boldsymbol{x})}{L(\hat{\theta}; \boldsymbol{x})} = \left(\frac{\max(\boldsymbol{x})}{\theta_0}\right)^n = \left(\frac{8}{10}\right)^5 = 0.32768.$$

(3) Now we need to determine the p-value, that is, $P(\Lambda \leq \lambda \mid H_0$ is true$)$.

In this case we can do this fairly easily since we know the distribution of $\max(\boldsymbol{X})$:

$$\begin{aligned}
P(\Lambda \leq \lambda) &= P((\max(\boldsymbol{X})/\theta_0)^n \leq \lambda) \\
&= P((\max(\boldsymbol{X})/\theta_0) \leq \sqrt[n]{\lambda}) \\
&= P(\max(\boldsymbol{X}) \leq \theta_0 \sqrt[n]{\lambda}) \\
&= P(\forall i \; X_i \leq \theta_0 \sqrt[n]{\lambda}) \\
&= (\sqrt[n]{\lambda})^n = \lambda = 0.32768.
\end{aligned}$$

(4) Draw a conclusion.

With such a large p-value, we do not have enough evidence to reject the null hypothesis. Our data are consistent with the hypothesis that $\theta = 10$. ◁

Although Example 5.2.1 demonstrates the method in a simple setting, it is quite atypical in a number of ways. First, it is not usually this easy to determine the sampling distribution of Λ; it is certainly very unusual that the likelihood ratio statistic and the p-value are equal. More importantly, in this example the **support** of X (possible values X can take on) depends on θ. Many of the theoretical results for likelihood ratio tests that make them so popular only apply when the support does not depend on the parameter value(s).

When our model is sufficiently nice, if we have no other means of deriving the exact sampling distribution of the likelihood ratio statistic, then we can use the following theorem to approximate the sampling distribution.

Theorem 5.2.1. *If the pdf or pmf of a one-parameter model is sufficiently smooth and if its support does not depend on the parameter θ of the model, then*

$$-2\log(\Lambda) = 2(l(\hat{\theta}) - l(\theta_0)) \approx \mathsf{Chisq}(1),$$

provided θ_0 is the true value of the parameter (that is, when H_0 is true in a test of $H_0 : \theta = \theta_0$ vs. $H_a : \theta \neq \theta_0$). □

We will not prove Theorem 5.2.1 nor state the smoothness conditions precisely. For sketches of the argument involved, see [**Ric88**]. For a more thorough treatment, consult, for example, [**CB01**] or [**Kal85a**]. Figure 5.5 provides some geometric intuition behind the likelihood ratio test statistic. Expressed in terms of the log-likelihood l, the likelihood ratio test statistic is twice the vertical distance between

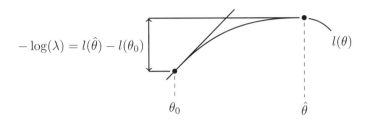

Figure 5.5. Geometric intuition behind the likelihood ratio test. $-\log(\lambda)$ is the vertical distance between the value of the likelihood function at the null hypothesis value θ_0 and the MLE $\hat{\theta}$. The slope of the tangent line is called the score.

$l(\hat{\theta})$, the maximum value of l, and $l(\theta_0)$, the likelihood of our data if the null hypothesis is true.

Example 5.2.2.

Q. Develop a likelihood ratio test to test $H_0 : \lambda = 1$ vs. $H_a : \lambda \neq 1$ assuming a population modeled by a $\mathsf{Pois}(\lambda)$-distribution. Apply this test to the data in Example 5.1.5.

A. From Example 5.1.5, we know that $\hat{\lambda} = \overline{x} = 1.4$. So our likelihood ratio statistic is

$$\lambda = \frac{L(1; \boldsymbol{x})}{L(\overline{x}; \boldsymbol{x})} = \prod_x \frac{\frac{e^{-1}1^x}{x!}}{\frac{e^{-\overline{x}}\overline{x}^x}{x!}} = \prod_x \frac{e^{-1}}{e^{-\overline{x}}\overline{x}^x} \ .$$

Thus

$$-2\ln(\lambda) = 2\sum_x \left[-\overline{x} + x\log(\overline{x}) - (-1) \right] = 2\left[-n\overline{x} + n\overline{x}\log(\overline{x}) + n \right] \ .$$

Plugging in the values of n and $\overline{x}$ from our data, we have

$$-2\ln(\lambda) = 2\left[-10 \cdot 1.4 + 10 \cdot 1.4\log(1.4) + 10 \right] = 1.4212 \ .$$

We can estimate a p-value two ways. Relying on asymptotics (but realizing that $n = 10$ may be too small for a very good approximation), we could estimate the p-value using the chi-squared distribution:

```
> x <- c(1,1,0,4,2,1,3,0,0,2); table(x)
x
0 1 2 3 4
3 3 2 1 1
> m <- mean(x); n <- length(x); m
[1] 1.4
> lrtStat <- function(x) {
+     n <- length(x); m <- mean(x)
+     2 * ( -n * m + n * m * log(m) + n)
+     }
> lrtStat(x)
[1] 1.4212
```

mle-pois

```
> pval <- 1 - pchisq(lrtStat(x), df=1); pval
[1] 0.2332
```

Alternatively, we could estimate the p-value by simulation. To facilitate this, we will again use the `statTally()` function, first introduced in Section 4.10.1, to tabulate our test statistic computed on a large number of random samples. With this function in hand, obtaining an empirical p-value is as easy as defining the test statistic as a function and simulating random samples assuming H_0 is true.

```
# generate 5000 samples of size 10                          mle-pois2
> rdata <- replicate(5000, rpois(10,1))
> statTally(x,rdata,lrtStat)
Test Stat function:

function (x)
{
    n <- length(x)
    m <- mean(x)
    2 * (-n * m + n * m * log(m) + n)
}

Test Stat applied to sample data = 1.421

Test Stat applied to random data:

   50%    90%    95%    99%
0.4297 3.0401 4.0414 6.7762
        Of the random samples
                3702 ( 74.04 % ) had test stats < 1.421
                270 ( 5.4 % ) had test stats = 1.421
                1028 ( 20.56 % ) had test stats > 1.421
```

Both the empirical p-value and the p-value based on the chi-squared distribution are sufficiently large that we would not reject the null hypothesis based on this (small) data set.

Note that

$$.2056 < 0.2332 = P(X^2 \geq 1.42) < 0.2056 + 0.054 \,.$$

So in this sense, our $\mathsf{Chisq}(1)$ p-value appears to have done as well as possible given the discrete nature of the exact sampling distribution. ◁

5.2.2. The Generalized Likelihood Ratio Test

The likelihood ratio can be used to test a wide range of hypotheses. The null hypotheses in the previous section are called **simple hypotheses** because they completely determine the null distribution. Many interesting null hypotheses do not completely determine the null distribution. For example, we might assume a normal model and want to test $H_0 : \mu = \mu_0$ vs. $H_a : \mu \neq \mu_0$ without assuming we know the standard deviation of the population. In this case, there are many distributions that fit the null hypothesis, namely the $\mathsf{Norm}(\mu_0, \sigma)$-distributions for any σ. If we do not assume H_0 is true, then a larger set of distributions is available, namely $\mathsf{Norm}(\mu, \sigma)$-distributions for any pair $\langle \mu, \sigma \rangle$.

The idea behind the generalized likelihood ratio test is to maximize the likelihood function twice and compute the ratio of the two maximum likelihood values. The numerator is computed by maximizing over parameter values that are allowable under the null hypothesis. The denominator is computed by maximizing over a larger set of possible parameter values. Formally, we define two sets of parameter values, Ω_0 and Ω_a, such that our hypotheses are

- $H_0 : \boldsymbol{\theta} \in \Omega_0$, and
- $H_a : \boldsymbol{\theta} \in \Omega_a$.

We then maximize the likelihood function twice:

- $L(\hat{\Omega}_0) = \max\{L(\boldsymbol{\theta}; \boldsymbol{x}) \mid \boldsymbol{\theta} \in \Omega_0\}$, and
- $L(\hat{\Omega}) = \max\{L(\boldsymbol{\theta}; \boldsymbol{x}) \mid \boldsymbol{\theta} \in \Omega\}$, where $\Omega = \Omega_0 \cup \Omega_a$.

The likelihood ratio is then

$$\lambda = \frac{L(\hat{\Omega}_0)}{L(\hat{\Omega})} .$$

In the case of a simple null hypothesis, Ω_0 contains only one value: $\Omega_0 = \{\boldsymbol{\theta}_0\}$, so $L(\hat{\Omega}_0) = L(\boldsymbol{\theta}_0; \boldsymbol{x})$. Similarly, $L(\hat{\Omega}) = L(\hat{\boldsymbol{\theta}})$, where $\hat{\boldsymbol{\theta}}$ is the MLE, and we get exactly the same likelihood ratio that we discussed above. The more general framework, however, allows us to handle a wide range of interesting situations.

The following theorem describes the asymptotic behavior of the generalized likelihood ratio.

Theorem 5.2.2. *If the pdf or pmf of a model is sufficiently smooth and its support does not depend on the parameters $\boldsymbol{\theta}$ of the model, then under the null hypothesis that $\boldsymbol{\theta} \in \Omega_0$,*

$$-2\log(\Lambda) = 2(l(\hat{\boldsymbol{\theta}}) - l(\hat{\boldsymbol{\theta}}_0)) \approx \mathsf{Chisq}(\dim \Omega - \dim \Omega_0) .$$

Here $\dim \Omega$ is the total number of free parameters in the model Ω, and $\dim \Omega_0$ is the number of free parameters in the restricted model that assumes $\boldsymbol{\theta} \in \Omega_0$.

Example 5.2.3.

Q. Assuming a normal population, use a likelihood ratio test to test $H_0 : \mu = 0$ vs. $H_a : \mu \neq 0$.

A. We already know from Example 5.1.9 that the MLE is

$$\hat{\boldsymbol{\theta}} = \langle \hat{\mu}, \hat{\sigma} \rangle = \langle \overline{x}, s\sqrt{\tfrac{n}{n-1}} \rangle .$$

Letting $\hat{\sigma} = s\sqrt{\tfrac{n}{n-1}}$, we get

$$L(\hat{\Omega}) = \prod_{i=1}^{n} \frac{1}{\hat{\sigma}\sqrt{2\pi}} e^{-\frac{(x-\overline{x})^2}{2\hat{\sigma}^2}} .$$

Using the fact that $\sum_i (x_i - \overline{x})^2 = n\hat{\sigma}^2$, this simplifies to

$$L(\hat{\Omega}) = (2e\pi\hat{\sigma}^2)^{-n/2} .$$

Similarly, if $\mu = \mu_0$, we must maximize

$$l_0(\sigma; \boldsymbol{x}) = \sum_i -\frac{1}{2}\log(2\pi\sigma^2) - \frac{(x_i - \mu_0)^2}{2\sigma^2} = -\frac{n}{2}\log(2\pi\sigma^2) - \frac{1}{2\sigma^2}\sum_i (x_i - \mu_0)^2 ,$$

from which we obtain

$$L(\hat{\Omega}_0) = \prod_{i=1}^{n} \frac{1}{\hat{\sigma}_0\sqrt{2\pi}} e^{\frac{(x - \mu_0)^2}{2\hat{\sigma}_0}} = (e2\pi\hat{\sigma}_0^2)^{-n/2} ,$$

where $\hat{\sigma}_0^2 = \frac{\sum_i (x_i - \mu_0)^2}{n}$.

The likelihood ratio is then

$$\lambda = \left(\frac{\hat{\sigma}}{\hat{\sigma}_0}\right)^n .$$

This is a reasonable statistic; it compares the "variance" of the data calculated two ways, once using the mean of the data and once using the mean from the null hypothesis. If these are very different, it suggests that the true mean is not μ_0.

Using the identity

$$\sum_i (x_i - \mu_0)^2 = \sum_i [(x_i - \overline{x}) + (\overline{x} - \mu_0)]^2$$

$$= \sum_i (x_i - \overline{x})^2 - 2(x_i - \overline{x})(\overline{x} - \mu_0) + (\overline{x} - \mu_0^2)^2$$

$$= n(\overline{x} - \mu_0)^2 + \sum_i (x_i - \overline{x})^2 ,$$

we can express this test another way. Namely, the test rejects $H_0 : \mu = \mu_0$ when

$$\frac{\sum_i (x - \overline{x})^2}{\sum_i (x - \mu_0)^2} = \frac{\sum_i (x - \overline{x})^2}{n(\overline{x} - \mu_0)^2 + \sum_i (x_i - \overline{x})^2} = \frac{1}{\frac{n(\overline{x} - \mu_0)^2}{\sum_i (x - \overline{x})^2} + 1}$$

is small, that is, when

$$\frac{n(\overline{x} - \mu_0)^2}{\sum_i (x - \overline{x})^2} = \frac{\frac{n}{n-1}(\overline{x} - \mu_0)^2}{\sum_i (x - \overline{x})^2/(n-1)} = \frac{1}{n-1} \cdot \frac{(\overline{x} - \mu_0)^2}{s^2/n}$$

is large. For fixed n this is proportional to the square of the statistic from the usual t-test, so we have derived the same test by a different method. Since the sampling distribution of the t statistic is known (under the model assumption that the population is normal), there is no reason to derive the sampling distribution of Λ since it will yield an equivalent test. The asymptotic distribution of Λ is $\mathsf{Chisq}(1)$ since there are 2 free parameters (μ and σ), in the unrestricted model but only 1 (σ) in the restricted model .   ◁

5.2.3. Numerical Likelihood Ratios

Example 5.2.4.

Q. Evaluate the Old Faithful eruption time data (Example 5.1.12) to see if there is sufficient evidence to reject the hypothesis that there is an equal proportion of "long" and "short" eruptions.

A. Our model has five parameters: $\boldsymbol{\theta} = \langle \alpha, \mu_1, \mu_2, \sigma_1, \sigma_2 \rangle$. The null hypothesis of interest is $H_0 : \alpha = 0.5$. The likelihood ratio statistic $-2 \log(\lambda)$ can be computed by maximizing over two sets of parameters:

$$\Omega = \{ \boldsymbol{\theta} \mid \alpha \in [0,1], \sigma_1 > 0, \sigma_2 > 0 \},$$

$$\Omega_0 = \{ \boldsymbol{\theta} \mid \alpha = \frac{1}{2}, \sigma_1 > 0, \sigma_2 > 0 \}.$$

The calculations can be done in R using `nlm()` twice, once to maximize over Ω and once to maximize over Ω_0.

```
# density function for mixture of normals                    lrt-faithful
> dmix <- function(x, alpha,mu1,mu2,sigma1,sigma2) {
+       if (alpha < 0) return (dnorm(x,mu2,sigma2))
+       if (alpha > 1) return (dnorm(x,mu1,sigma1))
+
+       alpha * dnorm(x,mu1,sigma1) + (1-alpha) * dnorm(x,mu2,sigma2)
+       }
>
# log-likelihood
> loglik <- function(theta, x) {
+       alpha <- theta[1]
+       mu1 <- theta[2]
+       mu2 <- theta[3]
+       sigma1 <- theta[4]
+       sigma2 <- theta[5]
+       density <- function (x) {
+           if (alpha < 0) return (Inf)
+           if (alpha > 1) return (Inf)
+           if (sigma1<= 0) return (Inf)
+           if (sigma2<= 0) return (Inf)
+           dmix(x,alpha,mu1,mu2,sigma1,sigma2)
+       }
+       sum( log ( sapply( x, density) ) )
+ }
>
> loglik0 <- function(theta, x) {
+       theta <- c(0.5,theta)
+       return(loglik(theta,x))
+ }
# seed the algorithm
> m <- mean(faithful$eruptions)
> s <- sd(faithful$eruptions)
>
> oldopt <- options(warn=-1)     # suppress warnings from log(0)
> mle <-  nlmax(loglik, p=c(0.5,m-1,m+1,s,s), x=faithful$eruptions)$estimate
> mle
[1] 0.34840 2.01861 4.27334 0.23562 0.43706
> loglik(mle,x=faithful$eruptions)
[1] -276.36
> mle0 <- nlmax(loglik0,p=c(m-1,m+1,s,s), x=faithful$eruptions)$estimate
> mle0
```

```
[1] 2.02837 4.28232 0.25104 0.42356
> loglik0(mle0,x=faithful$eruptions)
[1] -288.74
> stat <- 2 * (loglik(mle,x=faithful$eruptions)
+            - loglik0(mle0,x=faithful$eruptions)); stat
[1] 24.757
> 1 - pchisq(stat,df=1)          # p-value based on asymptotic distribution
[1] 6.5028e-07
> options(oldopt)
```

The small p-value gives strong evidence against an equal proportion of short and long eruption times. ◁

5.3. Confidence Intervals

Often it is more natural to express inference about parameters using a confidence interval rather than a hypothesis test. Fortunately, there is a natural duality between these two approaches that allows us to invert tests to form confidence intervals (and vice versa).

Example 5.3.1. Recall that the t-test for $H_0 : \mu = \mu_0$ vs. $H_a : \mu \neq \mu_0$ uses the test statistic

$$T = T(\boldsymbol{X}) = \frac{\overline{X} - \mu_0}{S/\sqrt{n}} \sim \mathrm{t}(n-1)$$

assuming the null hypothesis is true. Suppose we decide to reject H_0 if the p-value is at most α. That is, we reject if

$$T(\boldsymbol{X}) = \frac{|\overline{X} - \mu_0|}{S/\sqrt{n}} \geq t_* \,,$$

where t_* is the appropriate critical value. This test fails to reject precisely when

$$|\overline{X} - \mu_0| < t_* S/\sqrt{n} \,,$$

i.e., when

$$\overline{X} - t_* S\sqrt{n} < \mu_0 < \overline{X} + t_* S/\sqrt{n} \,.$$

So the $100(1-\alpha)\%$ confidence interval consists of precisely those values of μ_0 that would not be rejected by our test. ◁

The basic idea of Example 5.3.1 is generally true. Confidence intervals can be formed by "inverting" hypothesis tests. That is, confidence intervals can be thought of as the set of parameter values that would not be rejected by some test (and so are plausible values for $\boldsymbol{\theta}$). Before stating the duality theorems for confidence intervals and hypothesis tests, we first generalize the notion of confidence interval, which worked well for a single parameter, to the notion of a **confidence region**, which is more appropriate for multi-parameter models. We also introduce some terminology related to hypothesis tests.

Let Θ denote the set of all possible values of the parameter $\boldsymbol{\theta}$ in some model (i.e., in a family of distributions). A confidence region produces a subset of Θ that contains plausible values for $\boldsymbol{\theta}$.

Definition 5.3.1. Let $\boldsymbol{X}$ be random data drawn from a model with parameters $\boldsymbol{\theta}$. A 100p% **confidence region** for $\boldsymbol{\theta}$ is defined by a function C such that

- $C(\boldsymbol{X}) \subseteq \Theta$, and
- $\mathrm{P}(\boldsymbol{\theta}_0 \in C(\boldsymbol{X}) \mid \boldsymbol{\theta} = \boldsymbol{\theta}_0) = p$ for all $\boldsymbol{\theta}_0$.

That is, given random data $\boldsymbol{X}$, $C(\boldsymbol{X})$ is a random region in the space of possible parameters that has probability p of containing the correct parameter value $\boldsymbol{\theta}_0$.

Definition 5.3.2. Consider a hypothesis test of $H_0 : \boldsymbol{\theta} = \boldsymbol{\theta}_0$ at level α. The **rejection region** of this test is the following set:

$$R(\boldsymbol{\theta}_0) = \{\boldsymbol{x} \mid \text{the test rejects with data } \boldsymbol{x}\} \ .$$

Note that $\mathrm{P}[\boldsymbol{X} \in R(\boldsymbol{\theta}_0)] = \alpha$.

Typically we determine if a data set $\boldsymbol{x}$ is in the rejection region by computing a test statistic t and rejecting or not solely based on the value of the test statistic. In this case we will also refer to

$$
\begin{aligned}
t(R(\boldsymbol{\theta}_0)) &= t(\{\boldsymbol{x} \mid \text{the test rejects with data } \boldsymbol{x}\}) \\
&= \{t(\boldsymbol{x}) \mid \text{the test rejects with data } \boldsymbol{x}\} \\
&= \{t \mid \text{the test rejects with test statistic } t\}
\end{aligned}
$$

as the rejection region of the test.

The theorems that follow are simply formalizations of the same reasoning used in Example 5.3.1. The first shows that we can use likelihood ratio tests (or any other tests) to produce confidence regions.

Theorem 5.3.3. *Suppose that for every value of $\boldsymbol{\theta}_0 \in \Theta$ there is a test at level α of $H_0 : \theta = \theta_0$ vs. $H_a : \theta \neq \theta_0$. Define C by*

$$C(\boldsymbol{X}) = \{\boldsymbol{\theta} \mid \boldsymbol{X} \notin R(\boldsymbol{\theta})\} \ .$$

Then C defines a $100(1 - \alpha)\%$ confidence region for $\boldsymbol{\theta}$.

Proof. We must show that $\mathrm{P}[\boldsymbol{\theta}_0 \notin C(\boldsymbol{X})] = \alpha$ where $\boldsymbol{\theta}_0$ is the true value of the parameter. But this is immediate since

$$
\begin{aligned}
\mathrm{P}[\boldsymbol{\theta}_0 \notin C(\boldsymbol{X})] &= \mathrm{P}[\boldsymbol{X} \in R(\boldsymbol{\theta}_0)] \\
&= \alpha \ . \qquad \square
\end{aligned}
$$

We can also go in the other direction, producing hypothesis tests from confidence intervals.

Theorem 5.3.4. *Suppose that for every α, C_α defines a $100(1 - \alpha)\%$ confidence interval for $\boldsymbol{\theta}$ and that these confidence intervals are monotonic in the sense that $\alpha_1 > \alpha_2$ implies $C_{\alpha_1}(\boldsymbol{x}) \subseteq C_{\alpha_2}(\boldsymbol{x})$ for all $\boldsymbol{x}$. Let*

$$p(\boldsymbol{x}) = \sup\{\alpha \mid \boldsymbol{\theta}_0 \in C_\alpha(\boldsymbol{x})\} \ .$$

Then p defines a p-value for a test of $H_0 : \theta = \theta_0$ vs. $H_a : \theta \neq \theta_0$.

Proof. By definition,

$$\alpha < p(\boldsymbol{X}) \implies \theta_0 \in C_\alpha(\boldsymbol{X}) \text{ , and}$$
$$\alpha > p(\boldsymbol{X}) \implies \theta_0 \notin C_\alpha(\boldsymbol{X}) \text{ .}$$

Thus

$$\mathrm{P}\left[(p(\boldsymbol{X}) \le \alpha\right] = \mathrm{P}(\theta_0 \notin C_\alpha(\boldsymbol{X})) + \mathrm{P}\left[\alpha = p(\boldsymbol{X}) \text{ and } \theta_0 \in C_\alpha(\boldsymbol{X})\right]$$
$$= \alpha + \mathrm{P}\left[\alpha = p(\boldsymbol{X}) \text{ and } \theta_0 \in C_\alpha(\boldsymbol{X})\right]$$
$$\ge \alpha \text{ ,}$$

and

$$\mathrm{P}\left[p(\boldsymbol{X}) < \alpha\right] = \mathrm{P}(\theta_0 \notin C_\alpha(\boldsymbol{X})) - \mathrm{P}\left[\alpha = p(\boldsymbol{X}) \text{ and } \theta_0 \notin C_\alpha(\boldsymbol{X})\right]$$
$$= \alpha - \mathrm{P}\left[\alpha = p(\boldsymbol{X}) \text{ and } \theta_0 \notin C_\alpha(\boldsymbol{X})\right]$$
$$\le \alpha \text{ .}$$

So aside from some possible discreteness effects, $p(\boldsymbol{X}) \sim \mathsf{Unif}(0,1)$. That is precisely the definition of a p-value. $\qquad \square$

Example 5.3.2.

Q. Compute a 95% confidence interval for α in Example 5.2.4.

A. We'll let R do the calculations for us. First we define a function p such that $p(a)$ is the p-value for the test with $H_0 : \alpha = a$. Then we can use `uniroot()` to estimate when $p(a) = 0.05$. The two solutions are the endpoints of the confidence interval.

```
> oldopt <- options(warn=-1)     # suppress warnings from log(0)     faithful-ci
# loglik defined above
> mle <-  nlmax(loglik, p=c(0.5,m-1,m+1,s,s), x=faithful$eruptions)$estimate
> f <- function(a) {
+     loglik0 <- function(theta, x) {
+         theta <- c(a,theta)
+         return(loglik(theta,x))
+     }
+     mle0 <- nlmax(loglik0,p=c(m-1,m+1,s,s), x=faithful$eruptions)$estimate
+     stat <- 2 * (loglik(mle,x=faithful$eruptions)
+             - loglik0(mle0,x=faithful$eruptions)); stat
+     pval <- 1 - pchisq(stat,df=1)
+         return(pval)
+ }
> uniroot( function(a){f(a) - 0.05}, c(0,mle[1]))$root
[1] 0.29288
> uniroot( function(a){f(a) - 0.05}, c(1,mle[1]))$root
[1] 0.40692
> options(oldopt)
```

So our confidence interval is $(0.293, 0.407)$. $\qquad \triangleleft$

5.4. Goodness of Fit Testing

Goodness of fit testing approaches the question of how well a model fits the data from a different perspective. The hypothesis tests we have been discussing are focused on the accuracy of our parameter estimates. Goodness of fit testing is focused more on the overall shape of the resulting best-fit distribution and how well it matches the data. In this section we present a popular approach to goodness of fit testing.

5.4.1. Ad Hoc Methods

We begin by considering *ad hoc* methods in the context of an example.

Example 5.4.1. Recall Example 4.10.1 where we discussed the owners of a residence located along a golf course who collected 500 golf balls that landed on their property. Of the 500 golf balls collected, 486 of them bore a number between 1 and 4. The results are tabulated below:

number on ball	1	2	3	4
count	137	138	107	104

Q. What should the owners conclude?

A. We will follow our usual four-step hypothesis test procedure to answer this question.

(1) State the null and alternative hypotheses.
   - The null hypothesis is that all of the numbers are equally likely.
   - The alternative is that there is some other distribution (in the population).

(2) Calculate a test statistic.
       We need to devise a test statistic that will measure how well the data match the distribution of the null hypothesis. In Example 4.10.1, we used the range statistic $R = \max \text{count} - \min \text{count}$. For variety, let's use the maximum-count statistic M this time. When M is close to 122 (the smallest value it can have), all of the counts are nearly the same. A large value of M gives us evidence against the null hypothesis because at least one number is coming up more frequently than the others. The value of this test statistic in our data set is $m = 138$.

(3) Determine the p-value.
       Since $m = 138$, we are looking for

$$\text{p-value} = \text{P}(M \geq 138)$$

under the assumption that the null hypothesis is true. By simulating many samples of 486 golf balls under the assumption that each number is equally likely, we determined the proportion of these that have a test statistic at least as large as 138.

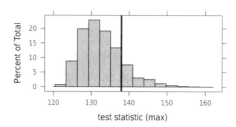

Figure 5.6. Empirical distribution of the maximum-count statistic for testing goodness of fit in the golf ball data set.

```
                                                                   golfballs-max
> plot <- statTally(golfballs, rgolfballs, max,
+                   xlab="test statistic (max)")
Test Stat function:

function (..., na.rm = FALSE)  .Primitive("max")

Test Stat applied to sample data = 138

Test Stat applied to random data:

50% 90% 95% 99%
132 140 143 149
        Of the random samples
                8095 ( 80.95 % ) had test stats < 138
                402 ( 4.02 % ) had test stats = 138
                1503 ( 15.03 % ) had test stats > 138
```

From this it appears that

$$\text{p-value } = P(M \geq 138) > 0.15 \ .$$

(See also Figure 5.6.)

(4) Interpret the p-value and draw a conclusion.

From the output above we see that our test statistic ($m = 138$) is not wildly unusual. Based on this statistic, the data are not enough to cause us to reject the null hypothesis. The fact that this p-value is larger than the p-value using the range statistic is an indication that this test is not as powerful as the one developed in Example 4.10.1. Intuitively, this makes sense since we are not using as much information from the data.   ◁

We can improve on and generalize the methods in Example 5.4.1 in a number of ways.

(1) We will develop better test statistics. Our new test statistics will be more powerful and have known asymptotic distributions.

(2) In this example, there were only 4 possible values for the number on a golf ball. We will use binning to apply our methods to situations where the number of possible values is large or where the underlying distribution is continuous.

(3) We will develop methods to deal with null hypotheses that do not completely specify the underlying distribution.

5.4.2. Pearson's Chi-Squared Statistic

Let's begin by developing a better test statistic for our golf ball example.

Example 5.4.2. Intuitively, the test statistic used above feels like it is not making full use of the evidence. For example, it would treat each of the following three samples the same, even though the first seems to offer the strongest evidence against the null hypothesis:

number on ball	1	2	3	4
sample 1	137	138	107	104
sample 2	120	138	107	121
sample 3	115	138	117	116

At a minimum, it seems like a good idea to use all of the cell counts. Let o be the vector of observed cell counts, and let $e_i = 486/4 = 121.5$ be the expected count in cell i if the null hypothesis is true. Then the following statistics (among others) come to mind:

$$A = \sum_i |o_i - e_i|, \tag{5.6}$$

$$B = \sum_i (o_i - e_i)^2, \tag{5.7}$$

$$C = \sum_i |o_i - \bar{o}|, \tag{5.8}$$

$$D = \sum_i (o_i - \bar{o})^2. \tag{5.9}$$

◁

Exercise 5.19 asks you investigate these (and any others that come to mind). The statistic traditionally used in this situation is similar to the statistic B in (5.7). The **Pearson chi-square statistic**, first proposed by Karl Pearson, has many applications. In each case, the form of the statistic is

$$X^2 = \sum_{i=1}^{n} \frac{(o_i - e_i)^2}{e_i} \tag{5.10}$$

where o_i is the observed count and e_i is the expected count for cell i. The effect of the denominator is to allow for more variability in cells where we expect larger counts. When all the expected counts are the same, as is the case in Example 5.4.1, (5.10) and (5.7) are equivalent since they are proportional to each other.

Example 5.4.3.

Q. Use simulations of the Pearson chi-squared statistic to evaluate the golf ball data from Example 5.4.1.

A. We can obtain an empirical p-value using `statTally()` as before.

```
+ E <- rep(486/4,4)                                    golfballs-pearson-sim
+ rgolfballs <- rmultinom(n=10000,size=486,prob=rep(0.25,4))
+ chisqstat <- function(x) { sum((x-E)^2 / E) }
+ plot3 <- statTally(golfballs, rgolfballs, chisqstat,
+                    xlab=expression(X^2))
```

Notice that this p-value is much smaller than the p-value we observed in Example 5.4.1. This is because the test using the Pearson chi-squared statistic is more powerful than the test used there. Based on the Pearson chi-squared statistic, we would reject the null hypothesis at the usual 5% level.   ◁

One advantage of the Pearson chi-squared statistic is that it can be shown that X^2 has an approximately chi-squared distribution when the null hypothesis is true and the sample size is large enough to expect sufficiently many observations in each cell.

The degrees of freedom depend a bit on the situation (more on this shortly), but for a table like the one we have been investigating where the null hypothesis and the sample size completely determine the expected cell counts of each of the n cells,

$$X^2 \approx \mathsf{Chisq}(n-1) \ .$$

Example 5.4.4.

Q. Use the asymptotic distribution of the Pearson chi-squared statistic to evaluate the golf ball data from Example 5.4.1.

A. We can use R to do the calculations "manually", or we can use `chisq.test()`.

```
# manual calculation                                   golfballs-pearson
> O <- golfballs; O
  1   2   3   4
137 138 107 104
> E <- rep(486/4,4); E
[1] 121.5 121.5 121.5 121.5
> X <- sum ( (O - E)^2 / E); X
[1] 8.4691
> 1 - pchisq(X,df=3)
[1] 0.037249
>
# repeated using built-in method
> chisq.test(O)

        Chi-squared test for given probabilities

data:  O
X-squared = 8.4691, df = 3, p-value = 0.03725
```

As we would expect, the p-value based on the asymptotic distribution and based on the simulations above are quite close. Keep in mind that both p-values are approximations. The empirical p-value is an approximation because it is based on a random sample from the sampling distribution rather than from the sampling distribution itself. This approximation becomes more accurate as the number of simulations increases. The asymptotic p-value is an approximation because the sampling distribution of our test statistic is only approximately $\mathsf{Chisq}(3)$. This approximation becomes more accurate as the sample size (total number of observations) increases. ◁

5.4.3. Likelihood Ratio Tests for Goodness of Fit

If we restate our goodness of fit problem as a likelihood ratio test, we get an alternative to the Pearson statistic.

Example 5.4.5. We could also have approached the golf ball problem using a likelihood ratio test where

$$\Omega_0 = \{\boldsymbol{\pi} \mid \boldsymbol{\pi} = \langle 0.25, 0.25, 0.25, 0.25 \rangle \} \text{ , and}$$

$$\Omega = \{\boldsymbol{\pi} \mid \pi_i \geq 0 \text{ and } \sum \pi_i = 1\} \text{ .}$$

From Example 5.1.10, we know that

$$\hat{\boldsymbol{\pi}} = \frac{\boldsymbol{o}}{n} = \langle \frac{o_1}{n}, \frac{o_2}{n}, \frac{o_3}{n}, \frac{o_4}{n} \rangle = \langle \frac{138}{486}, \frac{137}{486}, \frac{107}{486}, \frac{104}{486} \rangle \text{ .}$$

So our likelihood ratio statistic is

$$\lambda = \frac{L(\hat{\Omega}_0)}{L(\hat{\boldsymbol{\pi}})}$$

$$= \prod_i \left(\frac{(0.25)^{o_i}}{\left(\frac{o_i}{n}\right)^{o_i}} \right) = \prod_i \left(\frac{0.25n}{o_i} \right)^{o_i} \text{ ,}$$

and

$$-2\log \lambda = -2 \sum_i o_i \log \left(\frac{0.25n}{o_i} \right)$$

$$= 2 \sum_i o_i \log \left(\frac{o_i}{e_i} \right) \text{ .}$$

The likelihood ratio statistic is asymptotically $\mathsf{Chisq}(3)$ since Ω_0 has 3 free parameters and Ω_0 has none.

golfballs-lrt

```
+ # LRT calculation
+ O <- golfballs; O
+ E <- rep(486/4,4); E
+ G <- 2 * sum ( O * log(O/E)); G        # lrt Goodness of fit statistic
+ 1 - pchisq(G,df=3)
```

The results of the Pearson chi-squared statistic and the likelihood ratio chi-squared statistic are very similar. ◁

The likelihood ratio test statistic for goodness of fit is often denoted

$$G^2 = 2 \sum_i o_i \log \left(\frac{o_i}{e_i} \right) \text{ .}$$

This makes it easier to distinguish it from Pearson's test statistic, which is denoted

$$X^2 = \sum_i \frac{(o_i - e_i)^2}{e_i} \, ,$$

since both statistics have the same asymptotic distribution. This allows for succinct reporting of results that still makes it clear which test statistic was used. The results of Example 5.4.5, for example, could be reported as "$G^2 = 8.50$, $df = 3$, $p = 0.037$".

5.4.4. Complex Hypotheses

The hypotheses of the previous sections were all **simple hypotheses** in the sense that the null hypothesis completely determined the model (there were no remaining free parameters). Both Pearson's chi-squared statistic and the likelihood ratio test can also be used to test goodness of fit for **complex hypotheses**, that is, in situations where the null hypothesis restricts the model but still has free parameters.

Example 5.4.6. Continuing with our golf ball example, suppose that packages containing six golf balls typically contained two 1's, two 2's, one 3, and one 4 and that other packages (containing 4 or 12 golf balls) have an equal amount of each number. In this case it would be reasonable to expect an equal number of 1's and 2's and similarly an equal number of 3's and 4's, but without knowing how frequently golf balls are purchased in various quantities, there is no natural hypothesis about the individual percentages. In other words, a natural hypothesis to test in this situation is

- H_0: $\pi_1 = \pi_2$ and $\pi_3 = \pi_4$,
- H_a: $\pi_1 \neq \pi_2$ or $\pi_3 \neq \pi_4$.

The likelihood ratio test works equally well for a complex hypothesis like this. In this case

$$\Omega_0 = \{\boldsymbol{\pi} \mid \boldsymbol{\pi} = \langle \alpha, \alpha, \beta, \beta \rangle \text{ for some } \alpha \text{ and } \beta \text{ with } 2\alpha + 2\beta = 1\} \, , \text{ and}$$

$$\Omega = \{\boldsymbol{\pi} \mid \pi_i \geq 0 \text{ and } \sum \pi_i = 1\} \, .$$

We have already computed

$$L(\hat{\Omega}) = \left(\frac{138}{486}\right)^{138} \left(\frac{137}{486}\right)^{137} \left(\frac{107}{486}\right)^{107} \left(\frac{104}{486}\right)^{104} .$$

To compute $L(\hat{\Omega}_0)$, we must maximize

$$\alpha^{138} \alpha^{137} \beta^{107} \beta^{104} = \alpha^{275} \beta^{211}$$

subject to the constraint that $2\alpha + 2\beta = 1$. This maximum occurs when

$$l(\alpha) = \log\left(\alpha^{275} \left(\frac{1}{2} - \alpha\right)^{211}\right) = 275 \log(\alpha) + 211 \log(\frac{1}{2} - \alpha)$$

is maximized. Taking the derivative, we see that this occurs when

$$0 = \frac{275}{\alpha} - \frac{211}{\frac{1}{2} - \alpha} \, ,$$

$$\frac{275}{\alpha} = \frac{211}{\frac{1}{2} - \alpha} \, ,$$

$$\frac{275}{\alpha} = \frac{211}{\frac{1}{2} - \alpha} \, ,$$

$$\frac{275}{211} = \frac{\alpha}{\frac{1}{2} - \alpha} \, ,$$

$$\frac{275}{486} = \frac{\alpha}{\frac{1}{2}} = 2\alpha \, ,$$

that is, when the 2α is the sample proportion of balls bearing number 1 or 2.

The likelihood ratio statistic is therefore

$$\lambda = \frac{L(\hat{\Omega}_0)}{L(\hat{\pi})}$$

$$= \frac{\hat{\alpha}^{275} \, \hat{\beta}^{211}}{\left(\frac{138}{486}\right)^{138} \left(\frac{137}{486}\right)^{137} \left(\frac{107}{486}\right)^{107} \left(\frac{104}{486}\right)^{104}} \, .$$

The following R code calculates $-2 \log(\lambda)$ and the p-value. We use two degrees of freedom because there are 3 and 1 free parameters in the full and null models, respectively.

golfballs-complex

```
> O
[1] 137 138 107 104
> a <- sum(O[1:2]) / (2 * sum(O)); a
[1] 0.28292
> b <- sum(O[3:4]) / (2 * sum(O)); b
[1] 0.21708
> a+b                                # should equal 0.5
[1] 0.5
> lnum <- 275 * log(a) + 211 * log(b)
> ldenom <- sum( O * log (O/ sum(O)))
> X <- -2 * ( lnum - ldenom); X
[1] 0.046292
> 1 - pchisq(X,df=2)
[1] 0.97712
```

With such a large p-value, we certainly cannot reject the null hypothesis. In fact, our data look "good". Most random samples will not fit the null hypothesis as well, even when the null hypothesis is true.   ◁

The next few examples apply these ideas to test whether a proposed family of distributions is a good fit based on sample data. In each case, our null model will be a multinomial model with probabilities restricted by free parameters that capture the range of distributions in the family. The full model will again be the unrestricted multinomial model.

Example 5.4.7.

Q. As part of an experiment, a biology student counted the number of insects found on 50 plants. The data appear in the table below:

number of insects	0	1	2	3	4	5	6
plants observed	2	10	16	11	5	3	3

How well does a Poisson distribution fit the data?

A. Essentially we have a test of the following hypotheses:

- H_0: the probabilities in each cell are determined according to the $\mathsf{Pois}(\theta)$-distribution for some θ.

- H_a: the probabilities in each cell are not determined according to a Poisson distribution.

That is,

$$\Omega_0 = \{\boldsymbol{\pi}(\theta) \mid \theta > 0\}$$

where

$$\pi_i(\theta) = \mathrm{P}(\text{observation in cell } i \text{ given a } \mathsf{Pois}(\theta)\text{-distribution}) = \mathrm{dpois}(i, \theta) .$$

Our likelihood ratio statistic is

$$\lambda = \frac{L(\Omega_0)}{L(\Omega)} = \frac{L(\hat{\boldsymbol{\pi}}_0)}{L(\hat{\boldsymbol{\pi}})} ,$$

where $L(\hat{\boldsymbol{\pi}})$ is the unrestricted multinomial maximum likelihood (so $\hat{\pi}_i = \frac{o_i}{n}$) and $L(\hat{\boldsymbol{\pi}}_0)$ is the maximum likelihood subject to the constraint that the probabilities $\boldsymbol{\pi}$ agree with the probabilities of a $\mathsf{Pois}(\theta)$-distribution for some θ.

We have already observed that the MLE for the Poisson parameter is $\hat{\theta} = \overline{x}$, and $L(\Omega_0) = L(\boldsymbol{\pi}(\hat{\theta})) = L(\boldsymbol{\pi}(\overline{x}))$, so we obtain

$$\lambda = \frac{L(\boldsymbol{\pi}(\hat{\theta}))}{L(\hat{\boldsymbol{\pi}})} = \prod_i \left(\frac{\pi(\hat{\theta})_i}{\hat{\pi}_i} \right)^{o_i} ,$$

and

$$-2 \log \lambda = -2 \sum_i o_i \log \left(\frac{\pi(\hat{\theta})}{\hat{\pi}_i} \right)$$

$$= -2 \sum_i o_i \log \left(\frac{e_i/n}{o_i/n} \right)$$

$$= 2 \sum_i o_i \log \left(\frac{o_i}{e_i} \right) ,$$

where $e_i = n\pi(\hat{\theta})_i$. Notice that this takes exactly the same form as the test in the golf ball example. The only difference is how we determine e_i.

Alternatively, we could use the same e_i and o_i to calculate the Pearson chi-squared statistic,

$$X^2 = \sum_i \frac{(o_i - e_i)^2}{e_i} .$$

In either case, the degrees of freedom are computed in the usual way for a likelihood ratio test.

There are, however, a few wrinkles to smooth. The first wrinkle is that although we have only displayed seven cells in our table, there are actually infinitely many possible values for count data of this sort. Our formulas for the likelihood ratio statistics assume that the sums and products are taken of cells with non-zero observed counts. Moreover, for both statistics, we want to use a finite degrees of freedom. The typical solution is to restrict our attention to the cells with non-zero observed counts and to combine cells when cell counts are small (because the chi-squared approximation is not very good when cell counts are small).

In our example, our collapsed table might be

number of insects	0 or 1	2	3	4	5 or more
plants observed	12	16	11	5	6

This results in a test with $4 - 1 = 3$ degrees of freedom since there are 4 free parameters in the unrestricted model and 1 in the restricted model.

```
> o <- c(2,10,16,11,5,3,3)                                          fit-bugs-pois
> o.collapsed <- c(2+10,16,11,5,3+3)
> n <- sum(o)
> m <- sum(o * 0:6 ) / n       # mean count = MLE for lambda (full data)
> p <- dpois(0:6,m)
> p.collapsed <- c(p[1] + p[2], p[3:5], 1-sum(p[1:5]))   # collapsed probs
> e.collapsed <- p.collapsed * n
> print(cbind(o.collapsed,p.collapsed,e.collapsed))
     o.collapsed p.collapsed e.collapsed
[1,]          12     0.27520     13.7602
[2,]          16     0.25331     12.6656
[3,]          11     0.21616     10.8080
[4,]           5     0.13834      6.9171
[5,]           6     0.11698      5.8491
> lrt   <- 2 * sum(o.collapsed * log(o.collapsed/e.collapsed)); lrt
[1] 1.6409
> pearson <- sum( (o.collapsed-e.collapsed)^2/e.collapsed ); pearson
[1] 1.6416
> 1-pchisq(lrt, df=3)
[1] 0.65016
> 1-pchisq(pearson,df=3)
[1] 0.64999
```

The p-value indicates that we have no reason to doubt that these data come from a Poisson distribution. A small p-value would have indicated that the fit is poor.   ◁

More wrinkles

The method used in the previous example is the method that is nearly always used in practice without comment even though there still remain a couple of wrinkles to smooth. This situation is similar to the typical situation in that it is easier to estimate θ from the full data (using (5.2)) than to estimate from the collapsed data (which often must be done numerically). Furthermore, one might expect that using the full data would yield a more powerful test. As is pointed out in [**CL54**], "This is in fact the procedure recommended in many textbooks, particularly for the fitting of Poisson distributions, either as an approximation to the one with known theory ..., or more often without comment." The situation has not changed much since 1954.

Unfortunately, using the full data to derive maximum likelihood estimates of the underlying parameters and then applying these to obtain chi-squared statistics from the collapsed data makes the procedure anti-conservative (the p-values produced are too small).

The following theorem, which we state without proof, gives us some indication of how anti-conservative this approach might be. See [**CL54**] for further details.

Theorem 5.4.1. *Let $\hat{\boldsymbol{\theta}} = \langle \hat{\theta}_1, \ldots, \hat{\theta}_m \rangle$ be the maximum likelihood estimates for $\boldsymbol{\theta} = \langle \theta_1, \ldots, \theta_m \rangle$ based on the full sample and let X^2 be the likelihood ratio or Pearson chi-squared statistic computed using these estimates to determine the expected cell counts for a collapsed frequency table with k cells. Then*

$$\texttt{pchisq}(X^2, k-1-m) \leq \mathrm{P}(\chi^2 \geq X^2) \leq \texttt{pchisq}(X^2, k-1) \, . \qquad \square$$

Example 5.4.8.

Q. Using Theorem 5.4.1, give an interval containing the p-value in Example 5.4.7.

A. The true distribution lies somewhere between $\mathsf{Chisq}(4)$ and $\mathsf{Chisq}(3)$.

```
> 1-pchisq(pearson,df=5-1)
[1] 0.80129
> 1-pchisq(pearson,df=5-1-1)
[1] 0.64999
```

fit-bugs-pois2

◁

Although there is a substantial difference between the two probabilities calculated above, there is no change in the overall conclusion in this case. As long as both of the "p-values" calculated using Theorem 5.4.1 lead to the same decision, there is little cause for concern, but it is possible that the two values will straddle our stated significance level α, in which case it is not possible to know (without further analysis) whether or not we should reject at that level. It is important to know that the method most commonly seen in the literature is anti-conservative and thus overstates the strength of the evidence provided by the data. The magnitude of the effect depends on the underlying distribution and is typically quite modest for Poisson data. In the examples that follow, we will typically calculate only the anti-conservative p-value that is usually presented.

A final wrinkle is that typically the data collapse is determined *after* looking at the data but the methods assume that the number of cells was known in advance of collecting the data. Often this can be finessed because the researcher has some sense for what the data will be like prior to collecting them. Exercise 5.26 asks you to use simulations to see how much the choice of bins influences the resulting p-values.

5.4.5. Additional Examples

A general approach

The method presented in Example 5.4.7 generalizes to a wide variety of situations. The basic outline of the method is the following.

(1) Specify the hypotheses, paying attention to the number of free parameters in the null hypothesis.

(2) Calculate e and o.

It may be necessary to combine cells to avoid sparsely populated cells. For continuous data, a decision must be made about how to bin the data. (See Example 5.4.9.)

Various methods have been proposed for calculating the "expected" cell count. We have been using maximum likelihood estimators. Another method is to choose e_i to minimize the test statistic. This approach is more conservative than the MLE approach. (Exercise 5.25 asks you to explain why.)

(3) Use e and o to calculate the maximum likelihood or Pearson chi-squared statistic.

(4) Calculate the p-value by comparing the test statistic to the appropriate chi-squared distribution.

The number of free parameters in Ω is always $k-1$, where k is the number of cells in the table of observation counts. The number of free parameters under the null hypothesis will vary according to the hypothesis. Typically, the degrees of freedom will satisfy

degrees of freedom $= k - 1 -$ number of parameters estimated before calculating e

(but this may be anti-conservative if the full data are used to estimate expected cell counts that are analyzed after collapsing the frequency table to a smaller number of cells).

Pearson's chi-squared statistic and the likelihood ratio statistic are asymptotically equivalent. A Taylor series expansion argument shows that when H_0 is true and n is large, the two statistics are approximately equal (as they were in the previous example). Pearson's statistic is more commonly used in practice, largely because it is easier to calculate without a computer and so gained popularity before computers were easily available. Uses of Pearson's chi-squared test appear throughout the literature in a wide range of disciplines.

Example 5.4.9.

Q. How well does an exponential distribution fits the following data?

$$
\begin{array}{cccccccccc}
18.0 & 6.3 & 7.5 & 8.1 & 3.1 & 0.8 & 2.4 & 3.5 & 9.5 & 39.7 \\
3.4 & 14.6 & 5.1 & 6.8 & 2.6 & 8.0 & 8.5 & 3.7 & 21.2 & 3.1 \\
10.2 & 8.3 & 6.4 & 3.0 & 5.7 & 5.6 & 7.4 & 3.9 & 9.1 & 4.0
\end{array}
$$

A. Let θ be the rate parameter of the exponential distribution. The mean of the data is 7.98, so $\hat{\theta} = 1/7.98 = 0.125$. After selecting endpoints for data bins, we can use $\hat{\theta}$ to determine the expected count in each cell. The (anti-conservative) degrees of freedom is $4 - 1 - 1 = 2$ since there is one free parameter in the null model.

fit-exp

```
> data <- c(18.0,6.3,7.5,8.1,3.1,0.8,2.4,3.5,9.5,39.7,
+           3.4,14.6,5.1,6.8,2.6,8.0,8.5,3.7,21.2,3.1,
+           10.2, 8.3,6.4,3.0,5.7,5.6,7.4,3.9,9.1,4.0)
> n <- length(data)
> theta.hat <- 1/mean(data); theta.hat
[1] 0.12526
> cutpts <- c(0,2,4,7,12,Inf)
> bin.data <- cut(data, cutpts)
> p <- diff(pexp(cutpts,theta.hat))
> e <- n * p
> o <- table(bin.data)
> print(cbind(o,e))
           o      e
(0,2]      1 6.6482
(2,4]     10 5.1749
(4,7]      6 5.6939
(7,12]     9 5.8101
(12,Inf]   4 6.6730
> lrt  <- 2 * sum(o * log(o/e)); lrt
[1] 13.798
> pearson <- sum( (o-e)^2/e ); pearson
[1] 12.136
> 1-pchisq(lrt, df=2)            # df = (4 - 1) - 1 [anti-conservative]
[1] 0.0010087
> 1-pchisq(pearson,df=2)
[1] 0.0023157
> 1-pchisq(lrt, df=3)           # df = 4 - 1        [conservative]
[1] 0.0031931
> 1-pchisq(pearson,df=3)
[1] 0.0069312
```

The low p-value indicates that this data would be quite unusual for an exponential distribution. A comparison of observed and expected counts shows that compared with an exponential distribution, our data are too "bunched together". We may need to consider a different model for these data. (Perhaps a normal, gamma, or Weibull distribution will fit better. Exercise 5.20 asks you to find out.)   ◁

5.5. Inference for Two-Way Tables

The ideas of the previous section have been applied to the analysis of **two-way tables** – tables reporting the number of observational units with each possible combination of values for two categorical variables.

Example 5.5.1. A study of students was conducted in which subjects were asked whether or not they smoked and how many of their parents smoked. The results of the study were reported in [**Zag67**] and appear also in [**BM09**]. We can represent the data in a two-way table where rows correspond to the student smoking status and columns to the parental smoking status.

```
> xtabs(~Student+Parents,data=familySmoking) -> smokeTab      smoke-xtabs
> smokeTab
                 Parents
Student          NeitherSmokes OneSmokes BothSmoke
  DoesNotSmoke            1168      1823      1380
  Smokes                   188       416       400                        ◁
```

The general form of such a two-way table with I rows and J columns is

	1	2	3	$\ldots$	J	total
1	n_{11}	n_{12}	n_{13}	$\ldots$	n_{1J}	$n_{1\cdot}$
2	n_{21}	n_{22}	n_{23}	$\ldots$	n_{2J}	$n_{2\cdot}$
$\vdots$	$\vdots$	$\vdots$	$\vdots$	$\ddots$	$\vdots$	$\vdots$
I	n_{I1}	n_{I2}	n_{I3}	$\ldots$	n_{IJ}	$n_{I\cdot}$
total	$n_{\cdot1}$	$n_{\cdot2}$	$n_{\cdot3}$	$\ldots$	$n_{\cdot J}$	$n_{\cdot\cdot}$

This describes the data. Now we need to determine our hypotheses. These are more easily expressed in terms of proportions:

$$\pi_{ij} = \text{proportion of population in cell } \langle i,j \rangle .$$

In a test of independence, the null hypothesis is

$$H_0 : \pi_{ij} = \pi_{i\cdot}\pi_{\cdot j} \text{ for all } \langle i,j \rangle ,$$

where $\pi_{i\cdot} = \sum_j \pi_{ij}$ and $\pi_{\cdot j} = \sum_i \pi_{ij}$. That is, the probability of a random observation being in cell $\langle i,j \rangle$ is the product of the probabilities of being in row i and column j.

As in the previous section, we need to determine the expected cell counts e_{ij} under the null hypothesis (after estimating some parameters) and the appropriate degrees of freedom. Once again, the probabilities must sum to 1, so

$$\Omega = \left\{ \boldsymbol{\pi} \mid \pi_{ij} \geq 0 \text{ and } \sum_{i,j} \pi_{ij} = 1 \right\} , \text{ and}$$

$$\Omega_0 = \left\{ \boldsymbol{\pi} \mid \pi_{ij} = \pi_{i\cdot}\pi_{\cdot j} \text{ and } \sum_i \pi_{i\cdot} = \sum_j \pi_{\cdot j} = 1 \right\} .$$

Because of the constraints on $\boldsymbol{\pi}$, we see that there are $IJ - 1$ degrees of freedom for Ω and $(I-1) + (J-1)$ degrees of freedom for Ω_0. And since

$$IJ - 1 - [(I-1) + (J-1)] = (I-1)(J-1) ,$$

our test statistics will have the $\mathsf{Chisq}((I-1)(J-1))$-distribution when H_0 is true.

It can be shown that the MLE for $\hat{\boldsymbol{\pi}}$ in Ω_0 satisfies

$$\hat{\pi}_{ij} = \frac{n_{i\cdot}}{n_{\cdot\cdot}} \cdot \frac{n_{\cdot j}}{n_{\cdot\cdot}} .$$

Thus

$$e_{ij} = n_{\cdot\cdot}\hat{\pi}_{ij} = \frac{n_{i\cdot}n_{\cdot j}}{n_{\cdot\cdot}} = \frac{\text{row total} \cdot \text{column total}}{\text{grand total}} .$$

Example 5.5.2. We can easily calculate the test statistics and p-value for our smoking data set.

smoke-manual

```
> row.sum <- apply(smokeTab,1,sum)
> col.sum <- apply(smokeTab,2,sum)
> grandTotal <- sum(smokeTab)
> e <- outer(row.sum,col.sum)/grandTotal; e
            NeitherSmokes OneSmokes BothSmoke
DoesNotSmoke      1102.71    1820.78   1447.51
Smokes             253.29     418.22    332.49
> o <- smokeTab
> stat <- sum ( (e-o)^2/e); stat
[1] 37.566
> pval <- 1 - pchisq(stat,df=2); pval
[1] 6.9594e-09
```

Of course, R also provides a function to do the work for us.

smoke-test

```
> chisq.test(smokeTab)

        Pearson's Chi-squared test

data:  smokeTab
X-squared = 37.566, df = 2, p-value = 6.96e-09
```

The object returned by `chisq.test()` contains some additional information which we can access if we want more details.

smoke-test-attr

```
> attributes((chisq.test(smokeTab)))
$names
[1] "statistic" "parameter" "p.value"   "method"    "data.name" "observed"
[7] "expected"  "residuals"

$class
[1] "htest"
```

The `xchisq.test()` function in `fastR` accesses this information to produce a summary table listing four values for each cell:

- the observed count,
- the expected count,
- the contribution to the χ^2 statistic: $\dfrac{(o-e)^2}{e}$, and
- the residual: $\dfrac{(o-e)}{\sqrt{e}}$.

smoke-test2

```
> xchisq.test(smokeTab)

        Pearson's Chi-squared test

data:  smokeTab
X-squared = 37.566, df = 2, p-value = 6.96e-09
```

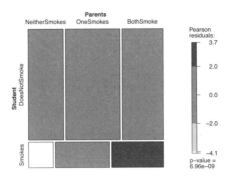

Figure 5.7. A mosaic plot of the family smoking data. Colors indicate cells with large positive (dark) and negative (light) residuals.

```
 1168.00     1823.00     1380.00
(1102.71)   (1820.78)   (1447.51)
[ 3.8655]   [ 0.0027]   [ 3.1488]
< 1.966>    < 0.052>    <-1.774>


  188.00      416.00      400.00
( 253.29)   ( 418.22)   ( 332.49)
[16.8288]   [ 0.0118]   [13.7086]
<-4.102>    <-0.109>    < 3.703>

key:
        observed
        (expected)
        [contribution to X-squared]
        <residual>
```

The residuals can be used to help us determine in what ways our data depart from the hypothesis of independence. In this case, we see that students are more likely to be smokers if both their parents smoke and less likely if neither parent smokes. We can visualize this using a **mosaic plot** (see Figure 5.7).

smoke-mosaic

```
> require(vcd)
> smoke.plot <- mosaic(~Student+Parents,familySmoking,shade=T)
```
◁

Example 5.5.3. In another study of smoking behavior [**CEF96**] asked 7th grade students to answer questions about their smoking exposure and smoking behavior. Students were classified as either "experimenters" or non-smokers. In one of the items to gauge exposure, students were asked to assess their exposure to advertisements in magazines as "Don't read magazines with cigarette ads", "Hardly ever/never see ads", or "See ads sometimes or a lot".

The results are tabulated below:

	Never	Hardly Ever	Sometimes or a lot
NonExperimenter	171	15	148
Experimenter	89	10	132

The analysis of such a table proceeds as before.

```
> smTab <- rbind(NonExperimenter=c(171,15,148),                    smoking-ads
+                Experimenter=c(89,10,132))
> colnames(smTab) = c('Never','Hardly Ever', 'Sometimes or a lot')
> smTab
                Never Hardly Ever Sometimes or a lot
NonExperimenter   171          15                148
Experimenter       89          10                132
> chisq.test(smTab)


        Pearson's Chi-squared test

data:  smTab
X-squared = 9.3082, df = 2, p-value = 0.009523
```

The small p-value indicates an association between magazine advertising exposure and smoking behavior. An analysis of residuals shows that experimenters are more likely to have higher exposure to magazine advertisements:

```
> xchisq.test(smTab)                                              smoking-ads2


        Pearson's Chi-squared test

data:  smTab
X-squared = 9.3082, df = 2, p-value = 0.009523

  171.00      15.00      148.00
 (153.70)   ( 14.78)    (165.52)
 [1.9474]   [0.0033]    [1.8549]
 < 1.396>   < 0.058>    <-1.362>

   89.00      10.00      132.00
 (106.30)   ( 10.22)    (114.48)
 [2.8158]   [0.0048]    [2.6820]
 <-1.678>   <-0.069>    < 1.638>

key:
        observed
        (expected)
        [contribution to X-squared]
        <residual>                                                       ◁
```

Example 5.5.4. Schooler *et al* [**CEF96**] also asked about exposure to billboard advertising. The results are tabulated below:

	Never	Hardly ever	Sometimes or a lot
NonExperimenter	34	4	296
Experimenter	15	3	213

This time the p-value does not suggest an association:

```
> smTab2 <- rbind(NonExperimenter=c(34,4,296),                    smoking-bbs
+                 Experimenter=c(15,3,213))
> colnames(smTab2) <- c('Never','Hardly ever','Sometimes or a lot')
> smTab2
               Never Hardly ever Sometimes or a lot
NonExperimenter   34           4                296
Experimenter      15           3                213
> chisq.test(smTab2)


        Pearson's Chi-squared test

data:  smTab2
X-squared = 2.3455, df = 2, p-value = 0.3095


Warning message:
In chisq.test(smTab2) : Chi-squared approximation may be incorrect
```

The warning message is due to the small expected cell counts in the middle column of the table. Since our p-value is not close to indicating significance, the fact that the chi-squared distribution is providing a poor approximation does not worry us too much. To be safe, we repeat the analysis with the middle column removed. (Another option would have been to combine those results with one of the other two columns.)

```
> smTab2[,-2]                                                     smoking-bbs2
               Never Sometimes or a lot
NonExperimenter   34                296
Experimenter      15                213
> chisq.test(smTab2[,-2])


        Pearson's Chi-squared test with Yates' continuity correction

data:  smTab2[, -2]
X-squared = 1.8928, df = 1, p-value = 0.1689
```

Although the p-value has changed a fair amount, the interpretation is unchanged. These data do not provide evidence that smoking behavior is associated with self-reported billboard exposure.   ◁

In the previous example students were asked about their and their parents' smoking status. Now let's consider a slightly different situation. Suppose the researchers randomly assign subjects to treatment groups and measure a categorical response for each subject. Such data can also be summarized in a two-way table, but this time the marginal probabilities and totals in one direction are known based on the study design. Although the argument is different, the same χ^2 statistic and degrees of freedom apply for this type of two-way table, too.

Example 5.5.5. One of the most famous clinical trials is the *Physicians' Health Study* [**The89**]. This was a large prospective study that used physicians as subjects. Physicians were used because they were considered well qualified to assess the risks and give informed consent and because they were able to report reliable physiological measurements without costing the researchers the time and expense to set up appointments with each subject.

Among the research questions was whether low doses of aspirin reduced the risk of heart attack. Physicians were randomly assigned to receive either an aspirin or placebo tablet. The unlabeled pills arrived by mail so that the subjects were unaware which they were taking. Here are the results of this component of the study presented in tabular form:

	Heart Attack	No Heart Attack
Aspirin	104	10,933
Placebo	189	10,845

The large number of participants was necessary in order to have a reasonably large expected number of heart attacks. Once again, we can use the chi-squared test to analyze these data.

```
> phs <- cbind(c(104,189),c(10933,10845))                              phs
> rownames(phs) <- c("aspirin","placebo")
> colnames(phs) <- c("heart attack","no heart attack")
> phs
        heart attack no heart attack
aspirin          104           10933
placebo          189           10845
> xchisq.test(phs)


        Pearson's Chi-squared test with Yates' continuity correction

data:  phs
X-squared = 24.429, df = 1, p-value = 7.71e-07

   104.00    10933.00
(  146.52) (10890.48)
 [12.34]     [ 0.17]
 <-3.51>    < 0.41>

   189.00    10845.00
(  146.48) (10887.52)
 [12.34]     [ 0.17]
 < 3.51>    <-0.41>

key:
        observed
        (expected)
        [contribution to X-squared]
        <residual>
```

The data so clearly indicate a beneficial effect of aspirin that this part of the study was terminated early. It was decided that it was no longer ethical to withhold this information from the subjects (and the population at large).   ◁

The **Yates' continuity correction** [Yat34] mentioned in the output above is a correction that is (optionally) applied in an attempt to achieve more accurate p-values. The chi-squared test compares a discrete test statistic to a continuous chi-squared distribution. Just as was the case when we approximated a binomial distribution with a normal distribution, this approximation in the case of 2×2 tables can be improved by using an adjustment of 0.5. The Yates' continuity correction uses

$$\sum \frac{(\mid o_i - e_i \mid - 0.5)^2}{e_i}$$

in place of

$$\sum \frac{\mid o_i - e_i \mid^2}{e_i}$$

as the test statistic. For large data sets such as the one from the *Physicians' Health Study*, the difference between the two methods will typically be small. The Yates' continuity correction was intended to improve p-value calculations for smaller data sets. For a 2×2 table we also have the option of using Fisher's exact test.

Example 5.5.6.

Q. Recall the twin study from Example 2.7.4. How do the p-values from Fisher's exact test and the chi-squared test with and without a continuity correction compare?

A. We can calculate the three p-values in R:

```
> convictions <- rbind(dizygotic=c(2,15), monozygotic=c(10,3))      chisq-twins
> colnames(convictions) <- c('convicted','not convicted')
> convictions
            convicted not convicted
dizygotic           2            15
monozygotic        10             3
> chisq.test(convictions,correct=FALSE)

        Pearson's Chi-squared test

data:  convictions
X-squared = 13.032, df = 1, p-value = 0.0003063

> chisq.test(convictions)$p.value
[1] 0.0012211
> fisher.test(convictions)$p.value
[1] 0.00053672
```

The uncorrected chi-squared test produces a p-value that is smaller than the p-value from Fisher's exact test. The Yates' correction, on the other hand, is conservative and produces a p-value that is a bit too large.   ◁

In many situations, it is not possible to assign treatments randomly. Researchers may instead collect independent samples in each of two or more categories of subject. A case-control study is one example of this. In a case-control study, researchers conduct separate samples of cases and controls rather than sampling from a larger population and checking to see which subjects are cases and which are controls. The primary advantage to this is that it is possible to get roughly equal numbers of cases and controls even if the incidence rate of the cases is quite small in the population at large. The challenge in collecting data for a case-control study is to ensure that the two samples are similar in all other ways except for case-control status, as they would be in a randomized design. Otherwise we can't be sure whether differences between the groups are associated with case-control status or with some other factor that differs between the groups.

The analysis in this situation is identical to that of the designed experiment with randomly assigned treatment groups since the marginal counts and percentages in one direction are fixed by the design of the study in either case.

Example 5.5.7. Genomewide association studies (GWAS) search for genes that contribute to diseases by genotyping large numbers of genetic markers in cases and controls. One famous early GWAS genotyped 116,204 single-nucleotide polymorphisms (SNPs) in 50 age-related macular degeneration (AMD) cases and 96 controls. The results for one of these SNPs (the C/G SNP rs380390) can be estimated from the numerical summaries and graphs in [**KZC**$^+$**05**]:

	CC	CG	GG
cases	26	17	6
controls	13	46	37

Because the cases are more likely to have the C allele than the controls, the C allele is sometimes called the "risk allele". There are three different models that we could fit to this data given the machinery currently at our disposal. We could do a 2-degree of freedom Chi-squared test to see if the the proportions of cases and controls in each genotype class are the same. We could also fit models that correspond to dominant or recessive genetic models (from the perspective of the C allele) by combining either the first two or the last two columns.

Q. For which of these models do we have the strongest evidence?

A. We can fit all three models in R and compare the p-values.

```
> amd <- rbind(cases=c(27,17,6), controls=c(13,46,37))        amd
> dom <- rbind(cases=c(27+17,6), controls=c(13+46,37))
> rec <- rbind(cases=c(27,17+6), controls=c(13,46+37))
> chisq.test(amd)

        Pearson's Chi-squared test

data:  amd
X-squared = 28.982, df = 2, p-value = 5.089e-07

> chisq.test(dom)

        Pearson's Chi-squared test with Yates' continuity correction

data:  dom
X-squared = 9.9059, df = 1, p-value = 0.001648
```

```
> chisq.test(rec)

        Pearson's Chi-squared test with Yates' continuity correction

data:  rec
X-squared = 25.059, df = 1, p-value = 5.56e-07
```

All of the p-values are small, but keep in mind that this is just one (the best one) out of more than 100,000 such tests that were performed, so we must adjust our threshold for significance. We would expect approximately 5,000 of these SNPs to produce p-values below 0.05 just by chance, for example. A crude adjustment (called the Bonferroni correction; see Section 7.3.7) is to use $0.05/100{,}000 = 5 \cdot 10^{-7}$ as our significance level for each test. Using this (conservative) adjustment, we see that the 2-degree of freedom model and the recessive model are borderline significant, but there is no evidence to support the dominant model. The authors used a slightly different model and reported a p-value of $4.1 \cdot 10^{-8}$.

As genotyping costs decreased, the number and size of genomewide association studies increased rapidly in the several years following the publication of this result. Another example is presented in Exercise 5.28. ◁

5.6. Rating and Ranking Based on Pairwise Comparisons

In this section we discuss some likelihood-based methods for rating or ranking based on pairwise comparisons. We will describe it in terms of contests between sports teams that result in a winner and a loser (there is also a way to generalize the method to accommodate ties), but the methods can be applied to a wide variety of other situations. For example, one could rate a number of colas by having a number of people taste two of them and express their preference. The question was originally posed by Zermelo, who was interested in methods for declaring a winner of a chess tournament that gets interrupted before all the matches have been played.

We can divide this problem into two phases:

(1) Specify the model.
    What are the parameters of the model and how are they related?

(2) Fit the model.
    How do we calculate the "best" values for the parameters from our data?

As we have seen, the method of maximum likelihood provides a general framework for fitting models once we have specified the parameters and distributions involved. Alternatively, we could minimize some measure of lack of fit (as is done in least squares methods) or use a method similar to the method of moments (by requiring for example that each team's expected winning percentage be their actual winning percentage). Our focus here will be primarily on specifying a reasonable model. We will fit the model using maximum likelihood, and we will let R do all the heavy lifting for us.

5.6.1. Specifying a Model

The natural parameters for a model are

$$\pi_{ij} = \text{probability team } i \text{ defeats team } j .$$

We will constrain these parameters so that $\pi_{ij} + \pi_{ji} = 1$. If we consider the outcome of a contest between teams i and j to be a $\mathsf{Binom}(1, \pi_{i,j})$ random variable and if we further consider the outcome of each game to be independent of the others, then the likelihood is a simple product of binomial pmfs, so it appears that we have all we need to use maximum likelihood to fit the model.

Unfortunately, this choice of parameters immediately leads to some (surmountable) difficulties:

- It is unclear how to convert these parameters into a ranking of teams from best to worst, since they may provide information that is not transitive.

  For example, if $\pi_{12} = \frac{2}{3}$, $\pi_{23} = \frac{2}{3}$, and $\pi_{31} = \frac{2}{3}$, which team do we consider to be the best among these three? One can argue that this is a feature rather than a problem: It may well be that team 1 has an edge against team 2, team 2 against team 3, and team 3 against team 1. But if our goal is to linearly order the teams, then this is awkward.

- In a typical sports application (and other settings, too) the number of pairs of teams is much larger than the number of games played.

  That is, most teams don't play most of the other teams. Having far more parameters than data makes estimation problematic.

A solution to both difficulties is to reparameterize the problem. For each team we introduce a parameter

$$R_i = \text{rating of team } i .$$

If we can relate these ratings to the π_{ij}'s, then we can use maximum likelihood to derive estimates. A number of methods have been proposed, including

- $\pi_{ij} = .5 + (R_i - R_j)$.

  Like the methods below, this satisfies $\pi_{ij} + \pi_{ji} = 1$ and has the property that larger ratings differences lead to higher probabilities that the higher rated team wins. One weakness of this method is that it is not constrained to produce probabilities. If $R_i - R_j > .5$, for example, then $\pi_{ij} > 1$. Nevertheless, this model has been used. It works best in situations where teams are relatively close in ability.

- $\pi_{ij} = \dfrac{R_i}{R_i + R_j}$.

  This avoids the problem of π_{ij}'s that are not probabilities but is not as commonly used as our third method.

- $\pi_{ij} = \dfrac{1}{1 + e^{-(R_i - R_j)}}$.

  Although this idea has been "discovered" repeatedly, including by Zermelo, it is generally called the **Bradley-Terry model**. This can be motivated several ways (and is probably the most commonly used method). The

Bradley-Terry model can be expressed several equivalent ways:

$$\pi_{ij} = \frac{1}{1 + e^{-(R_i - R_j)}} = \frac{e^{R_i}}{e^{R_i} + e^{R_j}} = \frac{e^{R_i - R_j}}{1 + e^{R_i - R_j}} = 1 - \frac{1}{1 + e^{(R_i - R_j)}} \, .$$

From this we see that this is just a logarithmic rescaling of the previous model. We can invert this relationship as follows:

$$\pi_{ij} = \frac{1}{1 + e^{-(R_i - R_j)}} \, ,$$

$$\frac{1}{\pi_{ij}} = 1 + e^{-(R_i - R_j)} \, ,$$

$$\frac{1}{\pi_{ij}} - 1 = e^{-(R_i - R_j)} \, ,$$

$$\log(\frac{1}{\pi_{ij}} - 1) = -(R_i - R_j) \, ,$$

$$\log(\frac{\pi_{ij}}{1 - \pi_{ij}}) = R_i - R_j \, .$$

This shows that the difference in ratings is the logarithm of the odds that the stronger team wins.

See Section 6.7 (logistic regression) for more information about this transformation.

Note that for the last two models, shifting all the ratings by a constant yields the same winning probabilities. This constant is generally chosen to achieve some desirable property. For example, we could give our favorite team a rating of 0. Better teams would have a positive rating and worse teams would have a negative rating. Alternatively, we could shift things so that the maximum rating is 100 or so all ratings are positive numbers.

Using any of these (or other) functions relating π_{ij} to R_i and R_j, we can use maximum likelihood to fit a model where the outcome of each game is considered to be a $\mathsf{Binom}(1, \pi_{ij})$-random variable. One nice feature of the Bradley-Terry model is that there are fast iterative methods for approximating the maximum likelihood estimates.

Home team advantage

A home team advantage can be modeled by adjusting the formulas that determine π_{ij} from R_i and R_j. In the Bradley-Terry model, this is typically taken to be

$$\pi_{ij} = \frac{1}{1 + e^{-(\alpha + R_i - R_j)}} = \frac{e^{\alpha + R_i - R_j}}{1 + e^{\alpha + R_i - R_j}} \, ,$$

where team i is the home team. This model gives the same advantage (generically called the **order effect**) to all home teams. The probability that the home team wins a contest between two evenly matched teams is then

$$\pi_{\text{home}} = \frac{1}{1 + e^{-\alpha}} = \frac{e^{\alpha}}{1 + e^{\alpha}} \, .$$

For games contested on a neutral site, α is simply omitted. More complicated models can be made that estimate separate home court advantages for each team.

In the context of chess, the home team advantage is replaced by the color of pieces a player uses. In a taste-test situation, the order effect can be used to measure the effect of the order in which samples are tasted.

5.6.2. Fitting the Bradley-Terry Model in R

The `BradleyTerry2` package in R makes it easy to fit a Bradley-Terry model from a data frame that lists the winner and loser of each contest. We need to do a little data management to transform data that lists home and away teams and their scores.

Example 5.6.1. We can apply this technique to the 2007 NFL season (including post-season). We begin with a little preprocessing of the data.

```
> nfl <- nfl2007                              # shorten name of data set   nfl-prep
> head(nfl,3)
          Date            Visitor VisitorScore               Home HomeScore
1 09/06/2007 New Orleans Saints           10 Indianapolis Colts        41
2 09/09/2007 Kansas City Chiefs            3      Houston Texans        20
3 09/09/2007      Denver Broncos           15       Buffalo Bills        14
  Line TotalLine
1  5.5      53.5
2  3.0      38.0
3 -3.0      37.0
> nfl$dscore <- nfl$HomeScore - nfl$VisitorScore
> w <- which(nfl$dscore > 0)
> nfl$winner <- nfl$Visitor; nfl$winner[w] <- nfl$Home[w]
> nfl$loser <- nfl$Home; nfl$loser[w] <- nfl$Visitor[w]
>
# did the home team win?
> nfl$homeTeamWon <- nfl$dscore > 0
> head(nfl,3)
          Date            Visitor VisitorScore               Home HomeScore
1 09/06/2007 New Orleans Saints           10 Indianapolis Colts        41
2 09/09/2007 Kansas City Chiefs            3      Houston Texans        20
3 09/09/2007      Denver Broncos           15       Buffalo Bills        14
  Line TotalLine dscore             winner              loser homeTeamWon
1  5.5      53.5     31 Indianapolis Colts New Orleans Saints        TRUE
2  3.0      38.0     17      Houston Texans Kansas City Chiefs        TRUE
3 -3.0      37.0     -1      Denver Broncos       Buffalo Bills       FALSE
```

The `BTm()` function fits a Bradley-Terry model if we provide it the data on winners and losers.

```
# fit Bradley-Terry model                                              nfl-bt
> require(BradleyTerry2)
> BTm(cbind(homeTeamWon,!homeTeamWon), Home, Visitor,
+              data=nfl, id='team') -> nfl.model
```

The pure Bradley-Terry model cannot be fit to the regular season alone since New England did not lose in the regular season. Fortunately (for the purposes of this example) they lost to the New York Giants in the Super Bowl. So if we include the post-season games, we can apply this method. In this case there is no MLE since the likelihood can always be increased by increasing New England's rating relative to the other teams' ratings. More generally, the Bradley-Terry model cannot be uniquely fit whenever the teams can be partitioned into two subsets A and B, so that all games between a team from A and a team from B were won by the team from A. Methods have been developed to deal with this situation, but they require some arbitrary choices.

With a little post-processing, we can compare the ratings from the Bradley-Terry model with the records of the teams.

nfl-post

```
> bta <- BTabilities(nfl.model)
> nflRatings<- data.frame(
+       team = rownames(bta),
+       rating = bta[,"ability"],
+       se = bta[,"s.e."],
+       wins = as.vector(table(nfl$winner)),
+       losses = as.vector(table(nfl$loser))
+       )
> rownames(nflRatings) = NULL
>
> nfl$winnerRating <- nflRatings$rating[as.numeric(nfl$winner)]
> nfl$loserRating <- nflRatings$rating[as.numeric(nfl$loser)]
> nfl$upset <- nfl$loserRating > nfl$winnerRating
> nflRatings[rev(order(nflRatings$rating)),]
                   team    rating      se wins losses
19 New England Patriots   4.49516 1.26441   18      1
9        Dallas Cowboys   2.90791 0.97269   13      4
21       New York Giants  2.71939 0.92199   14      6
12      Green Bay Packers  2.62748 0.91271   14      4
14      Indianapolis Colts 2.40009 0.93081   13      4
26     San Diego Chargers  2.22357 0.89032   13      6
15  Jacksonville Jaguars   1.74064 0.84936   12      6
32   Washington Redskins   1.55026 0.85807    9      8
24   Philadelphia Eagles   1.46789 0.89083    8      8
31       Tennessee Titans  1.23687 0.84567   10      7
18      Minnesota Vikings  1.18262 0.87524    8      8
6          Chicago Bears   1.01555 0.85668    7      9
11          Detroit Lions  1.00631 0.84158    7      9
13         Houston Texans  0.76722 0.84332    8      8
30  Tampa Bay Buccaneers   0.69610 0.76992    9      8
28      Seattle Seahawks   0.67183 0.72228   11      7
10        Denver Broncos   0.66205 0.85446    7      9
25   Pittsburgh Steelers   0.50686 0.78727   10      7
8       Cleveland Browns   0.43869 0.79201   10      6
5       Carolina Panthers  0.30337 0.78674    7      9
4          Buffalo Bills   0.29763 0.90231    7      9
20     New Orleans Saints  0.19494 0.77223    7      9
1       Arizona Cardinals  0.00000 0.00000    8      8
7      Cincinnati Bengals -0.36132 0.78382    7      9
16     Kansas City Chiefs -0.37729 0.90807    4     12
23       Oakland Raiders  -0.40157 0.93060    4     12
```

```
2          Atlanta Falcons  -0.63400 0.81582   4    12
27     San Francisco 49ers  -0.77368 0.74825   5    11
3        Baltimore Ravens   -0.80634 0.82005   5    11
22          New York Jets   -0.94156 0.95841   4    12
29          St Louis Rams   -1.39114 0.81833   3    13
17         Miami Dolphins   -2.51582 1.25246   1    15
```

And we can see how big an upset the Super Bowl was.

```
> require(faraway)          # for ilogit(), the inverse logit      nfl-post2
> nfl$pwinner <- ilogit(nfl$winnerRating - nfl$loserRating)
# how big an upset was the Super Bowl?
> nfl[nrow(nfl),]
        Date       Visitor VisitorScore          Home HomeScore
267 02/03/2008 New York Giants          17 New England Patriots        14
    Line TotalLine dscore       winner             loser homeTeamWon
267 12.5      54.5     -3 New York Giants New England Patriots       FALSE
    winnerRating loserRating upset pwinner
267       2.7194      4.4952  TRUE 0.14483
```

Even using the result that the New York Giants defeated New England in the Super
Bowl, our model would predict this to happen again less than 15% of the time if
they played a rematch.   ◁

Example 5.6.2. We can also fit a Bradley-Terry model to the 2009–2010 NCAA
Division I basketball season. Before fitting the model, we manipulate the data
into the proper format and we remove teams that have played only a few games.
Presumably these are Division II teams that played a small number of games against
Division I teams. Many of these teams do not have both a win and a loss, which is
problematic for the reasons discussed above.

```
> ncaa <- ncaa2010   # make a local copy (with shorter name)   ncaa2010-bt-prep
# at a neutral site?
> ncaa$neutralSite <- grepl('notes',ncaa$n,ignore.case=TRUE)
# did home team win?
> ncaa$homeTeamWon <- ncaa$hscore > ncaa$ascore
# remove teams that didn't play >= 5 at home and >=5 away
# (typically div II teams that played a few div I teams)
> h <- table(ncaa$home); a <- table(ncaa$away)
> deleteTeams <- c(names(h[h<=5]), names(a[a<=5]))
> ncaa <- ncaa[!( ncaa$home %in% deleteTeams |
+     ncaa$away %in% deleteTeams ), ]
# remove unused levels from home and away factors
> teams <- union(ncaa$home, ncaa$away)
> ncaa$home <- factor(ncaa$home, levels=teams)
> ncaa$away <- factor(ncaa$away, levels=teams)
```

Now we fit the model

```
# fit a Bradley-Terry model                              ncaa2010-bt-fit
> require(BradleyTerry2)
> ncaa.model <- BTm( cbind(homeTeamWon, 1-homeTeamWon),
+                     home, away, data=ncaa, refcat="Duke")
```

```
                                                              ncaa2010-bt-look
# look at top teams
> coef(ncaa.model)[rev(order(coef(ncaa.model)))[1:6]]
     ..Kansas      ..Kentucky      ..Syracuse ..West Virginia
       1.31689        0.78092        0.45011          0.22476
   ..Villanova     ..Kansas St.
      -0.16292       -0.18196
```

```
                                                             ncaa2010-bt-look2
```

```
# nicer output this way        Pittsburgh     -0.57280 0.73428
> ratings <-                    Texas A&M      -0.65481 0.74605
+       BTabilities(ncaa.model) Tennessee      -0.65819 0.76753
> ratings[                      Ohio St.       -0.65937 0.74158
+    rev(order(ratings[,1]))[1:30],]  Brigham Young -0.74902 0.81210
          ability    s.e.       Butler         -0.75424 0.81366
Kansas         1.31689 0.95579  Texas          -0.81704 0.73482
Kentucky       0.78092 0.95100  Wisconsin      -0.84290 0.71344
Syracuse       0.45011 0.80106  Maryland       -0.86076 0.66085
West Virginia  0.22476 0.74229  Vanderbilt     -0.93658 0.74012
Duke           0.00000 0.00000  Marquette      -0.96081 0.70536
Villanova     -0.16292 0.73933  Northern Iowa  -1.01536 0.80212
Kansas St.    -0.18196 0.77811  Michigan St.   -1.02432 0.72921
Purdue        -0.23129 0.78047  Xavier         -1.04813 0.72422
Temple        -0.27092 0.78729  Gonzaga        -1.05466 0.75151
New Mexico    -0.36386 0.83286  Notre Dame     -1.11335 0.71332
Georgetown    -0.37770 0.69641  Florida St.    -1.13394 0.66605
Baylor        -0.45184 0.76570  St. Mary's     -1.17581 0.80087
```

We can also fit a model that takes into account home court advantage. In this model a fixed (i.e., the same for all teams and games) home court advantage is added to the rating of the home team before determining the probability of winning. The ratings that result are the ratings for a game at a neutral site (and can be adjusted to give probabilities for a game where one team has a home court advantage).

```
                                                              ncaa2010-bt-hc-fit
> require(BradleyTerry2)
# home team gets advantage unless on neutral court
> ncaa$home <- data.frame(team=ncaa$home, at.home = 1 - ncaa$neutralSite)
> ncaa$away <- data.frame(team=ncaa$away, at.home = 0)
> ncaa.model2 <- update(ncaa.model, id='team', formula = ~ team + at.home)
```

```
                                                             ncaa2010-bt-hc-look
# the "order effect" is the coefficient on "at.home"
> coef(ncaa.model2)["at.home"] -> oe; oe
at.home
 0.6649
# expressed a multiplicative odds factor
> exp(oe)
at.home
 1.9443
# prob home team wins if teams are "equal"
> require(faraway); ilogit(oe)
at.home
0.66036
```

```
                                                       ┌─────────────────────────┐
                                                       │ ncaa2010-bt-hc-look2    │
                                                       └─────────────────────────┘
> ratings <-                        Brigham Young -0.463328 0.84876
+    BTabilities(ncaa.model2)       Pittsburgh    -0.477626 0.77024
> ratings[                          Ohio St.      -0.637364 0.78583
+    rev(order(ratings[,1]))[1:30],] Tennessee    -0.642515 0.78890
              ability    s.e.       Butler        -0.645638 0.82876
Kansas        1.527435 0.98592      Texas         -0.757237 0.77427
Kentucky      0.714602 0.98144      Xavier        -0.807881 0.76032
Syracuse      0.451811 0.83578      Marquette     -0.809803 0.74027
West Virginia 0.214972 0.76749      Richmond      -0.811680 0.75188
Duke          0.000000 0.00000      Wisconsin     -0.824389 0.73810
Villanova    -0.020024 0.78019      Cornell       -0.847105 0.94670
Kansas St.   -0.088619 0.80451      Maryland      -0.863548 0.68484
Temple       -0.127344 0.80936      Northern Iowa -0.928290 0.83222
New Mexico   -0.196177 0.87726      Gonzaga       -0.938791 0.79018
Purdue       -0.216995 0.80953      St. Mary's    -1.019850 0.82802
Georgetown   -0.338387 0.73128      Vanderbilt    -1.021330 0.77665
Baylor       -0.353114 0.79613      Notre Dame    -1.047816 0.75626
Texas A&M    -0.430974 0.77809
```

It is straightforward now to estimate the probability of one team defeating another at a neutral site.

```
                                         ┌──────────────────────┐
                                         │ ncaa2010-compare     │
                                         └──────────────────────┘
> compareTeams <-
+   function(team1,team2,model,abilities=BTabilities(model)) {
+     a <- abilities[team1,1]
+     b <- abilities[team2,1]
+     return(ilogit(a-b))
+ }
> compareTeams('Kansas','Kentucky',ab=ratings)
[1] 0.69271
> compareTeams('Michigan St.','Butler',ab=ratings)
[1] 0.40033
> compareTeams('Butler','Duke',ab=ratings)
[1] 0.34397
```

In the 2010 NCAA Basketball Tournament, Butler defeated Michigan State in the semi-finals, but lost to Duke in the championship game.   ◁

5.7. Bayesian Inference

To this point the approach we have been taking to inference about a parameter of a distribution is usually referred to as the **frequentist approach**. In this approach, we assume that the true value of the parameter θ is fixed – but unknown. The data are considered as a random sample from this unknown distribution, and we are able to make probability statements about the data:

- P-values are probabilities of obtaining data that produce a test statistic in a given range under some assumption about the true value of θ.

- Confidence intervals are random intervals (because they are computed based on random data) that have a particular probability of containing the parameter θ.

In both cases, the probability statements involved are about random samples – even though in practice we only have one sample and not a distribution of samples.

In this section we provide a brief introduction to another approach called the **Bayesian approach**. For simplicity we will only cover cases where a single parameter is estimated. The Bayesian approach is based on conditional probability and Bayes' Theorem, one simple version of which we state here.

Theorem 5.7.1.

$$P(H \mid E) = \frac{P(E \mid H) \, P(H)}{P(E)} . \tag{5.11}$$

Proof. Exercise 5.29. $\square$

We have used this reasoning before – in Example 2.2.22, for example. There we were interested in the probability of the hypothesis (H) that a patient had a disease given the evidence (E) of a positive laboratory test result. That is, we wanted to calculate $P(H \mid E)$. Bayes' Theorem tells us that we can do this by calculating $P(E \mid H)$, $P(H)$, and $P(E)$. In the Bayesian approach, we interpret these values as follows:

- $P(H)$ is the **prior probability**. That is, this is the probability assigned to H before gathering any evidence. Since this probability is not based on the evidence but on some prior knowledge or intuition, the Bayesian approach is also called the **subjectivist approach** because it always relies on some sort of prior.

- $P(E \mid H)$ is the probability of the evidence given the hypothesis. If the hypothesis provides sufficient information about the distribution of samples (evidence), then this probability can be calculated.

- $P(H \mid E)$ is called the **posterior probability**. The posterior probability represents how confident we are that H is true *after* we have seen the evidence. The Bayesian approach therefore quantifies how much the new evidence (E) modifies our estimation of whether the hypothesis (H) is true or false. It may well be that different researchers begin with different priors, but for each prior, the Bayesian approach quantifies the effect the new data should have.

- $P(E)$ may be calculated using

$$P(E) = P(E \cap H) + P(E \cap H_1) + P(E \cap H_2) + \cdots + P(E \cap H_n)$$
$$= P(H) \cdot P(E \mid H) + P(H_1) \cdot P(E \mid H_1) + \cdots + P(H_n) \cdot P(E \mid H_n)$$

  if $H_1, \ldots, H_n$ form an exhaustive set of mutually exclusive alternative hypotheses and the probabilities $P(E \mid H)$ are known (or assumed). In Chapter 2 we dealt mostly with the case where there was only one alternative to H (e.g., diseased vs. not diseased), but the reasoning generalizes easily to any (finite) number of alternatives.

Now let's consider the case where we are interested in a parameter θ that can take on infinitely many possible values. For each value of θ there is a corresponding hypothesis that it is the correct value. In this setting, a prior is a *distribution* telling which values of θ were how likely before gathering our evidence (the data). Similarly, the posterior will be another distribution of possible values for θ, updated to take into account the evidence our data provide.

If we replace probability statements with pdfs in (5.11), we obtain a Bayes' Theorem that applies to continuous distributions.

Theorem 5.7.2. *Let X and Y be jointly distributed continuous random variables. Then*

$$f_{X|Y}(x \mid y) = \frac{f_{Y|X}(y \mid x)\, f_X(x)}{\int f_{Y|X}(y \mid t)\, f_X(t)\, dt}\,. \tag{5.12}$$

Proof. This follows directly from our definitions of joint, conditional, and marginal probability:

$$\begin{aligned}
f_{X|Y}(x \mid y) &= \frac{f_{X,Y}(x,y)}{f_Y(y)} \\
&= \frac{f_{Y|X}(y \mid x)\, f_X(x)}{f_Y(y)} \\
&= \frac{f_{Y|X}(y \mid x)\, f_X(x)}{\int f_{Y|X}(y \mid t)\, f_X(t)\, dt}\,. \qquad \square
\end{aligned}$$

Now let's change notation slightly. Let's use θ in place of X, $\boldsymbol{X}$ in place of Y, and let's use f, g, and h for the three pdfs (so we can avoid the use of subscripts). Then (5.12) becomes

$$h(\theta \mid \boldsymbol{x}) = \frac{f(\boldsymbol{x} \mid \theta)\, g(\theta)}{\int f(\boldsymbol{x} \mid t)\, g(t)\, dt}\,. \tag{5.13}$$

Notice that

- $f(\boldsymbol{x} \mid \theta)$ is the likelihood function, and if the data are an iid random sample from a distribution with parameter θ, then

$$f(\boldsymbol{x} \mid \theta) = L(\theta; \boldsymbol{x}) = \prod f(x_i; \theta)\,.$$

- $g(\theta)$ is the prior pdf for θ and reflects the knowledge or belief about θ before examining the new evidence.

  If the amount of prior information is substantial, the prior may be very peaked and highly concentrated in a narrow portion of its support. On the other hand, if there is very little prior information, this is represented by a very wide, flat prior distribution.

- The denominator guarantees that our posterior $h(\theta \mid \boldsymbol{x})$ is a legitimate pdf.

It's time for a simple example where we put this all together.

Example 5.7.1.

Q. Use the Bayesian approach to make an inference about a population proportion π if the data $\boldsymbol{x}$ consist of a successes and b failures. Use a uniform prior.

A. If we have no knowledge shedding light on π before we collect our data, we can use an **uninformative prior** like a pdf for $\mathsf{Unif}(0,1)$, i.e.,

$$g(\pi) = 1 \text{ on } [0,1] \,.$$

This prior expresses that every value of π seemed equally likely prior to collecting evidence.

In this simple case, the likelihood function is simple to calculate:

$$f(\boldsymbol{x} \mid \pi) = \pi^a(1 - \pi)^b \,.$$

Applying (5.13), we obtain

$$h(\pi \mid \boldsymbol{x}) = \frac{\pi^a(1 - \pi)^b \cdot 1}{\int t^a(1 - t)^b \cdot 1 \, dt} \,. \tag{5.14}$$

This is immediately recognizable as the pdf of a $\mathsf{Beta}(a + 1, b + 1)$-distribution. (Ignore the constants and look at the kernel.) ◁

Now that we have calculated our first posterior distribution, what do we do with it? Back when we were using the frequentist approach, we computed point estimates and confidence intervals to estimate parameters. We can do a similar thing here. One natural point estimate is the expected value of the posterior distribution. We can obtain an interval estimate by forming an interval that contains the specified percentage of the posterior distribution. Such intervals are called **credibility intervals** to avoid confusion with the frequentist confidence intervals. They serve much the same purpose as confidence intervals, but the interpretation is quite different. Remember that in a frequentist approach, it was the interval itself that was considered to be random; the parameter was fixed and unknown. In the Bayesian approach, our knowledge of θ is expressed via the posterior distribution and it makes sense to form probability statements about the parameter based on this distribution. In both situations, the resulting interval gives us some indication of the precision of our point estimate.

Example 5.7.2.

Q. Returning to the previous example, compute a Bayesian point estimate and 95% credibility interval for π.

A. The mean of a $\mathsf{Beta}(a + 1, b + 1)$-distribution is $\frac{a+1}{a+b+2}$, so our Bayesian point estimate is $\hat{\pi} = \frac{a+1}{a+b+2}$. This is only slightly different from our usual estimator of $\frac{a}{a+b}$.

We can obtain a 95% credibility interval by eliminating the most extreme 2.5% from each end of a $\mathsf{Beta}(a + 1, b + 1)$-distribution. For example, if $a = 20$ and $b = 30$, then we calculate a 95% credibility interval as follows. For comparison, we also compute the usual 95% confidence interval.

```
> qbeta(c(0.025,0.975),20+1,30+1)        binom-credible
[1] 0.27584 0.53886
> binom.test(20,50)$conf.int          # for comparison
[1] 0.26408 0.54821
attr(,"conf.level")
[1] 0.95
```

```
> prop.test(20,50)$conf.int             # for comparison
[1] 0.26733 0.54795
attr(,"conf.level")
[1] 0.95                                                                      ◁
```

Now we consider the situation with a different prior.

Example 5.7.3.

Q. Returning to the previous example, compute a Bayesian point estimate and 95% credibility interval for π using a $\mathsf{Beta}(\alpha, \beta)$ prior.

A. We can simplify things if we observe that when working with (5.13), we can ignore the denominator and the scaling constants if we are satisfied with determining the kernel of $h(\theta \mid x)$. We could then subsequently integrate to determine the scaling constant if the resulting distribution is unfamiliar. In the current situation, the kernel of our posterior distribution is

$$f(\boldsymbol{x} \mid \theta) \, g(\theta) = \pi^a (1 - \pi)^b \pi^{\alpha-1} \pi^{\beta-1}$$
$$= \pi^{a+\alpha-1}(1 - \pi)^{b+\beta-1}$$

which we recognize as the pdf for a $\mathsf{Beta}(a + \alpha, b + \beta)$-distribution.

The expected value of the posterior distribution is $\frac{a+\alpha}{b+\beta}$. This has a natural interpretation. We can view our $\mathsf{Beta}(\alpha, \beta)$ prior as representing the information contained in a previous sample with α successes and β failures. Our posterior information is equivalent to pooling the two samples to make one large sample.

Given this interpretation, it is natural to ask about using a $\mathsf{Beta}(0, 0)$ prior in place of the $\mathsf{Unif}(0, 1) = \mathsf{Beta}(1, 1)$ prior. This is almost possible. The problem is that there is no $\mathsf{Beta}(0, 0)$ pdf. If we set $\alpha = \beta = 0$ in the pdf, the resulting function does not have a convergent integral. The method still works, however, because the righthand side of (5.13) is still defined. A prior that is not a pdf but for which the posterior can still be defined by (5.13) is called an **improper prior**. See Exercise 5.35 for another example using an improper prior.   ◁

Finally, we turn our attention to the Bayesian equivalent of p-values.

Example 5.7.4.

Q. A coin is flipped 100 times yielding 38 head and 62 tails. Is this enough evidence for a Bayesian to reject the coin as unfair?

A. The short answer, of course, is that it depends on the prior. We'll provide answers using the uniform prior.

First we calculate 95% credibility and confidence intervals.

```
> qbeta(c(0.025,0.975),38+1,62+1)                        binom-credible2
[1] 0.29091 0.47817
> binom.test(38,100)$conf.int             # for comparison
[1] 0.28477 0.48254
attr(,"conf.level")
[1] 0.95
> prop.test(38,100)$conf.int              # for comparison
[1] 0.28639 0.48294
```

```
attr(,"conf.level")
[1] 0.95
```

Since none of these intervals contains 0.50, both the frequentist and the Bayesian reject that value based on interval estimates.

Note that, using the Bayesian approach, it makes sense to calculate probabilities like

$$P(\pi = 0.50 \mid \boldsymbol{x})$$

and

$$P(\pi \geq 0.50 \mid \boldsymbol{x})$$

because we have a posterior distribution for π. The first probability above is clearly 0 because the distribution is continuous. The second probability can be used to calculate something akin to a p-value. Using a uniform prior, the second probability is given below.

```
> 1- pbeta(0.5,38+1,62+1)      # 1-sided Bayesian p-value   binom-bayes-pval
[1] 0.008253
> binom.test(38,100,alt="less")$p.value      # for comparison
[1] 0.010489
> prop.test(38,100,alt="less")$p.value       # for comparison
[1] 0.010724
```

Once again the value is very similar to that produced using a frequentist approach, but the interpretation is different. The Bayesian can reject the coin on the grounds that given the evidence, the probability that π is as large as 0.50 (or larger) is quite small. So to a Bayesian it makes sense to say that π *probably* isn't as large as 0.50. From a frequentist perspective, this is a nonsense statement. The frequentist claims only that data like this would be quite unlikely to occur by random chance if $\pi = 0.5$. Both agree, however, that $\pi = 0.50$ should be rejected as a plausible value of π.   ◁

In the examples above we see that the Bayesian methods have natural interpretations and that they give results consistent with those obtained by frequentist methods. Many people find the interpretation of the Bayesian results more intuitive and easier to understand. In general, when there is an obvious candidate for the prior, Bayesian inference provides accepted methods with a clear interpretation. The challenge arises when there is debate about what prior should be used. Since the resulting point estimates, credibility intervals, and p-values all depend on the prior, when there is disagreement about the prior, there can be (sometimes heated) disagreements about the analysis and its conclusions. This is the downside of the Bayesian approach and the reason it is called the subjective approach; the choice of a prior is often very subjective indeed.

Example 5.7.5.

Q. How does a Bayesian do inference for μ, the unknown mean of a normal population, assuming that variance of the population (σ^2) is known?

A. The first step is to decide upon a prior. If we had no previous knowledge, we might use an improper prior that is uniform on $(-\infty, \infty)$ or $[0, \infty)$. Here we will investigate using a $\mathsf{Norm}(\mu_0, \sigma_0)$ prior. The kernel of the posterior distribution for μ is then

$$h(\mu \mid \boldsymbol{x}) = \left(\prod e^{-\frac{1}{2}(x_i - \mu)^2 / \sigma^2} \right) e^{-\frac{1}{2}(\mu - \mu_0)^2 / \sigma_0^2} \tag{5.15}$$

$$= e^{-\frac{1}{2} \left[\frac{(x_1 - \mu)^2}{\sigma^2} + \frac{(x_2 - \mu)^2}{\sigma^2} + \cdots + \frac{(x_n - \mu)^2}{\sigma^2} + \frac{(\mu - \mu_0)^2}{\sigma_0^2} \right]}. \tag{5.16}$$

Expanding the portion of the exponent in square brackets, we see that

$$\left[\frac{(x_1 - \mu)^2}{\sigma^2} + \frac{(x_2 - \mu)^2}{\sigma^2} + \cdots + \frac{(x_n - \mu)^2}{\sigma^2} + \frac{(\mu - \mu_0)^2}{\sigma_0^2} \right]$$

$$= \frac{1}{\sigma^2} \left(\sum x_i^2 - 2\mu \sum x_i + n\mu^2 \right) + \frac{1}{\sigma_0^2} \left(\mu^2 - 2\mu\mu_0 + \mu_0^2 \right)$$

$$= \left(\frac{n}{\sigma^2} + \frac{1}{\sigma_0} \right) \mu^2 - 2(\frac{n\overline{x}}{\sigma^2} + \frac{\mu_0}{\sigma_0^2})\mu + A$$

$$= \left(\frac{n}{\sigma^2} + \frac{1}{\sigma_0} \right) \left[\mu^2 - 2 \frac{\frac{n\overline{x}}{\sigma^2} + \frac{\mu_0}{\sigma_0^2}}{\frac{n}{\sigma^2} + \frac{1}{\sigma_0^2}} \mu + B \right]$$

$$= \left(\frac{n}{\sigma^2} + \frac{1}{\sigma_0} \right) \left[\mu - \frac{n\overline{x} + \frac{\mu_0}{\sigma_0^2}}{\frac{n}{\sigma^2} + \frac{1}{\sigma_0^2}} \right]^2 + C$$

$$= \frac{1}{\sigma_1^2} \left[\mu - \mu_1 \right]^2 + C$$

where A, B, and C do not depend on μ,

$$\sigma_1^2 = \frac{1}{\frac{n}{\sigma^2} + \frac{1}{\sigma_0}}, \text{ and}$$

$$\mu_1 = \frac{\frac{n\overline{x}}{\sigma^2} + \frac{\mu_0}{\sigma_0^2}}{\frac{n}{\sigma^2} + \frac{1}{\sigma_0^2}}.$$

Substituting in (5.16), we see that our kernel is

$$h(\mu \mid \boldsymbol{x}) = e^{-\frac{1}{2} \left[\frac{1}{\sigma_1^2} (\mu - \mu_1)^2 + C \right]}$$

$$= e^{-\frac{1}{2} C / \sigma_1^2} \cdot e^{-\frac{1}{2} \left[\frac{1}{\sigma_1^2} (\mu - \mu_1)^2 \right]},$$

i.e., the kernel of a $\mathsf{Norm}(\mu_1, \sigma_1)$-distribution. Notice that the posterior mean μ_1 is a weighted average of $\overline{x}$ and μ_0 where the weights are the reciprocals of the prior variance (σ_0^2) and the variance of $\overline{x}$ (σ^2/n). Similarly, the posterior variance is the reciprocal of the sum of the reciprocals of these two variances. This can be expressed more simply if we define the **precision** to be the reciprocal of the variance (since a larger variance means less precision). Then the posterior precision is the sum of the prior precision and the precision of the data.   ◁

The assumption that σ is known is artificial but simplifies our example. Without this assumption, our prior and posterior distributions would be *joint* distributions of μ and σ, but the underlying ideas would be the same.

Example 5.7.6.

Q. Use the method of the previous example to estimate μ based on the following (sorted) data, assuming that the population is normal with mean $\sigma = 5$:

$$20 \quad 24 \quad 27 \quad 28 \quad 28 \quad 28 \quad 29 \quad 30 \quad 30 \quad 30 \quad 30 \quad 32 \quad 33 \quad 34 \quad 35 \quad 38$$

A. The function `posterior()` defined below calculates the mean μ_1 and standard deviation σ_1 of the posterior distribution. Using this function, we can easily estimate μ with various priors (μ_0 and σ_0). It is important to note that σ_0 is *not* an estimate for the population standard deviation σ. Rather σ_0 describes our prior certainty about our estimate μ_0 for μ. A relatively large value of σ_0 expresses relatively little prior knowledge about what μ might be.

```
                                                                    bayes-normal
> x <- c(20,24,27,28,28,28,29,30,30,30,30,32,33,34,35,38)
> mean(x)
[1] 29.75
> sd(x)
[1] 4.2817
> posterior <- function(x,mu0,sigma0,sigma=5) {
+       n <- length(x)
+       N <- (n*mean(x)/sigma^2 + mu0/sigma0^2)
+       D <- (n/sigma^2 + 1/sigma0^2)
+       mu1 <- N/D; sigma1 <- sqrt(1/D)
+       precision1 <- D
+       precision0 <- 1/sigma0^2
+       precision.data <- n/sigma^2
+       return(cbind(mu1,sigma1,precision1,precision0,precision.data))
+       }
> posterior(x,20,1)
         mu1  sigma1 precision1 precision0 precision.data
[1,] 23.805 0.78087       1.64          1           0.64
> posterior(x,20,4)
         mu1 sigma1 precision1 precision0 precision.data
[1,] 28.883 1.1931     0.7025     0.0625           0.64
> posterior(x,20,16)
         mu1 sigma1 precision1 precision0 precision.data
[1,] 29.691 1.2462     0.6439  0.0039062           0.64
> posterior(x,20,1000)
        mu1 sigma1 precision1 precision0 precision.data
[1,] 29.75   1.25       0.64      1e-06           0.64
> 5/sqrt(length(x))
[1] 1.25
```

Exercise 5.34 asks you to calculate a 95% credibility interval and a 95% confidence interval (using σ) from this data and to compare them.   ◁

In each of our examples so far, the posterior distribution has been in the same family as the prior distribution. This is not always the case, but when it is, the prior is said to be a **conjugate prior**.

5.8. Summary

5.8.1. Maximum Likelihood

In this chapter we presented several statistical methods based on the likelihood function. We began by considering the problem of parameter estimation. In Chapter 3, we introduced the **method of moments** as a way to obtain **point estimates** by equating the moments of the model distribution (as functions of the parameters) with the **sample moments**. A solution to this system of equations is a method of moments estimate.

Here we introduced a new method, called **maximum likelihood estimation**, which is the most frequently used method of **point estimation**. Maximum likelihood estimates are found by maximizing the **likelihood function** $L(\boldsymbol{\theta}; \boldsymbol{x}) = f(\boldsymbol{x}; \boldsymbol{\theta})$. Here we consider the joint pdf of $\boldsymbol{x}$ as a function of the parameters $\boldsymbol{\theta}$ for fixed data (our sample) instead of the other way around. Intuitively, the maximum likelihood estimate is the vector of parameter values that makes the observed data most likely.

The theory of likelihood allows us to do statistical inference based on maximum likelihood estimators. The **generalized likelihood ratio test** provides a flexible method for producing test statistics and p-values in a wide range of situations. The null and alternative hypotheses are expressed as sets of parameter vectors Ω_0 and Ω_a. That is, the hypotheses are represented by the statistical models determined by the parameters in Ω_0 and Ω_a, respectively. The test statistic is the **likelihood ratio**

$$\lambda = \lambda(\boldsymbol{x}) = \frac{L(\hat{\Omega}_0; \boldsymbol{x})}{L(\hat{\Omega}; \boldsymbol{x})}$$

where $\Omega = \Omega_0 \cup \Omega_a$ and $L(\hat{S}; \boldsymbol{x}) = \max\{L(\boldsymbol{\theta}; \boldsymbol{x}) \mid \boldsymbol{\theta} \in S\}$. Often the **asymptotic distribution** of $\Lambda = \lambda(\boldsymbol{X})$ is known when the null hypothesis is true. Under sufficient **regularity conditions**,

$$-2\log(\Lambda) \sim \mathsf{Chisq}(\nu) \quad \text{where} \quad \nu = \dim \Omega - \dim \Omega_0 \ ,$$

the difference in the number of free parameters in the unrestricted and restricted models. Geometrically, $-2\log(\lambda)$ is twice the difference between two maximal values of the likelihood function, once maximized in Ω and once in Ω_0.

Likelihood ratio tests are important because

- the method suggests a test statistic, so we are not forced to use *ad hoc* methods for each new situation;

- numerical methods exist that in many cases make it possible to obtain good numerical approximations of maximum likelihood estimates and likelihood ratios;

- often the asymptotic distribution is known, providing good approximate p-values when sample sizes are sufficiently large; and

- likelihood ratio tests satisfy certain optimality properties which indicate that they are often at least as powerful as any other test of the same hypotheses.

By the duality between p-values and confidence regions, we can invert the likelihood ratio test (or some other hypothesis test) to obtain **confidence regions** for $\boldsymbol{\theta}$.

In addition to inference about the parameter vector $\boldsymbol{\theta}$, **goodness of fit tests** can be used to assess how well a model fits the data. Two commonly used approaches are based on comparing observed cell counts ($\boldsymbol{o}$) to expected cell counts ($\boldsymbol{e}$) in an appropriate table summarizing the observed data. The **Pearson chi-squared statistic**

$$X^2 = \sum_i \frac{(o_i - e_i)^2}{e_i}$$

and the likelihood ratio statistic

$$G^2 = -2\log(\lambda) = 2\sum_i o_i \log\left(\frac{o_i}{e_i}\right)$$

are each asymptotically $\mathsf{Chisq}(\nu)$ with

$$\nu = k - 1 - (\text{number of parameters estimated before calculating } \boldsymbol{e}).$$

Historically, the Pearson statistic has been more frequently used, largely because it is computationally simpler to obtain. A small p-value resulting from a goodness of fit test is an indication that none of the distributions in the model is a particularly good fit with the data and suggests that barring additional information not contained in the data, one should at least consider alternative models.

5.8.2. Bayesian Inference

This chapter concludes with a brief introduction to **Bayesian inference**. Bayesian inference also uses the likelihood function, but it is a very different approach to inference. In Bayesian inference we begin with a **prior distribution** of the parameter(s) of the model. This distribution is intended to reflect our knowledge about the parameters prior to observing our data. The Bayesian approach then produces a new distribution called the **posterior distribution** of the parameter(s) of the model. The posterior distribution indicates our updated knowledge of the parameter(s) based on both the prior distribution and the data.

At its core, Bayesian inference is based on the following version of **Bayes' Theorem**:

$$P(H \mid E) = \frac{P(E \mid H)\,P(H)}{P(E)}. \tag{5.17}$$

If we interpret H as our hypothesis and E as some evidence (data) but shift our focus to distributions of random variables rather than events, this becomes (5.13):

$$h(\theta \mid \boldsymbol{x}) = \frac{f(\boldsymbol{x} \mid \theta)\,g(\theta)}{\int f(\boldsymbol{x} \mid t)\,g(t)\,dt}, \tag{5.18}$$

where f is the likelihood function, g is the prior distribution of θ, and h is the posterior distribution of θ. When a case can be made for an appropriate prior, Bayesian inference has the advantage that we can express our reasoning about the model in terms of probability statements about the parameters, like "the probability that π is between 0.45 and 0.55", and draw conclusions based on those probabilities.

The disadvantage of Bayesian inference is that it depends not only on the data but also on the prior, and there can be strong disagreements about what the prior should be. For this reason, Bayesian inference is sometimes referred to as the "subjectivist approach" in contrast the the "frequentist approach" we have been using to this point. For a more complete introduction to Bayesian inference see, for example, [**GCSR03**].

5.8.3. R Commands

Here is a table of important R commands introduced in this chapter. Usage details can be found in the examples and using the R help.

`nlm(f,p,x)`	Minimize `f` starting from point `p` .
`nlmin(f,p,x)`	Minimize `f` starting from point `p` [`fastR` wrapper for `nlm()`].
`nlmax(f,p,x)`	Maximize `f` starting from point `p` [`fastR` wrapper for `nlm()`].
`summary(nlmax(f,p,x))`	Summary output for `nlmax()`.
`oldopt <- options(warn=-1)`	Turn off warnings and save previous options.
`options(oldopt)`	Revert to old options.
`xhistogram(~x,data,...)`	Histogram with some extras [`fastR`].
`uniroot(f,interval,...)`	Numerically approximate a solution to $f(x) = 0$ for x within the interval specified by `interval`.
`nrow(x); ncol(x)`	The number of rows or columns in an object `x`.
`rbind(...)`	Bind together rowwise into a matrix.
`cbind(...)`	Bind together columnwise into a matrix.
`rownames(x)`	Access or set the row names of object `x`.
`colnames(x)`	Access or set the column names of object `x`.
`chisq.test(x,...)`	Perform a Pearson Chi-squared test; handles some simple goodness of fit testing and 2-way tables.
`xchisq.test(x,...)`	Perform a Pearson Chi-squared test and display some extra information [`fastR`].
`mosaic(...)`	Construct a mosaic plot [`vcd`].
`merge(x,y,...)`	Merge data frames `x` and `y`.
`BTm(...)`	Fit a Bradley-Terry model [`BradleyTerry2`].
`coef(model)`	Compute coefficients of a model.
`logit(x), ilogit()`	Logit and inverse logit functions [`faraway`].

Exercises

5.1. Let $\boldsymbol{X} \overset{\text{iid}}{\sim} \mathsf{Unif}(0, \theta)$ as in Example 5.1.4. Find the maximum likelihood estimate for θ using the data from Example 4.2.2. What advantage of the MLE over the method of moments estimator does this example illustrate?

5.2. Let X be a random variable with pdf
$$f(x; \theta) = (\theta + 1)x^\theta \text{ on } [0, 1] \, .$$

a) Derive the method of moments estimator for θ.

b) Derive the maximum likelihood estimator for θ.

c) Find the method of moments estimator for θ based on the sample data below.

d) Find the maximum likelihood estimator for θ based on the sample data below:

$$\begin{array}{ccccc} 0.90 & 0.078 & 0.93 & 0.64 & 0.45 \\ 0.85 & 0.75 & 0.93 & 0.98 & 0.78 \end{array}$$

5.3. Recall Example 5.1.6. Here is another sampling method to estimate the plant density λ. This time the biologist selects n non-overlapping regions and counts the number of plants of the species in each region. Let a_i be the area of region i (known in advance) and let x_i be the number of plants observed in region i.

a) Given this sampling method, derive the MLE for λ.

b) Use simulation to compare these two methods. Under what conditions does one seem to be preferred to the other? To simplify things, you can assume that the regions in the method just described all have the same area.

5.4. Justify (5.3) and show that this definition agrees with the usual definition of binomial coefficients when $k = 2$.

5.5. Let $\boldsymbol{X} \sim \mathsf{Multinom}(n, \boldsymbol{\pi})$. What is the conditional distribution of X_i given $X_j = k$?

Note: According to the R help pages, this is how the function `rmultinom()` is implemented.

5.6. Write your own function to replace `rmultinom()` using the following idea. Partition the unit interval into r subintervals of lengths $\pi_1, \pi_2, \ldots, \pi_r$. Now sample using `runif()` and interpret the result based on the subinterval which contains it.

5.7. Use Lagrange multipliers to verify (5.4).

5.8. Repeat Example 5.1.4 using numerical methods.

5.9. Repeat Example 5.1.7 using numerical methods.

5.10. Find data that can be considered a sample of proportions from some larger population. (There are many examples in sports, for example.) Use maximum likelihood to fit a beta model to your data and provide some informal analysis of how good you think the fit is. That is, answer the following two questions as best you can: Does the beta model seem like a good model for your data? What makes you think so?

5.11. Find data that can be considered a sample of lifetimes (time to failure) of some product. Use maximum likelihood to fit three models:

- $X \sim \mathsf{Exp}(\lambda)$.
- $X \sim \mathsf{Gamma}(\alpha, \lambda)$.
- $X \sim \mathsf{Weibull}(\alpha, \lambda)$.

Provide an informal analysis of how good you think the fits are. Which model do you like best? Why?

5.12. Let X and Y be continuous random variables. Define a new random variable Z according to the following process. First flip a biased coin with probability α of obtaining a head. If the result is heads, sample from X; otherwise sample from Y. Such a distribution is called a **mixture** of X and Y.

a) Derive a formula for the cdf of Z in terms of α and the cdfs of X and Y.

b) Derive a formula for the pdf of Z in terms of α and the pdfs of X and Y.

c) Let $W \sim 0.3 \cdot \mathsf{Norm}(8, 2) : 0.7 \cdot \mathsf{Norm}(16, 3)$ (that is, a 30-70 mix of two normal distributions). What is $P(W \leq 12)$?

5.13. Develop a likelihood ratio test to test $H_0 : \mu = \mu_0$ vs. $H_a : \mu \neq \mu_0$ when our model is $X \overset{\text{iid}}{\sim} \mathsf{Norm}(\mu, 1)$. How does it compare to other tests we have seen that do not assume a known value of σ?

5.14. The **Laplace distribution** (also called the double exponential distribution) with parameters θ and λ has a pdf with kernel

$$e^{-\lambda|x-\theta|} \ .$$

This resembles the pdfs for the normal and exponential distributions. Compared to the normal distribution, the Laplace distribution uses an absolute value in place of a square. This will cause the Laplace distribution to have heavier tails than a normal distribution. The Laplace distribution can also be viewed as a shifted and mirrored version of an exponential distribution: It is symmetric about θ and each half looks like an exponential distribution with parameter λ.

a) Compute the mean and variance of a Laplace random variable.

b) Write an R function `dlaplace()` to compute the pdf of a Laplace random variable.

c) Write an R function `plaplace()` to compute the cdf of a Laplace random variable.

d) Use your functions to determine the following:
- $P(X \leq 3)$ when $X \sim \mathsf{Laplace}(\lambda = 2, \theta = 1)$,
- $P(-3 \leq X \leq 3)$ when $X \sim \mathsf{Laplace}(\lambda = 2, \theta = 1)$.

5.15. Suppose

$$1.00 \quad -1.43 \quad 0.62 \quad 0.87 \quad -0.66 \quad -0.59 \quad 1.30 \quad -1.23 \quad -1.53 \quad -1.94$$

is a random sample from a Laplace distribution (see Exercise 5.14).

  a) Estimate the parameters θ and λ using maximum likelihood.

  b) Find the method of moments estimators from the same data.

5.16. Using the data from Exercise 5.14 and a likelihood ratio test, evaluate the evidence for $H_0 : \theta = 0$ vs. $H_a : \theta \neq 0$.

5.17. Suppose that X is a random variable that has a pdf that depends on a parameter θ with $\theta \geq 0$:

$$f(x; \theta) = \begin{cases} (\theta + 1)x^\theta & \text{if } 0 < x \leq 1, \\ 0 & \text{otherwise.} \end{cases}$$

We want to perform a hypothesis test of $H_0 : \theta = 0$ versus $H_1 : \theta > 0$. (Notice that H_0 says that the distribution is the uniform distribution on $[0, 1]$.) We're feeling pretty lazy so we only get a random sample of size 1 from this distribution. Call the value that results x.

  a) What should the form of the test be? (That is, should we reject when x is large, when x is small, or for both large and small values of x? We have seen all three kinds of tests in our study of hypothesis testing.) Be sure to explain your reasoning.

  b) What is the p-value of this test if $x = .2$?

  c) What is the p-value of this test if $x = .9$?

  d) Given $\alpha = .05$, for what values of x should we reject H_0?

  e) What is the power of this test (with $\alpha = .05$) if the true value of θ is 1?

5.18. You're feeling a bit less lazy now, and are willing at least to consider a larger sample size (n) in the preceding problem. It is less clear now what you should use for a test statistic, so you decide to use the likelihood ratio method. Determine the form of the likelihood ratio test statistic assuming a sample $x = \langle x_1, x_2, \ldots, x_n \rangle$. (You do not need to compute a p-value.)

5.19. Use simulation to investigate test statistics for the golf ball situation of Example 5.4.1. Compare the results of the various statistics.

5.20. Return to the data in Example 5.4.9. Try fitting normal, gamma, and Weibull distributions. Do any of these families provide a reasonable fit to the data?

5.21. Test the goodness of fit of the Hardy-Weinberg model to the data in Example 5.1.7. Is there any reason to worry that the assumption of Hardy-Weinberg equilibrium is being violated?

5.22. Derive a likelihood ratio test for $H_0 : \sigma = \sigma_0$ vs. $H_a : \sigma \neq \sigma_0$ for a sample from a normal distribution with unknown mean.

5.23. In 1958, Fisher examined data from a genetic study of two genes in the offspring of self-pollinated heterozygous plants. The results of scoring the offspring plants as either starchy or sugary and has having either a green or a white base leaf appear in the table below:

1) starchy-green	2) starchy-white	3) sugary-green	4) sugary-white
1997	906	904	32

According to a genetic model for these traits, the probability that a plant exhibits one of these trait combinations should be $0.25(2+\theta)$ for the first combination, $0.25(1-\theta)$ for the middle two, and 0.25θ for the last where θ is a parameter related to the linkage (closeness on the chromosome) between the two genes and measures the probability of inheritance following the Punnet square on the left vs. that on the right:

$$\theta \qquad\qquad\qquad 1-\theta$$

	AB	ab
AB	1	1
ab	1	4

	Ab	aB
Ab	2	1
aB	1	3

(The numbers in the Punnet squares correspond to the four phenotypes.)

a) Determine the MLE for θ.

b) Test the hypothesis that $\theta = 0.05$.

c) Test the hypothesis that $\theta = 0.03$.

d) Test the hypothesis that $\theta = 0.07$.

e) Perform a goodness of fit test to see how well the data fit the model.

f) Form a 95% confidence interval for θ.

5.24. A geneticist is interested in estimating the percentage of twins that are identical. She has access to birth records that allow her to count the number of twins born and that lists the sex of each twin but does not indicate whether the twins are identical or not. You have been asked to help her estimate the proportion θ of twins that are identical twins.

The geneticist provides you with the following biological assumptions that you may use:

• Identical twins are always same-sex twins.

• The rate of identical twins is the same among males and females.

• Males and females are equally likely.

• In non-identical twins, the sexes of the twins are independent.

Develop a method for estimating θ for data of the following form:

sex of twins	Male-Male	Male-Female	Female-Female
count	n_1	n_2	n_3

Bonus: Find data and apply this method.

5.25. Explain why calculating expected counts based on minimizing the test statistic (as discussed on page 287) is always at least as conservative as using maximum likelihood methods to determine expected cell counts.

5.26. Use simulations to investigate the effect of bin choices when doing goodness of fit testing. Using a variety of Poisson, normal, and exponential distributions (add others if you like) compute the goodness of fit p-values from simulated data sets using different numbers or sizes of cells and produce a scatterplot or other type of summary to compare the results.

Based on your experimental results, how much of a concern is this?

5.27. In one of his famous experiments [**Men65**], Gregor Mendel crossed plants that were heterozygous for two traits: color (yellow or green) and shape (round or wrinkled). According to his genetic model, the offspring of such a cross should have a $9 : 3 : 3 : 1$ ratio of yellow-round : yellow-wrinkled : green-round : green-wrinkled phenotypes. (See Example 2.2.21.)

The table below contains some data from this experiment:

Phenotype	Frequency
yellow-round	315
yellow-wrinkled	101
green-round	108
green-wrinkled	32

How well do these data agree with Mendel's model? Conduct a goodness of fit test and comment on the results.

5.28. The FUSION study of the genetics of type 2 diabetes [**SMS**[+]**04**] was introduced in Exercise 3.40. Genotypes for some of the FUSION subjects are available in the `fusion1` data set. The marker in `fusion1` is a single nucleotide polymorphism (SNP) in the region of *TCF7L2*, a gene that has repeatedly been associated with diabetes. These data were analyzed in a 2008 paper that appeared in *Science* [**SMB**[+]**07**].

Use this data to conduct an association test for type 2 diabetes (the `t2d` variable in `pheno`). To get started, you will need to merge the phenotype and genotype information into a single data set using the `merge()` function.

`fusion1-merge`

```
# merge fusion1 and pheno keeping only id's that are in both
> fusion1m <- merge(fusion1, pheno, by='id', all.x=FALSE, all.y=FALSE)
```

This allows us to cross-tabulate case-control status with genotype or dose (the number of G alleles).

`fusion1-xtabs-geno`

```
> xtabs(~t2d + genotype, fusion1m)
        genotype
t2d        GG  GT  TT
  case    737 375  48
  control 835 309  27
```

```
> xtabs(~t2d + Gdose, fusion1m)                    fusion1-xtabs-dose
         Gdose
t2d        0   1   2
  case    48 375 737
  control 27 309 835
```

a) Which allele (G or T) is the "risk allele"? That is, which allele is more common among cases than among controls?

b) Perform an appropriate Chi-squared test and interpret the results to see if there is an association between this SNP and type 2 diabetes.

c) This marker was just one part of a larger genomewide association study that tested over 300,000 SNPs for association with type 2 diabetes. How does knowing this affect how you interpret the results?

5.29. Prove Theorem 5.7.1.

5.30. Do Exercise 2.23.

5.31. Do Exercise 2.24.

5.32. Do Exercise 2.28.

5.33. In (5.13) we introduced f, g, and h as a way to avoid subscripts. For each function, say what the correct subscript would be.

5.34. Use the data in Example 5.7.6 to answer the following.

a) Compute a 95% confidence interval for μ (using the fact that σ is known) and a 95% credibility intervals for μ using two or three different priors.

b) Your credibility intervals should be narrower than the confidence interval. Explain why this is always the case when estimating μ this way.

c) When will the confidence intervals and credibility intervals be most similar?

5.35. Suppose we sample from a normal distribution with unknown mean μ but known standard deviation σ. The function $f(x) = 1$ on $(-\infty, \infty)$ can be used as an improper prior that is intended to represent no prior information about the mean (precision = 0).

a) Determine the general form of the posterior using this prior.

b) How do credibility intervals computed this way compare with confidence intervals computed using the known value of σ?

5.36. Suppose we sample from a Poisson distribution with unknown parameter λ. Using a Gamma(α, β)-distribution as the prior for λ, determine the general form of the posterior distribution.

5.37. Recall that the mean and variance of a Poisson random variable are equal. Intuitively, this means that something like

$$T = \frac{s_x^2}{\bar{x}} = \frac{\text{sample variance}}{\text{sample mean}}$$

ought to be a possible test statistic for testing whether sample data appear to be coming from a Poisson distribution. Of course, to use this sort of test statistic, we would need to know the sampling distribution of T under the null hypothesis that the population being sampled from is indeed Poisson. As it turns out, if we randomly sample from a Poisson distribution, then

$$(n-1)T \approx \mathsf{Chisq}(n-1) \, ,$$

with the approximation getting better as the sample size gets larger. The resulting test is called the Poisson dispersion test.

a) For what sorts of values of T will you reject the null hypothesis? (Large values? Small values? Large and small values?) Relate your answer to the alternative hypothesis of this test.

b) Use the Poisson dispersion test to decide if the data below (from a sample of size 15) appear to come from a Poisson distribution and interpret the results.

count	0	1	2	3	4
frequency	9	2	3	0	1

5.38. Suppose we sample from a normal distribution with mean 0 and unknown tolerance $\tau = 1/\sigma^2$. Using a $\mathsf{Gamma}(\alpha, \beta)$ prior for τ, determine the general form of the posterior distribution.

## Chapter 6

Introduction to Linear Models

Remember that all models are wrong; the practical question is how wrong do they have to be to not be useful.

George Box [**BD87**]

Most of the statistical models we have encountered thus far have been **univariate**. Univariate statistical models allow us to ask and answer questions about the distribution of a single variable or measurement. We developed methods to answer questions like the following:

- What is the mean value of the distribution?

- Is a Poisson distribution a good model for this variable?

- Is it reasonable to assume that the true proportion is 50%?

While these methods are important, many questions of interest are more easily expressed as questions about relationships between multiple variables. We might want to answer questions like the following:

- Which of two or more *treatments* leads to a better (average) *reduction in cholesterol* in patients?

- How is the *age* of a bridge related to its *strength*?

- What *factors* help determine the *selling price* of a home?

- How is the *risk of diabetes* related to various *genetic effects*, *health factors* (such as weight, blood pressure, etc.), and *environmental factors* (such as income, education, place of residence, etc.)?

We have already seen a few examples of multivariate models. Chi-squared tests for two-way tables (and Fisher's exact test for 2×2 tables), for example, can be used to answer questions about the relationship between two categorical variables. In this chapter and the next we will begin to develop a flexible, systematic approach to modeling a wide range of multivariate situations.

6.1. The Linear Model Framework

6.1.1. Modeling Relationships with Functions

Regression analysis is used for explaining or modeling the relationship between a variable Y (called the **response**, **output**, or **dependent** variable) and a set of one or more other variables $X_1, X_2, \ldots, X_p$ (called the **predictor**, **input**, **explanatory**, or **independent** variables). The relationship between the predictors and the response is modeled by a *function*. In its most general form, such a model could have the form

$$Y = f(X_1, X_2, \ldots, X_p) + \varepsilon \, .$$

Such a model says that Y is determined by the values $X_1, X_2, \ldots, X_p$ via the function f to within some "error" ε.

Typically we will assume that $\mathrm{E}(\varepsilon) = 0$, in which case there is another interpretation of the model. The function f gives the mean value of Y for a particular set of predictors:

$$\mathrm{E}(Y \mid X_1 = x_1, \ldots, X_p = x_p) = f(x_1, \ldots, x_p) \, .$$

Note that this model is implicitly partitioning the variability in Y into two components:

- Some of the variability is "explained" by the model.
  The values of the response Y vary (in part) because they are associated with different predictors $\boldsymbol{X}$.

- But the model does not predict perfectly, leaving some of the variation "unexplained" by the model.
  Even if two values of the response Y are associated with the same predictors $\boldsymbol{X}$, they may differ because the values of ε are different.

6.1.2. Linear Models

To keep things simpler (and hence more interpretable), we make two simplifying restrictions to this general approach:

(1) We will restrict the form of the function f to be *linear*. Such a model function with p predictors can be written as

$$f(x_1, x_2, x_3) = \beta_0 + \beta_1 x_1 + \beta_2 x_2 + \cdots + \beta_p x_p \, . \tag{6.1}$$

The vector $\boldsymbol{\beta} = \langle \beta_0, \beta_1, \beta_2, \ldots, \beta_p \rangle$ is the parameter vector for this model, and fitting such a model reduces to estimating these $p + 1$ parameters.

(2) We will not directly model variability (including measurement error) in the predictors. That is, the model considers the predictors to be constant rather than random variables. This assumption is closer to reality in some controlled

experiments than it is in most observational studies, and there are other models that do not make this assumption.

Linear models are much more flexible than (6.1) might lead you to believe. We will require that the function f be linear as a function of the *parameters* but not of the variables. So, for example, we will allow a function like

$$f(x_1, x_2, x_3) = \beta_0 + \beta_1 x_1 + \beta_2 \log(x_2) + \beta_3 x_3^2 + \beta_4 x_1 x_2 \sqrt{x_3} \,.$$

This makes sense, since expressions like $\log(x_2)$, $x_1 x_2 \sqrt{x_3}$ *are* variables – each takes on a value for each observational unit in our data set.

We will also allow transformations of Y. So a relationship like

$$Y = \varepsilon \beta_0 e^{\beta_1 x} \,,$$

which is equivalent to

$$\log(Y) = \log(\beta_0) + \beta_1 x + \log(\varepsilon) \,,$$

can also be handled within the linear model framework. Linear models are actually quite flexible and can express a wide variety of relationships between the explanatory and response variables.

6.1.3. Vector and Matrix Notation

It is convenient to represent the data involved in regression analysis using vectors and matrices.[1] A typical data set with n observations and p explanatory variables has the general form

$$
\begin{array}{ccccc}
y_1 & x_{11} & x_{12} & \cdots & x_{1p} \\
y_2 & x_{21} & x_{22} & \cdots & x_{2p} \\
\vdots & \vdots & \vdots & \ddots & \vdots \\
y_n & x_{n1} & x_{n2} & \cdots & x_{np}
\end{array}
$$

The use of subscripts, especially the double subscripts on the explanatory variables, makes expressions like

$$Y_i = \beta_0 + \beta_1 x_{i1} + \beta_2 x_{i2} + \cdots + \beta_p x_{ip} + \varepsilon_i \tag{6.2}$$

difficult to read. It will be simpler for both representation and manipulation to use vectors and matrices instead. Equation (6.2) then becomes

$$\boldsymbol{Y} = \beta_0 \boldsymbol{1} + \beta_1 \boldsymbol{x}_{.1} + \cdots + \beta_p \boldsymbol{x}_{.p} + \boldsymbol{\varepsilon} \,, \tag{6.3}$$

or even more succinctly

$$\boldsymbol{Y} = \boldsymbol{X\beta} + \boldsymbol{\varepsilon} \,, \tag{6.4}$$

where $\boldsymbol{Y} = [Y_1, \ldots, Y_n]^\top$, $\boldsymbol{1} = [1, 1, \ldots, 1]^\top$, $\boldsymbol{\varepsilon} = [\varepsilon_1, \ldots, \varepsilon_n]^\top$, $\boldsymbol{\beta} = [\beta_0, \ldots, \beta_p]^\top$, and $\boldsymbol{X}$ is the **model matrix**

$$
\boldsymbol{X} = \begin{bmatrix}
1 & x_{11} & x_{12} & \cdots & x_{1p} \\
1 & x_{21} & x_{22} & \cdots & x_{2p} \\
\vdots & \vdots & \vdots & \ddots & \vdots \\
1 & x_{n1} & x_{n2} & \cdots & x_{np}
\end{bmatrix} \,.
$$

---

[1] Appendix C covers the necessary linear algebra for this chapter.

The column of 1's in X allows us to incorporate the **intercept term** β_0. We will denote the columns of X as $x_{\cdot i}$ or even more succinctly as x_i for $0 \le i \le p$. (So $x_0 = 1$.) Note that in this context, X does not represent a random vector; rather it is a fixed matrix. As usual, we will replace Y with y when we are considering the values in a particular data set.

6.1.4. Expressing Relationships Using Linear Models

Let's begin by looking at something we already know how to do, but from the perspective of linear models.

Example 6.1.1. In Chapter 4 we developed methods of inference for an unknown population mean μ. Recall that these methods were based on the assumptions that the population was normally distributed and that our sample was an iid (or simple random) sample. That is, we assumed that

$$Y_i \overset{\text{iid}}{\sim} \text{Norm}(\mu, \sigma)$$

for some parameter values μ and σ.

We can express the assumptions in a different way, namely as

$$Y_i = \mu + \varepsilon_i$$

where $\varepsilon_i \overset{\text{iid}}{\sim} \text{Norm}(0, \sigma)$. Using our linear algebra notation, we can also write this as

$$Y = \mu \mathbf{1} + \varepsilon \ .$$

This is a simple linear model with only an intercept parameter: $\boldsymbol{\beta} = \langle \beta_0 \rangle = \langle \mu \rangle$; the model matrix X in this case would consist of a single column of 1's:

$$Y = \begin{bmatrix} Y_1 \\ Y_2 \\ \vdots \\ Y_n \end{bmatrix} = X\boldsymbol{\beta} + \varepsilon = \mu \begin{bmatrix} 1 \\ 1 \\ \vdots \\ 1 \end{bmatrix} + \begin{bmatrix} \varepsilon_1 \\ \varepsilon_2 \\ \vdots \\ \varepsilon_n \end{bmatrix} = \begin{bmatrix} \mu + \varepsilon_1 \\ \mu + \varepsilon_2 \\ \vdots \\ \mu + \varepsilon_n \end{bmatrix} . \qquad \triangleleft$$

To see the utility of the linear model framework, let's look at some additional examples.

Example 6.1.2. A physics student measures the distance a spring stretches under different load weights. She expects a linear relationship between stretch and weight. An appropriate model for this is

$$Y = \beta_0 \mathbf{1} + \beta_1 x_1 + \varepsilon$$

where Y is a vector of measured amounts of stretch and x_1 is a vector of measured weights. In this case β_0 indicates the length of the spring with no load and β_1 is a spring constant. $\qquad \triangleleft$

Example 6.1.3. Suppose that we measure the height of a sample of adults. It is reasonable to assume that the heights of men and women have different distributions. We can model this as

$$Y = \mu \mathbf{1} + \alpha x_1 + \varepsilon \ .$$

Here Y_i is the height of subject i, μ is the mean height of women in the population, α is the difference between the mean heights of men and women, and

$$x_{i1} = [\![\text{subject } i \text{ is male}]\!] = \begin{cases} 0 & \text{if subject } i \text{ is female,} \\ 1 & \text{if subject } i \text{ is male.} \end{cases}$$

The variable $\boldsymbol{x}_1$ is called an **indicator variable** since it indicates whether some condition is true or false.

Putting things into standard notation, we have the following model:

$$\boldsymbol{Y} = \boldsymbol{X}\boldsymbol{\beta} + \boldsymbol{\varepsilon}$$

where $\boldsymbol{\beta} = [\beta_0, \beta_1]^\top = [\mu, \alpha]^\top$. Measuring in inches and placing the women first and men after, our data set would look something like

$$\boldsymbol{y} = \begin{bmatrix} 65 \\ 70 \\ \vdots \\ 68 \\ 69 \\ 75 \\ \vdots \\ 72 \end{bmatrix}, \quad \boldsymbol{X} = \begin{bmatrix} 1 & 0 \\ 1 & 0 \\ \vdots & \vdots \\ 1 & 0 \\ 1 & 1 \\ 1 & 1 \\ \vdots & \vdots \\ 1 & 1 \end{bmatrix}, \quad \text{or, equivalently,} \quad [\boldsymbol{y} \mid \boldsymbol{X}] = \begin{bmatrix} 65 & 1 & 0 \\ 70 & 1 & 0 \\ \vdots & \vdots & \vdots \\ 68 & 1 & 0 \\ 69 & 1 & 1 \\ 75 & 1 & 1 \\ \vdots & \vdots & \vdots \\ 72 & 1 & 1 \end{bmatrix}.$$

Except for the column of 1's, this looks very much like how we might store the data in an R data frame. ◁

Example 6.1.4. Suppose we are interested in the relationship between age and cholesterol in a population of men. It is perhaps reasonable to assume that a scatterplot of cholesterol vs. age would follow a roughly parabolic shape. An appropriate model for this would be

$$\boldsymbol{Y} = \beta_0 + \beta_1 \boldsymbol{x}_1 + \beta_2 \boldsymbol{x}_1^2 .$$

The columns of our model matrix $\boldsymbol{X}$ would then consist of a column of 1's (for the intercept term), a column of ages, and a column of squared ages. ◁

6.1.5. Describing Linear Models in R

Linear models can be fit in R using the function `lm()`, which has a formula interface that mimics the way we have been describing our models. We will learn about `lm()` shortly. For now, we will focus only on the formula interface. The formulas will look familiar because they are very similar to the formulas used to describe plots in `lattice`. The general form (with three predictors) is

```
response ~ predictor1 + predictor2 + predictor3
```

The intercept term, the model parameters (coefficients), and the error term are assumed, but the intercept can be made explicit if we like:

```
response ~ 1 + predictor1 + predictor2 + predictor3
```

or removed if we don't want it in the model:

```
response ~ -1 + predictor1 + predictor2 + predictor3
```

The correspondence between models and R formulas for a simple model with one predictor is

model	R formula
$Y = \beta_0 \mathbf{1} + \beta_1 x + \varepsilon$	y ~ 1 + x or y ~ x
$Y = \beta_1 x + \varepsilon$	y ~ -1 + x or y ~ 0 + x

Example 6.1.5. We can describe the models of the previous section using R as follows:

- Example 6.1.1:
  ```
  y ~ 1              # note explicit intercept is the only term on right
  ```
- Example 6.1.2:
  ```
  length ~ weight        # using implicit intercept
  length ~ 1 + weight    # using explicit intercept
  ```
- Example 6.1.3:
  ```
  height ~ sex           # using implicit intercept
  height ~ 1 + sex       # using explicit intercept
  ```
- Example 6.1.4:
  ```
  cholesterol ~ age + I(age^2)
  ```

    A note of explanation is required here. The I() function "inhibits in-terpretation". As we will learn in Chapter 7, model formulas treat certain operators (including +, *, ^, and :) in special ways to simplify specifying common types of models. Wrapping age^2 in the I() function tells R not to treat exponentiation in a special formula-specific way but to do normal arith-metic instead. Alternatively, we could define an additional row in our data frame that contains the squared ages of our subjects. Another use of I() is illustrated in Example 6.1.6.   ◁

Example 6.1.6.

Q. A college administrator is investigating how well standardized test scores predict college grade point average. If SATM and SATV are the mathematics and verbal scores from the SAT test, what is the difference between the following two models?

```
gpa ~ SATM + SATV
gpa ~ I(SATM + SATV)
```

A. The first formula corresponds to the model

$$\text{gpa} = \beta_0 \mathbf{1} + \beta_1 \text{SATM} + \beta_2 \text{SATV} \,,$$

and the second to the model

$$\text{gpa} = \beta_0 \mathbf{1} + \beta_1 (\text{SATM} + \text{SATV}) = \beta_0 \mathbf{1} + \beta_1 \text{SATM} + \beta_1 \text{SATV} \,.$$

One way to interpret the difference between the two models is that the second model predicts college performance from the sum (equivalently, mean) of the two partial scores while the first model uses a weighted average of the two. The models are only the same if $\beta_1 = \beta_2$.   ◁

6.1.6. Questions for Linear Models

The examples from the previous section are only a small sample of the types of relationships that can be expressed within the linear model framework. The use of linear models gives rise to questions of the following types:

- How do we estimate the parameters? How do we interpret these estimates?

    We will be interested both in point estimates and in confidence regions for some or all of the parameters in our models. Selecting models where the parameters are meaningful and correctly interpreting their estimates is an important part of regression analysis.

- Is my model too complicated?

    Can I omit some of the terms in the model so that I have a simpler model? Simpler models are usually easier to interpret and may rely on fewer assumptions. More complicated models run the risk of "over-fitting".

- Is my model a good fit?

    Parameter estimation will find a "best-fit" model of the specified type, but is it a good fit or should I consider some different model?

- Which of several models is the best one to use?

    The flexibility of the linear model and the ease with which linear models can be analyzed in software makes it possible to investigate multiple models for the same situation. How do we decide which models to fit and then which model or models to use in our analysis?

- What does my model predict? How sure am I about the prediction?

    Often linear models are used to predict future observations under various conditions.

- Is there a causal relationship among the variables?

    Suppose we observe that diabetics are heavier than non-diabetics on average. How do we interpret this finding? Does obesity cause (increased risk for) diabetes? Does diabetes cause (increased risk for) obesity?

    While causality is an interesting and important question, at a fundamental level, statistical inference alone can never answer it. We can fit models to measure various relationships between variables, but the statistics will not tell us what (if anything) is causing what. Inferring a (likely, plausible, etc.) causal relationship must be the result of more than just the mathematics of statistical inference. Where possible, proper design of randomized experiments is one of the most important components in drawing conclusions about causality, and so is a non-statistical understanding of the underlying situation. This is one reason why statisticians often work collaboratively with researchers from other disciplines who can bring the appropriate and necessary expertise to the project.

We will develop methods to at least partially answer all of these questions.

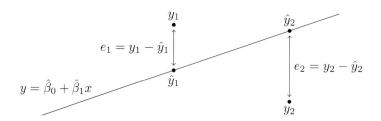

Figure 6.1. Graphically, a residual is the vertical distance on a scatterplot between an observation and the corresponding fitted value (i.e., the regression line).

6.1.7. Parameter Estimation for Linear Models

In this section we address the question of parameter estimation in a linear model. Given a linear model, what parameter values provide the best fit to our data?

It is useful to think of linear models as describing two components of the response:

$$\text{response} = \text{systematic variation} + \text{random variation} \,,$$
$$\boldsymbol{Y} = \boldsymbol{X}\boldsymbol{\beta} + \boldsymbol{\varepsilon} \,.$$

Since we don't know the true values of the parameters $(\boldsymbol{\beta})$ in our model, we will have to estimate them somehow. Our estimate for $\boldsymbol{\beta}$ will be denoted $\hat{\boldsymbol{\beta}}$. From $\hat{\boldsymbol{\beta}}$ we can compute fitted values for the predictors in our data set using

$$\hat{\boldsymbol{y}} = \boldsymbol{X}\hat{\boldsymbol{\beta}}$$

or fitted values for arbitrary values of the predictors using

$$\hat{\boldsymbol{y}}^{*} = \boldsymbol{x}^{*}\hat{\boldsymbol{\beta}} \,.$$

That is, the fitted values are determined by using our parameter estimates in place of the true (and unknown) parameter values $\boldsymbol{\beta}$. The difference between the observed response and the fitted response is called a **residual** and is denoted

$$e_i = y_i - \hat{y}_i \,.$$

The vector of all residuals is denoted

$$\boldsymbol{e} = \boldsymbol{y} - \hat{\boldsymbol{y}} \,.$$

Graphically, a residual is the vertical distance on a scatterplot between an observation and the corresponding fitted value on the regression line. (See Figure 6.1.) Our observed response can then be described in terms of fit and residual:

$$\text{response} = \text{fit} + \text{residual} \,,$$
$$\boldsymbol{y} = \hat{\boldsymbol{y}} + \boldsymbol{e} \,.$$

There is an important distinction between $\boldsymbol{e}$ and $\boldsymbol{\varepsilon}$:

$$\boldsymbol{\varepsilon} = \boldsymbol{Y} - (\beta_0 \boldsymbol{1} + \beta_1 \boldsymbol{x}) \,,$$
$$\boldsymbol{e} = \boldsymbol{y} - (\hat{\beta}_0 \boldsymbol{1} + \hat{\beta}_1 \boldsymbol{x}) \,.$$

We can calculate the values of e from our data (once we have calculated the parameter estimates $\hat{\boldsymbol{\beta}}$), but $\boldsymbol{\varepsilon}$ is a random variable used to describe the model and remains unknown because we don't know the values of β_0 and β_1.

As long as there is an intercept term in the model, we may assume that $\mathrm{E}(\boldsymbol{\varepsilon}) = \mathbf{0}$, since any non-zero mean could be absorbed into the intercept coefficient. Often we will further assume that the error terms are independent and normally distributed:

$$\varepsilon \overset{\text{iid}}{\sim} \mathsf{Norm}(0, \sigma) \, .$$

One method of fitting such a linear model comes to mind immediately: maximum likelihood. Indeed maximum likelihood is a useful method (and sometimes the preferred method) for fitting a linear model. Since $\boldsymbol{e} = \boldsymbol{y} - \hat{\boldsymbol{y}}$ is completely determined by the data and the (estimated) parameters of the model, the maximum likelihood method amounts to seeking the parameters under which the residuals are most likely, assuming the true distribution is normal (or follows some other specified distribution). The variance parameter σ^2 is typically unknown, so this method estimates both the coefficients $\boldsymbol{\beta}$ of the linear model and the parameter σ^2.

But there is another method that drives much of the development of regression analysis. When it is available (sometimes it is not), it is also computationally more efficient. Furthermore, it typically gives rise to the same estimators as the maximum likelihood method does. It can be motivated by considering the linear model geometrically.

6.1.8. The Principle of Least Squares

Figure 6.2 depicts a geometric interpretation of the generic linear model. The response vector $\boldsymbol{y}$ can be considered as a vector in $\mathbb{R}^n$, where n is the number of observations in the data. The fit vector $\hat{\boldsymbol{y}}$ can also be viewed as a vector in $\mathbb{R}^n$. Since (for a fixed $\boldsymbol{X}$) this vector is determined by specifying the $p + 1$ parameters (intercept plus coefficients on p predictors) of the model, this vector lives in a $(p + 1)$-dimensional subspace of $\mathbb{R}^n$ called the **model space**.

The residuals form another vector $e = \boldsymbol{y} - \hat{\boldsymbol{y}}$, and the length of this vector is one way to measure how well the model fits the data. If e is short, the fit is good because the y_i's are close to the $\hat{y}_i$'s. The length of this vector satisfies

$$|e|^2 = SSE = \sum e_i{}^2 = \sum (y_i - \hat{y}_i)^2 \, .$$

In SSE, SS stands for "sum of squares" and E for "error".

There is an implicit dependence of SSE on the estimates $\hat{\boldsymbol{\beta}}$ because $\hat{\boldsymbol{y}}$ depends on $\hat{\boldsymbol{\beta}}$. The goal of least squares estimation is to determine the values of $\hat{\boldsymbol{\beta}}$ that make the residual vector ($e = \boldsymbol{y} - \hat{\boldsymbol{y}}$) as short as possible.

The other vectors appearing in Figure 6.2 will become important in our analysis, too. Let $\overline{\boldsymbol{y}} = \overline{y}\,\mathbf{1}$ be the vector defined by

$$\overline{y}_i = \overline{y} \, .$$

That is, $\overline{\boldsymbol{y}}$ is a vector in $\mathbb{R}^n$ such that every component is equal to the mean of the observed responses. Because $\boldsymbol{y} - \hat{\boldsymbol{y}}$ is shorter than $\boldsymbol{y} - \tilde{\boldsymbol{y}}$ for any other vector $\tilde{\boldsymbol{y}}$,

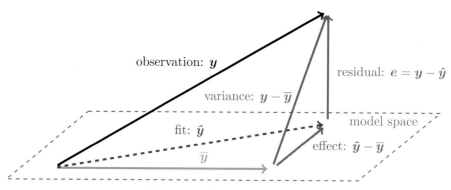

Figure 6.2. Generic linear model.

Box 6.1. The Principle of Least Squares

The **least squares estimates** of the parameters of a linear model are the parameters that minimize the sum of the squares of the residuals (SSE). This is equivalent to minimizing the length of the residual vector $e = y - \hat{y}$.

$y - \hat{y}$ will always be orthogonal to the model space. In particular, since $\hat{y}$ and $\overline{y}$ are both in the model space, $\hat{y} - \overline{y}$ is also in the model space, and thus

$$\hat{y} - \overline{y} \perp y - \hat{y} .$$

The vector $y - \overline{y}$ is related to the sample variance of the response values:

$$|y - \overline{y}|^2 = \sum_i (y_i - \overline{y})^2 = (n-1)s_y^2 ,$$

where s_y^2 is the variance of y. This variance vector can be partitioned into an **effect** component ($\hat{y} - \overline{y}$) and a **residual** component ($e = y - \hat{y}$):

$$y - \overline{y} = \hat{y} - \overline{y} + y - \hat{y} .$$

Moreover, since $\hat{y} - \overline{y} \perp y - \hat{y}$, we have the important identity

$$|y - \overline{y}|^2 = |\hat{y} - \overline{y}|^2 + |y - \hat{y}|^2 \tag{6.5}$$

as well. We will refer to (6.5) as a **Pythagorean decomposition** of $y - \overline{y}$.

A comparison of the lengths of these vectors indicates how much of the variance is explained by the model ($\hat{y} - \overline{y}$) vs. unexplained by the model ($y - \hat{y}$). Analysis of variance (ANOVA) methods are based on a comparison of the lengths of these two vectors. When $|\hat{y} - \overline{y}|^2$ is relatively large and $|y - \hat{y}|^2$ is relatively small, then our model is effective in explaining the variation in the response variable. This explains the use of the term *effect* vector. It is not meant to imply that there is necessarily a cause-and-effect relationship between the predictors and the response.

6.2. Simple Linear Regression

We will develop the linear regression methods outlined in the previous section first for **simple linear regression**. Simple linear regression handles the case where the linear model has only one explanatory variable and the errors are independent and normally distributed:

$$\boldsymbol{Y} = \beta_0 \boldsymbol{1} + \beta_1 \boldsymbol{X}_1 + \boldsymbol{\varepsilon}\,, \qquad \varepsilon \overset{\text{iid}}{\sim} \mathsf{Norm}(0, \sigma)\,.$$

This distinguishes it from **multiple linear regression**, when there are two or more explanatory variables.

6.2.1. Parameter Estimates via Least Squares

Least squares estimation can be approached either from the perspective of calculus or from the perspective of linear algebra and geometry. We begin with the calculus approach but will eventually find the other approach more useful.

Our goal is to choose $\boldsymbol{\beta}$ so that

$$S(\boldsymbol{\beta}) = \sum_{i=1}^{n} e_i^2(\boldsymbol{\beta}) = \sum_{i=1}^{n} (y_i - \hat{y}_i[\boldsymbol{\beta}])^2$$

is as small as possible. Here we are using $\hat{\boldsymbol{y}}[\boldsymbol{\beta}]$ to indicate the fit vector if we use $\boldsymbol{\beta}$ for our parameter estimates. In the case of simple linear regression (dropping the limits of summation for readability), this becomes

$$S(\boldsymbol{\beta}) = \sum (y_i - \beta_0 - \beta_1 x_{i1})^2\,.$$

Since there is only one explanatory variable, it is convenient to use $\boldsymbol{x}$ in place of $\boldsymbol{x}_{\cdot 1}$. Then

$$S(\boldsymbol{\beta}) = \sum (y_i - \beta_0 - \beta_1 x_i)^2\,.$$

We can minimize S by determining where the two partial derivatives vanish, that is, by solving the following system of equations:

$$
\begin{aligned}
0 = \frac{\partial S}{\partial \beta_0} &= \sum 2(y_i - \beta_0 - \beta_1 x_i)(-1) \\
&= (-2)[n\overline{y} - n\beta_0 - n\beta_1 \overline{x}] \\
&= (-2n)[\overline{y} - \beta_0 - \beta_1 \overline{x}]\,,
\end{aligned}
\tag{6.6}
$$

$$
\begin{aligned}
0 = \frac{\partial S}{\partial \beta_1} &= \sum 2(y_i - \beta_0 - \beta_1 x_i)(-x_i) \\
&= (-2) \sum \left[x_i y_i - \beta_0 x_i - \beta_1 x_i^2 \right]\,.
\end{aligned}
\tag{6.7}
$$

From (6.6) we see that

$$\overline{y} = \hat{\beta}_0 + \hat{\beta}_1 \overline{x}\,,$$

so the point $\langle \overline{x}, \overline{y} \rangle$ is always on the least squares regression line, and

$$\hat{\beta}_0 = \overline{y} - \hat{\beta}_1 \overline{x}\,.
\tag{6.8}$$

Substituting (6.8) into (6.7), we see that

$$0 = \sum \left[x_i y_i - (\overline{y} - \hat{\beta}_1 \overline{x}) x_i - \hat{\beta}_1 x_i^2 \right]$$
$$= \sum \left[x_i y_i - \overline{y} x_i + \hat{\beta}_1 \overline{x} x_i - \hat{\beta}_1 x_i^2 \right] ,$$

so

$$\hat{\beta}_1 = \frac{\sum (y_i - \overline{y}) x_i}{\sum x_i^2 - \sum x_i \overline{x}} = \frac{\sum (y_i - \overline{y}) x_i}{\sum x_i (x_i - \overline{x})} \tag{6.9}$$

$$= \frac{\sum (y_i - \overline{y})(x_i - \overline{x})}{\sum (x_i - \overline{x})^2} = \frac{S_{xy}}{S_{xx}} \tag{6.10}$$

$$= \left[\frac{1}{n-1} \sum_i \left(\frac{x_i - \overline{x}}{s_x} \cdot \frac{y_i - \overline{y}}{s_y} \right) \right] \cdot \frac{s_y}{s_x}$$

$$= r \frac{s_y}{s_x} . \tag{6.11}$$

We will see the quantities

$$S_{xx} = \sum (x_i - \overline{x})^2 , \tag{6.12}$$

$$S_{xy} = \sum (y_i - \overline{y})(x_i - \overline{x}) , \text{ and} \tag{6.13}$$

$$r = \frac{1}{n-1} \sum_i \left(\frac{x_i - \overline{x}}{s_x} \cdot \frac{y_i - \overline{y}}{s_y} \right) \tag{6.14}$$

again and will have more to say about the **correlation coefficient** r and its interpretation later. For now, the most important observation is that determining the values of $\hat{\beta}_0$ and $\hat{\beta}_1$ is a computationally simple matter.

6.2.2. Parameter Estimates via Maximum Likelihood

We can also estimate the regression parameters (β_0, β_1, and σ) using maximum likelihood. Here we need to make explicit use of the distributional assumption on the error term in the model:

$$\varepsilon = \boldsymbol{Y} - \beta_0 \boldsymbol{1} - \beta_1 \boldsymbol{x} \stackrel{\text{iid}}{\sim} \mathsf{Norm}(0, \sigma) .$$

From this it follows that the likelihood function is

$$L(\beta_0, \beta_1, \sigma; \boldsymbol{x}, \boldsymbol{y}) = \prod_i \frac{1}{\sigma \sqrt{2\pi}} e^{-(y_i - \beta_0 - \beta_1 x_i)^2 / 2\sigma^2}$$

and the log-likelihood is

$$l(\beta_0, \beta_1, \sigma; \boldsymbol{x}, \boldsymbol{y}) = \log(L(\beta_0, \beta_1, \sigma; \boldsymbol{x}, \boldsymbol{y}))$$

$$= \sum_i -\log(\sigma) - \frac{1}{2} \log(2\pi) - (y_i - \beta_0 - \beta_1 x_i)^2 / 2\sigma^2 .$$

We can maximize the log-likelihood function using methods from calculus:

$$\frac{\partial l}{\partial \beta_0} = \frac{1}{\sigma^2} \sum_i (y_i - \beta_0 - \beta_1 x_i) \,, \tag{6.15}$$

$$\frac{\partial l}{\partial \beta_1} = \frac{1}{\sigma^2} \sum_i (y_i - \beta_0 - \beta_1 x_i) x_i \,, \tag{6.16}$$

$$\frac{\partial l}{\partial \sigma} = \frac{-n}{\sigma} - \frac{1}{\sigma^3} \sum_i (y_i - \beta_0 - \beta_1 x_i)^2$$

$$= \frac{1}{\sigma^3} \left(n\sigma^2 - \sum_i (y_i - \beta_0 - \beta_1 x_i)^2 \right) = \frac{1}{\sigma^3} \left(n\sigma^2 - \sum_i e_i^2 \right) \,. \tag{6.17}$$

The solutions to (6.15) and (6.16) are the same as the solutions to (6.6) and (6.7), so the maximum likelihood estimate for $\boldsymbol{\beta}$ is the same as the least squares estimate.

As a side effect of the maximum likelihood approach, we also get an estimate for σ^2, namely

$$\frac{\sum_i e_i^2}{n} \,.$$

This is an intuitive estimator, but it is not typically used. The problem is that it is biased. As was the case for estimating the population variance based on a sample (Section 4.6), we will prefer a version of this estimator that has been corrected to be an unbiased estimator of σ^2. As we will show in Section 6.3.4, the unbiased estimator for σ^2 in the case of the simple linear regression model is

$$\hat{\sigma}^2 = \frac{\sum_i e_i^2}{n - 2} \,.$$

6.2.3. Parameter Estimates via Geometry and Linear Algebra

An alternative approach to determining the least squares estimates β_0 and β_1 relies on geometry rather than on calculus. Since

$$|\boldsymbol{e}|^2 = |\boldsymbol{y} - \hat{\boldsymbol{y}}|^2 = \sum (y_i - \hat{y})^2 \,,$$

the least squares estimate for $\boldsymbol{\beta}$ is the parameter value that makes $\boldsymbol{e}$ shortest. Since $\hat{\boldsymbol{y}}$ must lie in the model space (see Figure 6.2), and since the shortest vector from the model space to $\boldsymbol{y}$ must be orthogonal to the model space, we see that

$$\hat{\boldsymbol{y}} \perp \boldsymbol{y} - \hat{\boldsymbol{y}} \,.$$

This means that we can obtain $\hat{\boldsymbol{y}}$ from $\boldsymbol{y}$ by orthogonal projection into the model space. In the case of simple linear regression,

$$\hat{\boldsymbol{y}} = \hat{\beta}_0 \boldsymbol{1} + \hat{\beta}_1 \boldsymbol{x} \,,$$

so the model space is 2-dimensional. Combining this geometric view with tools from linear algebra allows us to solve for $\hat{\boldsymbol{\beta}}$.

A clever choice of basis for the model space will both simplify the algebra and provide additional insight. We will use the following basis for the model space:

$$\boldsymbol{v}_0 = \boldsymbol{1} = [1, 1, 1, \ldots, 1]^\top ,$$

$$\boldsymbol{v}_1 = \boldsymbol{x} - \overline{\boldsymbol{x}} = [x_1 - \overline{x}, x_2 - \overline{x}, \ldots, x_n - \overline{x}]^\top .$$

Notice that $|\boldsymbol{v}_1|^2 = S_{xx}$ from (6.12) and that $\boldsymbol{v}_0 \perp \boldsymbol{v}_1$ since

$$\boldsymbol{v}_0 \cdot \boldsymbol{v}_1 = \sum 1 \cdot (x_i - \overline{x}) = \sum x_i - n\overline{x} = 0 .$$

These vectors form a basis for the model space since

$$\alpha_0 \boldsymbol{v}_0 + \alpha_1 \boldsymbol{v}_1 = \alpha_0 + \alpha_1 (\boldsymbol{x}_1 - \overline{\boldsymbol{x}})$$
$$= (\alpha_0 - \alpha_1 \overline{\boldsymbol{x}}) + \alpha_1 \boldsymbol{x} .$$

In fact, we see that $\beta_1 = \alpha_1$ and $\beta_0 = \alpha_0 - \alpha_1 \overline{x}$. Essentially we have reparameterized the model as

$$\boldsymbol{Y} = \alpha_0 + \beta_1 (\boldsymbol{x} - \overline{\boldsymbol{x}}) + \varepsilon$$
$$= \alpha_0 \boldsymbol{v}_0 + \beta_1 \boldsymbol{v}_1 + \varepsilon .$$

We can compute $\hat{\boldsymbol{y}}$ by projecting in the directions of $\boldsymbol{v}_0$ and $\boldsymbol{v}_1$. Let $\boldsymbol{u}_i = \boldsymbol{v}_i / |\boldsymbol{v}_i|$ be the unit vector in the direction of $\boldsymbol{v}_i$. Then

(1) Since $\{\boldsymbol{v}_0, \boldsymbol{v}_1\}$ span the model space and are orthogonal

$$\hat{\boldsymbol{y}} = \mathrm{proj}(\boldsymbol{y} \to \boldsymbol{v}_0) + \mathrm{proj}(\boldsymbol{y} \to \boldsymbol{v}_1) \tag{6.18}$$

$$= (\boldsymbol{y} \cdot \boldsymbol{u}_0)\boldsymbol{u}_0 + (\boldsymbol{y} \cdot \boldsymbol{u}_1)\boldsymbol{u}_1 . \tag{6.19}$$

(2) $\mathrm{proj}(\boldsymbol{y} \to \boldsymbol{v}_0) = \overline{\boldsymbol{y}}$.

This follows by direct calculation:

$$\boldsymbol{y} \cdot \boldsymbol{u}_0 = \frac{1}{\sqrt{n}} \sum y_i = \sqrt{n}\, \overline{y} ,$$

so

$$\mathrm{proj}(\boldsymbol{y} \to \boldsymbol{v}_0) = (\boldsymbol{y} \cdot \boldsymbol{u}_0)\boldsymbol{u}_0$$

$$= \left(\sqrt{n}\overline{y} \right) \frac{1}{\sqrt{n}} = \overline{\boldsymbol{y}} .$$

(3) From this it follows that

$$\mathrm{proj}(\boldsymbol{y} \to \boldsymbol{v}_1) = (\boldsymbol{y} \cdot \boldsymbol{u}_1)\boldsymbol{u}_1 = \hat{\boldsymbol{y}} - \overline{\boldsymbol{y}} .$$

Thus projection in the direction of $\boldsymbol{v}_0$ gives the overall mean, and projection in the direction of $\boldsymbol{v}_1$ gives what we called the effect vector in Figure 6.2.

Since $\hat{\boldsymbol{y}} = \hat{\alpha}_0 \boldsymbol{v}_0 + \hat{\beta}_1 \boldsymbol{v}_1$, we can recover the coefficients $\hat{\boldsymbol{\beta}}$ as follows:

$$\hat{\beta}_1 = \frac{|\mathrm{proj}(\boldsymbol{y} \to \boldsymbol{v}_1)|}{|\boldsymbol{v}_1|} = \frac{\boldsymbol{y} \cdot \boldsymbol{u}_1}{|\boldsymbol{v}_1|} = \frac{\boldsymbol{y} \cdot \boldsymbol{v}_1}{|\boldsymbol{v}_1|^2} , \tag{6.20}$$

$$\hat{\alpha}_0 = \frac{|\mathrm{proj}(\boldsymbol{y} \to \boldsymbol{v}_0)|}{|\boldsymbol{v}_0|} = \frac{\boldsymbol{y} \cdot \boldsymbol{u}_0}{|\boldsymbol{v}_0|} = \frac{\boldsymbol{y} \cdot \boldsymbol{v}_0}{|\boldsymbol{v}_0|^2} , \tag{6.21}$$

$$\hat{\beta}_0 = \alpha_0 + \beta_1 \overline{\boldsymbol{x}} . \tag{6.22}$$

The hat matrix

We can re-express (6.19) in matrix form as follows:

$$\hat{y} = u_0 u_0^\top y + u_1 u_1^\top y = (u_0 u_0^\top + u_1 u_1^\top)y = Hy . \qquad (6.23)$$

The $n \times n$ matrix $H = u_0 u_0^\top + u_1 u_1^\top$, called the **hat matrix**, is the matrix that performs orthogonal projection onto our model space. We can obtain our least squares estimate $\hat{\beta}$ from H as follows:

$$X\hat{\beta} = \hat{y} = Hy ,$$

$$X^\top X \hat{\beta} = X^\top Hy ,$$

$$\hat{\beta} = (X^\top X)^{-1} X^\top Hy ,$$

provided $X^\top X$ is invertible. (Recall that for the simple linear model, X is a $2 \times n$ matrix and so cannot be invertible if $n \neq 2$. But $X^\top X$ is an $n \times n$ matrix and often *is* invertible.) With a bit more linear algebra, it can be shown that

$$H = X(X^\top X)^{-1} X^\top \qquad (6.24)$$

and

$$\hat{\beta} = (X^\top X)^{-1} X^\top y .$$

This approach to determining $\hat{\beta}$ is of theoretical interest more than of practical interest. The geometry and linear algebra leading to the definition of H motivate much of regression analysis and are important for this reason. But the matrix H can be very large, and numerical algorithms with these matrices can be inefficient and prone to round-off errors. In general it is wisest to use regression algorithms that have been carefully programmed to avoid both inefficiency and inaccuracy.

6.2.4. Fitting Linear Models in R

Linear models can be fit in R using the function `lm()`. We use a model formula to describe the model to `lm()`, which produces a model object that can be further manipulated to obtain information about the model. We begin with a very simple model and a very small data set.

Example 6.2.1.

Q. Use R to fit a simple linear model to the following data:

x	1	2	3	4
y	2	3	5	6

A. A linear model is fit by providing the model formula to `lm()`:

```
> x <- 1:4; y <- c(2,3,5,6)
> model <- lm(y~x)
```

lm-demo1

There are a number of functions (like `summary()`, `coef()`, and `fitted()` shown here) that extract information about a regression model.

```
> coef(model)            # the coefficients          lm-demo2
(Intercept)          x
       0.5         1.4
> fitted(model)          # y-hat values
   1   2   3   4
1.9 3.3 4.7 6.1
> summary(model)
< 8 lines removed >
Coefficients:
            Estimate Std. Error t value Pr(>|t|)
(Intercept)    0.500      0.387    1.29     0.33
x              1.400      0.141    9.90     0.01

Residual standard error: 0.316 on 2 degrees of freedom
Multiple R-squared: 0.98,       Adjusted R-squared: 0.97
F-statistic:   98 on 1 and 2 DF,  p-value: 0.0101
```

In addition, one can access the information stored in the lm object or its summary directly:

```
> names(model)                                       lm-demo3
 [1] "coefficients"   "residuals"    "effects"      "rank"
 [5] "fitted.values"  "assign"       "qr"           "df.residual"
 [9] "xlevels"        "call"         "terms"        "model"
> names(summary(model))
 [1] "call"           "terms"        "residuals"    "coefficients"
 [5] "aliased"        "sigma"        "df"           "r.squared"
 [9] "adj.r.squared"  "fstatistic"   "cov.unscaled"              ◁
```

Example 6.2.2.

Q. Fit the linear model from the previous example without using lm().

A. Although there is generally no reason to do so in practice, we can also fit this example "by hand" using the methods developed in the previous section. First we define the vectors v_0 and v_1, and from these we can compute $\hat{y}$, $\hat{\beta}_1$, $\hat{\alpha}_0$, and $\hat{\beta}_0$:

```
                                                     lm-demo-vector
> v0 <- rep(1,4); v0          # two ways to compute beta_1-hat
[1] 1 1 1 1                   > project(y,v1,type='l')/vlength(v1)
> u0 <- v0/vlength(v0); u0    [1] 1.4
[1] 0.5 0.5 0.5 0.5           > dot(y,v1)/(vlength(v1))^2 -> b1; b1
> v1 <- x - mean(x); v1       [1] 1.4
[1] -1.5 -0.5  0.5  1.5       #
> u1 <- v0/vlength(v1); u1    # two ways to compute alpha_0-hat
[1] 0.44721 0.44721 0.44721 0.44721  > project(y,v0,type='l')/vlength(v0)
#                             [1] 4
# projecting into the model space  > dot(y,v0)/(vlength(v0))^2 -> a0; a0
> project(y,v0)              [1] 4
[1] 4 4 4 4                   #
> project(y,v1)              # beta_1-hat
[1] -2.1 -0.7  0.7  2.1       > a0 - b1 * mean(x)
#                             [1] 0.5
```

Alternatively, we can use the matrix approach, defining the hat matrix $\boldsymbol{H}$:

lm-demo-matrix

```
> one <- rep(1,4)                          [,1]
> X <- cbind(one,x); X          [1,]   0.5
      one x                     [2,]   0.5
[1,]   1 1                      [3,]   0.5
[2,]   1 2                      [4,]   0.5
[3,]   1 3                      > v1 <- x - mean(x); v1
[4,]   1 4                              [,1]
> x <- as.matrix(x);x          [1,]  -1.5
       [,1]                     [2,]  -0.5
[1,]    1                       [3,]   0.5
[2,]    2                       [4,]   1.5
[3,]    3                       > u1 <- v1/sqrt(sum(v1^2)); u1
[4,]    4                                [,1]
> y <- as.matrix(y); y         [1,]  -0.67082
       [,1]                     [2,]  -0.22361
[1,]    2                       [3,]   0.22361
[2,]    3                       [4,]   0.67082
[3,]    5                       #
[4,]    6                       # the hat matrix
> v0 <- as.matrix(one); v0     > H <- u0 %*% t(u0) + u1 %*% t(u1)
       [,1]                     > H
[1,]    1                            [,1] [,2] [,3] [,4]
[2,]    1                       [1,]  0.7  0.4  0.1 -0.2
[3,]    1                       [2,]  0.4  0.3  0.2  0.1
[4,]    1                       [3,]  0.1  0.2  0.3  0.4
> u0 <- v0/sqrt(sum(v0^2));  u0 [4,] -0.2  0.1  0.4  0.7
```

The model matrix $\boldsymbol{X}$ can also be obtained using the utility function `model.matrix()`, and we can obtain $\boldsymbol{H}$ using (6.24):

lm-demo-matrix2

```
> X <- model.matrix(model); X   # hat matrix
   (Intercept) x                > X %*% solve(t(X) %*% X) %*% t(X)
1            1 1                     1    2   3    4
2            1 2                 1  0.7 0.4 0.1 -0.2
3            1 3                 2  0.4 0.3 0.2  0.1
4            1 4                 3  0.1 0.2 0.3  0.4
attr(,"assign")                 4 -0.2 0.1 0.4  0.7
[1] 0 1
```

Whichever way we compute $\boldsymbol{H}$, we can use it to obtain $\hat{\boldsymbol{y}}$ and $\hat{\boldsymbol{\beta}}$:

lm-demo-hat

```
# y-hat values:                 #
> H %*% y                       # beta-hat values:
       [,1]                     #
[1,]   1.9                      > solve(t(X) %*% X) %*% t(X) %*% y
[2,]   3.3                                   [,1]
[3,]   4.7                      (Intercept)  0.5
[4,]   6.1                      x            1.4
```

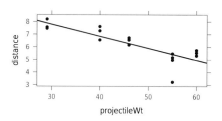

Figure 6.3. The distance traveled decreases as the weight of the projectile increases.

◁

Example 6.2.3. A counterweight trebuchet is a device developed in the Middle Ages as an improvement on the catapult. The improved range and accuracy of the new devices were an important military advantage. Trebuchet construction is also a common event in school science competitions.

The `trebuchet2` data set contains measurements from an experiment that involved firing projectiles with a small trebuchet under different conditions. In this data set, we can compare the distance traveled (in cm) to the weight of the projectile (in grams).

It is natural to assume that (at least within certain bounds) the lighter the projectile, the farther it will fly when released. The R code below fits a linear model and also generates a scatterplot with the least squares regression line (Figure 6.3).

```
> treb.model <- lm(distance~projectileWt,data=trebuchet2)          lm-trebuchet2
> coef(treb.model)
 (Intercept) projectileWt
   10.629388    -0.094604
> plot1 <- xyplot(distance~projectileWt,data=trebuchet2, type=c('p','r'))
```

The plot is important. Statistical packages will happily fit a least squares line to any data set, even if a line is not a good fit for the data.

The linear model seems to fit reasonably well with the exception of one measurement for a 55-gram projectile which seems not to have flown very far. Perhaps there was some malfunction of the trebuchet on that attempt. The slope coefficient tells us that for each additional gram the projectile weighs, we can expect roughly 9.5 cm decreased distance (on average). The intercept is not informative on its own since we are not interested in how far a weightless projectile will travel.

Notice that there are only 5 distinct projectile weights and that each has been repeated several times. Each weight corresponds to a particular projectile that was launched several times. The distribution of these weights is part of the design of the study. In controlled experiments, the researchers can often choose the values of the explanatory variables. In this case, these weights were determined by the projectiles at hand.

◁

Example 6.2.4. The `elasticband` data set in the `DAAG` package contains measurements from an experiment that involved stretching an elastic band over the

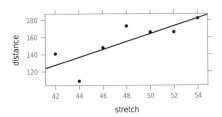

Figure 6.4. An elastic band travels farther if it is stretched farther before launch.

end of a ruler and then releasing the band. The distance stretched (in mm) and the distance traveled (in cm) were recorded.

It is natural to assume that (at least within certain bounds) the farther the band is stretched, the farther it will fly when released. The R code below fits a linear model and also generates a scatterplot with the least squares regression line (Figure 6.4).

```
> require(DAAG)                                          lm-elasticband
> eband.model <- lm(distance~stretch,data=elasticband); coef(eband.model)
(Intercept)     stretch
   -63.5714     4.5536
> plot1 <- xyplot(distance~stretch,data=elasticband, type=c('p','r'))
```

This is a small data set, so it is difficult to assess the quality of the fit. The slope coefficient tells us that for each additional mm the band is stretched before release, we can expect roughly 4.5 cm of increased distance (on average). The intercept is not informative on its own since we are not interested in how far an elastic band will travel if it is not stretched. Notice, too, that the stretch measurements are equally spaced along the range from 42 to 54 mm with one measurement at every 2 mm. This is part of the design of the study. The researchers decided to fire the elastic band once with each of those "treatments". ◁

6.3. Inference for Simple Linear Regression

The simple linear regression model has three parameters: β_0, β_1, and σ^2. Confidence intervals and hypothesis tests are based on the sampling distributions of our estimators. If we consider $\boldsymbol{x}$ to be fixed, then the standard linear model says that $\boldsymbol{Y}$ is random and satisfies

$$\boldsymbol{Y} = \beta_0 \boldsymbol{1} + \beta_1 \boldsymbol{x} + \varepsilon, \qquad \varepsilon \stackrel{\text{iid}}{\sim} \mathsf{Norm}(0, \sigma),$$

which we can also express as

$$\boldsymbol{Y} - (\beta_0 \boldsymbol{1} + \beta_1 \boldsymbol{x}) \stackrel{\text{iid}}{\sim} \mathsf{Norm}(0, \sigma).$$

Since our estimators are calculated from $\boldsymbol{x}$ and $\boldsymbol{Y}$, they too can be considered random variables. In this section we will use capital Roman notation for these estimators to emphasize that they are random variables rather than point estimates:

$$B_0 = \hat{\beta}_0 \,,$$

$$B_1 = \hat{\beta}_1 \,.$$

S^2 will be used to represent an unbiased estimator for σ^2. We will introduce this estimator shortly.

6.3.1. The Linear Algebra Toolkit

The sampling distributions for these random variables can be derived from the geometry of the linear model and the following important facts.

Lemma 6.3.1. *Let $Y_j \sim \mathsf{Norm}(\mu_j, \sigma_j)$ be independent random variables and let $\boldsymbol{u}$ and $\boldsymbol{v}$ be (constant) unit vectors. Then:*

(1) $\boldsymbol{u} \cdot \boldsymbol{Y} = \sum u_j Y_j \sim \mathsf{Norm}\left(\sum u_j \mu_j, \sqrt{\sum u_j^2 \sigma_j^2}\right) = \mathsf{Norm}\left(\boldsymbol{u} \cdot \boldsymbol{\mu}, \sqrt{\boldsymbol{u}^2 \cdot \boldsymbol{\sigma}^2}\right) .$

    (a) *If, in addition, $\sigma_j = \sigma$ for all j, then*

$$\boldsymbol{u} \cdot \boldsymbol{Y} \sim \mathsf{Norm}(\boldsymbol{u} \cdot \boldsymbol{\mu}, \sigma) \,.$$

    (b) *If, in addition, $\boldsymbol{u} \perp \mathbf{1}$ and for all j, $\sigma_j = \sigma$ and $\mu_j = \mu$, then*

$$\boldsymbol{u} \cdot \boldsymbol{Y} \sim \mathsf{Norm}(0, \sigma) \,.$$

(2) *If $\boldsymbol{u} \perp \boldsymbol{v}$, then $\boldsymbol{u} \cdot \boldsymbol{Y}$ and $\boldsymbol{v} \cdot \boldsymbol{Y}$ are independent.*

Proof. Part (2) follows immediately from Lemma 4.6.4. Part (1) is left as Exercise 6.7. □

Lemma 6.3.2. *For any vectors $\boldsymbol{a}$, $\boldsymbol{b}$, and $\boldsymbol{c}$, if $\boldsymbol{a} \perp \boldsymbol{c}$, then*

$$\boldsymbol{a} \cdot \boldsymbol{b} = \boldsymbol{a} \cdot (\boldsymbol{b} + \boldsymbol{c})$$
$$= \boldsymbol{a} \cdot (\boldsymbol{b} - \boldsymbol{c}) \,.$$

Proof. Exercise 6.8 □

6.3.2. The Sampling Distribution of $\hat{\beta}_1$

Now let $\{\boldsymbol{v}_0, \boldsymbol{v}_1, \ldots, \boldsymbol{v}_{n-1}\}$ be an orthogonal basis for $\mathbb{R}^n$ such that $\boldsymbol{v}_0 = \mathbf{1}$ and $\boldsymbol{v}_1 = \boldsymbol{x} - \overline{\boldsymbol{x}}$, and let $\{\boldsymbol{u}_0, \boldsymbol{u}_1, \ldots, \boldsymbol{u}_{n-1}\}$ be an orthonormal basis for $\mathbb{R}^n$ such that $\boldsymbol{u}_i = \boldsymbol{v}_i/|\boldsymbol{v}_i|$. So $\{\boldsymbol{u}_0, \boldsymbol{u}_1\}$ spans the model space and the remaining vectors span the error space.

We first turn our attention to the projection in the direction of $\boldsymbol{u}_1$ and the sampling distribution of B_1. By (6.20),

$$B_1 = \frac{1}{|\boldsymbol{x} - \overline{\boldsymbol{x}}|} \left(\boldsymbol{u}_1 \cdot \boldsymbol{Y}\right) , \tag{6.25}$$

from which it follows by Lemma 6.3.1 that B_1 is normally distributed with standard deviation $\sigma/|\boldsymbol{x} - \overline{\boldsymbol{x}}|$. It remains only to calculate the expected value of B_1.

$$\begin{aligned}
\mathrm{E}(B_1) \cdot |\boldsymbol{x} - \overline{\boldsymbol{x}}|^2 &= \mathrm{E}\left((\boldsymbol{x} - \overline{\boldsymbol{x}}) \cdot \boldsymbol{Y}\right) \\
&= (\boldsymbol{x} - \overline{\boldsymbol{x}}) \cdot (\beta_0 \mathbf{1} + \beta_1 \boldsymbol{x}) \\
&= \beta_1 (\boldsymbol{x} - \overline{\boldsymbol{x}}) \cdot \boldsymbol{x} \\
&= \beta_1 (\boldsymbol{x} - \overline{\boldsymbol{x}}) \cdot (\boldsymbol{x} - \overline{\boldsymbol{x}}) \\
&= \beta_1 |\boldsymbol{x} - \overline{\boldsymbol{x}}|^2 \,,
\end{aligned} \tag{6.26}$$

and so

$$B_1 \sim \mathsf{Norm}\left(\beta_1, \frac{\sigma}{|\boldsymbol{x} - \overline{\boldsymbol{x}}|}\right) \,. \tag{6.27}$$

Notice that this means that our least squares estimator B_1 is an unbiased estimator of β_1.

The presence of $|\boldsymbol{x} - \overline{\boldsymbol{x}}|$ in the denominator of the standard deviation of the sampling distribution of B_1 indicates that we can obtain more precise estimates of β_1 by increasing the number and variance of the x values. Taken to an extreme, an optimal (in this sense) design would select two values of x, widely separated, and equally partition the observations between these two predictor values. This extreme design has other problems, however. In particular, it is impossible to diagnose whether a linear fit is a good fit because there are no intermediate observations. When the researcher has control over the predictor values, it is generally wise to distribute them fairly evenly across the range of values being studied.

6.3.3. The Sampling Distribution of $\hat{\beta}_0$

We can apply a similar argument to derive the sampling distribution of B_0. Since $B_0 = \overline{Y} - B_1 \overline{x}$, we begin by deriving the distribution of $\overline{Y}$:

$$\overline{Y} = \frac{\mathbf{1} \cdot \boldsymbol{Y}}{n} = \frac{\mathbf{1} \cdot \boldsymbol{Y}}{|\mathbf{1}|^2} = \frac{1}{|\mathbf{1}|^2}\left(\boldsymbol{v}_0 \cdot \boldsymbol{Y}\right) = \frac{1}{|\mathbf{1}|}\left(\boldsymbol{u}_0 \cdot \boldsymbol{Y}\right) \,,$$

from which it follows by Lemma 6.3.1 that $\overline{Y}$ is normally distributed with standard deviation $\dfrac{\sigma}{|\mathbf{1}|} = \dfrac{\sigma}{\sqrt{n}}$. It remains only to calculate the expected value of $\overline{Y}$.

$$\begin{aligned}
\mathrm{E}(\overline{Y}) \, |\mathbf{1}|^2 &= \mathrm{E}(\mathbf{1} \cdot \boldsymbol{Y}) \\
&= \mathbf{1} \cdot (\beta_0 \mathbf{1} + \beta_1 \boldsymbol{x}) \\
&= n\beta_0 + n\beta_1 \overline{x} \\
&= n\left(\beta_0 + \beta_1 \overline{x}\right) \\
&= |\mathbf{1}|^2 \left(\beta_0 + \beta_1 \overline{x}\right) \,,
\end{aligned} \tag{6.28}$$

so

$$\overline{Y} \sim \mathsf{Norm}\left(\beta_0 + \beta_1 \overline{x}, \frac{\sigma}{|\mathbf{1}|}\right) \,.$$

Thus

$$B_0 = \overline{Y} - B_1\overline{x} \sim \mathsf{Norm}\left(\beta_0,\ \sigma\sqrt{\frac{1}{|\mathbf{1}|^2} + \frac{\overline{x}^2}{|\mathbf{x} - \overline{\mathbf{x}}|^2}}\right) = \mathsf{Norm}\left(\beta_0,\ \sigma\sqrt{\frac{1}{n} + \frac{\overline{x}^2}{S_{xx}}}\right)$$

because $\overline{Y}$ and B_1 are independent. (Independence follows from $\mathbf{u}_0 \perp \mathbf{u}_1$.) It is important to note that B_0 and B_1 are not independent.

6.3.4. Estimating σ

Because $\beta_0\mathbf{1} + \beta_1\mathbf{x}$ and $\hat{\beta}_0\mathbf{1} + \hat{\beta}_1\mathbf{x}$ lie in the model space, both are orthogonal to $\mathbf{u}_i$ when $i \geq 2$. Thus for $i \geq 2$,

$$\mathbf{u}_i \cdot \mathbf{Y} = \mathbf{u}_i \cdot (\mathbf{Y} - (\beta_0\mathbf{1} + \beta_1\mathbf{x})) = \mathbf{u}_i \cdot (\mathbf{Y} - (\hat{\beta}_0\mathbf{1} + \hat{\beta}_1\mathbf{x}))\ .$$

This is convenient because we know that

$$\mathbf{Y} - (\beta_0\mathbf{1} + \beta_1\mathbf{x}) = \boldsymbol{\varepsilon} \overset{\text{iid}}{\sim} \mathsf{Norm}(0, \sigma)\ ,$$

and so

$$\mathbf{u}_i \cdot (\mathbf{Y} - (\hat{\beta}_0\mathbf{1} + \hat{\beta}_1\mathbf{x})) = \mathbf{u}_i \cdot \mathbf{Y} \sim \mathsf{Norm}(0, \sigma)\ , \qquad i \geq 2\ .$$

Furthermore, each of these random dot products is independent of the others by Lemma 6.3.1. From this it immediately follows that $|\mathbf{E}|^2 = |\mathbf{Y} - \hat{\mathbf{Y}}|^2$ is the sum of the squares of $n - 2$ independent $\mathsf{Norm}(0, \sigma)$-random variables since

$$|\mathbf{E}|^2 = \sum_{i \geq 2}(\mathbf{u}_i \cdot \mathbf{Y})^2\ .$$

Thus

$$\frac{|\mathbf{E}|^2}{\sigma^2} = \frac{\sum E_i^2}{\sigma^2} = \frac{SSE}{\sigma^2} \sim \mathsf{Chisq}(n - 2)\ .$$

Among other things, this implies that

$$\mathrm{E}(\frac{SSE}{\sigma^2}) = n - 2\ ,$$

so

$$\mathrm{E}\left(\frac{SSE}{n - 2}\right) = \sigma^2\ ,$$

and we have found an unbiased estimator of σ^2:

$$\hat{\sigma}^2 = S^2 = \frac{SSE}{n - 2}\ . \tag{6.29}$$

This, rather than the maximum likelihood estimator, will be our standard estimator for σ^2 when performing least squares regression.

6.3.5. Confidence Intervals and Hypothesis Tests for β_i

Since $B_1 \sim \mathsf{Norm}\left(\beta_1, \frac{\sigma}{|\boldsymbol{x} - \overline{\boldsymbol{x}}|}\right)$ and $\frac{(n-2)S^2}{\sigma^2} \sim \mathsf{Chisq}(n-2)$,

$$\frac{B_1 - \beta_1}{\sigma/|\boldsymbol{x} - \overline{\boldsymbol{x}}|} \sim \mathsf{Norm}(0,1) \quad \text{and} \quad \frac{B_1 - \beta_1}{S/|\boldsymbol{x} - \overline{\boldsymbol{x}}|} \sim \mathsf{t}(n-2) .$$

This means that inference for β_1 can proceed just as it did for the 1-sample t-procedures after substituting the estimated standard error for the slope $(s/|\boldsymbol{x} - \overline{\boldsymbol{x}}|)$ for the standard error for the mean $(s/\sqrt{n})$ and adjusting the degrees of freedom. The estimated standard error for the slope can be rewritten a number of different ways:

$$SE(\hat{\beta}_1) = SE(\text{slope}) = \frac{s}{|\boldsymbol{x} - \overline{\boldsymbol{x}}|} = \frac{\sqrt{MSE}}{\sqrt{S_{xx}}} = \frac{SE(\text{residuals})}{\sqrt{S_{xx}}} .$$

An even more precise notation is to use either $\hat{\sigma}_{\hat{\beta}_1}$ or $s_{\hat{\beta}_1}$ to indicate that we are dealing with an estimated standard deviation of an estimator.

A confidence interval can be formed in the usual way:

$$\hat{\beta}_1 \pm t_* \, SE(\hat{\beta}_1)$$

where t_* is a critical value from the t-distribution with $n-2$ degrees of freedom. Similarly we can test the hypothesis $H_0 : \beta_1 = \beta_{10}$ using a t-statistic

$$t = \frac{\hat{\beta}_1 - \beta_{10}}{SE(\hat{\beta}_1)} .$$

The test of $H_0 : \beta_1 = 0$ vs. $H_a : \beta_1 \neq 0$ is especially interesting. If $\beta_1 = 0$, then the slope of the regression line is 0, which means that our model predicts the same mean value of y for any x. In other words, our predictor is of no value in predicting the response. If we fail to reject this null hypothesis, then our data do not provide any evidence that the model is useful. For this reason, this test is sometimes called the **model utility test**.

Exactly the same structure works for inference about β_0, although this is usually less interesting. We simply replace $SE(\hat{\beta}_1)$ with

$$SE(\hat{\beta}_0) = s\sqrt{\frac{1}{|\mathbf{1}|^2} + \frac{\overline{x}^2}{|\boldsymbol{x} - \overline{\boldsymbol{x}}|^2}}$$

and proceed as above. Fortunately, these standard errors are calculated for us by statistical software.

Example 6.3.1.

Q. Returning to Example 6.2.3, test the hypothesis $H_0 : \beta_1 = 0$ and give a 95% confidence interval for β_1.

A. Two-sided p-values for the tests of $\beta_i = 0$ are given in the standard summary table, as are the standard errors, from which we can compute confidence intervals. Alternatively, the function `confint()` will compute a confidence interval for each β_i. The default level of confidence is 95%.

```
> summary(treb.model)                                    lm-trebuchet2-summary
< 8 lines removed >
Coefficients:
            Estimate Std. Error t value Pr(>|t|)
(Intercept)  10.6294     0.8188   12.98 3.4e-09
projectileWt -0.0946     0.0171   -5.52 7.5e-05

Residual standard error: 0.743 on 14 degrees of freedom
Multiple R-squared: 0.685,        Adjusted R-squared: 0.663
F-statistic: 30.5 on 1 and 14 DF,  p-value: 7.5e-05
```

```
> -0.0946 + c(-1,1) * 0.01713 * qt(0.975,df=14)  # CI by hand   lm-trebuchet2-ci
[1] -0.13134 -0.05786
> confint(treb.model,"projectileWt")                # CI using confint()
            2.5 %   97.5 %
projectileWt -0.13134 -0.057872
```

In this case we see that we have enough evidence to conclude that the weight of
the projectile does matter, but our confidence interval for this effect is quite wide
(the largest value in the interval is more than twice the smallest value). Collecting
additional data might clarify things. It is important to note that while collecting
data might allow us to estimate the parameters more accurately, it will not reduce
the variability in the data that may be caused because the firing apparatus (or its
operator) is not very consistent. It may be that there would be a great deal of
variability in distance traveled even if the same projectile were used each time. In
fact, the scatterplot gives some indication that this is so.

It is instructive to look under the hood of `confint.lm()`, the function called
to compute confidence intervals from the object returned by `lm()`.

```
> stats:::confint.lm                                      confint-lm
function (object, parm, level = 0.95, ...)
{
    cf <- coef(object)
    pnames <- names(cf)
    if (missing(parm))
        parm <- pnames
    else if (is.numeric(parm))
        parm <- pnames[parm]
    a <- (1 - level)/2
    a <- c(a, 1 - a)
    fac <- qt(a, object$df.residual)
    pct <- format.perc(a, 3)
    ci <- array(NA, dim = c(length(parm), 2L), dimnames = list(parm,
        pct))
    ses <- sqrt(diag(vcov(object)))[parm]
    ci[] <- cf[parm] + ses %o% fac
    ci
}
<environment: namespace:stats>
```

vcov() returns a matrix of the variances and covariances of the estimates. The square roots of the diagonal entries in this matrix are the standard errors we want. The argument parm allows us to produce confidence intervals for any subset of the parameters. The default is all parameters. ◁

Example 6.3.2.

Q. Returning to Example 6.2.4, test the hypothesis $H_0 : \beta_1 = 0$ and give a 95% confidence interval for β_1.

A. Two-sided p-values for the tests of $\beta_i = 0$ are given in the standard summary table, as are the standard errors, from which we can compute confidence intervals. As in the previous example, the function confint() will compute a confidence interval for each β_i. The default level of confidence is 95%.

```
> summary(eband.model)                              lm-elasticband-ci
< 8 lines removed >
Coefficients:
            Estimate Std. Error t value Pr(>|t|)
(Intercept)   -63.57      74.33   -0.86    0.431
stretch         4.55       1.54    2.95    0.032

Residual standard error: 16.3 on 5 degrees of freedom
Multiple R-squared: 0.635,        Adjusted R-squared: 0.562
F-statistic: 8.71 on 1 and 5 DF,  p-value: 0.0319

> 4.554 + c(-1,1) * 1.543 * qt(0.975,df=5)          # CI by hand
[1] 0.58759 8.52041
> confint(eband.model,"stretch")                    # CI using confint()
        2.5 % 97.5 %
stretch 0.58656 8.5206
```

In this case we see that we have enough evidence to conclude that the distance the elastic band is stretched does matter, but our confidence interval for this effect is quite wide. ◁

6.3.6. ANOVA for Regression

There is another way to derive a hypothesis test for $H_0 : \beta_1 = 0$ that can be motivated directly from our geometric understanding of the simple linear model (see Figure 6.5) and will be important when we consider other models. Recall from (6.25) and (6.27) that

$$\boldsymbol{u}_1 \cdot \boldsymbol{Y} = |\boldsymbol{x} - \overline{\boldsymbol{x}}| B_1 \sim \mathsf{Norm}(|\boldsymbol{x} - \overline{\boldsymbol{x}}| \beta_1, \sigma)$$

and that for $i = 2, \ldots, n - 1$,

$$\boldsymbol{u}_i \cdot \boldsymbol{Y} \sim \mathsf{Norm}(0, \sigma) .$$

Thus if $\beta_1 = 0$, then

$$\frac{(\boldsymbol{u}_1 \cdot \boldsymbol{Y})^2}{\sigma^2} = \frac{|\hat{\boldsymbol{Y}} - \overline{\boldsymbol{Y}}|^2}{\sigma^2} \sim \mathsf{Chisq}(1) ,$$

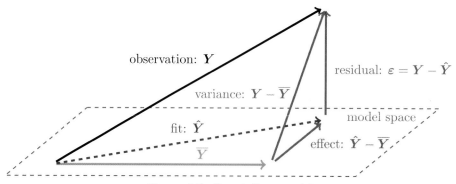

Figure 6.5. Generic linear model.

and since

$$\frac{SSE}{\sigma^2} \sim \mathsf{Chisq}(n-2)\,,$$

we have

$$F = \frac{|\hat{\mathbf{Y}} - \overline{\mathbf{Y}}|^2/1}{SSE/(n-2)} \sim \mathsf{F}(1, n-2)\,.$$

F has an interesting interpretation in terms of the variance of Y. Let

$$SST = |\mathbf{Y} - \overline{\mathbf{Y}}|^2 = \sum (Y_i - \overline{Y})^2 = (n-1)S_Y^2\,,$$

$$SSM = |\hat{\mathbf{Y}} - \overline{\mathbf{Y}}|^2 = \sum_i (\hat{Y}_i - \overline{Y}_i)^2\,,\text{ and}$$

$$SSE = |\mathbf{Y} - \hat{\mathbf{Y}}|^2 = \sum (Y_i - \hat{Y}_i)^2\,.$$

SST, SSM, and SSE are the total sum of squares (needed to compute the sample variance of the observed responses) and the portion of the sum of squares associated with the model and with error. We have already seen that

$$SST = SSM + SSE\,,$$

so we have partitioned the variance into two components. Now let

$$MST = \frac{SST}{n-1}\,,$$

$$MSM = \frac{SSM}{1}\,,\text{ and}$$

$$MSE = \frac{SSE}{n-2}\,.$$

MS stands for "mean square", and each MS term is the quotient of an SS term and the associated degrees of freedom. Notice that MST is our usual estimator for the variance of Y and that MSE is our unbiased estimator of σ^2 from (6.29). Under the null hypothesis, the ratio

$$F = \frac{MSM}{MSE} = \frac{SSM/1}{SSE/(n-2)} \sim \mathsf{F}(1, n-2)\,,$$

and the expected value of the numerator and denominator are both 1. But if $\beta_1 \neq 0$, then the expected value of the numerator of F will be larger than 1. This gives us an alternative way to test $H_0 : \beta_1 = 0$. Large values of F are evidence against the null hypothesis, and the $F(1, n-2)$-distribution can be used to calculate the p-value. This approach to inference is called analysis of variance or ANOVA. The anova() function in R presents the ANOVA results in a tabular form that goes back to Fisher.

Example 6.3.3. Here is the ANOVA table for our trebuchet example.

```
> anova(treb.model)                                          anova-trebuchet2
Analysis of Variance Table


Response: distance
             Df Sum Sq Mean Sq F value  Pr(>F)
projectileWt  1  16.84   16.84    30.5 7.5e-05
Residuals    14   7.73    0.55
```
◁

Often ANOVA tables are presented with one additional row giving the total degrees of freedom, SST and MST. The F-statistic, degrees of freedom, and p-value also appear in the output of summary(). By Lemma 4.7.3, the p-value for this test and for the t-test presented earlier will always be the same because $t^2 = F$ (Exercise 6.10).

We can also now explain the value of Multiple R-squared that appears in the output of summary():

$$r^2 = \frac{SSM}{SSM + SSE} = \frac{SSM}{SST} = \frac{|\hat{y} - \overline{y}|^2}{|y - \overline{y}|^2} \ .$$

So r^2 (the **coefficient of determinism**) gives the fraction of the total variation in the response that is explained by the model.

Example 6.3.4. In the case of our trebuchet experiment,

$$r^2 = \frac{16.8433}{16.8433 + 7.7277} = 0.685 \ ,$$

so the model explains roughly 69% of the variation in distance traveled.

```
> summary(treb.model)                                    lm-trebuchet2-summary
< 8 lines removed >
Coefficients:
             Estimate Std. Error t value Pr(>|t|)
(Intercept)   10.6294     0.8188   12.98 3.4e-09
projectileWt  -0.0946     0.0171   -5.52 7.5e-05


Residual standard error: 0.743 on 14 degrees of freedom
Multiple R-squared: 0.685,          Adjusted R-squared: 0.663
F-statistic: 30.5 on 1 and 14 DF,  p-value: 7.5e-05
```
◁

This interpretation of r^2 will hold for the more complicated linear models we will encounter later as well. It is important to note, however, that this interpretation does not hold for models without an intercept term. (See Exercise 6.5 for more about models without an intercept term.)

Example 6.3.5. Here is the ANOVA table for our elastic band example.

```
> anova(eband.model)                                        anova-elasticband
Analysis of Variance Table


Response: distance
          Df Sum Sq Mean Sq F value Pr(>F)
stretch    1   2322    2322    8.71  0.032
Residuals  5   1334     267
```

From this we can compute

$$r^2 = \frac{2322.3}{2322.3 + 1333.7} = 0.635 \ .$$

This value is also reported in the output presented in Example 6.3.2.   ◁

6.3.7. Confidence Intervals and Prediction Intervals for $\hat{y}$

We have already seen that the regression line provides us with a way of predicting the response for a given value of the explanatory variable. We would like to know something about the accuracy of this prediction. For this we need to know the sampling distribution of $\hat{Y}_* = B_0 + B_1 x_*$, our predicted response when the value of the explanatory variable is x_*. This distribution is easily determined if we first perform a bit of algebraic manipulation:

$$\hat{Y}_* = B_0 + B_1 x_*$$
$$= \overline{Y} - B_1 \overline{x} + B_1 x_*$$
$$= \underbrace{\overline{Y}}_{\mathsf{Norm}(\beta_0 + \beta_1 \overline{x}, \frac{\sigma}{|\mathbf{1}|})} + \underbrace{B_1(x_* - \overline{x})}_{\mathsf{Norm}(\beta_1 x_* - \beta_1 \overline{x}, \frac{\sigma(x_* - \overline{x})}{|\boldsymbol{x} - \overline{\boldsymbol{x}}|})}$$
$$\sim \mathsf{Norm}\left(\beta_0 + \beta_1 x_*, \sigma \sqrt{\frac{1}{|\mathbf{1}|^2} + \frac{(x_* - \overline{x})^2}{|\boldsymbol{x} - \overline{\boldsymbol{x}}|^2}}\right) \ .$$

(Once again we have used the fact that $\overline{Y}$ and B_1 are independent.) Thus

$$\hat{Y}_* - (\beta_0 + \beta_1 x_*) \sim \mathsf{Norm}\left(0, \sigma \sqrt{\frac{1}{|\mathbf{1}|^2} + \frac{(x_* - \overline{x})^2}{|\boldsymbol{x} - \overline{\boldsymbol{x}}|^2}}\right),$$

and a confidence interval for the mean response when $x = x_*$ is given by

$$\hat{\beta}_0 + \hat{\beta}_1 x_* \pm t_* \ s \ \sqrt{\frac{1}{|\mathbf{1}|^2} + \frac{(x_* - \overline{x})^2}{|\boldsymbol{x} - \overline{\boldsymbol{x}}|^2}} \ ,$$

where t_* is the appropriate critical value from the $\mathsf{t}(n-2)$-distribution.

It is important to note that this is a confidence interval for the *mean* response. We can also give an interval (typically called a prediction interval) for an individual response. For this we consider the sampling distribution of $Y_* - \hat{Y}_*$ where $Y_* \sim \mathsf{Norm}(\beta_0 + \beta_1 x_*, \sigma)$ is a random variable modeling a new (i.e., independent

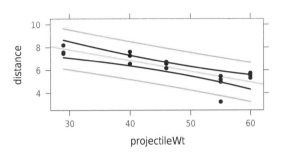

Figure 6.6. A plot showing confidence intervals (darker) and prediction intervals (lighter) for the trebuchet experiment.

from the data used to produce the regression line) observation of the response. Since

$$Y_* - \hat{Y}_* \sim \mathsf{Norm}\left(0, \sigma\sqrt{1 + \frac{1}{|\mathbf{1}|^2} + \frac{(x_* - \overline{x})^2}{|\boldsymbol{x} - \overline{\boldsymbol{x}}|^2}}\right),$$

a prediction interval is given by

$$\hat{\beta}_0 + \hat{\beta}_1 x_* \pm t_*\, s\, \sqrt{1 + \frac{1}{|\mathbf{1}|^2} + \frac{(x_* - \overline{x})^2}{|\boldsymbol{x} - \overline{\boldsymbol{x}}|^2}}\,.$$

The R function `predict()` can be used to make point and interval predictions of the response for specified values of the predictors.

Example 6.3.6. Returning again to Example 6.2.3, we use R to calculate point and interval predictions.

```
> predict(treb.model,newdata=data.frame(projectileWt=44))
    1
6.4668
> predict(treb.model,newdata=data.frame(projectileWt=44),
+    interval='confidence')
   fit    lwr    upr
1 6.4668 6.0575 6.8761
> predict(treb.model,newdata=data.frame(projectileWt=44),
+    interval='prediction')
   fit    lwr    upr
1 6.4668 4.8216 8.112
```

Figure 6.6 shows the confidence and prediction intervals graphically for x_* within the range of the data and was generated using the following code:

```
> trebuchet2.bandplot <- xyplot(distance~projectileWt,
+    data=trebuchet2, ylim=c(2.5,10.5),
+    panel=panel.lmbands,
+    conf.lty=1,
+    pred.lty=1
+    )
```

◁

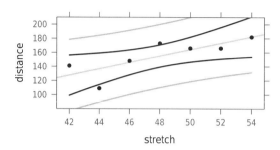

Figure 6.7. A plot showing confidence intervals (darker) and prediction intervals (lighter) for the elastic band experiment.

Example 6.3.7. Returning again to Example 6.2.4, we use R to calculate point and interval predictions.

```
> predict(eband.model,newdata=data.frame(stretch=30))       elasticband-predict
     1
73.036
> predict(eband.model,newdata=data.frame(stretch=30),interval='confidence')
     fit     lwr     upr
1 73.036 -0.11229 146.18
> predict(eband.model,newdata=data.frame(stretch=30),interval='prediction')
     fit     lwr     upr
1 73.036 -11.304 157.38
```

Figure 6.7 shows the confidence and prediction intervals graphically for x_* within the range of the data.   ◁

There are several things to note about these confidence intervals and prediction intervals.

(1) Both intervals are narrower when $|x_* - \bar{x}|$ is smaller.

This reflects the fact that we are more confident about predictions made near the center of the range of our data and explains why the confidence bands in Figures 6.6 and 6.7 are curved.

(2) Prediction intervals for an individual response are wider than confidence intervals for the mean response.

(3) Confidence intervals are more robust against departures from normality than are prediction intervals.

We have not yet discussed the robustness of our regression methods, but the Central Limit Theorem and our previous experience with t-distributions should lead us to hope that confidence intervals for the mean response will be reasonably robust. But since we only consider one new observation when we make a prediction interval, the Central Limit Theorem does not help us. In general, prediction intervals are quite sensitive to the assumptions made about the shape of the underlying distribution.

6.4. Regression Diagnostics

The linear model is based on several assumptions. A number of procedures have been developed to check the appropriateness of these assumptions. Generally speaking, **regression diagnostics** are looking for one or more of the following types of potential problems:

(1) Problems with the errors.

   The linear model assumes that the errors (ε) are
   (a) independent,
   (b) normally distributed, and
   (c) homoskedastic[2] (i.e., that they have equal variance).
   Any or all of these assumptions may be violated in a given situation.

(2) Problems with the fit.

   The structural part of the model assumes that the true relationship between the response and predictors is linear, but it may be that the true relationship is of another shape.

(3) Problems due to unusual observations.

   It is possible that one or a small number of the observations in the data have a disproportionately large influence on the fit of the model. This sensitivity to certain observations may give us a mistaken sense for the strength of the relationship we think we are observing or may mask patterns present in the rest of the data.

(4) Failure to include important variables in the model.

   We will cover models with multiple predictors in Chapter 7. But our diagnostics for the simpler models covered in this chapter may lead us to suspect that additional variables will be required to obtain a sufficiently good model.

Diagnostic procedures may be graphical or numerical. Graphical procedures can often help us quickly identify major problems, and it is wise to get into the habit of always making use of diagnostic plots. Numerical procedures tend to focus on a narrower range of problems than do graphical procedures, but because they are numerical, they can be easier to automate and may generalize more readily to multiple regression, where certain types of plots become more difficult to generate or interpret. Often regression diagnostics will not only identify potential problems but also suggest possible solutions. In practice, one may fit several models to the same data and use various diagnostics to select the most appropriate model.

6.4.1. Residual Plots

The distributional assumptions in the linear model are about the distribution of the error term. For this reason it is important to look at the residuals to see that the assumptions of the model do not appear to be grossly violated.

---

[2]also spelled homoscedastic.

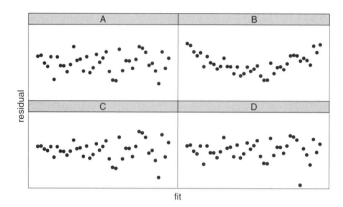

Figure 6.8. Examples of residuals plots.

The sum of the residuals (and hence their mean) will always be 0. This can be demonstrated algebraically, but a geometric proof is much simpler: Since $\mathbf{1}$ lies in the model space and $\mathbf{e}$ is orthogonal to the model space,

$$\mathbf{1} \cdot \mathbf{e} = \sum_i e_i = 0 \, .$$

Several plots involving residuals can help assess if the other assumptions of the model are reasonable. In most of these plots we are looking to *not* see any pattern. We can begin by plotting the residuals against the predictor $\mathbf{x}$. It is actually more standard to plot the residuals versus $\hat{\mathbf{y}}$ rather than $\mathbf{x}$ since this works as well for multiple regression as it does for simple linear regression. In the case of simple linear regression, the two plots will look identical (except for scale).

If the model assumptions are met, the vertical variation about the horizontal line representing a residual value of 0 should look roughly the same across the width of the plot. The plots in Figure 6.8 are illustrative. Panel A was generated from simulated data that meet the model assumptions precisely. The residuals show no particular pattern. In panel B there is a clear bowing to the plot. The residuals at the ends tend to be positive. In the middle the residuals tend to be negative. This shows lack of independence (if a residual is positive, another residual corresponding to the same or similar value of the predictors is more likely to also be positive). A typical cause for this sort of pattern is that the relationship in the data is not linear. In this case, at the ends the linear fit is too small and in the center it is too large.

Panel C demonstrates a different problem. There is clearly more variation in the residuals at the right end of the plot. This violates the assumption that all the errors come from a normal distribution with the same variance. It is not uncommon that when the values are larger, the variance is also larger. Sometimes a transformation of the data can be used to combat this problem. Another option is to use another analysis method, such as weighted least squares.

The problem in panel D is more subtle. There is one residual that is noticeably smaller than all the rest. This indicates one data value that does not fit the pattern of the rest of the data very well. Outliers can have a large effect on the regression model. Because the parameters are estimated by least squares, there is a large penalty for large residuals. This in turn can cause the parameter estimates to be heavily influenced by a small number of (atypical) data points. As always, once we detect an outlier, we would like to know why it doesn't fit the general pattern.

Similar plots can be made of the residuals vs. the order of the observations (if this is known). Unwanted patterns in these plots could be due to a number of things:

- If the variance decreases over time, there may be a training effect.

  The researchers may have become more accurate in making measurements as they became familiar with the procedure. If the variance eventually stabilizes, the first portion of the data may need to be ignored (because it is less reliable) in analysis.

- If the variance increases over time, there may be a fatigue effect.

  Either the equipment (in a laboratory setting) or the researchers may be less accurate later than they were at the beginning of data collection.

- Time of measurement may be correlated with some important **lurking variable**.

  A lurking variable is an unmeasured variable that is (potentially) correlated with the recorded variables. Depending on the context of the study, such things as temperature, season, news events, weekend effects, etc., may be lurking variables that affect the distribution of the residuals.

Mysterious patterns in the residuals are always a cause for concern in regression analysis.

6.4.2. Standardized Residuals

It is important to emphasize that large residuals are not necessarily a problem. This may simply be an indication that σ is large, not an indication that the model is inappropriate. The underlying linear trend may still be of interest even if there is a good deal of variability about the linear pattern. Residual plots often use **standardized residuals** to avoid the distraction of absolute size and make it easier to compare residuals across data sets of different scales.

To complicate matters, there are two different rescalings of residuals that go by the name of standardized residuals. One way to compute standardized residuals (e') is the following:

$$e'_i = \frac{e_i}{s}$$

where s^2 is the unbiased estimate of σ^2 from (6.29). But even if $\varepsilon \overset{\text{iid}}{\sim} \mathsf{Norm}(0, \sigma)$, the variance of e_i is not σ^2. The variance of e_i is given by

$$\mathrm{Var}(e_i) = \sigma^2 (1 - h_i) \,,$$

where h_i is the ith diagonal element of the hat matrix $\boldsymbol{H}$, which can be shown to satisfy

$$h_i = \frac{1}{n} + \frac{(x_i - \bar{x})^2}{\sum (x_i - \bar{x})^2} .$$

Thus a more appropriate standardization of the residuals is

$$e_i' = \frac{e_i}{s\sqrt{1 - h_i}} .$$

This latter adjustment to the residuals also goes by the name of **studentized residuals**. Note that

- the studentized residual is always larger than e_i/s,
- when n is large, the two are nearly equal, and
- the difference between the two is larger when x_i is farther from $\bar{x}$.

We will not give a full theoretical justification for the studentized residuals here, but we can hint at the intuition behind the studentized residuals. The key observations are that even if the errors $\boldsymbol{\varepsilon}$ are independent and have equal variance as the model demands, the residuals $\boldsymbol{e}$ are not independent and do not have equal variance. In fact, in addition to the identity

$$\sum e_i = 0 ,$$

(6.7) implies that the residuals must also satisfy

$$\sum e_i x_i = 0 .$$

So if we know $n - 2$ of the residuals, the other two are determined. This clearly demonstrates lack of independence. The denominator of the studentized residual replaces our estimate for $\mathrm{Var}(\varepsilon_i)$ with an estimate of $\mathrm{Var}(e_i)$. In this way, the studentized residuals give a less distorted picture of how we should expect our residuals to behave.

6.4.3. Checking Normality

The normality assumption can be assessed by looking at a normal-quantile plot. It must be emphasized that it is difficult to verify normality (by any method) if the number of observations is small. Nevertheless, at a minimum, a normal-quantile plot of residuals should be examined to see if there are reasons to be concerned about the normality assumption.

A variety of hypothesis tests have been developed to test for normality. Most of them only have reasonable power to detect certain types of departures from normality. Detecting departures from normality is also complicated by the so-called *supernormality* of residuals: In situations where $\boldsymbol{\varepsilon}$ is not normal, the distribution of $\boldsymbol{e}$ is more like a normal distribution than is the distribution of $\boldsymbol{\varepsilon}$.

6.4.4. Outliers, Leverage, and Influence

In regression there are two types of outliers. An outlier in the response variable is a value that is not predicted well by the model and may be an indication of a problem with either the data or the model. Such outliers result in residuals with large absolute values and will likely be detected in our examination of the residuals.

Outliers in the predictors are called **leverage points** because they have the potential to exert undue influence on the fit of the model. Recall that

$$\text{Var}(e_i) = \sigma^2 (1 - h_i)$$

and that h_i is large when $|x_i - \bar{x}|$ is large. When h_i is large, $\text{Var}(e_i)$ will be small, because the fit will be forced to make $\hat{y}_i$ close to y_i. For this reason, h_i is taken to be a numerical measurement of **leverage**. It can be shown that $\sum h_i = 2p$ where p is the total number of parameters in the model ($p = 2$ for the simple linear model), so a standard rule of thumb says that any observations with leverage substantially greater than $2p/n$ should be looked at more closely. Large leverages are a result of extreme values of the predictors. In situations where the values of the predictors are controlled by the researchers, it is best to avoid outliers in the predictors. In observational studies, such outliers may be unavoidable. (Of course, one should always check to be sure there is no error in recording the data.)

So how serious a problem are (response) outliers and leverage points? To some extent, the answer to this question depends on the reasons for fitting the linear model in the first place. One way of assessing this is via **leave-one-out** analysis. The motivation behind leave-one-out analyses was summarized by Frank Anscombe [**Ans73**]:

> We are usually happier about asserting a regression relation if the relation is appropriate after a few observations (any ones) have been deleted – that is, we are happier if the regression relation seems to permeate all the observations and does not derive largely from one or two.

For leave-one-out analysis it is handy to introduce the following notation. Let $\hat{\beta}_j(i)$ and $\hat{\sigma}(i) = s(i)$ denote the parameter estimates that result from the data set with observation i removed, and let $\hat{\boldsymbol{y}}(i) = \boldsymbol{X}\hat{\boldsymbol{\beta}}(i)$ denote the vector of fitted values that results after the ith observation has been removed. If we are primarily interested in the fitted values derived from the model, we could look to see how much the fits change when any one observation is deleted. That is, we can compare $\hat{\boldsymbol{y}}$ with $\hat{\boldsymbol{y}}(i)$ for each value of i to see which observations exert the most influence on the fitted values.

In 1979, Cook [**Coo79**], proposed the following measure of influence, now known as **Cook's distance**. Cook's distance is defined for each observation i by

$$D_i = \frac{|\hat{\boldsymbol{y}} - \hat{\boldsymbol{y}}(i)|^2}{ps^2}$$

where p is the number of parameters in the model ($p = 2$ for simple linear regression). An algebraically equivalent expression for Cook's distance is

$$D_i = \frac{(e_i')^2}{p} \cdot \frac{h_i}{1 - h_i}$$

where e_i' is the studentized residual. This formulation shows that Cook's distance is the product of a residual component and a leverage component.

Plots of Cook's distance are included in the suite of diagnostic plots available using `plot(model)` (or `xplot(model)` from `fastR`) where `model` is an object returned by `lm()`. With this and the other measures we will introduce shortly, there is not an obvious cut-off to use when determining if a data set has problems that need to be addressed. The denominator of D_i is chosen so that D_i has a distribution similar to that of an $F(p, n - p)$-distribution, so values greater than 1 usually warrant a closer look. Some authors suggest that any values that exceed the median of the relevant F-distribution (which is often smaller than 1) should be flagged. Our primary method will be to look for values that are especially large relative to the others. That is, we will look for the most influential observations in our data set.

Another measure of influence on fit is DFFITS, introduced by Belsley, Kuh, and Welsch [**BKW80**]. DFFITS differs from Cook's distance primarily in two respects. First DFFITS looks only at the difference in the fitted values for the ith observation rather than at the full vector of differences, so DFFITS gives a more localized sense of influence. Second, DFFITS uses $s(i)$ in place of s to estimate the variance. The idea is that an unusual observation may inflate our estimate of σ and thereby mask its own influence as measured by D_i. The drawback, of course, is that a different estimate of the variance is used for each observation:

$$\text{DFFITS}_i = \frac{\hat{y}_i - \hat{y}(i)_i}{s_{(i)} \sqrt{h_i}} \, .$$

Similarly, we can look at the influence of the ith observation on a particular parameter in the model. For example,

$$\text{DFBETA}_{j,i} = \hat{\beta}_j - \hat{\beta}_j(i)$$

and $\text{DFBETAS}_{j,i}$ is a rescaling of $\text{DFBETA}_{j,i}$ that has variance approximately 1. The scaling factor depends on which parameter is being estimated, but it is easily computed with a bit of linear algebra. (The 'S' in DFFITS and DFBETAS stands for 'scaled'; 'DF' stands for 'difference'.) The primary purpose for the rescaling is to provide values that are more easily compared across data sets.

R provides functions `cooks.distance()`, `dffits()`, `dfbeta()`, `dfbetas()`, and `lm.influence()` to compute these diagnostic measures. For efficiency, these functions use algebraically equivalent formulations of these measures to calculate them without actually fitting n different linear models. All of these tools will become more valuable in the context of multiple regression where a simple scatterplot may not reveal important issues with the model.

The next example demonstrates the use of leverage and influence as diagnostics.

Example 6.4.1.

Q. Using the data in the `star` data set of the package `faraway`, investigate the relationship between the surface temperature and the light intensity of 47 stars in the cluster CYG OB1. (Note: Both temperature and light values have already been transformed to a logarithmic scale.)

A. Looking at the scatterplot (Figure 6.9), we see a cluster of four stars that seem very different from the others. The regression line fit without these four stars is very different from the regression line computed from the full data set. In fact, even the direction of the effect changes when we remove these four stars.

```
> require(faraway); require(grid)                          lm-star
> star.plot1 <- xyplot(light~temp,star)
> hotstar <- star[star$temp > 3.7,]        # select all but 4 coolest stars
> star.model1 <- lm(light~temp,star)
> star.model2 <- lm(light~temp,hotstar)
> star.plot2 <- xyplot(light~temp,star,
+     panel = function(x,y,...){
+         panel.abline(reg=star.model1, lwd=2, lty=1,
+             col=trellis.par.get('superpose.line')$col[2])
+         panel.abline(reg=star.model2, lwd=2, lty=1,
+             col=trellis.par.get('superpose.line')$col[1])
+         panel.xyplot(x,y,...)
+         ids <- which(star$temp < 4.0)
+         grid.text(x=x[ids] + 0.04, y=y[ids],
+             as.character(ids),
+             default.units="native",gp=gpar(cex=0.7))
+     })
```

Figure 6.10 shows the diagnostic plots produced by one of the following lines

```
plot(model,w=1:6)
xplot(model,w=1:6)
```

where `model` is an object produced by `lm()` and `w` indicates which of six possible diagnostic plots are desired. We will be primarily interested in the first two (top row, left and center), which show residuals vs. fits and a normal-quantile plot of the residuals. Plots 4–6 (bottom row) can help identify outliers based on leverage, Cook's distance, and residuals. Plot 5, labeled "Residuals vs. Leverage" overlays level curves for Cook's distance on a scatterplot of residuals vs. leverage and shows how these measures are related.

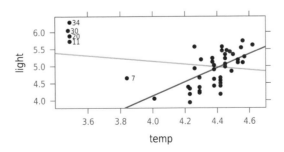

Figure 6.9. The regression lines with (lighter) and without (darker) the cluster of four observations are dramatically different.

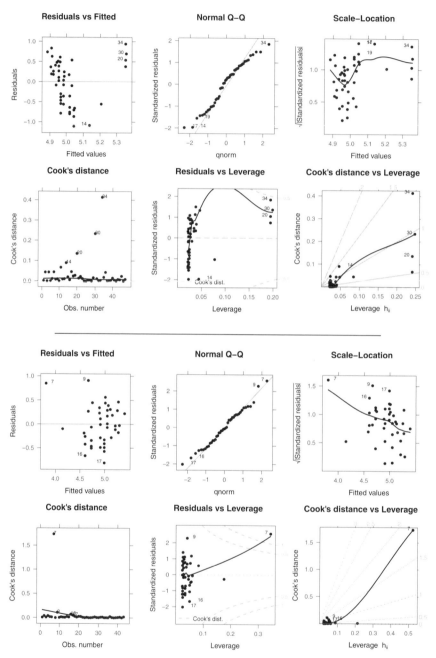

Figure 6.10. Regression diagnostic plots: xplot(star.model1) (above) and xplot(star.model2) (below).

In general, the leave-one-out methods may not be useful in detecting a cluster of observations that do not fit the overall pattern of the rest of the data, but in this case, three of the four unusual observations have Cook's distance values that although not particularly large in absolute value, do stand out among the other

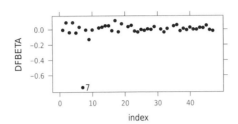

Figure 6.11. A plot of $\text{DFBETA}_{1,i}$ vs. i after removing the four giants from the `star` data set.

values in the data set (Figure 6.10). As it turns out those four stars are so-called giants and so are astronomically different from the others. In such a case, it makes sense to remove these stars from the data set and to focus our analysis on the remaining non-giant stars in the cluster.

After removing the four giants, Cook's distance reveals that there is still one observation with unusually large influence. Before proceeding, we may need to consult with an astronomer to find out if this star is unusual in any way. This example points out an important feature of regression diagnostics: they depend on the model fit and the data used. An outlier in one model may not be in another, and fitting a model again after removing all outliers may simply reveal new outliers. If every time the problematic observations are removed new problematic observations are revealed, this can be an indication of more systematic problems.

Finally, we demonstrate the use of `dfbeta()`. Assuming we are primarily interested in the slope (β_1), we can plot the values of $\text{DFBETA}_{1,i}$ and look for unusual values. (See Figure 6.11.) The value of $\text{DFBETA}_{1,7}$ tells us that removing observation 7 decreases the slope by approximately 0.8. If that magnitude of change in slope is meaningful in our analysis, we want to be quite certain about the quality of observation 7 before proceeding.

`lm-star-dfbeta`

```
> star.plot3 <- xyplot(dfbeta(star.model2)[,'temp']~index,
+     data=hotstar,
+     ylab="DFBETA",
+     panel=function(x,y,...) {
+         ids <- which(abs(y) > 0.5)
+         panel.xyplot(x,y,...)
+         grid.text(
+             x=x[ids]+1.5, y = y[ids],
+             as.character(ids), default.units="native")
+     })
> coef(lm(light~temp,hotstar))
(Intercept)        temp
    -4.0565      2.0467
> coef(lm(light~temp,hotstar[-7,]))
(Intercept)        temp
    -7.4035      2.8028
```

The functions `dfbetas()` and `dffits()` can be used in a similar manner. Observation 7 also stands out if we look at the intercept or at DFFITS. $\triangleleft$

6.5. Transformations in Linear Regression

The utility of linear models is greatly enhanced through the use of various transformations of the data. There are several reasons why one might consider a transformation of the predictor or response (or both).

- To correspond to a theoretical model.

    Sometimes we have *a priori* information that tells us what kind of nonlinear relationship we should anticipate. As a simple example, if we have an apparatus that can accelerate objects by applying a constant (but unknown) force F, then since $F = ma$, we would expect the relationship between the mass of the objects tested and the acceleration measured to satisfy

    $$a = \frac{F}{m}.$$

    This might lead us to fit a model with no intercept (see Exercise 6.5) after applying an inverse transformation to the predictor m:

    $$a = \beta_1 \cdot \frac{1}{m}.$$

    The parameter β_1 corresponds to the (unknown) force being applied and could be estimated by fitting this model.

    Alternatively, we could use a logarithmic transformation

    $$\log(a) = \beta_0 + \beta_1 \log(m),$$
    $$a = e^{\beta_0} m^{\beta_1}.$$

    In this model, e_0^{β} corresponds to the unknown force and we can test whether $\beta_1 = -1$ is consistent with our data.

    Many non-linear relationships can be transformed to linearity. Exercise 6.16 presents several examples and asks you to determine a suitable transformation.

- To obtain a better fit.

    If a scatterplot or residual plot shows a clearly non-linear pattern to the data, then there is no reason to fit a line. In the absence of a clear theoretical model, we may select transformations based on the shape of the relationship as revealed in a scatterplot. Section 6.5.1 provides some guidance for selecting transformations in this situation.

- To obtain better residual behavior.

    Some transformations are used to improve the agreement between the data and the assumptions about the error terms in the model. For example, if the variance in the response appears to increases as the predictor increases, a logarithmic or square root transformation of the response will decrease the disparity in variance.

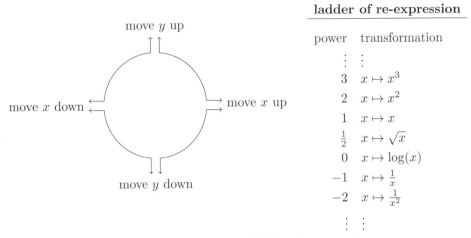

Figure 6.12. Bulge rules and ladder of re-expression.

In practice, all three of these issues are intertwined. A transformation that improves the fit, for example, may or may not have a good theoretical interpretation. Similarly, a transformation performed to achieve **homoskedasticity** (equal variance; the opposite is called **heteroskedasticity**) may result in a fit that does not match the overall shape of the data very well. Despite these potential problems, there are many situations where a relatively simple transformation is all that is needed to greatly improve the model.

6.5.1. The Ladder of Re-expression

In the 1970s, Mosteller and Tukey introduced what they called the **ladder of re-expression** and **bulge rules** [**Tuk77, MT77**] that can be used to suggest an appropriate transformation to improve the fit when the relationship between two variables (x and y in our examples) is monotonic and has a single bend. Their idea was to apply a power transformation to x or y or both – that is, to work with x^a and y^b for an appropriate choice of a and b. Tukey called this ordered list of transformations the ladder of re-expression. The identity transformation has power 1. The logarithmic transformation is a special case and is included in the list associated with a power of 0. The direction of the required transformation can be obtained from Figure 6.12, which shows four bulge types, represented by the curves in each of the four quadrants. A bulge can potentially be straightened by applying a transformation to one or both variables, moving up or down the ladder as indicated by the arrows. More severe bulges require moving farther up or down the ladder. A curve bulging in the same direction as the one in the first quadrant of Figure 6.12, for example, might be straightened by moving up the ladder of transformations for x or y (or both), while a curve like the one in the second quadrant, might be straightened by moving up the ladder for y or down the ladder for x.

This method focuses primarily on transformations designed to improve the overall fit. The resulting models may or may not have a natural interpretation. These transformations also affect the shape of the distributions of the explanatory

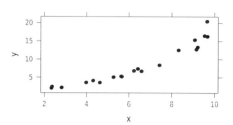

Figure 6.13. A scatterplot illustrating a non-linear relationship between x and y.

and response variables and, more importantly, of the residuals from the linear model (see Exercise 6.18). When several transformations lead to reasonable linear fits, these other factors may lead us to prefer one over another.

Example 6.5.1.

Q. The scatterplot in Figure 6.13 shows a curved relationship between x and y. What transformations of x and y improve the linear fit?

A. This type of bulge appears in quadrant IV of Figure 6.12, so we can hope to improve the fit by moving up the ladder for x or down the ladder for y. As we see in Figure 6.14, the fit generally improves as we move down and to the right – but not too far, lest we over-correct. A log-transformation of the response ($a = 1$, $b = 0$) seems to be especially good in this case. Not only is the resulting relationship quite linear, but the residuals appear to have a better distribution as well.   ◁

Example 6.5.2. Some physics students conducted an experiment in which they dropped steel balls from various heights and recorded the time until the ball hit the floor. We begin by fitting a linear model to this data.

```
> ball.model <- lm(time~height,balldrop)                          balldrop
> summary(ball.model)
< 8 lines removed >
Coefficients:
            Estimate Std. Error t value Pr(>|t|)
(Intercept)  0.19024    0.00430    44.2   <2e-16
height       0.25184    0.00552    45.7   <2e-16

Residual standard error: 0.0101 on 28 degrees of freedom
Multiple R-squared: 0.987,        Adjusted R-squared: 0.986
F-statistic: 2.08e+03 on 1 and 28 DF,  p-value: <2e-16

> ball.plot <- xyplot(time~height,balldrop,type=c('p','r'))
> ball.residplot <- xplot(ball.model,w=1)
```

At first glance, the large value of r^2 and the reasonably good fit in the scatterplot might leave us satisfied that we have found a good model. But a look at the residual plot (Figure 6.15) reveals a clear curvilinear pattern in this data. A knowledgeable physics student knows that (ignoring air resistance) the time should

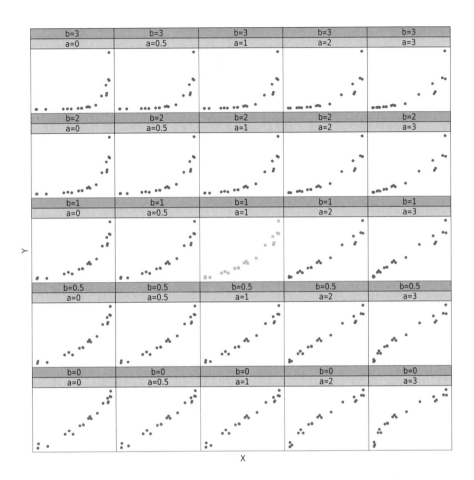

Figure 6.14. Using the ladder of re-expression to find a better fit.

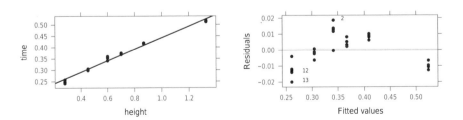

Figure 6.15. A scatterplot and a residual plot for the `balldrop` data set.

be proportional to the *square root* of the height. This transformation agrees with Tukey's ladder of re-expression, which suggests moving down the ladder for `height` or up the ladder for `time`.

```
> ball.modelT <- lm(time ~ sqrt(height),balldrop)
> summary(ball.modelT)
```

balldrop-trans

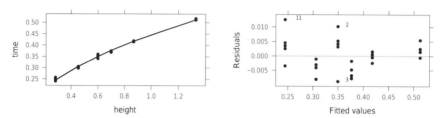

Figure 6.16. Using a square root transformation on `height` in the `balldrop` data improves the fit but reveals other problems.

```
< 8 lines removed >
Coefficients:
             Estimate Std. Error t value Pr(>|t|)
(Intercept)   0.01608    0.00408    3.94    5e-04
sqrt(height)  0.43080    0.00486   88.58   <2e-16

Residual standard error: 0.00523 on 28 degrees of freedom
Multiple R-squared: 0.996,        Adjusted R-squared: 0.996
F-statistic: 7.85e+03 on 1 and 28 DF,  p-value: <2e-16

> ball.plotT <- xyplot(time~height,balldrop,
+                 panel=panel.lm,model=ball.modelT)
> ball.residplotT <- xplot(ball.modelT,w=1)
```

This model does indeed fit better, but the residual plot in Figure 6.16 indicates that there may be some inaccuracy in the measurement of the height. In this experiment, the apparatus was set up once for each height and then several observations were made. So any error in this set-up affected all time measurements for that height in the same way. This could explain why the residuals for each height are clustered the way they are since it violates the assumption that the errors are *independent*. (See Example 6.5.3 for a simple attempt to deal with this problem.)   ◁

Example 6.5.3. One simple way to deal with the lack of independence in the previous example is to average all the readings made at each height. (This works reasonably well in our example because we have nearly equal numbers of observations at each height.) We pay for this data reduction in a loss of degrees of freedom, but it may be easier to justify that the errors in average times at each height are independent (if we believe that the errors in the height set-up are independent and not systematic).

balldrop-avg

```
> balldropavg <- aggregate(balldrop$time, by=list(balldrop$height), mean)
> names(balldropavg) <- c('height','time')
> ball.modelA <- lm(time ~ sqrt(height),balldropavg)
> summary(ball.modelA)
< 8 lines removed >
Coefficients:
```

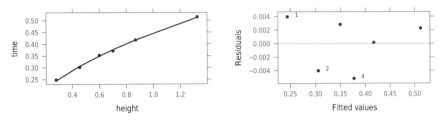

Figure 6.17. Using a square root transformation on averaged `height` measurements in the `balldrop` data gives a similar fit but a very different residual plot. The interpretation of this model is also different.

```
             Estimate Std. Error t value Pr(>|t|)
(Intercept)   0.01608    0.00740    2.17    0.096
sqrt(height)  0.43080    0.00882   48.86  1.0e-06


Residual standard error: 0.00424 on 4 degrees of freedom
Multiple R-squared: 0.998,      Adjusted R-squared: 0.998
F-statistic: 2.39e+03 on 1 and 4 DF,  p-value: 1.05e-06


> ball.plotA <- xyplot(time~height,balldropavg,
+                 panel=panel.lm,model=ball.modelA)
> ball.residplotA <- xplot(ball.modelA,w=1)
```

Notice that the parameter estimates are essentially the same as in the preceding example. The estimate for σ has decreased some. This makes sense since we are now estimating the variability in *averaged* measurements rather than in individual measurements.

Of course, we've lost a lot of degrees of freedom, and as a result, the standard error for our parameter estimate is about twice as large as before. This might have been different; had the mean values fit especially well, our standard error might have been smaller despite the reduced degrees of freedom.

One disadvantage of the data reduction is that it is hard to interpret the residuals (because there are fewer of them). At first glance there appears to be a downward trend in the residuals, but this is largely driven by the fact that the largest residual happened to be for the smallest fit. ◁

Example 6.5.4.

Q. Rex Boggs of Glenmore State High School in Rockhampton, Queensland, had an interesting hypothesis about the rate at which bar soap is used in the shower. He writes:

> I had a hypothesis that the daily weight of my bar of soap [in grams] in my shower wasn't a linear function, the reason being that the tiny little bar of soap at the end of its life seemed to hang around for just about ever. I wanted to throw it out, but I felt I shouldn't do so until it became unusable. And that seemed to take weeks.
>
> Also I had recently bought some digital kitchen scales and felt I needed to use them to justify the cost. I hypothesized that the daily

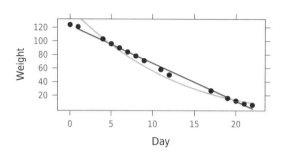

Figure 6.18. Comparing untransformed (darker) and transformed (lighter) fits to soap use data.

weight of a bar of soap might be dependent upon surface area, and hence would be a quadratic function . . . .

The data ends at day 22. On day 23 the soap broke into two pieces and one piece went down the plughole.

The data indicate that although Rex showered daily, he failed to record the weight for some of the days.

What do the data say in regard to Rex's hypothesis?

A. Rex's assumption that weight should be a (quadratic) function of time does not actually fit his intuition. His intuition corresponds roughly to the differential equation

$$\frac{\partial W}{\partial t} = kW^{2/3},$$

for some negative constant k since the rate of change should be proportional to the surface area remaining. (We are assuming that the bar shrinks in such a way that its shape remains proportionally unaltered.) Solving this equation (by separation of variables) gives

$$W^{1/3} = kt + C.$$

We can fit untransformed and transformed models (Weight^(1/3) ~ Day) to this data and compare.

```
> soap.model1 <- lm(Weight~Day,soap)                          lm-soap
> summary(soap.model1)
< 8 lines removed >
Coefficients:
            Estimate Std. Error t value Pr(>|t|)
(Intercept)  123.141     1.382    89.1   <2e-16
Day           -5.575     0.107   -52.2   <2e-16

Residual standard error: 2.95 on 13 degrees of freedom
Multiple R-squared: 0.995,       Adjusted R-squared: 0.995
F-statistic: 2.72e+03 on 1 and 13 DF,  p-value: <2e-16
```

The high value of r^2 (and the scatterplot in Figure 6.18, darker line) indicates that the untransformed model is already a good fit.

```
> soap.model2 <- lm(I(Weight^(1/3))~Day,soap)        lm-soap-trans
> summary(soap.model2)
< 8 lines removed >
Coefficients:
            Estimate Std. Error t value Pr(>|t|)
(Intercept)  5.29771    0.08381    63.2  < 2e-16
Day         -0.14698    0.00648   -22.7  7.7e-12


Residual standard error: 0.179 on 13 degrees of freedom
Multiple R-squared: 0.975,      Adjusted R-squared: 0.973
F-statistic:  515 on 1 and 13 DF,  p-value: 7.67e-12
```

The transformed model in this case actually fits worse. The higher value of r^2 for the untransformed model is an indication that it performs better. Figure 6.18 shows a scatterplot with both fits. The data do not support Rex's assumption that a transformation is necessary. We can also fit a quadratic model of the form Weight~I(Day^2), but this model is worse still. Fitting a full quadratic model requires two predictors (Day and Day^2) and so will have to wait until our discussion of multiple linear regression. The scatterplot and especially the residual plot both show that the residuals are mostly positive near the ends of the data and negative near the center. Part of this is driven by a flattening of the pattern of dots near the end of the measurement period. Perhaps as the soap became very small, Rex used slightly less soap than when the soap was larger. Exercise 6.15 asks you to remove the last few observations and see how that affects the models.

Finally, since a linear model appears to fit at least reasonably well (but see Exercise 6.15), we can give a confidence interval for β_1, the mean amount of soap Rex uses each shower.

```
> confint(soap.model1)                               lm-soap-ci
             2.5 %  97.5 %
(Intercept) 120.1547 126.127
Day          -5.8055  -5.344                                      ◁
```

6.6. Categorical Predictors

So far we have focused our attention on the situation where the response and the (single) predictor have been quantitative. Methods dealing with categorical predictors generally go by the name **ANOVA** and are treated in Chapter 7. Here we handle only the simplest case where there is one categorical predictor with only two levels. We described such a model in Example 6.1.3. We also discuss the more common alternative approach, the **2-sample** *t*-test.

6.6.1. Using the Linear Model to Compare Two Groups

Consider a categorical predictor with only two levels. The linear model framework can be used for this situation if we code the two levels of our categorical predictor as 0 and 1, as we did in Example 6.1.3. The linear model then becomes

$$Y = \beta_0 \mathbf{1} + \beta_1 x + \varepsilon, \qquad \varepsilon \overset{iid}{\sim} \text{Norm}(0, \sigma),$$

or

$$Y = \begin{cases} \beta_0 + \varepsilon, & x = 0, \\ \beta_0 + \beta_1 + \varepsilon, & x = 1. \end{cases}$$

In other words, the model describes two groups, distinguished by the variable x. In the first group, $Y \sim \text{Norm}(\beta_0, \sigma)$ and in the other group $Y \sim \text{Norm}(\beta_0 + \beta_1, \sigma)$. If the two groups are distinguished by a treatment, then β_0 can be interpreted as a baseline response, and β_1 can be interpreted as a treatment effect. The null hypothesis $\beta_1 = 0$ states that there is no treatment effect, or more generally that the mean response is the same for both groups.

Example 6.6.1. In testing food products for palatability, General Foods employed a 7-point scale from -3 (terrible) to $+3$ (excellent) with 0 representing "average". Their standard method for testing palatability was to conduct a taste test with 50 testers: 25 men and 25 women. The scores from these 50 people are then added to give an overall score.

The data in the `tastetest` data set, originally described in [**SC72**], contains the results from a study comparing different preparation methods for taste test samples. The samples were prepared for tasting in one of two ways, using either a coarse screen or a fine screen, to see whether this affects the scores given. Eight groups of testers were given samples prepared each way.

Q. Does the coarseness of the screen used to prepare the samples matter?

A. We begin by looking at a numerical summary of the data

```
> summary(score~scr, data=taste1, fun=favstats)          taste-favstats
score     N=16


+-------+------+--+--+-----+----+-----+----+------+------+-------+
|       |      |N |0%|25%  |50% |75%  |100%|mean  |sd    |var    |
+-------+------+--+--+-----+----+-----+----+------+------+-------+
|scr    |coarse| 8|16|23.25|37.0|44.25| 77 |38.875|20.629| 425.55|
|       |fine  | 8|64|79.25|90.0|98.75|129 |90.375|21.159| 447.70|
+-------+------+--+--+-----+----+-----+----+------+------+-------+
|Overall|      |16|16|38.00|64.5|88.00|129 |64.625|33.388|1114.78|
+-------+------+--+--+-----+----+-----+----+------+------+-------+
```

From this we see that in our data set the samples prepared with the fine screen were scored more highly than those prepared with a coarse screen. We can use R's `lm()` function just as before.

```
> taste.model <- lm(score~scr, data=taste1)              lm-taste
> summary(taste.model)
< 8 lines removed >
Coefficients:
```

```
           Estimate Std. Error t value Pr(>|t|)
(Intercept)   38.88      7.39    5.26  0.00012
scrfine       51.50     10.45    4.93  0.00022


Residual standard error: 20.9 on 14 degrees of freedom
Multiple R-squared: 0.634,       Adjusted R-squared: 0.608
F-statistic: 24.3 on 1 and 14 DF,  p-value: 0.000222
```

Things to note:

(1) The `taste` data frame codes `scr` as a factor with levels `coarse` and `fine`. It is not necessary to convert these to 0's and 1's (although one could). By default, R assigns 0 to the alphabetically first level and 1 to the other.

(2) The p-value for the test of $H_0 : \beta_1 = 0$ is small, so these data allow us to reject the hypothesis that coarseness does not matter at any usual level of significance.

(3) $\hat{\beta}_1 = 51.5$ is an estimate for how much the mean score improves when switching from a coarse screen to a fine screen.
   We can calculate a 95% confidence interval for this:

lm-taste-ci

```
> confint(taste.model)
             2.5 % 97.5 %
(Intercept) 23.030 54.720
scrfine     29.092 73.908
```

Someone from General Foods would have to tell us whether differences in this range are important. It may be useful to present the information on the scale of a per person difference.

lm-taste-rescale

```
> confint(taste.model,"scrfine") / 50
        2.5 % 97.5 %
scrfine 0.58183 1.4782
```

(4) The standard deviations of the two groups are quite similar. Informally, this is an indication that our equal variance assumption is justified.

(5) The estimate of σ (i.e., 20.9) given by the regression is much smaller than the overall sample standard deviation (33.38). These are estimating two different things. The former estimates the amount of variation in each group (assuming that amount is equal for the two groups). The latter estimates the variation in the combined population (ignoring that they are in two groups).

(6) We have not directly examined the residuals here, but they do not reveal any cause for concern. The fact that the two groups have nearly identical standard deviations is a partial indicator of this. A normal-quantile plot also reveals no cause for concern. ◁

Example 6.6.2. Gosset (a.k.a Student) did an experiment to test whether kiln dried seeds have better yield (lbs/acre) than regular seeds. Kiln dried seeds and regular seeds were each planted in 11 different plots and the yield in these plots was measured at harvest. Here is an analysis of his data. (For a better analysis, see Exercise 6.33.)

```
# the corn data frame has an inconvenient "shape"
# (each type of corn is in its own column)
> head(corn,3)
   reg kiln
1 1903 2009
2 1935 1915
3 1910 2011
# this puts all the yields in one column and type of seed in another
> corn2 <- stack(corn)
> corn2[c(1,2,12,13),]
     values  ind
1      1903  reg
2      1935  reg
12     2009 kiln
13     1915 kiln
# the default variable names aren't great, so we rename them
> names(corn2) <- c('yield','treatment')
> corn2[c(1,2,12,13),]
    yield treatment
1    1903       reg
2    1935       reg
12   2009      kiln
13   1915      kiln
> summary(yield~treatment,data=corn2,fun=favstats)
yield     N=22


+---------+----+--+----+------+----+------+----+------+------+------+
|         |    |N |0%  |25%   |50% |75%   |100%|mean  |sd    |var   |
+---------+----+--+----+------+----+------+----+------+------+------+
|treatment|kiln|11|1443|1538.5|1925|2066.5|2463|1875.2|332.85|110789|
|         |reg |11|1316|1561.5|1910|2010.5|2496|1841.5|342.74|117469|
+---------+----+--+----+------+----+------+----+------+------+------+
|Overall  |    |22|1316|1536.8|1920|2047.8|2496|1858.3|330.14|108992|
+---------+----+--+----+------+----+------+----+------+------+------+
> corn.model <- lm(yield~treatment,data=corn2)
> summary(corn.model)
< 8 lines removed >
Coefficients:
              Estimate Std. Error t value Pr(>|t|)
(Intercept)     1875.2      101.9   18.41  5.2e-14
treatmentreg     -33.7      144.1   -0.23     0.82

Residual standard error: 338 on 20 degrees of freedom
Multiple R-squared: 0.00273,      Adjusted R-squared: -0.0471
F-statistic: 0.0548 on 1 and 20 DF,  p-value: 0.817
```

The p-value for the model utility test is large, so we cannot (by this analysis) reject the hypothesis that kiln drying the corn has no effect on yield. In the actual experiment, there were 11 plots and each treatment was used in each plot. The rows of the original data set correspond to the different plots. Given this design, a paired test would be a better analysis. You are asked to do that analysis in Exercise 6.33. ◁

6.6.2. Eliminating the Equal Variance Assumption

The method just outlined for comparing the means of two groups is not the most common method in practice. The reason for this is that it assumes that the two groups have the same variance. In many situations, this is an unjustified or unnecessary assumption.

If we draw samples of sizes n_1 and n_2 independently from two normal populations with means μ_1 and μ_2 and variances σ_1^2 and σ_2^2, then

$$\overline{Y}_1 - \overline{Y}_2 \sim \mathsf{Norm}\left(\mu_1 - \mu_2, \sqrt{\frac{\sigma_1^2}{n_1} + \frac{\sigma_2^2}{n_2}}\right).$$

Typically, we will not know σ_1 and σ_2, so it is natural to ask what distribution we would get if we replaced these with sample estimates. Unfortunately, even if the populations are exactly normally distributed, it is not true that the resulting distribution is a t-distribution, but a t-distribution serves as a good approximation [**Sat46**]:

$$\frac{\overline{Y}_1 - \overline{Y}_2}{\sqrt{\frac{s_1^2}{n_1} + \frac{s_2^2}{n_2}}} \approx \mathsf{t}(\nu),$$

where

$$\nu = \frac{(s_1^2/n_1 + s_2^2/n_2)^2}{\frac{(s_1^2/n_1)^2}{n_1-1} + \frac{(s_2^2/n_2)^2}{n_2-1}}.$$

Some algebra shows the following:

- If $n_1 = n_2$ and $s_1 = s_2$, then $\nu = n_1 + n_2 - 2$.
- $\nu \leq n_1 + n_2 - 2$ (sum of degrees of freedom).
- $\nu \geq \min(n_1 - 1, n_2 - 1)$ (smaller degrees of freedom).

In general, even if $\sigma_1 = \sigma_2$, the 2-sample t-procedures, which do not make this assumption, are nearly as powerful as those from the linear model, which assumes equal variances. On the other hand, when $\sigma_1 \neq \sigma_2$, then the 2-sample t-procedures yield more accurate p-values and confidence intervals. For this reason, the 2-sample t-procedures are generally preferred over a linear model approach for this type of comparison.

Example 6.6.3.

Q. Analyze the data from Example 6.6.2 using the 2-sample t-procedure.

A. R provides two interfaces to `t.test()` for handling 2-sample tests.

```
> t.test(corn$kiln,corn$reg)        # 2-vector interface

        Welch Two Sample t-test
```

t-corn

```
data:  corn$kiln and corn$reg
t = 0.2341, df = 19.983, p-value = 0.8173
alternative hypothesis: true difference in means is not equal to 0
95 percent confidence interval:
 -266.77  334.23
sample estimates:
mean of x mean of y
   1875.2    1841.5


> t.test(yield~treatment,data=corn2)      # formula interface

        Welch Two Sample t-test

data:  yield by treatment
t = 0.2341, df = 19.983, p-value = 0.8173
alternative hypothesis: true difference in means is not equal to 0
95 percent confidence interval:
 -266.77  334.23
sample estimates:
mean in group kiln  mean in group reg
           1875.2             1841.5
```
◁

Example 6.6.4.

Q. Analyze the data from Example 6.6.1 using the 2-sample t-procedure.

A.

t-test-taste

```
> t.test(score~scr,data=taste1)

        Welch Two Sample t-test

data:  score by scr
t = -4.9293, df = 13.991, p-value = 0.0002223
alternative hypothesis: true difference in means is not equal to 0
95 percent confidence interval:
 -73.910 -29.090
sample estimates:
mean in group coarse    mean in group fine
            38.875                90.375
```
◁

6.6.3. Calculating the Power of t-Tests

Recall that the power of a statistical test against a particular alternative at specified significance level is the probability of obtaining a p-value small enough to reject the null hypothesis when the particular alternative is true. R provides a function to determine any one of power, sample size, effect size, standard deviation, or significance if all the others are specified.

Example 6.6.5.

Q. Suppose we want to compare two groups to test $H_0 : \mu_1 = \mu_2$. How large must the samples be if we want an 80% chance of detecting a difference of 5 between the two means assuming $\sigma = 10$ in each group?

A. If we set `delta=5`, `sd=10`, and `power=0.8`, then `power.t.test()` will determine the desired sample size assuming the default significance level (`sig.level=0.05`).

```
> power.t.test(delta=5,sd=10,power=0.8)                     power-t-test01a

        Two-sample t test power calculation

              n = 63.766
          delta = 5
             sd = 10
      sig.level = 0.05
          power = 0.8
    alternative = two.sided

 NOTE: n is number in *each* group
```

It is only the ratio of `delta` to `sd` that matters, as the following example indicates:

```
> power.t.test(delta=0.5,power=0.8)                        power-t-test01b

        Two-sample t test power calculation

              n = 63.766
          delta = 0.5
             sd = 1
      sig.level = 0.05
          power = 0.8
    alternative = two.sided

 NOTE: n is number in *each* group
```

Thus it suffices to express `delta` as a fraction of the standard deviation. Either way, we see that we need approximately 64 measurements from each group. `power.t.test()` does not handle the cases where the two populations have different variances or the two samples are of different sizes. (See Exercise 6.37.)   ◁

Example 6.6.6.

Q. How much would power decrease if we used samples of size 50 in the previous example? What if the difference in means was only 1/4 of the standard deviation?

A. Again we let R do the heavy lifting.

```
> power.t.test(delta=0.5,n=50)                             power-t-test01c

        Two-sample t test power calculation

              n = 50
          delta = 0.5
```

Power of a 2-sample test (n=50)

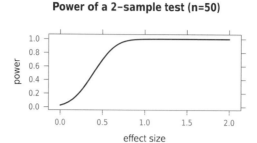

Figure 6.19. The power of a 2-sample t-test.

```
           sd = 1
    sig.level = 0.05
        power = 0.69689
  alternative = two.sided

NOTE: n is number in *each* group

> power.t.test(delta=0.25,n=50)

    Two-sample t test power calculation

            n = 50
        delta = 0.25
           sd = 1
    sig.level = 0.05
        power = 0.23509
  alternative = two.sided

NOTE: n is number in *each* group
```

Typically rather than reporting power under many scenarios, researchers proposing a study will present a graph that shows the relationship between power and other quantities of interest. In Figure 6.19 we plot power vs. effect size (`delta`) for a fixed sample size.

```
> pow <- function(effect) {                                    power-t-test01d
+     power.t.test(delta=effect,n=50)$power
+ }
> effect = seq(0,2,by=0.05)
> plot <- xyplot(pow(effect) ~ effect, type='l',
+     ylab="power", xlab="effect size",
+     main="Power of a 2-sample test (n=50)")                          ◁
```

Example 6.6.7. As the examples below show, we can use `power.t.test()` to make power calculations for 1-sample and paired t-tests as well.

```
> power.t.test(delta=0.5,power=0.8,type='one.sample')          power-t-test02

    One-sample t test power calculation
```

```
              n = 33.367
          delta = 0.5
             sd = 1
      sig.level = 0.05
          power = 0.8
    alternative = two.sided

> power.t.test(delta=0.5,power=0.8,type='paired')

      Paired t test power calculation

              n = 33.367
          delta = 0.5
             sd = 1
      sig.level = 0.05
          power = 0.8
    alternative = two.sided

NOTE: n is number of *pairs*, sd is std.dev. of *differences* within pairs
```

Note that although the interpretations of `delta` and `sd` are different, the output is identical since a paired t-test is a 1-sample t-test using paired differences. In the case of a 1-sample test, `sd` refers to the standard deviation of the single measured quantitative variable. In a paired test, it refers to the standard deviation of the *differences* of two measured quantitative variables. A similar statement holds for `delta`. ◁

6.6.4. Categorical Predictors with More Than Two Levels

Suppose we had a categorical predictor that had more than $k > 2$ levels. If we coded the different levels using the integers $0, 1, 2, \ldots, k - 1$, we would be back to familiar territory with a single quantitative predictor, and we could fit the standard linear model.

This is not, however, the usual way to handle a categorical predictor. (Exercise 6.43 asks you to think about why this model is not typically used.) In the next chapter we will return to the topic of fitting linear models using categorical predictors with more than two levels.

6.7. Categorical Response (Logistic Regression)

Categorical responses can be handled in a number of ways. Here we present an introduction to **logistic regression**, a first example of a more general framework known as **generalized linear models** that was developed in the 1970s to provide a unified approach to a wide range of statistical models (including simple linear regression). In this section we will use logistic regression to handle the special case where there is one (quantitative) predictor and the response has only two levels (success or failure).

Logistic regression is a bit different since we are not attempting to predict responses (success or failure) directly. Instead we predict the probability of success $\pi(x)$ for a given value of the explanatory variable x. One obvious way to do this would be with a model of the form

$$\pi(\boldsymbol{x}) = \beta_0 + \beta_1 \boldsymbol{x} + \varepsilon \,,$$

and in some situations, this is precisely the model that is used. But this model has certain drawbacks. In particular, $\pi(x)$ must be between 0 and 1, but the range of $\beta_0 + \beta_1 x$ is $(-\infty, \infty)$.

6.7.1. The logit Transformation

One solution to this problem is to use a transformation f such that the range of $f(\pi(x))$ is $(-\infty, \infty)$. There are many such transformations, but ideally we would like a transformation that yields a model in which the parameters have natural interpretations. The logit transformation can be thought of as the composition of two transformations.

- First we replace the probability of success π with the **odds** of success:

  odds: $\dfrac{\pi}{1 - \pi} \in [0, \infty)$ [assuming $\pi \neq 1$]

- Then we use a logarithmic transformation:

  log odds: $\log\left(\dfrac{\pi}{1 - \pi}\right) \in (-\infty, \infty)$ [assuming $\pi \neq 0$ and $\pi \neq 1$]

The full transformation

$$\operatorname{logit}(\pi) = \log\left(\frac{\pi}{1 - \pi}\right)$$

is known as the **logit** transformation, and regression that uses this transformation is called **logistic regression**. Other transformations are also used. See Exercise 6.41 for one example.

We can express the logistic model in the following two equivalent forms (with error term omitted):

$$\log\left(\frac{\pi(\boldsymbol{x})}{1 - \pi(\boldsymbol{x})}\right) = \beta_0 + \beta_1 \boldsymbol{x} \,,$$

$$\pi(\boldsymbol{x}) = \frac{e^{\beta_0 + \beta_1 \boldsymbol{x}}}{1 + e^{\beta_0 + \beta_1 \boldsymbol{x}}} = \frac{1}{e^{-(\beta_0 + \beta_1 \boldsymbol{x})} + 1} \,.$$

6.7.2. Fitting the Logistic Regression Model

Logistic regression models are not fit using a least squares method. Instead they are fit using the maximum likelihood method assuming that

$$Y_i \sim \operatorname{Binom}\left(1, \frac{e^{\beta_0 + \beta_1 x_i}}{1 + e^{\beta_0 + \beta_1 x_i}}\right) \,.$$

In the 1970s, Nelder and Wedderburn introduced **generalized linear models** as a general framework that "can be used to obtain maximum likelihood estimates of the parameters with observations distributed according to some exponential family and systematic effects that can be made linear by a suitable transformation" [**NW72**]. A generalized linear model is specified by giving a distribution (family) for the response and a transformation (called the **link function**) that relates the response to the linear model. We won't discuss details of the generalized linear model framework here, but for logistic regression in R, we need to specify the family (`binomial`) and the link (`logit`) and use `glm()` in place of `lm()`.[3] This is demonstrated in the following example using some of the data that was analyzed after the first space shuttle explosion.

Example 6.7.1. Following the explosion of the Space Shuttle Challenger in January 1986, an investigation was launched to determine the cause of the accident and to suggest changes in NASA procedures to avoid similar problems in the future. It was determined that damage to the O-rings used to seal one of the solid rocket boosters allowed pressurized hot gas from within the solid rocket motor to reach the adjacent attachment hardware and external fuel tank. This led to structural failure of the external tank and aerodynamic forces then broke up the shuttle.

Part of the ensuing investigation looked at the number of damaged O-rings for previous shuttle launches and compared this to the temperature at launch. There are several versions of this data in circulation, including the `orings` data set in the `faraway` package. A single O-ring failure is not catastrophic because of redundancies in the system, and there were damaged O-rings in some of the previous successful missions. But even a casual glance at the data suggests that damage is more likely at colder temperatures. We can fit a logistic model to this data to predict the probability of damage at various temperatures.

`glm-orings`

```
# select the version of this data set in the faraway package
> data(orings,package="faraway")
> orings$failure <- orings$damage != 0   # convert to binary response
> orings.model <-
+     glm(failure~temp,data=orings,family=binomial(link=logit))
> summary(orings.model)
< 8 lines removed >

Coefficients:
            Estimate Std. Error z value Pr(>|z|)
(Intercept)   15.043      7.379    2.04    0.041
temp          -0.232      0.108   -2.14    0.032


(Dispersion parameter for binomial family taken to be 1)

    Null deviance: 28.267  on 22  degrees of freedom
Residual deviance: 20.315  on 21  degrees of freedom
AIC: 24.32

Number of Fisher Scoring iterations: 5
```

---

[3]Actually, the logit link is the default for the binomial family, so it is not necessary to specify the link function in this case.

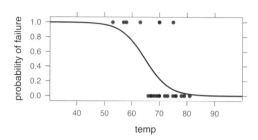

Figure 6.20. A logistic regression fit to pre-1986 space shuttle O-ring damage data.

Figure 6.20 shows the resulting fit graphically. We can use this model to predict the probability of at least one damaged O-ring if the temperature at launch is $31°$ F.

```
> predict(orings.model,newdata=data.frame(temp=31)) -> r; r      glm-orings-predict
      1
7.8459
> ilogit(r)                              # inverse logit transformation
      1
0.9996
> predict(orings.model,newdata=data.frame(temp=31),type='response')->p; p
      1
0.9996
```

Our model predicts a failure rate of 99.96%. This prediction should be considered with care, however. The prior launches were all at much warmer temperatures, so we are extrapolating well beyond temperatures for which we have any data. Furthermore, our choice of the logit link constrains the type of fit we obtain. This is an especially important consideration when extrapolating since there is no data with which to check the fit for such cold temperatures. Finally, we have not yet discussed inference for the model predictions or how to test goodness of fit, so it is unclear how reliable the estimate is, even if the model has been selected appropriately.

Nevertheless, this model does predict a very high probability of some O-ring damage. In fact, if one makes the further assumption that the number of damaged O-rings in a system with six O-rings should have a binomial distribution (i.e., that damage to each O-ring is equally likely and independent of the others), then the probability that any number of O-rings are damaged can also be estimated.

```
                                                            glm-orings-predict2
> 1 - (1-p)^(1/6) -> q; q          # P(damage to particular O-ring)
      1
0.72956
> 1 - dbinom(0,6,q)                # P(damage to >0 O-rings)
[1] 0.9996
> cbind(0:6, dbinom(0:6,6,q))      # table of all probabilities
      [,1]        [,2]
[1,]     0 0.00039122
[2,]     1 0.00633229
[3,]     2 0.04270638
```

```
[4,]    3 0.15361138
[5,]    4 0.31079678
[6,]    5 0.33537320
[7,]    6 0.15078875
```

This is likely a conservative estimate; it is reasonable to assume that there is some dependence among the six O-rings, since if one O-ring fails, it may be an indication that there are external conditions placing stress on the O-rings.   ◁

Example 6.7.2.

Q. How well does runs margin (number of runs scored minus number of runs allowed) predict winning percentage for major league baseball teams?

A. We can fit a logistic model to data from the 2004 major league baseball season contained in `mlb2004`.

runswins-look

```
> head(mlb2004,4)
        Team League  W  L   G   R  OR Rdiff   AB    H DBL TPL  HR  BB   SO
1    Anaheim    AL 92 70 162 836 734   102 5675 1603 272  37 162 450  942
2     Boston    AL 98 64 162 949 768   181 5720 1613 373  25 222 659 1189
3  Baltimore    AL 78 84 162 842 830    12 5736 1614 319  18 169 528  949
4  Cleveland    AL 80 82 162 858 857     1 5676 1565 345  29 184 606 1009
   SB CS    BA   SLG   OBA
1 143 46 0.282 0.429 0.341
2  68 30 0.282 0.472 0.360
3 101 41 0.281 0.432 0.345
4  94 55 0.276 0.444 0.351
```

The format of the data is a little different this time. The data frame contains one row for each team and lists that team's wins and losses. It would be a nuisance to have to reformat this data so that there is one row for each game (as there was one row for each shuttle launch in the previous example). Fortunately, `glm()` can handle data summarized in this way. The left-hand side of the model formula can contain a matrix with two columns – a column of success counts and a column of failure counts. In the example below we use `cbind()` to create this matrix.

runswins-glm

```
> bb <- mlb2004
> bb$runmargin = (bb$R - bb$OR) / (bb$G)
>
# data frame has summarized data for each team, so different syntax here:
> glm(cbind(W,L)~runmargin,data=bb,family='binomial') -> glm.bb
> summary(glm.bb)
< 8 lines removed >
Coefficients:
            Estimate Std. Error z value Pr(>|z|)
(Intercept) -0.00118    0.02906   -0.04     0.97
runmargin    0.44754    0.04165   10.75   <2e-16

(Dispersion parameter for binomial family taken to be 1)

    Null deviance: 131.689  on 29  degrees of freedom
Residual deviance:  12.649  on 28  degrees of freedom
```

```
AIC: 182.1

Number of Fisher Scoring iterations: 3
```

Since in this case we know the actual winning percentages, we can compare the fitted predictions with the actual winning percentages for each of the thirty teams.

runswins-glm2

```
> bb$winP <- bb$W/bb$G
> bb$predWinP <- predict(glm.bb,
+     newdata=data.frame(runmargin=bb$runmargin),type='response')
> bb$winPdiff <- bb$winP - bb$predWinP
> bb[rev(order(abs(bb$winPdiff)))[1:5],c(1,22:24)]
        Team    winP predWinP  winPdiff
27 Cincinnati 0.46914  0.39296  0.076179
8    New York 0.62346  0.56087  0.062587
5     Detroit 0.44444  0.48797 -0.043523
20    Chicago 0.54938  0.58453 -0.035144
28   New York 0.43827  0.46729 -0.029021
> bb.plot1 <- xyplot(winP~predWinP,data=bb,
+     panel=function(x,y,...) {
+         panel.xyplot(x,y,...)
+         panel.abline(0,1)
+     })
>
> rm <- seq(-5,5,by=0.1)
> wp <- predict(glm.bb, newdata=data.frame(runmargin=rm),type='response')
> bb.plot2 <- xyplot(winP~runmargin, data=bb,xlim=c(-2.5,2.5), ylim=c(0,1),
+     panel = function(x,y,...){
+         panel.xyplot(x,y,...)
+         panel.xyplot(rm,wp,type='l',col='gray50')
+     })
```

One interpretation of the lefthand plot in Figure 6.21 is that teams located above the diagonal were "lucky" because they won more games than their run production would have predicted. Similarly, those below the diagonal were "unlucky".   ◁

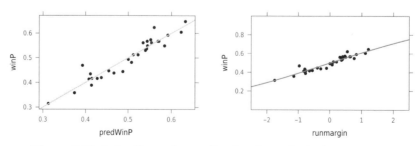

Figure 6.21. Left: Comparing predicted and actual winning percentages. Right: Predicted (curve) and actual (dots) winning percentages vs. run margin. Over this range of values, the logistic fit is nearly linear.

6.7.3. Interpreting the Parameters in Logistic Regression

Once again, the more interesting parameter is β_1. According to the logistic regression model,

$$\beta_1 = \log\left(\frac{\pi(x+1)}{1 - \pi(x+1)}\right) - \log\left(\frac{\pi(x)}{1 - \pi(x)}\right) = \log\left(\frac{\frac{\pi(x+1)}{1 - \pi(x+1)}}{\frac{\pi(x)}{1 - \pi(x)}}\right) = \log \text{ odds ratio}$$

and

$$e^{\beta_1} = \text{odds ratio}.$$

That is, for each increase of 1 unit for the explanatory variable, the odds increase by a factor of e^{β_1}.

Example 6.7.3. We can use the output from `summary()` to construct a confidence interval for this odds ratio in our O-rings example. Rather than copying the values by hand, we can access the necessary information from the summary object with a little bit of additional programming.

```
> s <- summary(orings.model)                                      glm-orings-ci
> sqrt(diag(s$cov.unscaled)) -> st.err; st.err
(Intercept)        temp
   7.37863     0.10824
> coef(orings.model)[2] + c(-1,1) * st.err[2] * qnorm(0.975)
[1] -0.444302 -0.020023
> exp(coef(orings.model)[2] + c(-1,1) * st.err[2] * qnorm(0.975))
[1] 0.64127 0.98018
```

From this we obtain the following confidence intervals:

$$95\% \text{ CI for } \beta_1: \quad (-0.0444, -0.020),$$

$$95\% \text{ CI for odds ratio } (e^{\beta_1}): \quad (e^{-0.0444}, e^{-0.020}) = (0.641, 0.980).$$

Note that the confidence interval for the odds ratio is not symmetric about the point estimate. The `confint()` function produces a different confidence interval (based on profile likelihood) that is not symmetric even on the log odds scale.

```
> confint(orings.model,parm='temp')                              glm-orings-ci2
Waiting for profiling to be done...
    2.5 %     97.5 %
-0.515472 -0.060821
```
◁

One reason for selecting the logistic regression model (as opposed to some competing transformation) is that e^{β_1} can be interpreted as an odds ratio. Unfortunately, most people do not have a good sense for what an odds ratio measures. Recall from Section 2.7.3 that if we have two probabilities p_1 and p_2 (that don't necessarily sum to 1), then

$$\text{odds ratio} = \frac{\left(\frac{p_1}{1 - p_1}\right)}{\left(\frac{p_2}{1 - p_2}\right)}, \text{ and}$$

$$\text{relative risk} = \frac{p_1}{p_2}.$$

Notice that the odds ratio satisfies

$$\frac{\left(\frac{p_1}{1-p_1}\right)}{\left(\frac{p_2}{1-p_2}\right)} = \frac{p_1}{p_2} \cdot \frac{1-p_2}{1-p_1},$$

so that relative risk and odds ratio are nearly the same when $\frac{1-p_1}{1-p_2} \approx 1$. This is often the case when p_1 and p_2 are small – rates of diseases, for example. Since relative risk is easier to interpret (I'm 3 times as likely ...) than an odds ratio (my odds are 3 times greater ...), we can often use relative risk to help us (or those to whom we communicate our results) get a feeling for odds ratios.

In addition to generalizing the type of predictors and responses (from quantitative to categorical), we also want to allow for multiple predictors. This will be discussed in Chapter 7.

6.8. Simulating Linear Models to Check Robustness

Simulating a linear model in R is straightforward. We simply select some values for our predictor and the parameters of the model, use `rnorm()` (or some other distribution) to simulate the error term, and combine. By repeating this process, we can check the coverage rate of the confidence intervals or the distribution of p-values empirically.

Example 6.8.1. We begin with a simulation where all the assumptions of the simple linear model are met exactly. First, we simulate a single data set and construct 95% confidence intervals for the parameters.

```
> b0 <- 3; b1 <- 5; sigma <- 2      # set model parameters     lm-simulate1
> x <- rep(1:5,each=4)              # 4 observations at each of 5 values
> e <- rnorm(length(x),sd=sigma)   # error term in the model
> y <- b0 + b1*x + e               # build response according to model
> model <- lm(y~x); summary(model)
< 8 lines removed >
Coefficients:
            Estimate Std. Error t value Pr(>|t|)
(Intercept)   1.4364     1.1119   1.292    0.213
x             5.4500     0.3352  16.257 3.33e-12 ***
---
Signif. codes:  0 *** 0.001 ** 0.01 * 0.05 . 0.1   1

Residual standard error: 2.12 on 18 degrees of freedom
Multiple R-squared: 0.9362,        Adjusted R-squared: 0.9327
F-statistic: 264.3 on 1 and 18 DF,  p-value: 3.329e-12

> confint(model)
                2.5 %    97.5 %
(Intercept) -0.8995973 3.772348
x            4.7457018 6.154346
```

Notice that each of these confidence intervals correctly covers the associated parameter. This will not always be the case, but we would expect that 95% of these confidence intervals will cover the associated parameter since we are simulating data that match the model assumptions exactly.

If we put all this into a function that performs one simulation, then we can use the `replicate()` function to repeat the simulation as many times as we like. From this we can compute an empirical coverage rate for the 95% confidence interval for β_1.

lm-simulate2

```
> sim <- function(b0=3,b1=5,sigma=2,
+                  x=rep(1:5,each=4)      # 4 observations at each of 5 values
+      ){
+      e <-rnorm(length(x),sd=sigma)
+      y <- b0 + b1*x + e
+      model <- lm(y~x)
+      ci <- confint(model,2)
+      return(b1 > ci[1] && b1 < ci[2])
+ }
> t <- table(replicate(10000,sim())); t / 10000

 FALSE    TRUE
0.0516 0.9484
> prop.test(t[2],sum(t),p=0.95)

        1-sample proportions test with continuity correction

data:  t[2] out of sum(t), null probability 0.95
X-squared = 0.5058, df = 1, p-value = 0.477
alternative hypothesis: true p is not equal to 0.95
95 percent confidence interval:
 0.9438375 0.9526142
sample estimates:
      p
0.9484
```

As the binomial test shows, a 95% coverage rate is fully consistent with our simulated data (the confidence interval contains 95%, and the p-value is large).   ◁

We can check robustness of the linear model against the assumption of normally distributed errors by considering different distributions of the errors.

Example 6.8.2. We can use an exponential distribution instead of a normal distribution to simulate the error terms. Since we prefer to have the mean of the errors be 0, we shift the exponential distribution by its mean. In the simulation below we see that the coverage rate of confidence intervals for β_1 is still close to the 95% nominal rate.

lm-simulate3

```
> sim <- function(b0=3,b1=5,lambda=1,
+                  x=rep(1:5,each=4)      # 4 observations at each of 5 values
+      ){
+      # shift to give a mean of 0.
+      e <- rexp(length(x),rate=1/lambda) - lambda
+      y <- b0 + b1*x + e
+      model <- lm(y~x)
```

```
+      ci <- confint(model,2)
+      return(b1 > ci[1] && b1 < ci[2])
+ }
> t <- table(replicate(10000,sim())); t / 10000

 FALSE   TRUE
0.0444 0.9556
> prop.test(t[2],sum(t),p=0.95)


        1-sample proportions test with continuity correction

data:  t[2] out of sum(t), null probability 0.95
X-squared = 6.4847, df = 1, p-value = 0.01088
alternative hypothesis: true p is not equal to 0.95
95 percent confidence interval:
 0.9513327 0.9595131
sample estimates:
     p
0.9556                                                                    ◁
```

Finally we check for robustness against heteroskedasticity.

Example 6.8.3. In the simulation below, the standard deviation of the error terms is proportional to the predictor, which ranges in size from 1 to 5. The result is a coverage rate that is too low.

```
> sim <- function(b0=3,b1=5,lambda=1,                        │ lm-simulate4 │
+                  x=rep(1:5,each=4)     # 4 observations at each of 5 values
+      ){
+      e <- x * rnorm(length(x))
+      y <- b0 + b1*x + e
+      model <- lm(y~x)
+      ci <- confint(model,2)
+      return(b1 > ci[1] && b1 < ci[2])
+ }
> t <- table(replicate(10000,sim())); t / 10000

 FALSE   TRUE
0.0642 0.9358
> prop.test(t[2],sum(t),p=0.95)


        1-sample proportions test with continuity correction

data:  t[2] out of sum(t), null probability 0.95
X-squared = 42.1521, df = 1, p-value = 8.444e-11
alternative hypothesis: true p is not equal to 0.95
95 percent confidence interval:
 0.9307749 0.9404869
sample estimates:
     p
0.9358                                                                    ◁
```

Similar simulations can be done to test other behaviors of the linear model.

6.9. Summary

Linear models provide a flexible framework for investigating relationships between **response** and **explanatory** variables. Linear models have two components: a systematic component (the pattern or signal) and a random component (error or noise). The basic form of a linear model is

$$\underbrace{Y}_{\text{data}} = \underbrace{f(X)}_{\text{signal}} + \underbrace{\varepsilon}_{\text{noise}} \,,$$

where f is a linear function, for example, $f(x) = \beta_0 + \beta_1 x$.

Typically the errors are assumed to be independent, normally distributed, with a common variance σ^2. (There are methods to deal with other distributional assumptions, too.) Regression diagnostics, including scatterplots of the data, various **residual plots**, and calculations of various measures of **leverage** and **influence** (e.g., Cook's distance, DFFITS, and DFBETAS) can be used to assess whether these assumptions of the model are reasonable and to detect the presence of observations that may be driving the fit. For small data sets, it can be difficult to justify the assumptions or to detect problems based on the data alone. **Transformations** of the explanatory and/or response variables can be used either to improve the overall fit or to make the assumptions about errors more reasonable.

Linear models can be fit by the method of **least squares** or by maximum likelihood to determine estimates $\hat{\boldsymbol{\beta}}$ for the parameters $\boldsymbol{\beta}$. In the case of simple linear regression where there is a single explanatory variable and a continuous response, these two methods yield the same estimators. Least squares estimates can be interpreted geometrically and derived using projections. The assumption that the errors are independent and normally distributed allows us to derive sampling distributions for the parameters of the model. The sampling distributions for each of the estimators in the Table 6.2 are normal with the indicated mean and standard deviation (i.e., standard error).

Since

$$\frac{(n-2)S^2}{\sigma^2} \sim \mathsf{Chisq}(n-2) \,,$$

our inference procedures replace σ with s and use the $\mathsf{t}(n-2)$-distribution in place of a normal distribution. The necessary standard errors are displayed in the summary of regression objects produced by `lm()`. Confidence intervals for the parameters β_0 and β_1 are also available via `confint()`, and confidence and prediction intervals for the response via `predict()`. As with inference for a mean, the procedures (except for prediction intervals) are fairly robust against lack of normality when the sample sizes are large enough. Prediction intervals remain sensitive to normality assumptions for every sample size.

The ANOVA table introduced by Fisher provides an alternative way of summarizing a linear model based on the geometry of linear models (see Figure 6.2). If $\beta_1 = 0$, then

$$SSE = \frac{|Y - \hat{Y}|^2}{\sigma^2} \sim \mathsf{Chisq}(n-2) \,,$$

and

$$SSM = \frac{|\hat{\boldsymbol{Y}} - \overline{\boldsymbol{Y}}|^2}{\sigma^2} \sim \mathsf{Chisq}(1)\,,$$

so

$$F = \frac{SSM/(1)}{SSE/(n-2)} \sim \mathsf{F}(1, n-2)\,.$$

This gives an equivalent way of testing $H_0 : \beta_1 = 0$ vs. $H_a : \beta_1 \neq 0$. Large values of F are evidence against H_0. This approach is called ANOVA (for analysis of variance), since

$$|\boldsymbol{y} - \overline{\boldsymbol{y}}|^2 = SST = SSM + SSE\,.$$

(This equality follows from the orthogonality of the vectors involved.) Thus we can partition the variation in the observations of the response into the portion explained by the model and the portion not explained by the model. The proportion of variation explained by the model is given by

$$r^2 = \frac{SSM}{SST}$$

and appears in most regression output. "Good" values of r^2 depend heavily on the context of the data.

We can incorporate categorical predictors with two levels into our model by coding the two levels as 0 and 1. The resulting model assumes there are two groups, each normally distributed with possibly different means but the same variance. The parameters of this model have a natural interpretation: The mean of the first group is given by β_0, and the mean of the second group by $\beta_0 + \beta_1$. The assumption that the errors are independent and normally distributed implies that the two samples should be independent. This is in contrast to the **paired t-test** that we derived earlier.

Table 6.2. Important normal distributions associated with regression. The sampling distribution for each of the estimators is normal with the indicated mean and standard deviation (i.e., standard error).

estimator	mean	standard error	
$\hat{\beta}_1$	β_1	$\sigma \dfrac{1}{|\boldsymbol{x} - \overline{\boldsymbol{x}}|}$	
$\hat{\beta}_0$	β_0	$\sigma \sqrt{\dfrac{1}{|\mathbf{1}|^2} + \dfrac{\overline{x}^2}{|\boldsymbol{x} - \overline{\boldsymbol{x}}|^2}}$	
$\hat{Y}_* = \hat{\beta}_0 + \beta_1 x_*$	$\beta_0 + \beta_1 x_*$	$\sigma \sqrt{\dfrac{1}{|\mathbf{1}|^2} + \dfrac{(x_* - \overline{x})^2}{|\boldsymbol{x} - \overline{\boldsymbol{x}}|^2}}$	(mean response)
$\hat{Y}_* = \hat{\beta}_0 + \beta_1 x_* + \varepsilon$	$\beta_0 + \beta_1 x_*$	$\sigma \sqrt{1 + \dfrac{1}{|\mathbf{1}|^2} + \dfrac{(x_* - \overline{x})^2}{|\boldsymbol{x} - \overline{\boldsymbol{x}}|^2}}$	(individual response)

The assumption that the two groups have the same standard deviation can be removed using a **2-sample** *t*-**test**, and since this procedure is nearly as powerful as the linear model approach even when the two groups have the same variance and yields more accurate results when the variances differ, this is the approach that is usually taken to investigate the difference in the means of two populations. Inference is based on the fact that

$$\overline{Y}_1 - \overline{Y}_2 \sim \mathsf{Norm}\left(\mu_1 - \mu_2, \sqrt{\frac{\sigma_1^2}{n_1} + \frac{\sigma_2^2}{n_2}}\right)$$

and that

$$\frac{\overline{Y}_1 - \overline{Y}_2}{\sqrt{\frac{s_1^2}{n_1} + \frac{s_2^2}{n_2}}} \approx \mathsf{t}(\nu), \quad \text{where } \nu = \frac{(s_1^2/n_1 + s_2^2/n_2)^2}{\frac{(s_1^2/n_1)^2}{n_1-1} + \frac{(s_2^2/n_2)^2}{n_2-1}}.$$

Two-sample *t*-procedures are quite robust.

The simple linear model described in this chapter can be extended in a number of important ways, some of which will be covered in Chapter 7.

(1) **ANOVA** allows for categorical predictors (with more than two levels).

(2) **Multiple regression** allows for more than one (quantitative) predictor.

(3) **ANCOVA** (analysis of covariance) allows for both quantitative and categorical predictors.

(4) The **generalized linear model** framework brings a wide range of models into a common framework.

6.9.1. Logistic Regression

Logistic regression uses a generalized linear model to accommodate two-level categorical responses. The model is fit by maximum likelihood assuming that

$$Y_i \sim \mathsf{Binom}\left(1, \frac{e^{\beta_0 + \beta_1 x_i}}{1 + e^{\beta_0 + \beta_1 x_i}}\right),$$

where Y_i is either 0 or 1. This is the same thing as a model where the probability of success $\pi(x)$ is related to the predictor x by

$$\log\left(\frac{\pi(x)}{1 - \pi(x)}\right) = \beta_0 + \beta_1 x.$$

In this context the **logit transformation**

$$\pi \mapsto \log\left(\frac{\pi}{1 - \pi}\right)$$

is called the **link function**. Generalized linear models can be fit in R using `glm()` by specifying

- a *model formula* indicating the systematic component of the model, which is a linear function (in this case $\beta_0 + \beta_1 x$),

- a *family of distributions* that describe the random component of the model (in this case the binomial distribution), and

- a *link function* (in this case the logit transformation).

Other generalized linear models are obtained by using different values for these three components of the model. The logit transformation is also used to fit **Bradley-Terry** models, which can be used to estimate ratings and rankings based on pairwise comparisons. (See Section 5.6.)

6.9.2. R Commands

Here is a table of important R commands introduced in this chapter. Usage details can be found in the examples and using the R help.

`lm(y~x,...)`	Fit a linear model.

`glm(y~x,`
`   family=binomial(link=logit),`
`   ...)`
	Fit a logistic regression model.

`glm(cbind(successes,failures)~x,`
`   family=binomial(link=logit),`
`   ...)`
	Fit a logistic regression model using tabulated data.

`I(...)`	Inhibit interpretation in model formulas. Several arithmetic operators have special meanings in the context of a formula. Surrounding them with `I()` causes them to them take on their usual arithmetic meaning.
`summary(model)`	Print a numerical summary of a model (output from `lm()` or `glm()`).
`anova(model)`	Print an ANOVA table.
`plot(model,...)`	Generate some diagnostic plots.
`xplot(model,...)`	Generate some diagnostic plots (using lattice plots) [`fastR`].
`confint(model,...)`	Compute confidence intervals for parameters.
`predict(model,...)`	Predict responses expressed as point estimate (default), confidence interval (`interval='confidence'`), or prediction interval (`interval='prediction'`). Use `newdata=data.frame(...)` to specify the explanatory variables for the predictions.

Exercises

6.1. Show that the estimates $\hat{\beta}_1$ and $\hat{\beta}_0$ given in (6.20) and (6.22) are the same as those given in Section 6.2.1.

6.2. Let $\boldsymbol{Y} = \langle Y_1, Y_2, Y_3 \rangle$ and let $\boldsymbol{v} = \langle 1, 2, -3 \rangle$. Suppose $Y_i \overset{\text{iid}}{\sim} \text{Norm}(5, 2)$. Show your work as you answer the following questions.

   a) What is $|\boldsymbol{v}|$?

   b) Is $\boldsymbol{v} \perp \mathbf{1}$?

   c) What is the distribution of $\boldsymbol{v} \cdot \boldsymbol{Y}$?

6.3. Collect your own elastic band data set and analyze the results. In what ways can you improve on the design used in the `elasticband` data set from DAAG analyzed in Example 6.2.4?

6.4. The `rubberband` data set contains the results of an experiment similar to the one in Example 6.2.4.

   a) In what ways is the design of this experiment better than the one discussed in Example 6.2.4?

   b) Analyze the `rubberband` data using `lm()` and compare the results to those presented in Example 6.2.4.

6.5. It is also possible to fit a linear model without the intercept term. The form of that model is

$$\boldsymbol{Y} = \beta_1 \boldsymbol{x} + \boldsymbol{\varepsilon}, \qquad \varepsilon \overset{\text{iid}}{\sim} \text{Norm}(0, \sigma) .$$

   a) Use the method of least squares to estimate β_1.

   b) Use the method of maximum likelihood to estimate β_1 and σ.

   c) Use the methods derived in parts a) and b) to fit such a model to the data in Example 6.2.1. R can fit this model, too. You can check your answer in R using one of the following:
```
lm(y~0+x)
lm(y~-1+x)
```

6.6. Show that the expressions in (6.9) and (6.10) are equivalent.

6.7. Prove part (a) of Lemma 6.3.1.

6.8. Prove Lemma 6.3.2.

6.9. Derive (6.25) directly from (6.9).

6.10. Prove that in the simple linear regression model, $t^2 = F$, where t and F are the test statistics used to test the null hypothesis $H_0 : \beta_1 = 0$.

   By Lemma 4.7.3, this implies that these two tests of H_0 are equivalent.

6.11. In a model without an intercept, it is still the case that

$$r^2 = \frac{SSM}{SST} \ .$$

But it is no longer the case that this represents the proportion of the variability in the response variable explained by the model.

What does r^2 represent in this situation? When is it a useful indicator of how well a model fits the data? The following may assist your exploration:

  a) For a model without an intercept, express SSM and SSE in terms of the lengths of vectors involving the response variable.

  b) For a model with an intercept, when is $r^2 = 0$?

  c) For a model without an intercept, when is $r^2 = 0$?

6.12. Use the methods of Section 6.3.7 to develop an interval estimate for the mean of m new observations when $x = x_*$.

How will the robustness of this interval depend on m?

6.13. Prove that $SSE = \sum_i e_i = 0$ without (directly) using the fact that $e \perp 1$.

6.14. In Example 6.5.2, we applied a square root transformation to the height. Is there another transformation that yields an even better fit?

6.15. Remove the last few days from the `soap` data set and refit the models in Example 6.5.4. How much do things change? Do the residuals look better, or is there still some cause for concern?

6.16. For each of the following relationships between a response y and an explanatory variable x, if possible find a pair of transformations f and g so that $g(y)$ is a linear function of $f(x)$:

$$g(y) = \beta_0 + \beta_1 f(x) \ .$$

For example, if $y = ae^{bx}$, then $\log(y) = \log(a) + bx$, so $g(y) = \log(y)$, $f(x) = x$, $\beta_0 = \log(a)$, and $\beta_1 = b$.

  a) $y = ab^x$.

  b) $y = ax^b$.

  c) $y = \frac{1}{a+bx}$.

  d) $y = \frac{x}{a+bx}$.

  e) $y = ax^2 + bx + c$.

  f) $y = \frac{1}{1 + e^{a+bx}}$.

  g) $y = \frac{100}{1 + e^{a+bx}}$.

6.17. What happens to the role of the error terms (ε) when we transform the data? For each transformation from Exercise 6.16, start with the form

$$g(y) = \beta_0 + \beta_1 f(x) + \varepsilon$$

and transform back into a form involving the untransformed y and x to see how the error terms are involved in these transformed linear regression models.

It is important to remember that when we fit a linear model to transformed data, the usual assumptions of the model are that the errors in the (transformed) linear form are additive and normally distributed. The errors may appear differently in the untransformed relationship.

6.18. The transformations in the ladder of re-expression also affect the shape of the distribution.

a) If a distribution is symmetric, how does the shape change as we move up the ladder?

b) If a distribution is symmetric, how does the shape change as we move down the ladder?

c) If a distribution is left skewed, in what direction should we move to make the distribution more symmetric?

d) If a distribution is right skewed, in what direction should we move to make the distribution more symmetric?

6.19. By attaching a heavy object to the end of a string, it is easy to construct pendulums of different lengths. Some physics students did this to see how the period (time in seconds until a pendulum returns to the same location) depends on the length (in meters) of the pendulum. The students constructed pendulums of lengths varying from 10 cm to 16 m and recorded the period length (averaged over several swings of the pendulum). The resulting data are in the `pendulum` data set.

Find a suitable transformation and apply a linear model to this data. What can you say about the apparent relationship between the pendulum length and period based on this fit?

6.20. The `pressure` data set contains data on the relation between temperature in degrees Celsius and vapor pressure in millimeters (of mercury). Using temperature as the predictor and pressure as the response, use transformations (if necessary) to obtain a good fit. Make a list of all the models you considered and explain how you chose your best model. What does your model say about the relationship between pressure and temperature?

6.21. The `cornnit` data set in the package `faraway` contains data from a study investigating the relationship between corn yield (bushels per acre) and nitrogen (pounds per acre) fertilizer application in Wisconsin. Using nitrogen as the predictor and corn yield as the response, use transformations (if necessary) to obtain a good fit. Make a list of all the models you considered and explain how you chose your best model.

6.22. The `anscombe` data set has four pairs of explanatory (x1, x2, x3, and x4) and response variables (y1, y2, y3, and y4). These data were constructed by Anscombe [**Ans73**].

a) For each pair, use R to fit the linear model and compare the results. Use, for example,

```
lm(y1~x1,data=anscombe) -> model1;  summary(model1)
```

b) Now make residual plots for each model.

c) For each pair, make a scatterplot that includes the regression line.

d) Comment on these results. Why do you think Anscombe created this data?

6.23. Verify that the standard error formulas

$$s = \sqrt{\frac{SSE}{n-2}},$$

$$SE(\hat{\beta}_1) = \frac{s}{|\boldsymbol{x} - \bar{\boldsymbol{x}}|}, \quad \text{and}$$

$$SE(\hat{\beta}_0) = s\sqrt{\frac{1}{|\boldsymbol{1}|^2} + \frac{\bar{x}^2}{|\boldsymbol{x} - \bar{\boldsymbol{x}}|^2}}$$

match the values given by the `summary()` function for a simple linear model by computing these values directly in R and comparing to the output in Example 6.3.1 (using the `trebuchet2` data set).

6.24. The object returned by `lm()` includes a vector named `effects`. (If you call the result `model`, you can access this vector with `model$effects`.) What are the values in this vector?

(Hint: Think geometrically, make a reasonable guess, and then do some calculations to confirm your guess. You may use one of the example data sets from this chapter or design your own data set if that helps you figure out what is going on.)

6.25. Use the R output below to create an ANOVA table.

```
> summary(lm(y~x,someData))                                  anova-from-lm
< 8 lines removed >
Coefficients:
            Estimate Std. Error t value Pr(>|t|)
(Intercept)    4.202      1.559    2.70    0.015
x              0.741      0.470    1.58    0.132


Residual standard error: 2.97 on 18 degrees of freedom
Multiple R-squared: 0.121,        Adjusted R-squared: 0.0724
F-statistic: 2.48 on 1 and 18 DF,  p-value: 0.132
```

6.26. The data set `actgpa` contains the ACT composite scores and GPAs of some randomly selected seniors at a Midwest liberal arts college.

  a) Give a 95% confidence interval for the mean ACT score of seniors at this school.

  b) Give a 95% confidence interval for the mean GPA of seniors at this school.

  c) What fraction of the variation in GPAs is explained by a student's ACT score?

  d) Use the data to estimate with 95% confidence the average GPA for a student who scored 25 on the ACT.

  e) Suppose you know a high school student who scored 30 on the ACT. Estimate with 95% confidence his GPA as a senior in college.

  f) Are there any reasons to be concerned about the analyses you have just done? Explain.

6.27. In the absence of air resistance, a dropped object will continue to accelerate as it falls. But if there is air resistance, the situation is different. The drag force due to air resistance depends on the velocity of an object and operates in the opposite direction of motion. Thus as the object's velocity increases, so does the drag force until it eventually equals the force due to gravity. At this point the net force is 0 and the object ceases to accelerate, remaining at a constant velocity called the terminal velocity.

Now consider the following experiment to determine how terminal velocity depends on the mass (and therefore on the downward force of gravity) of the falling object. A helium balloon is rigged with a small basket and just the right ballast to make it neutrally buoyant. Mass is then added and the terminal velocity is calculated by measuring the time it takes to fall between two sensors once terminal velocity has been reached.

The `drag` data set contains the results of such an experiment conducted by some undergraduate physics students. Mass is measured in grams and velocity in meters per second. (The distance between the two sensors used for determining terminal velocity is given in the `height` variable.)

Determine which of the following "drag laws" matches the data best:

- Drag is proportional to velocity.
- Drag is proportional to the square of velocity.
- Drag is proportional to the square root of velocity.
- Drag is proportional to the logarithm of velocity.

6.28. Construct a plot that reveals a likely systematic problem with the `drag` (see Exercise 6.27) data set. Speculate about a potential cause for this.

6.29. Exercise 6.28 suggests that some of the data should be removed before analyzing the `drag` data set. Redo Exercise 6.27 after removing this data.

6.30. The `spheres` data set contains measurements of the diameter (in meters) and mass (in kilograms) of a set of steel ball bearings. We would expect the mass to be proportional to the cube of the diameter. Fit a model and see if the data reflect this.

6.31. The `spheres` data set contains measurements of the diameter (in meters) and mass (in kilograms) of a set of steel ball bearings. We would expect the mass to be proportional to the cube of the diameter. Using appropriate transformations fit two models: one that predicts mass from diameter and one that predicts diameter from mass. How do the two models compare?

6.32. The `utilities` data set was introduced in Exercise 1.21. Fit a linear model that predicts `thermsPerDay` from `temp`.

a) What observations should you remove from the data before doing the analysis? Why?

b) Are any transformations needed?

c) How happy are you with the fit of your model? Are there any reasons for concern?

d) Interpret your final model (even if it is with some reservations listed in part c)). What does it say about the relationship between average monthly temperature and the amount of gas used at this residence? What do the parameters represent?

6.33. Recall Example 6.6.2. A number of factors (rainfall, soil quality, etc.) affect the yield of corn. In Gosset's experiment with regular and kiln dried corn, he planted corn of each type in plots with different conditions to block for some of these effects.

Present an analysis of his data using a paired t-test instead of a 2-sample comparison.

6.34. Tire tread wear can be measured two different ways: by weighing the tire and by measuring the depth of the grooves. Data from a 1953 study [**SRM53**] that estimated tire wear (in thousands of miles) of a set of tires by each of these two methods is available in the `tirewear` data set.

Turn the following informal questions into formal questions and present an analysis.

a) Do these methods give comparable results?

b) Does one of the two methods tend to give a higher estimate than the other?

c) How well can we predict one measurement from the other?

Be sure to discuss any places where you had more than one analysis option and how you made your choice.

6.35. Two varieties of oats were compared in an experiment to determine which variety had the higher yield. Since soil type also affects yield, the experimenter blocked out its effect by planting each variety of oats in seven different types of soil. With the data paired by soil types as given below, does it appear that variety A has the higher mean yield?

	Yield	
Soil type	A	B
1	71.2	65.2
2	72.6	60.7
3	47.8	42.8
4	76.9	73.0
5	42.5	41.7
6	49.6	56.6
7	62.8	57.3

6.36. Without using `power.test.test()`, compute the power of a 2-sample t-test if each sample has size 10, each population has standard deviation 3, and the true difference in means is 4.

6.37. Write your own version of `power.t.test()` that can handle samples from populations with different variances and use samples of different sizes.

6.38. Show how to recover the probability π if you know the odds o.

6.39. Return to Example 6.7.2 and fit a simple linear model instead of a logistic regression model. The predictor is again `runmargin` and the response is winning percentage.

How does this model compare with the logistic regression model?

6.40. Buckthorn is an invasive shrub that can grow to heights of 20 meters with trunk diameters of up to 10 cm and has become a significant problem on the campus of Calvin College and in much of the U.S. Midwest. Removing buckthorn from an infested area is challenging. Students and faculty at Calvin College have conducted experiments to learn how best to get rid of buckthorn. In one experiment, students cut down buckthorn plants to near ground level and painted different concentrations of glyphosate (RoundUp Herbicide) on the stump to see how the concentration of glyphosate affects the proportion of plants that die. Students came back later and counted the number of new shoots growing from the old stump. Any stumps with no shoots were considered "dead". The data from 2006 are available in the `buckthorn` data set.

   a) Fit a logistic regression model to this data and prepare a plot that shows the logistic regression fit (as a curve) and the proportion of dead plants at each concentration used (as dots).

   b) Interpret the parameter estimates of the model in a way that would be useful for a biologist.

   c) Do a follow-up goodness of fit test. How many degrees of freedom should you use?

   d) Is there any reason to be concerned about the use of this model?

   e) How might the design of the study be improved?

6.41. The logit transformation works in part because

$$\text{logit} : (0, 1) \to (-\infty, \infty) \, .$$

Other transformations with this property could also be used. The cdf F of a continuous random variable with support all of $\mathbb{R}$ has the property that

$$F : (-\infty, \infty) \to (0, 1) \, .$$

Thus

$$F^{-1} : (0, 1) \to (-\infty, \infty)$$

could be used in place of the logit transformation. The **probit transformation** uses the inverse of the cdf of the standard normal distribution. We can use `glm()` to fit a probit model by using the argument `family=binomial(link=probit)`.

   a) Graph $\pi(x) = f(\beta_0 + \beta_1 x)$ vs. x for several values of β_0 and β_1 where f is either the inverse logit or the inverse probit transformation. Based on your graphs, how do the logit and probit models differ?

   b) Fit a probit model to the data in Example 6.7.2 and compare the results to those from logistic regression.

   c) How does a probit model compare with the logit model on the O-rings data from Example 6.7.1?

One could, of course, play this game backwards and derive a cdf by inverting the logit transformation. The resulting distribution is called the **logistic distribution**. The usual R functions for working with logistic distributions are called `dlogis()`, `plogis()`, `qlogis()`, and `rlogis()`.

6.42. There is yet another way to interpret the correlation coefficient R geometrically. We defined r^2 as

$$r^2 = \frac{SSE}{SST} \ .$$

This implies that

$$|R| = \frac{\sqrt{SSE}}{\sqrt{SST}} \ .$$

The numerator and denominator of this expression can be interpreted as lengths of vectors. Using this intuition, express R in terms of trigonometric functions and angles.

6.43. Suppose we had a categorical predictor that had more than $k > 2$ levels. If we coded the different levels using the integers $0, 1, 2, \ldots, k - 1$, we would be back to familiar territory with a single quantitative predictor, and we could fit the standard linear model.

This is not, however, the usual way to handle a categorical predictor. Why not? That is, what constraints does this model impose that we often do not want?

6.44. What is the difference between a paired test and a 2-sample test? In what situations should one be preferred over the other? In what situations will a paired test be more powerful than a 2-sample test? In what situations is a 2-sample test more powerful?

## Chapter 7

More Linear Models

To call in the statistician after the experiment is done may be no more than asking him to perform a postmortem examination: he may be able to say what the experiment died of.

R. A. Fisher [**Fis38**]

In this chapter we present several extensions to the linear models presented in the previous chapter:

- multiple predictors,
- categorical predictors with more than two levels,
- interactions between predictors,
- a wider range of testable hypotheses and confidence intervals, and
- more examples of generalized linear models.

In addition we will discuss how to interpret various models we encounter and how to select a model when several models are possible. Along the way, we will see that there is a tight connection between the design of an experiment or observational study and the analysis of the resulting data.

7.1. Additive Models

7.1.1. Multiple Quantitative Predictors

The **general additive model** with k quantitative explanatory variables (also called **predictors** or **regressors**) has the form

$$\boldsymbol{Y} = \beta_0 + \beta_1 \boldsymbol{x}_1 + \beta_2 \boldsymbol{x}_2 + \cdots + \beta_k \boldsymbol{x}_k + \boldsymbol{\varepsilon}, \qquad \varepsilon \overset{\text{iid}}{\sim} \mathsf{Norm}(0, \sigma) \ .$$

Interpreting these parameters (and especially the statistical inferences involving them) is a bit more subtle than it was for simple linear regression. For example, β_1 tells how much the response increases (on average) if x_1 is increased by 1 *and the other predictors are not changed*. It is an assumption of the additive model that this effect does not depend on the particular values of any other predictors in the model. Statisticians say that the effects of the predictors $x_1, x_2, \ldots, x_k$ are *additive*. We will learn about other models that do not make this additivity assumption later.

The interpretation of the model parameters is even trickier when it really isn't possible to adjust one predictor without simultaneously adjusting other predictors. Imagine, for example, an economic model that includes predictors like inflation rate, interest rates, unemployment rates, etc. The government can take action to affect a parameter like interest rates, but that may result in changes to the other predictors.

In the next section we will discuss how an additive model is fit. But before we do that, let's take a look at an example. We'll see that as a first approximation we can transfer much of what we learned from the simple linear model.

Example 7.1.1. Investigators at the Virginia Polytechnic Institute and State University studied physical characteristics and ability in 13 football punters [**Pun83**]. Punting performance was based on ten punts from which they calculated the mean distance (in feet) and mean hang time (time until the punt returns to the ground after being kicked, in seconds). The investigators also recorded five measures of strength and flexibility for each punter: right leg strength (pounds), left leg strength (pounds), right hamstring muscle flexibility (degrees), left hamstring muscle flexibility (degrees), and overall leg strength (foot-pounds).

Q. How well do right leg strength and flexibility predict the distance a punter can punt?

A. We can fit an additive model with two predictors using `lm()`. The output is very similar to the output we saw for simple linear models.

```
> punting.lm <- lm(distance~rStrength+rFlexibility,punting)    punting-lm
> summary(punting.lm)
< 4 lines removed >
Residuals:
   Min    1Q Median    3Q    Max
-20.23 -11.76  -2.35  12.20  19.78

Coefficients:
             Estimate Std. Error t value Pr(>|t|)
(Intercept)   -76.064     57.458   -1.32    0.215
rStrength       0.475      0.302    1.58    0.146
rFlexibility    1.610      0.882    1.83    0.098

Residual standard error: 15.1 on 10 degrees of freedom
Multiple R-squared: 0.72,        Adjusted R-squared: 0.664
F-statistic: 12.8 on 2 and 10 DF,  p-value: 0.00173
```

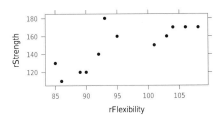

Figure 7.1. A scatterplot shows that leg strength and leg flexibility are positively correlated for the punters.

We're fitting a model of the form

$$\text{distance} = \beta_0 \mathbf{1} + \beta_1 \text{ rStrength} + \beta_2 \text{ rFlexibility} + \varepsilon.$$

Beginning at the bottom of the output, we see that a p-value for the model utility test ($H_0 : \beta_1 = \beta_2 = 0$) is quite small, suggesting that leg strength and flexibility are useful in predicting the distance of punts. Moving up, we see that $r^2 = 0.72$, so this model is explaining approximately 72% of the variability in the distances the punters punt. So knowing leg strength and flexibility is better than not knowing them.

Interestingly, none of the p-values for the individual parameters has a very small p-value. These tests of the null hypotheses

$$H_0 : \beta_j = 0$$

are in the context of the rest of the model. So knowing leg flexibility does not add much predictive power over just knowing leg strength and vice versa. One possible explanation for this is that the punters with the strongest legs tend to also have the most flexible legs, so little additional information is gained from the second measurement. We can check if this is the case by making a scatterplot of strength vs. flexibility (Figure 7.1) or by fitting a linear model to see how well one predicts the other.

```
> summary(lm(rFlexibility~rStrength, punting))
< 8 lines removed >
Coefficients:
            Estimate Std. Error t value Pr(>|t|)
(Intercept)  56.5630     9.7357    5.81  0.00012
rStrength     0.2649     0.0652    4.06  0.00187


Residual standard error: 5.15 on 11 degrees of freedom
Multiple R-squared:  0.6,        Adjusted R-squared: 0.564
F-statistic: 16.5 on 1 and 11 DF,  p-value: 0.00187


> punt.plot <- xyplot(rStrength~rFlexibility, punting)
# if all we want is the correlation coefficient, we can get it directly
> r <- cor(punting$rStrength, punting$rFlexibility); r
[1] 0.77468
> r^2
[1] 0.60012
```

punting2

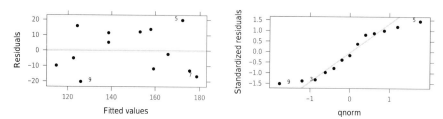

Figure 7.2. Residuals plots for a model predicting distance from leg strength and flexibility show a roughly symmetric distribution with tails that are somewhat shorter than the tails of a normal distribution.

In this case we see that leg strength and flexibility are indeed positively associated. In Sections 7.1.3 and 7.2.1 we will discuss ways to compare this additive model to models that include only one of the two predictors.

Finally, a look at the summary of the residuals suggests that the normality assumption of the model is not met as well as we might like – the residuals appear to be too spread out in the middle relative to the tails of the distribution. We have not yet discussed robustness, so we don't yet know how serious a problem this is. But we can make our diagnostic plots just as we did for simple linear regression using `plot()` or `xplot()`.

```
> resid.plots <- xplot(punting.lm)
```
punting-resid

The residual plots in Figure 7.2 confirm our suspicion from the numerical output – the distribution of the residuals is roughly symmetric but with tails that are somewhat shorter than the tails of a normal distribution. As our experience with other procedures based on normal distributions suggests, this is less problematic than some other types of departure from normality.

Point and interval predictions based on our model can be calculated using `predict()` just as for the simple linear model.

```
> newdata <- data.frame(rStrength=175,rFlexibility=100)
> predict(punting.lm, new=newdata, interval='confidence')
     fit    lwr    upr
1 168.15  152.2 184.10
> predict(punting.lm, new=newdata, interval='prediction')
     fit    lwr    upr
1 168.15 131.00 205.31
```
punting-predict

$\triangleleft$

In the next several sections we will look "under the hood" of the linear model engine to see what is going on behind the scenes when we use `lm()`.

7.1.2. Fitting the Additive Model

The additive model can be fit using either least squares or maximum likelihood. Although we will focus primarily on using R to fit these models and on interpreting the results, it will be important to have a good understanding of what is going on behind the scenes when these computations are being made. Before turning

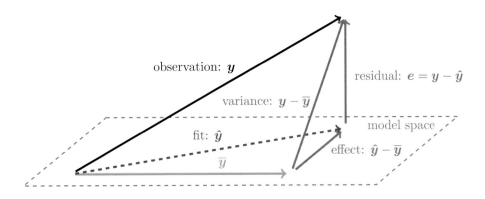

Figure 7.3. Generic linear model.

to a concrete example, it is useful to recall our generic picture of linear models (Figure 7.3) and to see how it applies in this new situation.

Our visualization is again limited in the number of dimensions available for our drawings, so we begin by enumerating the degrees of freedom for this model. Our data live in an n-dimensional space ($\mathbb{R}^n$). Our model space has $k+1$ dimensions where k is the number of predictors: 1 degree of freedom for the overall mean and k degrees of freedom for the k predictors. This leaves $n-k-1$ degrees of freedom for the residuals. Figure 7.4 shows a portion of this figure when $k=2$. In this figure the two degrees of freedom associated with the predictors are indicated by dashed and dotted green vectors. We will refer to this as the recentered model because it can be obtained by subtracting $\overline{y}$ from all of the observed response values. Alternatively, we can think of rotating the view so that the **1** vector is pointing directly at the observer (and so is invisible).

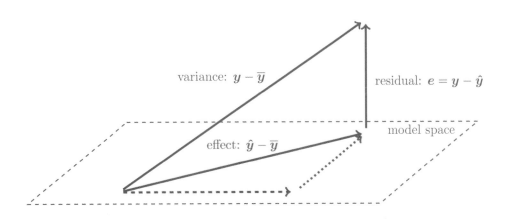

Figure 7.4. Recentered linear model.

Now we look at a concrete example.

Example 7.1.2. The `Devore6` package contains data sets used in [**Dev03**]. The data set `xmp13.13` contains data from an experiment testing the properties of concrete made with different proportions of ingredients. The original report on this study appeared in [**SH96**]. The variables of interest are compressive strength (measured in MPa) of the concrete after 28 days (`strength`), the percentage of limestone (`x1`), and water-cement ratio (`x2`). After creating a new data frame in which we rename the variables `x1` and `x2`, we fit the additive linear model.

```
> require(Devore6); data(xmp13.13, package='Devore6')          concrete-lm1
> concrete <- data.frame(
+      limestone=xmp13.13$x1,
+      water=xmp13.13$x2,
+      strength=xmp13.13$strength)
> concrete.lm1 <- lm(strength ~ limestone + water, concrete)
> concrete.lm1
< 5 lines removed >
(Intercept)     limestone          water
     84.817         0.164        -79.667
```

We can confirm that these parameter estimates are the least squares estimates by doing the linear algebra manually using the formulas we derived for simple linear regression. The only change is that we need a third vector to span the model space now. The model space is now the span of the following three vectors:

$$v_0 = 1 \,,$$
$$v_1 = x_1 - \overline{x_1} \,,$$
$$v_2 = x_2 - \overline{x_2} \,.$$

```
> y <- concrete$strength                                       concrete-lm1a
> n <- length(y); v0 <- rep(1,n)
> v1 <- concrete$limestone - mean(concrete$limestone)
> v2 <- concrete$water - mean(concrete$water)
> project(y,v0,type='v')
[1] 39.317 39.317 39.317 39.317 39.317 39.317 39.317 39.317 39.317
> mean(y)
[1] 39.317
> project(y,v1,type='l') / vlength(v1)
[1] 0.16429
> project(y,v2,type='l') / vlength(v2)
[1] -79.667
```

By adding the three projections together, we obtain the model fits ($\hat{y}$), which we could also obtain using `predict()` or `fitted()`. (If `newdata` is not provided to `predict()`, it provides fits to the predictor values in the original data set.)

```
> y <- concrete$strength                                       concrete-lm1b
> ef0 <- project(y,v0,type='v')
> ef1 <- project(y,v1,type='v')
> ef2 <- project(y,v2,type='v')
```

```
> ef0 + ef1 + ef2
[1] 36.483 44.450 34.183 42.150 41.617 37.017 31.350 47.283 39.317
> predict(concrete.lm1)
     1      2      3      4      5      6      7      8      9
36.483 44.450 34.183 42.150 41.617 37.017 31.350 47.283 39.317
> fitted(concrete.lm1)
     1      2      3      4      5      6      7      8      9
36.483 44.450 34.183 42.150 41.617 37.017 31.350 47.283 39.317
```

Actually, the methods used above will not work in general. Our example is special because $v_1 \perp v_2$:

```
> dot(v1,v2)                                              concrete-lm1c
[1] 0
```

So in this case the dotted and dashed components of Figure 7.4 correspond exactly to the projections in the directions of v_1 and v_2. This will only be the case in carefully designed experiments. As we will see, orthogonal designs make statistical inference and interpretation much simpler. Unfortunately, orthogonal designs are not always possible. ◁

We can explore the effects of orthogonality by removing one observation from our concrete data set.

Example 7.1.3.

Q. Suppose we remove the first observation from the `concrete` data frame. How does this affect our analysis of the concrete data?

A. We begin by imitating our method from the previous example.

```
# modify data by dropping first observation                concrete-minus
> concretemod <- concrete[-1,]
> concrete.lmmod <- lm(strength ~ limestone + water, concretemod)
> coef(concrete.lmmod)
(Intercept)    limestone        water
   80.52949      0.21264    -72.89744
> y <- concretemod$strength
> n <- length(y); v0 <- rep(1,n)
> v1 <- concretemod$limestone - mean(concretemod$limestone)
> v2 <- concretemod$water - mean(concretemod$water)
> project(y,v0,type='v')
[1] 40.038 40.038 40.038 40.038 40.038 40.038 40.038 40.038
> mean(y)
[1] 40.038
> project(y,v1,type='l') / vlength(v1)
[1] 0.2665
> project(y,v2,type='l') / vlength(v2)
[1] -75.977
> ef0 <- project(y,v0,type='v')
> ef1 <- project(y,v1,type='v')
> ef2 <- project(y,v2,type='v')
> ef0 + ef1 + ef2
[1] 45.460 34.131 41.729 43.527 36.065 32.198 47.394 39.796
```

```
> predict(concrete.lmmod)
     2      3      4      5      6      7      8      9
44.901 34.635 41.924 42.745 36.791 32.478 47.058 39.768
```

Now nothing matches quite right. This is because $v_1 \not\perp v_2$. Notice, however, that v_1 and v_2 remain orthogonal to v_0. (This is true in general and you are asked to prove this in Exercise 7.1.)

concrete-minus-dot

```
> dot(v0,v1)
[1] 0
> dot(v0,v2)
[1] 0
> dot(v1,v2)
[1] -0.39375
```

The fix is to adjust v_1 and v_2 so that we get an orthogonal pair that span the same space. We can do this by subtracting from v_1 or v_2 the portion that is the direction of the other, leaving only an orthogonal component remaining.

concrete-minus-adj

```
> w1 <- v1 - project(v1,v2)
> w2 <- v2 - project(v2,v1)
> dot(v0,w1)
[1] -1.3323e-15
> dot(v0,w2)
[1] -1.1276e-17
> dot(v1,w2)
[1] -1.2143e-16
> dot(w1,v2)
[1] 7.546e-17
```

We can fit the model one of two ways. We can use v_0, v_1, and w_2 or we can use v_0, w_1, and v_2. Each of these orthogonal sets of vectors spans the same space as v_0, v_1, and v_2.

concrete-minus2

```
> y <- concretemod$strength
# make fits using v1 and w2
> ef0 <- project(y,v0,type='v')
> ef1 <- project(y,v1,type='v')
> ef2 <- project(y,w2,type='v')
> ef0 + ef1 + ef2
[1] 44.901 34.635 41.924 42.745 36.791 32.478 47.058 39.768
# now try w1 and v2
> ef0 <- project(y,v0,type='v')
> ef1 <- project(y,w1,type='v')
> ef2 <- project(y,v2,type='v')
> ef0 + ef1 + ef2
[1] 44.901 34.635 41.924 42.745 36.791 32.478 47.058 39.768
# should match what lm() produces
> predict(concrete.lmmod)
     2      3      4      5      6      7      8      9
44.901 34.635 41.924 42.745 36.791 32.478 47.058 39.768
```

Now we turn our attention to estimating the coefficients of the model. The following computations reveal an important feature of the fit for this model.

`concrete-minus3`

```
# using v1 gives coefficient in model with only limestone as a predictor
> project(y,v1,type='l') / vlength(v1)
[1] 0.2665
> coef(lm(strength ~ limestone, concretemod))
(Intercept)    limestone
    36.5397       0.2665
# using v2 gives coefficient in model with only water as a predictor
> project(y,v2,type='l') / vlength(v2)
[1] -75.977
> coef(lm(strength ~ water, concretemod))
(Intercept)        water
     85.149      -75.977
# using w1 and w2 gives coefficients in the additive model
> project(y,w1,type='l') / vlength(w1)
[1] 0.21264
> project(y,w2,type='l') / vlength(w2)
[1] -72.897
> coef(concrete.lmmod)
(Intercept)    limestone          water
   80.52949      0.21264      -72.89744
```

As this shows, the coefficients are fit using $\boldsymbol{w}_1$ and $\boldsymbol{w}_2$. (See Lemma C.2.4 for the reason why.) Since $\boldsymbol{w}_1$ depends on $\boldsymbol{v}_2$ and $\boldsymbol{w}_2$ depends on $\boldsymbol{v}_1$, this means that the coefficient on one predictor depends on which other predictors are in a model. ◁

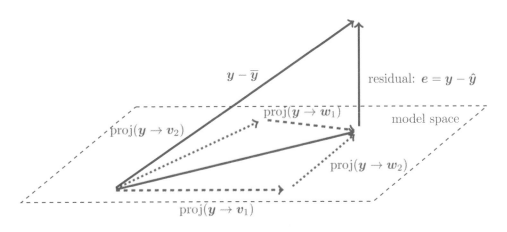

Figure 7.5. Non-orthogonal predictors. In this case $\mathrm{proj}(\boldsymbol{y} \to \boldsymbol{v}_i) \neq \mathrm{proj}(\boldsymbol{y} \to \boldsymbol{w}_i)$.

7.1.3. Model Comparison Tests

Most of the inferences for the linear model can be thought of as examples of the **model comparison test**. As its name implies, the model comparison test performs a test by comparing two models. One of these models must be a submodel of the other, by which we mean that its model space is a subspace of the larger model's model space. The goal of the test is to tell us whether we should prefer the larger model Ω over the smaller model ω.

The null hypothesis associated with a model comparison test describes how to restrict the larger model so that it becomes equivalent to the smaller model. In the first two examples below, ω is obtained by simply deleting terms from Ω (i.e., setting parameters equal to 0), but in the third and fourth examples ω is obtained from Ω by making a different kind of restriction on the model.

Example 7.1.4. A number of interesting tests can be expressed using model comparisons. Here are a few examples.

(1) Testing a single parameter.

$$\Omega : \ \mathrm{E}(Y) = \beta_0 + \beta_1 x_1 + \beta_2 x_2 \,,$$
$$\omega : \ \mathrm{E}(Y) = \beta_0 + \beta_1 x_1 \,.$$

This is equivalent to testing $H_0 : \beta_2 = 0$ in the larger model.

(2) Model utility test.

$$\Omega : \ \mathrm{E}(Y) = \beta_0 + \beta_1 x_1 + \beta_2 x_2 \,,$$
$$\omega : \ \mathrm{E}(Y) = \beta_0 \,.$$

This is the **model utility test** of $H_0 : \beta_1 = 0$ and $\beta_2 = 0$ in the larger model. If H_0 is true, then the predictors x_1 and x_2 do not actually help us predict the response.

(3) Testing a constraint on the model.

$$\Omega : \ \mathrm{E}(Y) = \beta_0 + \beta_1 x_1 + \beta_2 x_2 \,,$$
$$\omega : \ \mathrm{E}(Y) = \beta_0 + \beta_1 x_1 + \beta_1 x_2$$
$$= \beta_0 + \beta_1 (x_1 + x_2) \,.$$

Comparing these models tests the hypothesis $H_0 : \beta_1 = \beta_2$ in the larger model.

(4) A complex hypothesis.

$$\Omega : \ \mathrm{E}(Y) = \beta_0 + \beta_1 x_1 + \beta_2 x_2 + \beta_3 x_3 \,,$$
$$\omega : \ \mathrm{E}(Y) = \beta_0 + \beta_1 x_1 + \beta_1 x_2$$
$$= \beta_0 + \beta_1 (x_1 + x_2) \,.$$

Comparing these models tests the hypothesis $H_0 : \beta_1 = \beta_2$ and $\beta_3 = 0$ in the larger model.   ◁

The dimension of a model is the dimension of the model space as a vector space and is equal to the number of free parameters in the model. The dimensions of the models in Example 7.1.4 are given in the Table 7.1.

Table 7.1. Dimensions of models and submodels from Example 7.1.4.

example	$\dim \Omega$	$\dim \omega$	$\dim \Omega - \dim \omega$
1	3	2	1
2	3	1	2
3	3	2	1
4	4	2	2

Since the smaller model has a smaller model space, its best fit is not as good as the best fit in the larger model, where we have more room to maneuver and get close to the observed response vector $\boldsymbol{y}$. This means that SSE_Ω (the sum of the squares of the residuals in the larger model) will be smaller than SSE_ω (the sum of the squares of the residuals in the smaller model). If the the difference $SSE_\omega - SSE_\Omega$ is small, then the larger model does not fit much better than the smaller, and we prefer the simpler model. On the other hand, if this difference is large, the more complicated model is justified because of its much better fit. This is illustrated in Figure 7.6.

But how large is large? To answer this, we observe that the same reasoning used in Section 6.3.6 demonstrates that

$$\frac{SSE_\Omega}{\sigma^2} \sim \mathsf{Chisq}(n - \dim \Omega) , \tag{7.1}$$

and when H_0 is true, then

$$\frac{SSE_\omega - SSE_\Omega}{\sigma^2} \sim \mathsf{Chisq}(\dim \Omega - \dim \omega) . \tag{7.2}$$

So when H_0 is true,

$$F = \frac{(SSE_\omega - SSE_\Omega)/(\dim \Omega - \dim \omega)}{SSE_\Omega/(n - \dim \Omega)} \sim \mathsf{F}(\dim \Omega - \dim \omega, n - \dim \Omega) . \tag{7.3}$$

We can express (7.3) in a slightly different way that corresponds more directly to the output R produces when we perform a model comparison test. First, we will use RSS (residual sum of squares) in place of SSE. Then, since the residual degrees of freedom (rdf) of a model M satisfy

$$\mathrm{rdf}\, M = n - \dim M ,$$

(7.3) is equivalent to

$$F = \frac{(RSS_\omega - RSS_\Omega)/(\mathrm{rdf}\,\omega - \mathrm{rdf}\,\Omega)}{RSS_\Omega/\mathrm{rdf}\,\Omega} \sim \mathsf{F}(\mathrm{rdf}\,\omega - \mathrm{rdf}\,\Omega, \mathrm{rdf}\,\Omega) . \tag{7.4}$$

If this reminds you of a likelihood ratio test, it should. This test is algebraically equivalent to one derived as a likelihood ratio test. The advantage of the geometric approach is that it provides us with the exact distribution of the test statistic (assuming the model assumptions are true).

It's time for another concrete example of all this.

Example 7.1.5. Returning to the original concrete data set from Example 7.1.2 (the one with orthogonal predictor vectors), we can use the `anova()` function to compute model comparison tests.

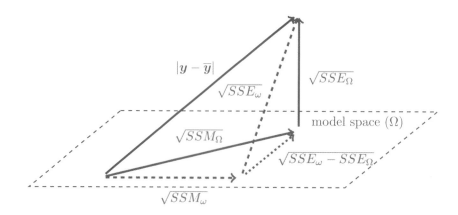

Figure 7.6. Model comparison tests. The length of each vector is indicated.

First we define several models.

```
> concrete.lm0 <- lm(strength ~ limestone + water, concrete)      concrete-models
> concrete.lm1 <- lm(strength ~ limestone, concrete)
> concrete.lm2 <- lm(strength ~ water, concrete)
> concrete.lm3 <- lm(strength ~ 1, concrete)
> concrete.lm4 <- lm(strength ~ I(limestone + water), concrete)
> concrete.lm5 <- lm(strength ~ -1 + limestone + water, concrete)
```

We are required to use I() in model 4 to distinguish it from model 1. In general, any arithmetic that could be interpreted as part of a model formula must be wrapped with I().

The submodel relationships among these models can be displayed graphically as follows:

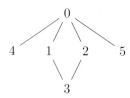

We can perform a model comparison test between any nested pair. Here are some examples.

```
> anova(concrete.lm1,concrete.lm0)                               concrete-mct
Analysis of Variance Table

Model 1: strength ~ limestone
Model 2: strength ~ limestone + water
  Res.Df   RSS Df Sum of Sq     F Pr(>F)
1      7 262.7
2      6  72.3  1       190 15.8 0.0073
> anova(concrete.lm2,concrete.lm0)
Analysis of Variance Table
```

```
Model 1: strength ~ water
Model 2: strength ~ limestone + water
  Res.Df  RSS Df Sum of Sq     F Pr(>F)
1      7 88.1
2      6 72.3  1       15.9 1.32   0.29
> anova(concrete.lm3,concrete.lm0)
Analysis of Variance Table

Model 1: strength ~ 1
Model 2: strength ~ limestone + water
  Res.Df   RSS Df Sum of Sq    F Pr(>F)
1      8 278.5
2      6  72.3  2      206 8.56  0.017
> anova(concrete.lm4,concrete.lm0)
Analysis of Variance Table

Model 1: strength ~ I(limestone + water)
Model 2: strength ~ limestone + water
  Res.Df   RSS Df Sum of Sq    F Pr(>F)
1      7 263.4
2      6  72.3  1      191 15.9 0.0072
> anova(concrete.lm5,concrete.lm0)
Analysis of Variance Table

Model 1: strength ~ -1 + limestone + water
Model 2: strength ~ limestone + water
  Res.Df RSS Df Sum of Sq  F  Pr(>F)
1      7 650
2      6  72  1      578 48 0.00045
```

Each of these model comparison tests is summarized in a table with the following structure:

model	Res. Df	RSS	df	Sum of Sq	F	Pr(>F)
ω	rdf ω	RSS_ω				
Ω	rdf Ω	RSS_Ω	rdf ω $-$ rdf Ω	$RSS_\omega - RSS_\Omega$	F	1 - pf$(F, \cdot, \cdot)$

We are now in a position to explain more fully the p-values that appear in the output that the summary() function produces from an additive model.

concrete-summary

```
> summary(concrete.lm0)
< 8 lines removed >
Coefficients:
            Estimate Std. Error t value Pr(>|t|)
(Intercept)   84.817     12.242    6.93  0.00045
limestone      0.164      0.143    1.15  0.29467
water        -79.667     20.035   -3.98  0.00731

Residual standard error: 3.47 on 6 degrees of freedom
Multiple R-squared: 0.741,       Adjusted R-squared: 0.654
F-statistic: 8.56 on 2 and 6 DF,  p-value: 0.0175
```

Each of the four p-values appearing in this output corresponds to a model comparison test where Ω is the full model and ω is a submodel obtained by deleting one term as in part (1) of Example 7.1.4. Since $\dim \Omega - \dim \omega = \mathrm{rdf}\,\omega - \mathrm{rdf}\,\Omega = 1$, R reports the t statistic with $\mathrm{rdf}\,\Omega$ degrees of freedom rather than the F statistic with 1 and $\mathrm{rdf}\,\Omega$ degrees of freedom. Recall that $t^2 = F$.

Here are some additional details about the output of `summary()`.

(1) This small data set has 9 observations. So the degrees of freedom are partitioned into 1 for the intercept, $k = 2$ for the two predictors, and $n - k - 1 = 6$ for the residuals (in Ω).

(2) The residual standard error is computed using $\mathrm{rdf}\,\Omega = n - k - 1$ degrees of freedom. Just as was the case for the simple linear model,

$$\frac{|\boldsymbol{E}|^2}{\sigma^2} = \frac{\sum E_i^2}{\sigma^2} = \frac{SSE}{\sigma^2} = \frac{RSS}{\sigma^2} \sim \mathsf{Chisq}(\mathrm{rdf}\,\Omega) \, . \tag{7.5}$$

Among other things, this implies that

$$\mathrm{E}\left(\frac{SSE}{\mathrm{rdf}\,\Omega}\right) = \frac{\sigma^2}{\mathrm{rdf}\,\Omega}\,\mathrm{E}(SSE/\sigma^2) = \sigma^2 \, ,$$

so we have found an unbiased estimator of σ^2:

$$\hat{\sigma}^2 = S^2 = \frac{SSE}{\mathrm{rdf}\,\Omega} \, .$$

This is the "residual standard error on 6 degrees of freedom" reported by R.

(3) For each coefficient (the intercept coefficient and coefficients associated with the `limestone` and `water` variables) we see the estimate, standard error, and results of the hypothesis tests

$$H_0 : \beta_i = 0 \text{ vs. } H_a : \beta_i \neq 0 \, .$$

We can compute confidence intervals for the parameters of the model using the standard errors listed in the R output:

$$\hat{\beta}_i \pm t_* SE \, .$$

We use $\mathrm{rdf}_\Omega = n - k - 1$ degrees of freedom to determine t_*, since that is the degrees of freedom associated with the residuals used to estimate σ. Alternatively, we can let R calculate the confidence intervals for us using `confint()`.

```
> confint(concrete.lm0)                                          concrete-confint1
                2.5 %     97.5 %
(Intercept)   54.86270 114.77064
limestone     -0.18588   0.51445
water       -128.69036 -30.64298
```

One must interpret these tests and confidence intervals in the context of the models involved. A small p-value for $H_0 : \beta_i = 0$ indicates that *even when the other predictors are included in the model*, β_i still has more influence on the response than would be expected by chance alone. Conversely, failure to reject H_0 does not necessarily mean that a predictor is not useful, but may indicate that the predictive power of this variables is largely contained in the other predictors in the model. We'll come back to this important idea in Section 7.2.4.

In our concrete example, both the confidence intervals and the p-values suggest that water is an important predictor but that limestone may not add much information to a model that already includes water as a predictor.

(4) $r^2 = \dfrac{SSM}{SSE + SSM}$.

In our example, $r^2 = 0.7406$, so our model (additive effects of limestone and water) explains approximately 74% of the batch-to-batch variation in the strength of concrete.

SSE and SSM are computed exactly as in simple linear regression:

$$SSE = \sum(y_i - \hat{y}_i)^2 = |\boldsymbol{y} - \hat{\boldsymbol{y}}|^2 \text{ , and}$$

$$SSM = \sum(\hat{y}_i - \overline{y}_i)^2 = |\hat{\boldsymbol{y}} - \overline{\boldsymbol{y}}|^2 \text{ .}$$

(5) The F statistic and p-value come from the **model utility test**.

In our example, the p-value for this test is 0.01746, giving us fairly strong evidence against the null hypothesis. Our model does appear to have at least some utility in predicting concrete strength. That is, it appears that we can do better using our model and knowing the amount of limestone and water used than if we did not know the amount of limestone and water used.   ◁

7.2. Assessing the Quality of a Model

In this section we present several diagnostic tools for assessing models with multiple predictors. Several of these build on the diagnostics we have already used for the simple linear regression model.

7.2.1. Adjusted r^2 and Akaike's Information Criterion

The coefficient of determination: r^2

As in simple linear regression, $r^2 = \frac{SSM}{SST}$ measures the fraction of total variation explained by the model. A larger value of r^2 indicates that more of the variation is explained by the model, so generally bigger is better. But there are some difficulties in interpreting this number:

- The absolute size of a "good" r^2 is highly dependent on the area of application. If there is a lot of variation in the population at each possible setting of the explanatory variables, then it is impossible to get high r^2 values.

- If we add parameters to a model, r^2 will always increase. (Technically it could stay the same, but it can't decrease.) So when comparing models with different numbers of parameters, the model with more parameters has an edge.

The `adjusted R-squared` shown in R output is one way to allow more equitable comparisons among models with different numbers of parameters. Recall that

$$r^2 = \frac{SSM}{SST} = 1 - \frac{SSE}{SST} = 1 - \frac{SSE/n}{SST/n} \; .$$

The adjusted r^2 replaces n in right-hand expression above with the appropriate degrees of freedom:

$$\texttt{Adjusted R-squared} = r_a^2 = 1 - \frac{SSE/(n-k-1)}{SST/(n-1)} = 1 - \frac{SSE/DFE}{SST/DFT} = 1 - \frac{MSE}{MST}$$

where $k + 1$ is the number of parameters in the model. This expression is not as easily interpreted as the original, and it is possible for this expression to be negative, but it allows for better comparison of models with different numbers of parameters.

AIC: Akaike's information criterion

Akaike's information criterion (AIC) is a measure that was first proposed in the 1970s for the purpose of comparing different models that are fit to the same data [**Aka74**]. Akaike's information criterion is defined by

$$AIC = 2k + n\log(RSS) + C \tag{7.6}$$

where k is the number of parameters, RSS is the residual sum of squares (SSE), and C is a constant that depends on the data set but not on the model. Different authors and software prefer different choices for C.

When interpreting AIC, smaller is better. There are theoretical reasons for this particular formula (based on likelihood methods), but notice that the first addend increases with each parameter in the model and the second decreases (because the fit improves); so this forms a kind of balance between the expected gains for adding new parameters against the costs of complication and potential for over-fitting. The scale of AIC is only meaningful relative to a fixed data set, so some authors will shift AIC by a constant intended to make such comparisons simpler.

The `AIC()` and `extractAIC()` functions will compute AIC for models fit with `lm()`. The values returned by these functions do not agree with each other or with (7.6) because they each select a different value of C, but the differences between values for different models fit using the same data set do agree, and that suffices for the intended use of AIC.

```
# these two methods give different numerical values          concrete-aic
> AIC(concrete.lm0)
[1] 52.287
> AIC(concrete.lm1)
[1] 61.903
> extractAIC(concrete.lm0)
[1]  3.000 24.746
> extractAIC(concrete.lm1)
[1]  2.000 34.363
# and neither agrees with our definition
> aic0 <- 2 * 3 + 9 * log(sum(resid(concrete.lm0)^2)); aic0
[1] 44.521
```

```
> aic1 <- 2 * 2 + 9 * log(sum(resid(concrete.lm1)^2)); aic1
[1] 54.138
# but differences between models are equivalent
> aic0 - aic1
[1] -9.6162
> AIC(concrete.lm0) - AIC(concrete.lm1)
[1] -9.6162
> extractAIC(concrete.lm0)[2] - extractAIC(concrete.lm1)[2]
[1] -9.6162
```

AIC is the method used by `step()` for comparing models for the purpose of stepwise regression (see Section 7.6.2). Several other criteria have also been proposed for comparing models.

7.2.2. Checking Model Assumptions

Looking at residuals

In models with two (or more) predictors, residuals become especially important in checking model assumptions. In the case of two predictors, we could conceivably attempt to plot our data in three dimensions (a 3-dimensional scatterplot is sometimes called a **cloud plot** and the `lattice` function `cloud()` will generate such a plot). If we did this, we would be looking to see if our data appear to be randomly scattered about some *plane*. It is easier, however, to inspect the residuals using various 2-dimensional plots.

Example 7.2.1. The `plot()` and `xplot()` functions can automate various residual plots for us. Alternatively, we can plot the residuals ourselves using functions like `resid()` and `fitted()` applied to our model.

concrete-plot1

```
> plot1 <- xplot(concrete.lm0,which=2)
> plot2 <- xplot(concrete.lm0,which=3)
> plot3 <- xplot(concrete.lm0,which=5,add.smooth=FALSE)
> plot4 <- xyplot(resid(concrete.lm1) ~ fitted(concrete.lm1),
+               main = "residuals vs fits",
+               ylab = "residuals",
+               xlab = "fitted values",
+               sub  = "lm(strength ~ limestone + water)" )
```

These residual plots appear in Figure 7.7. Observation 4 does not fit as well as we would like, but it is difficult to assess these plots with such a small data set. Perhaps we should at least check that there were no errors in recording observation 4. Alternatively, we could fit the model again without that observation to see how much things change. Ideally, we would like more data at each combination of water and limestone tested.   ◁

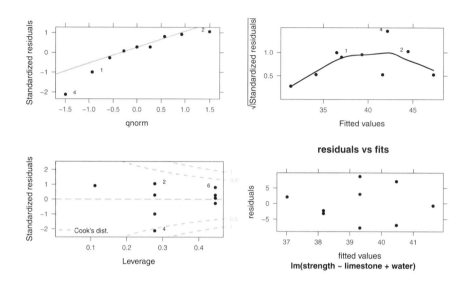

Figure 7.7. Residual plots for the model `lm(strength limestone+water,concrete)`

Additivity

The additive model also makes the assumption that the various effects of the model are additive. One good way to assess the additivity assumption is to fit a non-additive model as well and compare. We will learn how to do that soon. But there are graphical methods that are also useful and that illustrate what the additivity assumption entails.

Consider a model like

$$E(Y) = \beta_0 + \beta_1 x_1 + \beta_2 x_2 \ .$$

For a fixed value of x_1, this becomes

$$E(Y) = (\beta_0 + \beta_1 x_1) + \beta_2 x_2 \ ,$$

where the parenthesized expression is a constant. For different values of x_1, the dependence on x_2 is always linear *with the same slope*. The intercept varies with x_1. In other words, according to the additive model, observed responses should cluster about a set of *parallel lines*, one for each possible value of x_1. Similar statements occur with the roles of the variables reversed. The following simulation illustrates this graphically.

Example 7.2.2. In the code below, we begin by generating some random data according to the model

$$E(Y) = 1 + 2x_1 - 4x_2 \ .$$

Then we generate plots (Figure 7.8) showing first the model relationship and then simulated data with separate regression lines for each value of one of the predictors.

```
> x1 = rep(rep(1:4, each=4),4);   x2 = rep(rep(1:4, times=4),4)     additive
> fit = 1 + 2 * x1 - 4 * x2; e = rnorm(4*4*4,sd=1)
```

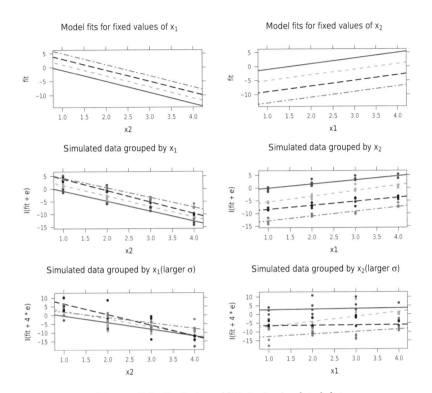

Figure 7.8. Exploring additivity in simulated data.

```
> plot1 <- xyplot(fit~x2,groups=x1,type='r',
+       main=expression(paste("Model fits for fixed values of ",x[1])))
> plot2 <- xyplot(fit~x1,groups=x2,type='r',
+       main=expression(paste("Model fits for fixed values of ",x[2])))
> plot3 <- xyplot(I(fit+e)~x2,groups=x1,type=c('p','r'),pch=16,
+       main=expression(paste("Simulated data grouped by ",x[1])))
> plot4 <- xyplot(I(fit+e)~x1,groups=x2,type=c('p','r'),pch=16,
+       main=expression(paste("Simulated data grouped by ",x[2])))
> plot5 <- xyplot(I(fit+4*e)~x2,groups=x1,type=c('p','r'),pch=16,
+       main=expression(paste("Simulated data grouped by ",x[1],
+               "(larger ", sigma,")")))
> plot6 <- xyplot(I(fit+4*e)~x1,groups=x2,type=c('p','r'),pch=16,
+       main=expression(paste("Simulated data grouped by ",x[2],
+               "(larger ", sigma,")")))
```

As we see, this sort of plot becomes more difficult to interpret as σ becomes larger because we cannot fit the linear relationship as accurately. Graphical assessment is also challenging if there are not repeated observations with the same predictors. Nevertheless, it is important to keep in mind what the additivity assumption is when fitting an additive model.   ◁

7.2.3. Partial Residual Plots

Partial residual plots are one way of determining whether we should try adding another predictor to our model. The idea is very simple, we plot the residuals of our model against a potential predictor not already in the model.

```
xyplot( resid(model) ~ somePredictor )
```

If the plot reveals a pattern, this is an indication that the new predictor may be useful in improving the fit of the model, since it is able to predict when the fitted value of the model is too large or too small. The shape of the pattern can also suggest transformations of the new predictor.

Example 7.2.3. The `utilities` data set was introduced in Exercise 1.21. It contains records of utilities bills at a residence over a number of years. The `utilities2` data set includes some additional calculated variables, including

$$\text{kwhpday} = \text{kwh/billingDays} .$$

In Exercise 6.32 you were asked to fit a model predicting `thermsPerDay` from `temp`. We will remove the first few months (where there was an erroneous meter reading) and the warm months (when the home is not heated) before fitting the model.

```
                                                                    utilities-lm1
# subset the data:
#   * remove first few months where there appears to have been a bad
#       meter reading (year == 2000 & month <= 6)
#   * remove months where there is little need to heat (temp > 60)
> ut <- subset(utilities2, subset=(year > 2000 | month > 6) & temp <= 60)
> ut.lm1 <- lm(thermsPerDay ~ temp, ut)
> summary(ut.lm1)
< 8 lines removed >
Coefficients:
            Estimate Std. Error t value Pr(>|t|)
(Intercept)  9.10698    0.14575    62.5   <2e-16
temp        -0.14041    0.00374   -37.6   <2e-16

Residual standard error: 0.458 on 68 degrees of freedom
Multiple R-squared: 0.954,        Adjusted R-squared: 0.953
F-statistic: 1.41e+03 on 1 and 68 DF,  p-value: <2e-16

> ut.plot <- xplot(ut.lm1)
```

The residual plots in Figure 7.9 suggest that there may be some heteroskedasticity and a mild right skew.

Now we want to see if we can improve the model by adding another variable, `kwhpday`. We begin my making a partial residual plot (Figure 7.10).

```
                                                                    utilities-presid
# partial residual plot
> ut.presid <- xyplot(resid(ut.lm1) ~ kwhpday, ut, type=c('p','r'))
```

There is a reasonably strong linear relationship between the residuals of the original model and `kwhpday`. This indicates that `kwhpday` explains some of the unexplained variation of the original model, so we proceed to fit a model that includes `kwhpday` as a predictor.

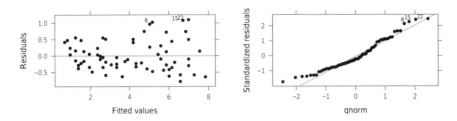

Figure 7.9. Residual plots for a model predicting `thermsPerDay` from `temp`.

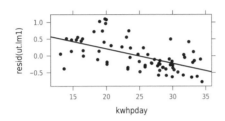

Figure 7.10. A partial residual plot.

```
> ut.lm2 <- lm(thermsPerDay ~ temp + kwhpday, ut)
> summary(ut.lm2)
< 8 lines removed >
Coefficients:
            Estimate Std. Error t value Pr(>|t|)
(Intercept) 10.37881    0.24501   42.36  < 2e-16
temp        -0.14443    0.00312  -46.27  < 2e-16
kwhpday     -0.04538    0.00764   -5.94  1.1e-07

Residual standard error: 0.373 on 67 degrees of freedom
Multiple R-squared: 0.97,        Adjusted R-squared: 0.969
F-statistic: 1.08e+03 on 2 and 67 DF,  p-value: <2e-16

> ut.plot2 <- xplot(ut.lm2)
```

utilities-kwh

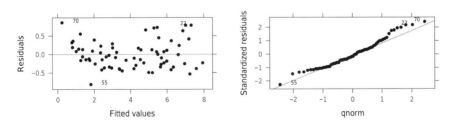

Figure 7.11. Residual plots for an additive model predicting `thermsPerDay` from `temp` and `kwhpday`.

The resulting fit seems better in every regard than the original but there is still some skew to the residuals. Both r^2 and adjusted r^2 have increased substantially, and the heteroskedasticity has been much improved. The negative coefficient on `kwhpday` indicates that among months with the same mean temperature, less gas is used in months when more electricity is used. ◁

7.2.4. More Examples

Before considering categorical predictors, let's take a look at some more examples with multiple quantitative predictors.

Example 7.2.4. The `gpa` data set contains SAT scores, ACT scores, and college GPAs for 271 students at a 4-year institution. We begin by fitting an additive model for GPA with the other variables as predictors.

```
                                                                          gpa1
> gpa.lm <- lm(gpa~satm+satv+act,gpa)
> summary(gpa.lm)
< 8 lines removed >
Coefficients:
            Estimate Std. Error t value Pr(>|t|)
(Intercept) 1.229589   0.196803    6.25  1.6e-09
satm        0.000691   0.000432    1.60   0.1104
satv        0.001223   0.000472    2.59   0.0100
act         0.033960   0.012511    2.71   0.0071

Residual standard error: 0.422 on 267 degrees of freedom
Multiple R-squared: 0.326,      Adjusted R-squared: 0.318
F-statistic:   43 on 3 and 267 DF,  p-value: <2e-16
```

Interestingly, SATV (the verbal component of the SAT score) and ACT seem to be useful predictors, but SATM does not. But keep in mind that these p-values must be interpreted in the context of the larger model. By itself, SATM does appear to have some predictive value.

```
                                                                          gpa2
> gpa.lm1<- lm(gpa~satm,gpa)
> summary(gpa.lm1)
< 8 lines removed >
Coefficients:
            Estimate Std. Error t value Pr(>|t|)
(Intercept) 1.703622   0.194149    8.77  < 2e-16
satm        0.002647   0.000309    8.58  7.7e-16

Residual standard error: 0.454 on 269 degrees of freedom
Multiple R-squared: 0.215,      Adjusted R-squared: 0.212
F-statistic: 73.6 on 1 and 269 DF,  p-value: 7.7e-16
```

In fact, the p-value here is very small. Does this mean that SATM is a better predictor than the larger model? No, the p-values must be interpreted correctly and are not directly comparable in this way. Notice that the larger model has a larger adjusted R^2 value and has a smaller p-value for the model utility test. These are indicators that the larger model does a better job of predicting GPA.

In this case it appears that SATV and ACT can improve our predictions even when the other variables are in the model. The case for adding SATM to the model *when the other variables are available* is much weaker. But if SATM is all we have, we are better off using it than ignoring it.

So why might it be that a variable that is useful by itself is no longer useful in the context of a larger model? One reason is that the information contained in that variable may be contained in the other variables. As the output below indicates, ACT is a good predictor for SATM, even if SATV is also in the model.

gpa3

```
> gpa.lm2<- lm(satm~satv+act,gpa); summary(gpa.lm2)
< 8 lines removed >
Coefficients:
            Estimate Std. Error t value Pr(>|t|)
(Intercept) 184.4413    25.4722    7.24  4.7e-12
satv         -0.0119     0.0667   -0.18    0.86
act          16.0959     1.4726   10.93  < 2e-16

Residual standard error: 59.7 on 268 degrees of freedom
Multiple R-squared: 0.558,       Adjusted R-squared: 0.555
F-statistic:  169 on 2 and 268 DF,  p-value: <2e-16
> gpa.lm3<- lm(satm~satv,gpa); summary(gpa.lm3)
< 8 lines removed >
Coefficients:
            Estimate Std. Error t value Pr(>|t|)
(Intercept) 267.012     29.195     9.15   <2e-16
satv          0.579      0.047    12.33   <2e-16

Residual standard error: 71.7 on 269 degrees of freedom
Multiple R-squared: 0.361,       Adjusted R-squared: 0.359
F-statistic:  152 on 1 and 269 DF,  p-value: <2e-16
> gpa.lm4<- lm(satm~act,gpa); summary(gpa.lm4)
< 8 lines removed >
Coefficients:
            Estimate Std. Error t value Pr(>|t|)
(Intercept) 183.020     24.145     7.58  5.6e-13
act          15.884      0.862    18.43  < 2e-16

Residual standard error: 59.6 on 269 degrees of freedom
Multiple R-squared: 0.558,       Adjusted R-squared: 0.556
F-statistic:  340 on 1 and 269 DF,  p-value: <2e-16
```

Linear relationships among the predictors is referred to as **collinearity** (or multicollinearity) and can make interpretation of the model difficult. It is important to understand the context in which various tests are conducted. It does not make sense to say our null hypothesis is $H_0 : \beta_i = 0$ without specifying the larger context, since in one context a predictor may be highly significant, and in another not significant at all.

ACT scores do not do as good a job predicting SATV scores, so SATV is useful additional information even when ACT scores (and SATM scores) are in the model.

```
> gpa.lm5<- lm(gpa~act+satv,gpa); summary(gpa.lm5)            gpa4
< 8 lines removed >
Coefficients:
            Estimate Std. Error t value Pr(>|t|)
(Intercept) 1.357092   0.180508    7.52  8.4e-13
act         0.045087   0.010436    4.32  2.2e-05
satv        0.001215   0.000473    2.57    0.011

Residual standard error: 0.423 on 268 degrees of freedom
Multiple R-squared: 0.319,        Adjusted R-squared: 0.314
F-statistic: 62.9 on 2 and 268 DF,  p-value: <2e-16
> gpa.lm6<- lm(satv~act,gpa); summary(gpa.lm6)
< 8 lines removed >
Coefficients:
            Estimate Std. Error t value Pr(>|t|)
(Intercept)  119.611     22.095    5.41  1.4e-07
act           17.872      0.789   22.66  < 2e-16

Residual standard error: 54.5 on 269 degrees of freedom
Multiple R-squared: 0.656,        Adjusted R-squared: 0.655
F-statistic:  513 on 1 and 269 DF,  p-value: <2e-16
```

Exercise 7.6 asks you to do some further analysis of this data set. The **students** data set contains a larger set of similar data. (See Exercise 7.8.)   ◁

Example 7.2.5. The **pheno** data set contains phenotype information for 2333 subjects from the Finland-United States Investigation of NIDDM Genetics (FUSION) study of type 2 diabetes [**SMS+04**]. Among the variables are **waist** (waist circumference in cm), **height** (in cm), and **weight** (in kg). If we model people as approximately cylindrical, that suggests

$$\texttt{weight} = K \cdot \texttt{waist}^2 \cdot \texttt{height}$$

for some constant K (related to the density of people and how elliptical their waists are). We can test this conjecture by fitting an additive model with logarithmic transformations:

$$\log(\texttt{weight}) = \beta_0 + \beta_1 \log(\texttt{waist}) + \beta_2 \log(\texttt{height}) .$$

In this model, $K = e^{\beta_0}$, and β_1 and β_2 are the exponents on **waist** and **height** in the untransformed model. Our question is answered by seeing if $\beta_1 = 2$ and $\beta_2 = 1$ are reasonable, given the data.

```
> pheno.lm <- lm(log(weight) ~ log(waist) + log(height), pheno)    pheno-weight
> summary(pheno.lm)
< 8 lines removed >
Coefficients:
            Estimate Std. Error t value Pr(>|t|)
(Intercept)  -5.1155     0.1458   -35.1   <2e-16
log(waist)    1.0754     0.0121    89.2   <2e-16
log(height)   0.8942     0.0302    29.6   <2e-16

Residual standard error: 0.0726 on 2237 degrees of freedom
```

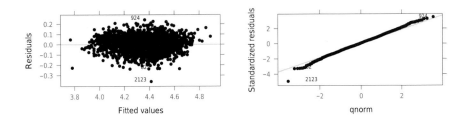

Figure 7.12. Residual plots for the model in Example 7.2.5.

```
(93 observations deleted due to missingness)
Multiple R-squared: 0.842,        Adjusted R-squared: 0.842
F-statistic: 5.95e+03 on 2 and 2237 DF,  p-value: <2e-16
```

The p-values displayed in the output of `summary()` are not the p-values we are interested in. But we can use the output to get the p-values we want since the estimate, standard error, and degrees of freedom are all listed (see Exercise 7.7). Alternatively, we can construct confidence intervals using `confint()`.

```
> confint(pheno.lm)                          pheno-weight-confint
              2.5 %    97.5 %
(Intercept) -5.40139 -4.82953
log(waist)   1.05179  1.09906
log(height)  0.83502  0.95346
```

As was already obvious from a casual glance at the estimates and standard errors, our intuition was not a very good fit for this data. The exponent on waist appears to be only slightly larger than 1 and the exponent on height slightly less than one.

Before leaving this example, we should inspect the residuals

```
> pheno.resid <- xplot(pheno.lm,w=1:2)        pheno-weight-resid
```

The plots are shown in Figure 7.12. Except possibly for a few values in the tails of the distribution, the residual plots look very good.   ◁

7.3. One-Way ANOVA

Regression with a quantitative response and a single categorical predictor is called one-way ANOVA. In ANOVA the predictors are often called **factors**, and especially in experimental situations, the **levels** (possible values) of the factors are often called **treatments**. Individuals with the same treatment are called a **group** or treatment group. This language is easily understood in the context of a medical trial where subjects are put into groups and each group is given a different medical treatment, but the language is used in other contexts as well.

We are primarily interested in whether the different treatments have different effects. That is, we want to know whether all the group means are the same and, if not, how they differ. We begin by exploring some data graphically and numerically.

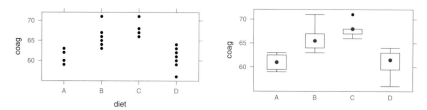

Figure 7.13. Scatterplot and boxplots for `coagulation` data (Example 7.3.1). The scatterplot works better for such a small sample.

Example 7.3.1. The `coagulation` data set in the `faraway` package contains data from an experiment exploring the effects of diet on coagulation in animals. In that experiment, discussed in [**BHH78**], four different diets were fed to the animals and the coagulation time was measured for each animal.

In our data set there is a fair amount of variation in each group, but the amount of variation is quite similar from group to group. Groups A and D have very similar responses. Average coagulation time is higher for both groups B and C, more so for C than for B. All of the coagulation times in group C are longer than any of those in groups A or D. Group B has some overlap with each of the other groups.

```
> data(coagulation,package="faraway")                                    coag-look
> summary(coag~diet,data=coagulation,fun=favstats)
coag      N=24


+-------+-+--+---+---+----+------+------+
|       | |N |min|max|mean|median|sd    |
+-------+-+--+---+---+----+------+------+
|diet   |A| 4|59 |63 |61  |61.0  |1.8257|
|       |B| 6|63 |71 |66  |65.5  |2.8284|
|       |C| 6|66 |71 |68  |68.0  |1.6733|
|       |D| 8|56 |64 |61  |61.5  |2.6186|
+-------+-+--+---+---+----+------+------+
|Overall| |24|56 |71 |64  |63.5  |3.8448|
+-------+-+--+---+---+----+------+------+
> coag.xyplot <- xyplot(coag~diet,coagulation)
> coag.bwplot <- bwplot(coag~diet,coagulation)
```

Both a scatterplot and boxplots appear in Figure 7.13, but there is no reason to use a boxplot with such a small data set. In one of the groups, the boxplot is based on a 5-number summary of 4 observations. In situations like this, it is best to simply display all the data rather than a summary. But for larger data sets, side-by-side boxplots can be a very good way to visualize the relationships between the group means and the amount of variability in the data.

We don't yet have methods to decide whether the differences between some groups are larger than we would expect by chance, even if all the animals were given the same diet. The one-way ANOVA model will help us decide.   ◁

Box 7.2. Notation for One-Way ANOVA

The following notation is used for our discussion of one-way ANOVA:

- n = the total sample size,
- I = the number of groups,
- n_i = the number of observations from group i,
- y_{ij} or Y_{ij} = the jth response from group i,
- $\overline{y}_{i\cdot}$ or $\overline{Y}_{i\cdot}$ = the mean of the sample responses from group i,
- $\overline{y}_{\cdot\cdot}$ or $\overline{Y}_{\cdot\cdot}$ = the mean of all responses in the sample (the grand mean),
- μ_i = the mean population response for group i,
- μ = the overall mean response in the population.

7.3.1. The One-Way ANOVA Model

We will describe the one-way ANOVA model using the notation in Box 7.2. The one-way ANOVA model assumes that the data are independently sampled from populations (one per treatment) that are each normally distributed with the same standard deviation but perhaps with different means. That is,

$$Y_{ij} \overset{\text{iid}}{\sim} \mathsf{Norm}(\mu_i, \sigma) . \tag{7.7}$$

We can bring this model into our linear model framework by introducing a numerical coding scheme for the categorical predictor. For example, if we let

$$x_{1i} = [\![\mathrm{Group} = \mathrm{B}]\!] = \begin{cases} 1 & \text{if observation } i \text{ in group B,} \\ 0 & \text{otherwise,} \end{cases}$$

$$x_{2i} = [\![\mathrm{Group} = \mathrm{C}]\!] = \begin{cases} 1 & \text{if observation } i \text{ in group C,} \\ 0 & \text{otherwise,} \end{cases}$$

$$x_{3i} = [\![\mathrm{Group} = \mathrm{D}]\!] = \begin{cases} 1 & \text{if observation } i \text{ in group D,} \\ 0 & \text{otherwise,} \end{cases}$$

then the model

$$\boldsymbol{Y} = \beta_0 + \beta_1 \boldsymbol{x}_1 + \beta_2 \boldsymbol{x}_2 + \beta_3 \boldsymbol{x}_2 + \boldsymbol{\varepsilon}, \qquad \varepsilon \overset{\text{iid}}{\sim} \mathsf{Norm}(0, \sigma),$$

is equivalent to (7.7), and $\mu_1 = \beta_0$, $\mu_2 = \beta_0 + \beta_1$, $\mu_3 = \beta_0 + \beta_2$, and $\mu_4 = \beta_0 + \beta_3$. This is not the only possible coding scheme, but it is the default coding scheme that `lm()` uses when the predictor is a factor (or something that R coerces to be a factor, as it does with character data).

Example 7.3.2. Returning to Example 7.3.1, we can use `lm()` to fit the model.

```
> coag.model <- lm(coag~diet,coagulation)
> summary(coag.model)
< 8 lines removed >
```

coag-lm

```
Coefficients:
             Estimate Std. Error t value Pr(>|t|)
(Intercept) 6.10e+01   1.18e+00   51.55  < 2e-16
dietB       5.00e+00   1.53e+00    3.27  0.00380
dietC       7.00e+00   1.53e+00    4.58  0.00018
dietD       2.18e-15   1.45e+00    0.00  1.00000

Residual standard error: 2.37 on 20 degrees of freedom
Multiple R-squared: 0.671,        Adjusted R-squared: 0.621
F-statistic: 13.6 on 3 and 20 DF,  p-value: 4.66e-05
```

Notice that the parameter estimates agree with the four group sample means:

$$\hat{\beta}_0 = 61; \qquad \hat{\beta}_0 + \hat{\beta}_1 = 66; \qquad \hat{\beta}_0 + \hat{\beta}_2 = 68; \qquad \hat{\beta}_0 + \hat{\beta}_3 = 61. \qquad \triangleleft$$

The p-values given in the output above are from tests of $H_0 : \beta_i = 0$ vs. $H_a : \beta_i \neq 0$. In many cases these are not the tests we are most interested in. In the present example, we know that $\beta_0 \neq 0$, so the first, although highly "significant", is not of interest. The other three tests test whether groups B, C, and D differ from group A (on average). In a situation where group A represents a control, these may be important tests, but certainly there are other relationships between the groups that are of interest (e.g., are treatments C and D equivalent?).

7.3.2. The Model Utility Test

The usual first test done in these situations is the model utility test (sometimes called the **omnibus test** in the ANOVA setting). Recall that the null hypothesis for the model utility test is that all the non-intercept parameters in the model are 0:

$$H_0 : \beta_1 = \beta_2 = \cdots \beta_{I-1} = 0 .$$

This is equivalent to testing

$$H_0 : \mu_1 = \mu_2 = \cdots = \mu_I \quad \text{vs.} \quad H_a : \mu_i \neq \mu_j \text{ for some } i \text{ and } j .$$

That is, we begin by asking the question: *Is there evidence that there are any differences in mean response?*

Once we have established that there are differences that are unlikely to be due solely to random variation, we can do follow-up analysis to determine which groups are different from which others. (There are also methods for testing other hypotheses of potential interest; stay tuned.)

The model utility test is a model comparison test between the following two models:

$$\Omega : \text{E}(Y) = \beta_0 + \beta_1 x_1 + \beta_2 x_2 + \cdots + \beta_{I-1} x_{i-1} ,$$
$$\omega : \text{E}(Y) = \beta_0 .$$

This means we already know how to have R perform the test for us.

Example 7.3.3. The output below confirms that the F-test in the `summary()` output in Example 7.3.2 is indeed this model comparison test.

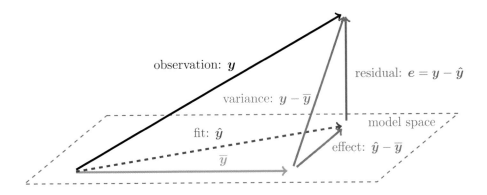

Figure 7.14. Generic linear model.

```
coag-mct
```

```
> coag <- coagulation
> coag$x1 <- coag$diet=='B'
> coag$x2 <- coag$diet=='C'
> coag$x3 <- coag$diet=='D'
> coag.model <- lm(coag~x1+x2+x3,coag)
> coag.model1 <- lm(coag~1,coag)
> anova(coag.model1,coag.model)
Analysis of Variance Table

Model 1: coag ~ 1
Model 2: coag ~ x1 + x2 + x3
  Res.Df RSS Df Sum of Sq    F  Pr(>F)
1     23 340
2     20 112  3       228 13.6 4.7e-05
```

The small p-value indicates that we have strong evidence that the four groups do not all have the same mean *in the population.*

R provides another way to report the results of this test in a tabular form that has become traditional for one-way ANOVA tests. Using `anova()` avoids the necessity of defining auxiliary variables by hand and building an additional model. It also labels things in a more informative way.

```
coag-anova
```

```
> coag.model <- lm(coag~diet,coagulation)
> anova(coag.model)
Analysis of Variance Table

Response: coag
          Df Sum Sq Mean Sq F value  Pr(>F)
diet       3    228    76.0    13.6 4.7e-05
Residuals 20    112     5.6
```

◁

7.3.3. A second look

Before leaving the coagulation data, let's look at the geometry of the model utility test (see Figure 7.14). The model space is I-dimensional, where I is the number of groups. Once again, our first basis element is the unit vector $\boldsymbol{u}_0$ in the direction of $\boldsymbol{v}_0 = \mathbf{1}$. Since the model specifies a mean μ_i for each group, when there are four groups, the model space is spanned by either of the two sets of vectors in Figure 7.15. From the first basis (let's denote it $\{\boldsymbol{w}_1, \boldsymbol{w}_2, \boldsymbol{w}_3, \boldsymbol{w}_4\}$) it follows that

$$\hat{y}_{ij} = \overline{y}_{i\cdot}$$

since

$$\mathrm{proj}(\boldsymbol{y} \to \boldsymbol{w}_i) = \frac{\boldsymbol{y} \cdot \boldsymbol{w}_i}{|\boldsymbol{w}_i|^2} \boldsymbol{w}_i = \frac{n_i \overline{y}_{i\cdot}}{n_i} \boldsymbol{w}_i = \overline{y}_{i\cdot} \boldsymbol{w}_i$$

and the vectors $\boldsymbol{w}_1, \boldsymbol{w}_2, \boldsymbol{w}_3, \boldsymbol{w}_4$ are independent. This gives us all the information we need to develop the F-test for our null hypothesis directly. If there are I groups and H_0 is true, then

$$\frac{|\hat{\boldsymbol{Y}} - \overline{\boldsymbol{Y}..}|^2}{\sigma^2} = \frac{SSM}{\sigma^2} \sim \mathsf{Chisq}(I-1) \text{ and}$$

$$\frac{|\boldsymbol{Y} - \hat{\boldsymbol{Y}}|^2}{\sigma^2} = \frac{SSE}{\sigma^2} \sim \mathsf{Chisq}(n-I) \text{ , so}$$

$$F = \frac{SSM/(I-1)}{SSE/(n-I)} = \frac{MSM}{MSE} \sim \mathsf{F}(I-1, n-I) \text{ .}$$

Furthermore, the quantities above are quite natural:

$$SSM = \sum_{i=1}^{n} (\overline{y}_{i\cdot} - \overline{y}..)^2 = \sum_{i=1}^{n} (\text{group mean} - \text{grand mean})^2 \text{ ,}$$

$$SSE = \sum_{i=1}^{n} (y_{ij} - \overline{y}_{i\cdot})^2 = \sum_{i=1}^{n} (\text{response} - \text{group mean})^2 \text{ .}$$

So SSM measures how much variability there is *between* groups and SSE measures how much variability there is *within* groups. When there is a lot of variability between groups relative to the amount of variability within groups, we can reject the null hypothesis.

Example 7.3.4. We can calculate SSM and SSE by hand and compare the results with output from `anova()`:

```
> stats <- function(x) {                                    coag-geom
+          c( mean = mean(x),
+             SS = sum( (x - mean(x))^2 )
+          )}
> summary(coag~diet, data=coagulation, fun=stats) -> s; s
coag     N=24
```

$$\left\{ \begin{bmatrix} 1 \\ 1 \\ 1 \\ 1 \\ 0 \\ 0 \\ 0 \\ 0 \\ 0 \\ 0 \\ 0 \\ 0 \\ 0 \\ 0 \\ 0 \\ 0 \\ 0 \\ 0 \\ 0 \\ 0 \\ 0 \\ 0 \\ 0 \\ 0 \\ 0 \\ 0 \end{bmatrix}, \begin{bmatrix} 0 \\ 0 \\ 0 \\ 0 \\ 1 \\ 1 \\ 1 \\ 1 \\ 1 \\ 1 \\ 0 \\ 0 \\ 0 \\ 0 \\ 0 \\ 0 \\ 0 \\ 0 \\ 0 \\ 0 \\ 0 \\ 0 \\ 0 \\ 0 \\ 0 \\ 0 \end{bmatrix}, \begin{bmatrix} 0 \\ 0 \\ 0 \\ 0 \\ 0 \\ 0 \\ 0 \\ 0 \\ 0 \\ 0 \\ 1 \\ 1 \\ 1 \\ 1 \\ 1 \\ 1 \\ 0 \\ 0 \\ 0 \\ 0 \\ 0 \\ 0 \\ 0 \\ 0 \\ 0 \\ 0 \end{bmatrix}, \begin{bmatrix} 0 \\ 0 \\ 0 \\ 0 \\ 0 \\ 0 \\ 0 \\ 0 \\ 0 \\ 0 \\ 0 \\ 0 \\ 0 \\ 0 \\ 0 \\ 0 \\ 1 \\ 1 \\ 1 \\ 1 \\ 1 \\ 1 \\ 1 \\ 1 \\ 1 \\ 1 \end{bmatrix} \right\} = \left\{ \begin{bmatrix} 1 \\ 0 \\ 0 \\ 0 \end{bmatrix}, \begin{bmatrix} 0 \\ 1 \\ 0 \\ 0 \end{bmatrix}, \begin{bmatrix} 0 \\ 0 \\ 1 \\ 0 \end{bmatrix}, \begin{bmatrix} 0 \\ 0 \\ 0 \\ 1 \end{bmatrix} \right\},$$

$$\left\{ \begin{bmatrix} 1 \\ 1 \\ 1 \\ 1 \\ 1 \\ 1 \\ 1 \\ 1 \\ 1 \\ 1 \\ 1 \\ 1 \\ 1 \\ 1 \\ 1 \\ 1 \\ 1 \\ 1 \\ 1 \\ 1 \\ 1 \\ 1 \\ 1 \\ 1 \\ 1 \\ 1 \end{bmatrix}, \begin{bmatrix} 0 \\ 0 \\ 0 \\ 0 \\ 1 \\ 1 \\ 1 \\ 1 \\ 1 \\ 1 \\ 0 \\ 0 \\ 0 \\ 0 \\ 0 \\ 0 \\ 0 \\ 0 \\ 0 \\ 0 \\ 0 \\ 0 \\ 0 \\ 0 \\ 0 \\ 0 \end{bmatrix}, \begin{bmatrix} 0 \\ 0 \\ 0 \\ 0 \\ 0 \\ 0 \\ 0 \\ 0 \\ 0 \\ 0 \\ 1 \\ 1 \\ 1 \\ 1 \\ 1 \\ 1 \\ 0 \\ 0 \\ 0 \\ 0 \\ 0 \\ 0 \\ 0 \\ 0 \\ 0 \\ 0 \end{bmatrix}, \begin{bmatrix} 0 \\ 0 \\ 0 \\ 0 \\ 0 \\ 0 \\ 0 \\ 0 \\ 0 \\ 0 \\ 0 \\ 0 \\ 0 \\ 0 \\ 0 \\ 0 \\ 1 \\ 1 \\ 1 \\ 1 \\ 1 \\ 1 \\ 1 \\ 1 \\ 1 \\ 1 \end{bmatrix} \right\} = \left\{ \begin{bmatrix} 1 \\ 1 \\ 1 \\ 1 \end{bmatrix}, \begin{bmatrix} 0 \\ 1 \\ 0 \\ 0 \end{bmatrix}, \begin{bmatrix} 0 \\ 0 \\ 1 \\ 0 \end{bmatrix}, \begin{bmatrix} 0 \\ 0 \\ 0 \\ 1 \end{bmatrix} \right\}.$$

Figure 7.15. Two spanning sets for the model space of an ANOVA model in Example 7.3.1. The first set is orthogonal. The second set (R's default) is not orthogonal, but includes the **1** vector.

```
+-------+-+--+----+---+
|       | |N |mean|SS |
+-------+-+--+----+---+
|diet   |A| 4|61  | 10|
|       |B| 6|66  | 40|
|       |C| 6|68  | 14|
|       |D| 8|61  | 48|
+-------+-+--+----+---+
|Overall| |24|64  |340|
+-------+-+--+----+---+
> s <- unclass(s)  # now we can access as a matrix
> grandMean <- mean(coagulation$coag); grandMean
[1] 64
> groupMean <- s[coagulation$diet,2]; groupMean
 A  A  A  A  B  B  B  B  B  B  C  C  C  C  C  C  D  D  D  D  D  D  D  D
61 61 61 61 66 66 66 66 66 66 68 68 68 68 68 68 61 61 61 61 61 61 61 61
> SST <- sum((coagulation$coag - grandMean)^2); SST  # total variation
[1] 340
> SSE <- sum((coagulation$coag - groupMean)^2); SSE  # w/in group variation
[1] 112
> SSM <- sum((groupMean - grandMean)^2 ); SSM         # b/w group variation
[1] 228
```

```
> coag.model <- lm(coag~diet,coagulation)                    coag-anova
> anova(coag.model)
Analysis of Variance Table

Response: coag
          Df Sum Sq Mean Sq F value  Pr(>F)
diet       3    228    76.0    13.6 4.7e-05
Residuals 20    112     5.6
```

We also see how $s = \hat{\sigma}$ (labeled residual standard error) was computed in the summary() output.

```
> coag.model <- lm(coag~diet,coagulation)                     coag-lm2
> summary(coag.model)
< 15 lines removed >
Residual standard error: 2.37 on 20 degrees of freedom
Multiple R-squared: 0.671,        Adjusted R-squared: 0.621
F-statistic: 13.6 on 3 and 20 DF,  p-value: 4.66e-05
```

That value is $\sqrt{MSE} = \sqrt{5.6} = 2.366$. There are $20 = 24 - 4$ degrees of freedom associated with this estimate, so the $t(20)$ distribution is used to calculate the p-values in the summary() table and other places where we use s^2 as an estimate for σ^2. ◁

An alternative model – and why we don't use it

In Figure 7.15 we presented two different bases for the model space. The upper spanning set is the default basis for lm(). The command

```
model.matrix()
```

can be used to reveal the details of the coding scheme that R is using if there is ever a doubt. This is important since the parameter estimates and tests reported depend on the basis that is being used. In principle, any basis may be used. As long as the basis includes a vector in the direction of **1**, the resulting F-test will be equivalent.

We can fit the model using the upper basis in Figure 7.15 if we like by removing the intercept term. But this basis does not include a vector in the direction of **1**, and the resulting F-test is not equivalent to the usual one.

coag-alt

```
> coag.altmodel <- lm(coag~diet-1,coagulation)
> summary(coag.altmodel)
< 8 lines removed >
Coefficients:
      Estimate Std. Error t value Pr(>|t|)
dietA   61.000      1.183    51.5   <2e-16
dietB   66.000      0.966    68.3   <2e-16
dietC   68.000      0.966    70.4   <2e-16
dietD   61.000      0.837    72.9   <2e-16

Residual standard error: 2.37 on 20 degrees of freedom
Multiple R-squared: 0.999,       Adjusted R-squared: 0.999
F-statistic: 4.4e+03 on 4 and 20 DF,  p-value: <2e-16
```

The command

```
model.matrix(coag.altmodel);
```

shows that the coding scheme is as we claim. This gives us our four group means directly, but the tests summarized are not the ones we want. The four t-tests are now testing whether each β_i $(= \mu_i)$ is 0. The F-test is testing $H_0 : \mu_1 = \mu_2 = \mu_3 = \mu_4 = 0$. In a typical situation, none of these tests is of any interest. So although this seems at first like a natural coding scheme for a categorical predictor, it is not as useful for inference.

7.3.4. A Small Example

Our next example uses a very small data set. This data set is borrowed from *Statistical Methods: A Geometric Primer* by Saville and Wood [**SW96**] which presents numerous examples of linear models from a geometric perspective. In practice, data sets this small are almost never desirable. The advantage of such a small data set is that we can inspect and display all of the vectors involved. They are color-coded to match our model diagrams.

Example 7.3.5. The data below come from a study of air pollution and consist of 6 measurements of air pollution, 2 each at 3 locations in a metropolitan area:

	pollution	location
1	124.00	Hill Suburb
2	110.00	Hill Suburb
3	107.00	Plains Suburb
4	115.00	Plains Suburb
5	126.00	Urban City
6	138.00	Urban City

airp-summary

```
> summary(pollution~location,data=airpollution,fun=mean)
pollution     N=6


+--------+-------------+-+---------+
|        |             |N|pollution|
+--------+-------------+-+---------+
|location|Hill Suburb  |2|117      |
|        |Plains Suburb|2|111      |
|        |Urban City   |2|132      |
+--------+-------------+-+---------+
|Overall |             |6|120      |
+--------+-------------+-+---------+
```

In our data, the largest values occur in the central city, but perhaps that is just due to chance. We can use the method just described to test the null hypothesis that there is no difference in air quality between the 3 locations. This can be done easily in R.

airp-anova

```
> airp.model <- lm(pollution~location,airpollution)
> anova(airp.model)
Analysis of Variance Table

Response: pollution
          Df Sum Sq Mean Sq F value Pr(>F)
location   2    468   234.0    3.48   0.17
Residuals  3    202    67.3
```

In this case the evidence is not strong enough to reject the hypothesis that air quality is the same at all three locations. This isn't too surprising given such a small sample size.

It is instructive to perform a **Pythagorean decomposition** of our response vector $\boldsymbol{y}$. That is, we would like to express $\boldsymbol{y}$ as the sum of 6 orthogonal vectors, the first three of which lie in the model space. In addition, we would like the first vector to be parallel to $\boldsymbol{1}$. A suitable set of vectors to use as a basis for $\mathbb{R}^6$ in this case is

$$\{\boldsymbol{v}_0, \boldsymbol{v}_1, \boldsymbol{v}_2, \boldsymbol{v}_3, \boldsymbol{v}_4, \boldsymbol{v}_5\} = \left\{ \begin{bmatrix} 1 \\ 1 \\ 1 \\ 1 \\ 1 \\ 1 \end{bmatrix}, \begin{bmatrix} 1 \\ 1 \\ -1 \\ -1 \\ 0 \\ 0 \end{bmatrix}, \begin{bmatrix} 1 \\ 1 \\ 1 \\ 1 \\ -2 \\ -2 \end{bmatrix}, \begin{bmatrix} 1 \\ -1 \\ 0 \\ 0 \\ 0 \\ 0 \end{bmatrix}, \begin{bmatrix} 0 \\ 0 \\ 1 \\ -1 \\ 0 \\ 0 \end{bmatrix}, \begin{bmatrix} 0 \\ 0 \\ 0 \\ 0 \\ 1 \\ -1 \end{bmatrix} \right\}.$$

The reason for this particular choice of basis will become clear shortly. We would, of course, normalize to unit length to get our vectors $\boldsymbol{u}_0, \ldots, \boldsymbol{u}_5$. A little matrix algebra gives

$$
\boldsymbol{y} = \begin{bmatrix} 124 \\ 110 \\ 107 \\ 115 \\ 126 \\ 138 \end{bmatrix} = \begin{bmatrix} 120 \\ 120 \\ 120 \\ 120 \\ 120 \\ 120 \end{bmatrix} + \begin{bmatrix} 3 \\ 3 \\ -3 \\ -3 \\ 0 \\ 0 \end{bmatrix} + \begin{bmatrix} -6 \\ -6 \\ -6 \\ -6 \\ 12 \\ 12 \end{bmatrix} + \begin{bmatrix} 7 \\ -7 \\ 0 \\ 0 \\ 0 \\ 0 \end{bmatrix} + \begin{bmatrix} 0 \\ 0 \\ -4 \\ 4 \\ 0 \\ 0 \end{bmatrix} + \begin{bmatrix} 0 \\ 0 \\ 0 \\ 0 \\ -6 \\ 6 \end{bmatrix} .
$$

Since these are orthogonal vectors, we see that

$$
\left| \begin{bmatrix} 124 \\ 110 \\ 107 \\ 115 \\ 126 \\ 138 \end{bmatrix} - \begin{bmatrix} 120 \\ 120 \\ 120 \\ 120 \\ 120 \\ 120 \end{bmatrix} \right|^2 = \left| \begin{bmatrix} 3 \\ 3 \\ -3 \\ -3 \\ 0 \\ 0 \end{bmatrix} \right|^2 + \left| \begin{bmatrix} -6 \\ -6 \\ -6 \\ -6 \\ 12 \\ 12 \end{bmatrix} \right|^2 + \left| \begin{bmatrix} 7 \\ -7 \\ 0 \\ 0 \\ 0 \\ 0 \end{bmatrix} \right|^2 + \left| \begin{bmatrix} 0 \\ 0 \\ -4 \\ 4 \\ 0 \\ 0 \end{bmatrix} \right|^2 + \left| \begin{bmatrix} 0 \\ 0 \\ 0 \\ 0 \\ -6 \\ 6 \end{bmatrix} \right|^2 ,
$$

$$
670 \quad = \quad 36 \quad + \quad 432 \quad + \quad 98 \quad + \quad 32 \quad + \quad 72 \, ,
$$

$$
670 \quad = \quad 468 \quad + \quad 202 \, .
$$

Notice how these results correspond to the R output above. ◁

7.3.5. Contrasts

The method of contrasts gives us a way of testing a wide range of potentially interesting hypotheses that can be expressed as linear combinations of the group means. Contrasts are technically somewhat simpler to work with when the number of observations in each group is the same, so we will begin by considering contrasts in the context of our air pollution example (Example 7.3.5). Specifically, we will consider the hypotheses

- $H_1 : \mu_1 - \mu_2 = 0$ (roughly: two suburbs have the same air quality), and
- $H_2 : \mu_1 + \mu_2 - 2\mu_3 = 0$ (roughly: air quality is the same in suburbs and city).

Each of these hypotheses can be expressed using a special type of linear combination of the group means called a contrast.

Definition 7.3.1. A **contrast** C is a linear combination of the group means

$$
C = \sum_{i=1}^{I} c_i \mu_i
$$

such that $\sum_i c_i = 0$. (That is, $\boldsymbol{c} \perp \boldsymbol{1}$.) □

The hypotheses above can be expressed as

- $H_1 : C_1 = 0$ and
- $H_2 : C_2 = 0$,

where

- $C_1 = \mu_1 - \mu_2$ and
- $C_2 = \mu_1 + \mu_2 - 2\mu_3$.

Testing contrasts via model comparison

A contrast C (or, more precisely, the hypothesis $H_0 : C = 0$) can be tested using a model comparison test where a submodel ω is formed by constraining the larger model Ω so that the contrast is equal to 0 in ω.

Example 7.3.6. We can test the contrasts

- $C_1 = \mu_1 - \mu_2$ and

- $C_2 = \mu_1 + \mu_2 - 2\mu_3$

from Example 7.3.5 using `anova()` if we can determine how to describe the submodels corresponding to each contrast. Recall that in the model

$$\boldsymbol{Y} = \beta_0 + \beta_1 \boldsymbol{x}_1 + \beta_2 \boldsymbol{x}_2 + \boldsymbol{\varepsilon}, \qquad \varepsilon \overset{\text{iid}}{\sim} \mathsf{Norm}(0, \sigma),$$

$\mu_1 = \beta_0$, $\mu_2 = \beta_0 + \beta_1$, and $\mu_3 = \beta_0 + \beta_2$.

Since $C_1 = 0$ means $\mu_1 = \mu_2$, our submodel has the form

$$\begin{aligned} \mathsf{E}(Y) &= \mu_1 [\![x = 1]\!] + \mu_1 [\![x = 2]\!] + \mu_3 [\![x = 3]\!] \\ &= \beta_0 + \beta_2 [\![x = 3]\!], \end{aligned}$$

where $x = 1$, 2, or 3 to indicate the location, $\beta_0 = \mu_1 = \mu_2$, $\beta_1 = 0$, and $\beta_2 = \mu_3 - \mu_1$. That is, the submodel has an intercept term and a term for the indicator variable $[\![x = 3]\!]$.

airp-modcomp1

```
# convert location to a numeric variable for convenience
> airp <- airpollution; airp$loc <- as.numeric(airp$location); airp
  pollution       location loc
1       124    Hill Suburb   1
2       110    Hill Suburb   1
3       107  Plains Suburb   2
4       115  Plains Suburb   2
5       126     Urban City   3
6       138     Urban City   3
> model <- lm(pollution~location, airp)
> model2 <- lm(pollution~ 1 + (loc==3), airp)
> anova(model2,model)
Analysis of Variance Table

Model 1: pollution ~ 1 + (loc == 3)
Model 2: pollution ~ location
  Res.Df RSS Df Sum of Sq    F Pr(>F)
1      4 238
2      3 202  1        36 0.53   0.52
```

For C_2, we proceed similarly. $C_2 = 0$ is equivalent to

$$\mu_3 = \frac{1}{2}\mu_1 + \frac{1}{2}\mu_2 \,,$$

$$\mu_3 - \mu_1 = -\frac{1}{2}\mu_1 + \frac{1}{2}\mu_2 \,,$$

$$\beta_2 = -\frac{1}{2}\beta_0 + \frac{1}{2}(\beta_0 + \beta_1) \,,$$

$$\beta_2 = \frac{1}{2}\beta_1 \,.$$

We can now express our submodel in two ways – in terms of the β_i's:

$$\mathrm{E}(Y) = \mu_1 [\![x = 1]\!] + \mu_2 [\![x = 2]\!] + \frac{1}{2}(\mu_1 + \mu_2)[\![x = 3]\!]$$

$$= \beta_0 [\![x = 1]\!] + (\beta_0 + \beta_1)[\![x = 2]\!] + (\beta_0 + \frac{1}{2}\beta_1)[\![x = 3]\!]$$

$$= \beta_0 \mathbf{1} + \beta_1 ([\![x = 2]\!] + \frac{1}{2}[\![x = 3]\!]) \,, \tag{7.8}$$

or in terms of the μ_i's:

$$\mathrm{E}(Y) = \mu_1 [\![x = 1]\!] + \mu_2 [\![x = 2]\!] + \frac{1}{2}(\mu_1 + \mu_2)[\![x = 3]\!]$$

$$= \mu_1 \left([\![x = 1]\!] + \frac{1}{2}[\![x = 3]\!] \right) + \mu_2 \left([\![x = 2]\!] + \frac{1}{2}[\![x = 3]\!] \right) \,. \tag{7.9}$$

Using (7.8) suggests defining a new variable $[\![x = 2]\!] + \frac{1}{2}[\![x = 3]\!]$, which then makes the model easy to describe in R.

```
# build a variable that makes the model easier to describe       airp-modcomp2
> airp$x <- with( airp, ((loc==2) + 0.5*(loc==3)) )
> model3 <- lm(pollution~ 1 + x, airp)
> anova(model3,model)
Analysis of Variance Table

Model 1: pollution ~ 1 + x
Model 2: pollution ~ location
  Res.Df RSS Df Sum of Sq     F Pr(>F)
1      4 634
2      3 202  1       432  6.42  0.085
```

On the other hand, (7.9) suggests defining two new variables and fitting the model without an intercept term.

```
# build two variables that make the model easier to describe    airp-modcomp2a
> airp$x1 <- with( airp, ((loc==1) + 0.5*(loc==3)) )
> airp$x2 <- with( airp, ((loc==2) + 0.5*(loc==3)) )
> model3 <- lm(pollution~ -1 + x1 + x2, airp)
> anova(model3,model)
Analysis of Variance Table

Model 1: pollution ~ -1 + x1 + x2
Model 2: pollution ~ location
```

```
   Res.Df RSS Df Sum of Sq     F Pr(>F)
1      4 634
2      3 202  1       432 6.42  0.085
```
$\triangleleft$

The geometry of contrasts

It is instructive to look at the geometry of contrasts. Suppose we had a unit direction vector $\boldsymbol{u}_C$ in the model space with the property that when $C =,$

$$\boldsymbol{u}_C \cdot \boldsymbol{Y} = |\operatorname{proj}(\boldsymbol{Y} \to \boldsymbol{u}_C)| \sim \mathsf{Norm}(0, \sigma) \,.$$

For such a vector, when $C = 0$,

$$F = \frac{(\boldsymbol{u}_C \cdot \boldsymbol{Y})^2}{MSE} \sim F(1, n - I) \,.$$

In order for our test to have power to detect when $C \neq 0$, we would like $\mathrm{E}(\boldsymbol{u}_C \cdot \boldsymbol{Y})$ to be proportional to the contrast C. That way, the more dramatically $C = 0$ fails to be true, the more likely we will be to obtain a small p-value when testing $H_0 : C = 0$. The following lemma tells us how to construct such a vector.

Lemma 7.3.2. *Suppose* $\boldsymbol{Y}_{ij} - \mu_i \overset{iid}{\sim} \mathsf{Norm}(0, \sigma)$ *as in the ANOVA model. Let* $C = \sum_{i=1}^{I} c_i \mu_i$ *be a contrast, and define*

$$\alpha_{ij} = \frac{c_i}{n_i} \,,$$
$$\boldsymbol{v}_C = \langle \alpha_{11}, \alpha_{12}, \dots \alpha_{1n_1}, \alpha_{21}, \dots \alpha_{2n_2}, \dots \alpha_{I1}, \dots \alpha_{In_I} \rangle \,,$$
$$\kappa = \frac{1}{|\boldsymbol{v}_C|} \,, \quad and$$
$$\boldsymbol{u}_C = \kappa \boldsymbol{v}_C \,.$$

Then

(a) $\boldsymbol{v}_C$ *and* $\boldsymbol{u}_C$ *lie in the model space,*

(b) $\boldsymbol{v}_C \perp \mathbf{1}$ *and* $\boldsymbol{u}_C \perp \mathbf{1}$,

(c) $|\boldsymbol{u}_C| = 1$, *and*

(d) $\boldsymbol{u}_C \cdot \boldsymbol{Y} = |\operatorname{proj}(\boldsymbol{Y} \to \boldsymbol{u}_C)| \sim \mathsf{Norm}(\kappa C, \sigma)$.

Proof.   (a) The vectors $\boldsymbol{v}_C$ and $\boldsymbol{u}_C$ lie in the model space because their coefficients are the same within groups.

(b) $\boldsymbol{v}_C \cdot \mathbf{1} = \sum_{i=1}^{I} \sum_{j=1}^{n_i} \alpha_{ij} \cdot 1 = \sum_{i=1}^{I} \sum_{j=1}^{n_i} \frac{c_i}{n_i} = \sum_{i=1}^{I} c_i = 0$ because C is a contrast.

(c) $|\boldsymbol{u}_C| = 1$ by the definition of κ.

(d) By Lemma 6.3.1, since $\boldsymbol{u}_C \perp \mathbf{1}$ and $|\boldsymbol{u}_C| = 1$, $\boldsymbol{u}_C \cdot \boldsymbol{Y}$ is normally distributed with a standard deviation of σ. The mean can be computed directly:

$$
\begin{aligned}
\mathrm{E}(\boldsymbol{u}_C \cdot \boldsymbol{Y}) &= \sum_{i=1}^{I} \sum_{j=1}^{n_i} \kappa \alpha_{ij} \, \mathrm{E}(Y_i) \\
&= \sum_{i=1}^{I} \sum_{j=1}^{n_i} \frac{\kappa c_i}{n_i} \mu_i \\
&= \kappa \sum_{i=1}^{I} c_i \mu_i \\
&= \kappa C \, .
\end{aligned}
$$
$\qquad\square$

It is important to note that in Lemma 7.3.2, κ depends on the coefficients c_i of the contrast C, but not on the means μ_i. Because of this dependence on C, it would be more precise to denote κ as κ_C. When the contrast is clear from context, however, we prefer to avoid the subscript.

Example 7.3.7. Returning to our small air pollution example, the direction vectors for our two contrasts are

$$
\boldsymbol{v}_{C_1} = \begin{bmatrix} 1/2 \\ 1/2 \\ -1/2 \\ -1/2 \\ 0 \\ 0 \end{bmatrix} = \frac{1}{2} \begin{bmatrix} 1 \\ 1 \\ -1 \\ -1 \\ 0 \\ 0 \end{bmatrix}, \quad \text{and} \quad \boldsymbol{v}_{C_2} = \begin{bmatrix} 1/2 \\ 1/2 \\ 1/2 \\ 1/2 \\ -2/2 \\ -2/2 \end{bmatrix} = \frac{1}{2} \begin{bmatrix} 1 \\ 1 \\ 1 \\ 1 \\ -2 \\ -2 \end{bmatrix},
$$

so

$$
\kappa_{C_1} = \frac{1}{|\boldsymbol{v}_{C_1}|} = 1, \quad \kappa_{C_2} = \frac{1}{|\boldsymbol{v}_{C_2}|} = \frac{1}{\sqrt{3}},
$$

$$
\boldsymbol{u}_{C_1} = \frac{1}{2} \begin{bmatrix} 1 \\ 1 \\ -1 \\ -1 \\ 0 \\ 0 \end{bmatrix}, \quad \text{and} \quad \boldsymbol{u}_{C_2} = \frac{1}{\sqrt{12}} \begin{bmatrix} 1 \\ 1 \\ 1 \\ 1 \\ -2 \\ -2 \end{bmatrix}.
$$

We can check that these vectors have the properties enumerated in Lemma 7.3.2. Properties (a)–(c) are immediate. We can check (d) by direct calculation. For example,

$$
\begin{aligned}
\mathrm{E}(\boldsymbol{u}_{C_2} \cdot \boldsymbol{Y}) &= \mathrm{E}\left(\frac{1}{\sqrt{12}} (Y_1 + Y_2 + Y_3 + Y_4 - 2Y_5 - 2Y_6) \right) \\
&= \frac{1}{\sqrt{12}} \, \mathrm{E}(2\overline{Y}_{.1} + 2\overline{Y}_{.2} - 4\overline{Y}_{.3}) \\
&= \frac{2}{\sqrt{12}} (\mu_1 + \mu_2 - 2\mu_3) = \frac{1}{\sqrt{3}} (\mu_1 + \mu_2 - 2\mu_3) \, ,
\end{aligned}
$$

as desired. A similar statement holds for $\boldsymbol{u}_{C_1}$. $\qquad\triangleleft$

Testing a contrast using vectors

Conducting an F-test for a contrast C is relatively simple once the direction vector u_C and constant κ have been determined. If H_0 is true, then

$$\boldsymbol{u}_C \cdot \boldsymbol{Y} \sim \mathsf{Norm}(\kappa C, \sigma) = \mathsf{Norm}(0, \sigma) , \qquad (7.10)$$

so

$$F = \frac{SSC/1}{SSE/(n-I)} = \frac{SSC}{MSE} = \frac{(\boldsymbol{u}_C \cdot \boldsymbol{Y})^2}{MSE} \sim \mathsf{F}(1, n - I) .$$

Equivalently,

$$\frac{(\boldsymbol{u}_C \cdot \boldsymbol{Y})}{\sqrt{MSE}} \sim \mathsf{t}(n - I) .$$

The hypothesis test can now be conducted by comparing the test statistic to the appropriate null distribution.

Example 7.3.8. In our air pollution example (Example 7.3.5), our basis included vectors parallel to $\boldsymbol{u}_{C_1}$ and $\boldsymbol{u}_{C_2}$, so so we can conduct the two tests by simply looking up the values we previously calculated. The following ANOVA table is a useful way to record the information:

	df	Sum Sq	Mean Sq	F value	Pr($>$F)
$H_1 : \mu_1 - \mu_2 = 0$	1	36.00	36.00	0.53	0.5176
$H_2 : \mu_1 + \mu_2 - 2\mu_3 = 0$	1	432.00	432.00	6.42	0.0852
Residuals	3	202.00	67.33		

Alternatively, we can compute the t-statistic directly:

$$\boldsymbol{u}_{C_1} \cdot \boldsymbol{Y} = \frac{1}{2} (1 \cdot 124 + 1 \cdot 110 - 1 \cdot 107 - 1 \cdot 115)$$
$$= 6 ,$$

so $t_{C_1} = \frac{6}{\sqrt{202/3}} = 0.422$, and

$$\boldsymbol{u}_{C_2} \cdot \boldsymbol{Y} = \frac{1}{\sqrt{12}} (1 \cdot 124 + 1 \cdot 110 + 1 \cdot 107 + 1 \cdot 115 - 2 \cdot 126 - 2 \cdot 138)$$
$$= -36/\sqrt{3} ,$$

so $t_{C_2} = \frac{-36/\sqrt{3}}{\sqrt{202/3}} = -2.53$.

Notice that the squares of these t-statistics are the F-statistics above, and that the p-values below match those in our table above:

```
> u1 <- 1/2 * c(1,1,-1,-1,0,0)
> u2 <- 1/sqrt(12) * c( 1,1,1,1,-2,-2)
> dot(airpollution$pollution, u1)
[1] 6
> dot(airpollution$pollution, u2)
[1] -20.785
> t1 <- dot(airpollution$pollution, u1) / sqrt(202/3); t1
[1] 0.7312
```

airp-vectors

```
> t2 <- dot(airpollution$pollution, u2) / sqrt(202/3); t2
[1] -2.5330
> t1^2
[1] 0.53465
> t2^2
[1] 6.4158
> 2 * pt(- abs(t1), df=3 )
[1] 0.51759
> 2 * pt(- abs(t2), df=3 )
[1] 0.085204                                                      ◁
```

Confidence intervals

Let $\hat{C}$ be the sample contrast

$$\hat{C} = \sum_{i=1}^{I} c_i \overline{Y}_{i \cdot} \, .$$

Then by Lemma 7.3.2,

$$\kappa \hat{C} = \boldsymbol{u}_C \cdot \boldsymbol{Y} \sim \mathsf{Norm}(\kappa C, \sigma) \, .$$

From this it follows that

$$\frac{\kappa \hat{C} - \kappa C}{\sigma} \sim \mathsf{Norm}(0, 1) \, , \text{ and}$$

$$\frac{\kappa \hat{C} - \kappa C}{S} \sim \mathsf{t}(n - I) \, , \text{ or, equivalently,}$$

$$\frac{\hat{C} - C}{S/\kappa} \sim \mathsf{t}(n - I) \, , \tag{7.11}$$

where $s^2 = MSE = SSE/(n - I)$ is the residual-based estimate for σ^2 on $n - I$ degrees of freedom. So a confidence interval for C is given by

$$\hat{C} \pm t_* \frac{s}{\kappa} \, , \tag{7.12}$$

where t_* is the appropriate critical value from the $\mathsf{t}(n - I)$-distribution. The expression s/κ is the estimated standard error for the contrast.

Recall that although s and t_* are the same for each contrast tested from the same data set, κ depends on the contrast and so should rightly be denoted κ_C.

Example 7.3.9.

Q. Calculate confidence intervals for each of the contrasts in Example 7.3.6.

A. We begin by handling the contrast $C_1 = \mu_1 - \mu_2$. We have already seen that $\kappa = 1$, but we could also compute κ by determining the constant of proportionality between $\mathrm{E}(\boldsymbol{u}_1 \cdot \boldsymbol{Y})$ and C:

$$\mathrm{E}(\boldsymbol{u}_1 \cdot \boldsymbol{Y}) = \mathrm{E}\left(\frac{Y_1 + Y_2 - Y_3 - Y_4}{\sqrt{4}}\right)$$
$$= 1(\mu_1 - \mu_2)\,,$$

so $\kappa = 1$. The estimated standard error for this contrast is $SE = s/1 = 8.21$ and a 95% confidence interval for C_1 is

$$\bar{y}_{1.} - \bar{y}_{2.} \pm t_* SE = (123 - 117) \pm t_*(8.21)$$
$$= 6 \pm 3.182(8.21)$$
$$= (-20.1, 32.1)\,.$$

Notice that 0 lies well within this interval, as we would expect since the p-value for the corresponding hypothesis test was 0.5176.

Now we turn our attention to $C_2 = \mu_1 + \mu_2 - 2\mu_3$. We already know that $\kappa = \frac{1}{\sqrt{3}}$, but this can also be seen from the following calculation:

$$\mathrm{E}(\boldsymbol{u}_2 \cdot \boldsymbol{Y}) = \mathrm{E}\left(\frac{Y_1 + Y_2 + Y_3 + Y_4 - 2Y_5 - 2Y_6}{\sqrt{12}}\right)$$
$$= \frac{2\mu_1 + 2\mu_2 - 4\mu_3}{\sqrt{12}}$$
$$= \frac{\mu_1 + \mu_2 - 2\mu_3}{\sqrt{3}}\,,$$

so $\kappa = \frac{1}{\sqrt{3}}$. Thus the estimated standard error for this contrast is $SE = s\sqrt{3} = 8.21\sqrt{3} = 14.21$, and a 95% confidence interval for C_2 is

$$\bar{y}_{1.} + \bar{y}_{2.} - 2\bar{y}_{3.} \pm t_* SE = 117 + 111 - 2(132) \pm t_*(14.21)$$
$$= -36 \pm 3.182 \cdot 14.21$$
$$= (-81.23, 9.23)\,.$$

This interval also contains 0, but a 90% confidence interval should not. Indeed, a 90% confidence interval is

$$-36 \pm 2.35 \cdot 14.21 = (-69.45, -2.55)\,.$$

The contrast $C_2' = \frac{1}{2}\mu_1 + \frac{1}{2}\mu_2 - 1\mu_3$ has a nicer interpretation since it is the difference between the mean of group 3 and the mean of the combination of groups 1 and 2. (The previous contrast was twice this amount.) The confidence intervals for C_2' are formed by halving everything. So, for example, a 90% confidence interval is

$$-18 \pm 2.35 \cdot 7.11 = (-34.72, -1.27)\,. \qquad \triangleleft$$

Orthogonal contrasts

Contrasts are said to be orthogonal if their corresponding direction vectors are orthogonal. The advantage of orthogonal contrasts is that the resulting tests are

---

Box 7.3. Key Properties of Contrasts

(1) A **contrast** C is a linear combination of the group means

$$C = \sum_{i=1}^{I} c_i \mu_i$$

such that $\sum_i c_i = 0$. (That is, $\boldsymbol{c} \perp \boldsymbol{1}$.)

(2) The **sample contrast** estimated from data is

$$\hat{C} = \sum_{i=1}^{I} c_i \overline{Y}_{i\cdot}$$

and is an unbiased estimator for C.

(3) The corresponding **direction vector** $\boldsymbol{u}_C$ satisfies the following:

(a) The coefficients of $\boldsymbol{u}_C$ are the same for all terms corresponding to the same group; that is, $\boldsymbol{u}_C$ can be written

$$[\alpha_1, \ldots, \alpha_1, \alpha_2, \ldots, \alpha_2, \ldots, \alpha_I, \ldots, \alpha_I]^{\top}$$

if the data are sorted by group. In other words, $\boldsymbol{u}_C$ lies in the ANOVA model space.

(b) $\boldsymbol{u}_C \perp \boldsymbol{1}$.

(c) $|\boldsymbol{u}_C| = 1$.

(d) $\mathrm{E}(\boldsymbol{u}_C \cdot \boldsymbol{Y}) = \kappa C$ for some constant κ.

The constant κ depends on the coefficients c_i of the contrast C but not on the means μ_i. This implies that the length $\boldsymbol{u}_C \cdot \boldsymbol{Y}$ is a measure of the evidence against the hypothesis that $C = 0$ since the expected value is 0 if and only if $C = 0$.

An explicit formula for κ is given by

$$\frac{1}{\kappa} = \sqrt{\sum c_i^2 n_i} \ .$$

(4) Let S^2 be our unbiased estimator of σ^2 on $\nu = n - I$ residual degrees of freedom. Then

$$\frac{\hat{C} - C}{S/\kappa} \sim \mathsf{t}(\nu) \ ,$$

so a confidence interval for C has the form

$$\hat{C} \pm t_* \frac{s}{\kappa} \ . \tag{7.13}$$

(5) Under the null hypothesis that $C = 0$,

$$\frac{(\boldsymbol{u}_C \cdot \boldsymbol{Y})^2}{SSE/(n - I)} = \frac{MSC}{MSE} \sim \mathsf{F}(1, \nu) \ .$$

independent and the sums of squares of a full set of orthogonal contrasts will sum to SST. In particular, if C_1 and C_2 are orthogonal, then we can test

$$H_0 : C_1 = 0 \text{ and } C_2 = 0$$

using

$$F = \frac{(SSC_1 + SSC_2)/2}{SSE/(n - I)} = \frac{(\boldsymbol{u}_{C_1} \cdot \boldsymbol{Y})^2 + (\boldsymbol{u}_{C_2} \cdot \boldsymbol{Y})^2}{MSE} \sim \mathsf{F}(2, n - I) .$$

Similar statements hold for three or more orthogonal contrasts. In our air pollution example, when $C_1 = 0$ and $C_2 = 0$, then $\mu_1 = \mu_2 = \mu_3$ (equivalently $\beta_1 = \beta_2 = 0$), and testing whether $C_1 = 0$ and $C_2 = 0$ is equivalent to the model utility test.

The two contrasts in our air pollution example were orthogonal, but there are many important situations where non-orthogonal contrasts are used.

7.3.6. Pairwise Comparisons

Comparing two groups

Often we are interested in comparing two groups. Suppose we want to compare groups i and j. For ease of notation, let's assume that the data have been sorted so that these two groups come first. The contrast associated with this is

$$C = \mu_i - \mu_j .$$

If group i has n_i observations and group j has n_j observations, then

$$\boldsymbol{v}_C = \left[\underbrace{\frac{1}{n_i}, \frac{1}{n_i}, \ldots, \frac{1}{n_i}}_{n_i}, \underbrace{\frac{-1}{n_j}, \frac{-1}{n_j}, \ldots, \frac{-1}{n_j}}_{n_j}, 0, \ldots, 0 \right]^\top .$$

So

$$|\boldsymbol{v}_C| = \frac{1}{\kappa} = \sqrt{\frac{1}{n_i} + \frac{1}{n_j}} ,$$

$$SE = \frac{s}{\kappa} = s\sqrt{\frac{1}{n_i} + \frac{1}{n_j}} ,$$

and our confidence interval has the form

$$(\bar{y}_{i\cdot} - \bar{y}_{j\cdot}) \pm t_* s\sqrt{\frac{1}{n_i} + \frac{1}{n_j}} . \tag{7.14}$$

The degrees of freedom used for t_* is the degrees of freedom used to estimate σ, namely $n - k$, where k is the number of groups.

This confidence interval should also remind you of the confidence interval from a 2-sample t-test from Section 6.6.2. There is an important difference, however. In (7.14), s is estimated using data from all of the groups, not just from the two groups being compared.

As the following example illustrates, the standard error and p-values that appear in the non-intercept rows in the summary output of the object returned by lm() are calculated using (7.14).

Example 7.3.10. Recall Example 7.3.1,

coag-lm

```
> coag.model <- lm(coag~diet,coagulation)
> summary(coag.model)
< 8 lines removed >
Coefficients:
            Estimate Std. Error t value Pr(>|t|)
(Intercept) 6.10e+01   1.18e+00   51.55  < 2e-16
dietB       5.00e+00   1.53e+00    3.27  0.00380
dietC       7.00e+00   1.53e+00    4.58  0.00018
dietD       2.18e-15   1.45e+00    0.00  1.00000


Residual standard error: 2.37 on 20 degrees of freedom
Multiple R-squared: 0.671,       Adjusted R-squared: 0.621
F-statistic: 13.6 on 3 and 20 DF,  p-value: 4.66e-05
```

We see that $s = 2.366$, so the last three standard errors are

$$2.366\sqrt{\frac{1}{4} + \frac{1}{6}} = 1.53 \ ,$$

$$2.366\sqrt{\frac{1}{4} + \frac{1}{6}} = 1.53 \ , \text{ and}$$

$$2.366\sqrt{\frac{1}{4} + \frac{1}{8}} = 1.45 \ .$$

These correspond to comparisons between group A and the other three groups. We can compare other pairs of groups if we like as well. For example, if we compare group C with group D, the standard error is $SE = 2.366\sqrt{\frac{1}{6} + \frac{1}{8}} = 1.28$. A 95% confidence interval for this difference is

$$7 \pm 2.08(1.28) = (2.06, 11.94) \ .$$

Since this interval does not contain 0, we know that the p-value for the test of $H_0 : \mu_3 = \mu_4$ vs. $H_a : \mu_3 \neq \mu_4$ is less than 0.05. ◁

7.3.7. Multiple Comparisons

The methods of the previous section are fine if we have only one contrast in mind *before we look at our data*. Sometimes this is the case. Comparing the suburbs to the city might have occurred to us as a natural thing to do, even if there had been measurements made at several urban and several suburban locations. In this case, a confidence interval for a contrast comparing the suburbs to the city is a meaningful thing to construct. Often, however, we don't know which groups we want to compare until after collecting the data. Or put differently, we want to compare all pairs of groups, looking to see which (if any) might have different means. Or perhaps one group represents a control and we want to compare all the other groups to the control group.

If we have 5 groups ($I = 5$), there are 10 different pairs of means we could compare. Clearly these can't all be independent comparisons since the model space has only 5 dimensions. But suppose for a moment that we make 10 independent 95% confidence intervals. Then the chance that at least one of them will fail to cover the true parameter value is much larger than 5%, namely $1 - (0.95)^{10} = 0.401$. That is, we make a mistake in 5% of our comparisons but in 40.1% of our data sets. We refer to 40.1% as the **familywise error rate**. This problem only gets worse as the number of confidence intervals increases. The same issue arises when doing multiple hypothesis tests instead of multiple confidence intervals and is generally referred to as the **multiple comparisons problem**.

If we want our entire testing procedure to have a familywise error rate of 5% *per data set*, it is clear that we need to make some adjustments. Several adjustment methods have been proposed. We will look at two of the simplest.

The Bonferroni correction

Suppose we want to construct m confidence intervals or perform m hypothesis tests. Let E_i be the event that interval (or test) i has a coverage (or type I) error. Then

$$\text{P(at least one error)} \leq \text{P}(E_1) + \text{P}(E_2) + \cdots + \text{P}(E_m) .$$

So if we ensure that $\text{P}(E_i) \leq \alpha/m$ for each test, then the familywise error rate will be at most $m\frac{\alpha}{m} = \alpha$. This is the Bonferroni correction.

Note that the inequality above is an equality only in the case that no data set can have two or more errors. So this method tends to "over-adjust". Statisticians say such a procedure is **conservative** because it makes intervals wider and p-values larger than they should be. Said another way, the coverage rate for confidence intervals will be higher than stated; the probability of an "extreme test statistic" will be lower. This means that our claims are better than advertised. This also reduces the power of the procedure, so a better correction method would be desirable.

Example 7.3.11.

Q. Use the Bonferroni correction to compare the three locations in the pollution data set from Example 7.3.5.

A. There are 3 comparisons, so we compute three 98.33% confidence intervals, since $0.9833 = 1 - \frac{0.05}{3}$. Our three groups each have 2 observations, so our Bonferroni corrected confidence intervals all have the form

$$\bar{y}_{i\cdot} - \bar{y}_{j\cdot} \quad \pm \quad t_* \sqrt{MSE} \sqrt{\frac{1}{2} + \frac{1}{2}} ,$$
$$\bar{y}_{i\cdot} - \bar{y}_{j\cdot} \quad \pm \quad 4.8567\sqrt{67.4}\sqrt{1} ,$$
$$\bar{y}_{i\cdot} - \bar{y}_{j\cdot} \quad \pm \quad 39.87 .$$

Our intervals are wider than the unadjusted intervals because of the new value of t_*.

Alternatively, we could conduct three hypothesis tests and only reject when the p-values are below $0.05/3 = 0.0167$. By the duality of confidence intervals and hypothesis tests, we will consider the differences between groups significant (at the $\alpha = 0.05$ level) only if the group means differ by at least 39.87.   ◁

Tukey's honest significant differences

An improved correction method must necessarily take into account the correlation structure between the tests or confidence intervals, since Bonferroni is optimal in the worst case. One of the first such methods was proposed by Tukey, and it is still commonly used to follow up a statistically significant ANOVA model utility test by testing all possible pairs of groups.

We'll outline how this is done in the simplest case, namely when all of the groups have the same size $r = n/k$. Notice that in this case the *unadjusted* margin of error of the confidence interval will be the same for each pair, namely

$$t_* s \sqrt{\frac{1}{r} + \frac{1}{r}} = \left(t_* \sqrt{\frac{1}{r} + \frac{1}{r}} \right) \sqrt{MSE} \;.$$

We will widen these intervals so that the margin of error is

$$D \sqrt{MSE}$$

for some constant D. We want to choose D so that the probability that one or more of the confidence intervals fails to cover the associated difference in means is α. That is, we let E_{ij} be the event that

$$|(\overline{Y}_{i\cdot} - \overline{Y}_{j\cdot}) - (\mu_i - \mu_j)| > D \sqrt{MSE} \;,$$

and we want to choose D so that

$$\mathrm{P}(\text{at least one error}) = \mathrm{P}(\bigcup E_{ij}) = \alpha \;.$$

The clever idea is to introduce new variables $W_i = \overline{Y}_{i\cdot} - \mu_i$. Then

$$E_{ij} \text{ is the event that } |W_i - W_j| > D \sqrt{MSE} \;,$$

and we want to determine D such that

$$\mathrm{P}(\bigcup_i E_i) = \mathrm{P}(\max_i W_i - \min_i W_i > D \sqrt{MSE}) = \alpha \;.$$

Since $W_i \sim N(0, \sigma/\sqrt{r})$ and MSE/r is an estimate for σ^2/r, we'll rewrite this one more time as

$$\mathrm{P}\left(\frac{\max_i W_i - \min_i W_i}{\sqrt{MSE}/\sqrt{r}} > \sqrt{r} D \right) = \alpha \;.$$

The quantity on the left can be expressed as

$$\max_i \frac{W_i}{\sqrt{MSE/r}} - \min_i \frac{W_i}{\sqrt{MSE/r}} = Q \;.$$

Q is the range of the studentized values of normally distributed W_{ij}. The distribution of Q has been studied and is called the **studentized range distribution**. This distribution has 2 parameters: $k = $ sample size for the numerator (i.e., the number of groups) and $\nu = $ degrees of freedom for the estimator in the denominator. We can get critical values for this distribution using `qtukey()` in R.

We want

$$P\left(Q > \sqrt{r}D\right) = \alpha \, ,$$

so we need $\sqrt{r}D$ to be the $1 - \alpha$ quantile of the distribution of Q. The number D that we are seeking is

$$D = \texttt{qtukey}(1 - \alpha, k, n - k)/\sqrt{r} \, .$$

Example 7.3.12. Let's return yet again to our air pollution data set. In that example, $k = 3$ (3 groups), and $\nu = 3$ (3 degrees of freedom for the error space), so our Tukey adjusted 95% confidence intervals look like

$$\bar{y}_{i\cdot} - \bar{y}_{j\cdot} \quad \pm \quad \frac{\texttt{qtukey(0.95,3,3)}}{\sqrt{2}} \sqrt{MSE}$$

$$= \bar{y}_{i\cdot} - \bar{y}_{j\cdot} \quad \pm \quad \left(\frac{7.66}{\sqrt{2}}\right)(8.21)$$

$$= \bar{y}_{i\cdot} - \bar{y}_{j\cdot} \quad \pm \quad 34.3 \, .$$

This is only slightly narrower than the Bonferroni adjusted intervals because we only made three comparisons. The difference between the two methods becomes more striking as the the number of groups increases.   ◁

Since all of the differences in our means were less than 34.3, all of the pairwise confidence intervals will contain 0, and none of these differences is significant at the $\alpha = 0.05$ level. This is not surprising since the global test of $H_0 : \mu_1 = \mu_2 = \mu_3$ was not significant either. In fact, typically one only does a Tukey pairwise comparison after first rejecting the null hypothesis that all the means are the same.

Example 7.3.13. Of course, R can do the multiple comparisons calculations for us, too:

```
> airp.aov <- aov(pollution~location,airpollution)
> TukeyHSD(airp.aov)
  Tukey multiple comparisons of means
    95% family-wise confidence level

Fit: aov(formula = pollution ~ location, data = airpollution)

$location
                          diff     lwr    upr   p adj
Plains Suburb-Hill Suburb   -6 -40.290 28.290 0.76435
Urban City-Hill Suburb      15 -19.290 49.290 0.30192
Urban City-Plains Suburb    21 -13.290 55.290 0.16009
```

HSD stands for **honest significant differences**. Note that `tukeyHSD()` displays both confidence intervals and p-values and requires a new function for fitting the model (`aov()`). This function computes and stores different information from `anova(lm())`.

Alternatively, we can use some utilities from the `multcomp` (multiple comparisons) package. The `glht()` (generalized linear hypothesis test) function requires a model (output from `lm()` or `glm()`, for example) and a description of the desired set of contrasts. The `mcp()` (multiple comparisons) function provides an easy way to specify all of the contrasts needed for pairwise comparisons.

```
> airp.cint <- confint(glht(airp.model,mcp(location="Tukey")))
> airp.cint


            Simultaneous Confidence Intervals

Multiple Comparisons of Means: Tukey Contrasts


Fit: lm(formula = pollution ~ location, data = airpollution)


Quantile = 4.175
95% family-wise confidence level



Linear Hypotheses:
                               Estimate lwr      upr
Plains Suburb - Hill Suburb == 0  -6.000  -40.260  28.260
Urban City - Hill Suburb == 0     15.000  -19.260  49.260
Urban City - Plains Suburb == 0   21.000  -13.260  55.260


> plot(TukeyHSD(airp.aov)); plot(airp.cint)   # plots
```

airp-glht

The `plot()` function can be used to produce a graphical depiction of the intervals produced either way (see Figure 7.16). ◁

Example 7.3.14. Returning to Example 7.3.1, we see that although we have only developed a method for handling balanced designs, R can also handle groups of different sizes.

```
> coag.aov <- aov(coag~diet,coagulation); coag.aov
Call:
   aov(formula = coag ~ diet, data = coagulation)

Terms:
                diet Residuals
Sum of Squares   228      112
Deg. of Freedom    3       20

Residual standard error: 2.3664
Estimated effects may be unbalanced
> TukeyHSD(coag.aov)
  Tukey multiple comparisons of means
    95% family-wise confidence level

Fit: aov(formula = coag ~ diet, data = coagulation)

$diet
    diff      lwr     upr   p adj
B-A    5  0.72455  9.2754 0.01833
C-A    7  2.72455 11.2754 0.00096
D-A    0 -4.05604  4.0560 1.00000
```

coag-TukeyHSD

```
C-B    2  -1.82407   5.8241 0.47660
D-B   -5  -8.57709  -1.4229 0.00441
D-C   -7 -10.57709  -3.4229 0.00013
```

As the documentation from `?TukeyHSD` warns:

> Technically the intervals constructed in this way would only apply to balanced designs where there are the same number of observations made at each level of the factor. This function incorporates an adjustment for sample size that produces sensible intervals for mildly unbalanced designs.

The methods in the `multcomp` package are somewhat better for unbalanced designs and employ multivariate *t*-distributions.

coag-glht

```
> require(multcomp)
> coag.glht <- glht(coag.model,mcp(diet="Tukey"))
> summary(coag.glht)


            Simultaneous Tests for General Linear Hypotheses

Multiple Comparisons of Means: Tukey Contrasts


Fit: lm(formula = coag ~ diet, data = coagulation)

Linear Hypotheses:
             Estimate Std. Error t value Pr(>|t|)
B - A == 0   5.00e+00   1.53e+00    3.27   0.0181
C - A == 0   7.00e+00   1.53e+00    4.58   <0.001
D - A == 0   2.18e-15   1.45e+00    0.00   1.0000
C - B == 0   2.00e+00   1.37e+00    1.46   0.4748
D - B == 0  -5.00e+00   1.28e+00   -3.91   0.0045
D - C == 0  -7.00e+00   1.28e+00   -5.48   <0.001
(Adjusted p values reported -- single-step method)


> plot(TukeyHSD(coag.aov)); plot(confint(coag.glht))   # plots
```

The graphical presentation of these simultaneous confidence intervals appears in Figure 7.16. The p-values reported by `summary(coag.glht)` are adjusted to take into account the multiple comparisons. The reported p-value is the smallest familywise error rate at which each test would still be significant.   ◁

Testing other contrasts using glht()

The `glht()` (general linear hypothesis test) function from the `multcomp` package can test a wide range of models and contrasts and provides a pleasant syntax for specifying contrasts. The model supplied can be the object returned by any of `lm()`, `aov()`, and `glm()`.

The `glht()` function can test other sets of contrasts, too, making the necessary adjustments to p-values and confidence intervals to account for the multiple comparisons. The `glht()` function requires a model (the output of `lm()`, `glm()`,

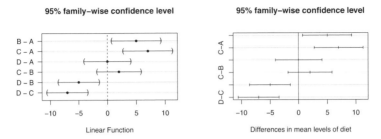

Figure 7.16. A graphical presentation of the simultaneous confidence intervals produced using `confint.glht()` (left) and `TukeyHSD()` (right).

or `aov()`) and a description of the contrast or contrasts to be tested. The output of `glht()` can be passed to `summary()` to produce p-values or to `confint()` to produce confidence intervals. In Example 7.3.14 we used `mcp()` to specify all of the contrasts involved in pairwise comparisons. We can specify other contrasts by providing the coefficients on each of the non-intercept terms in the model.

Example 7.3.15. We demonstrate the use of `glht()` with our air pollution data from Example 7.3.5. In the standard parameterization of the model,

$$E(Y) = \beta_0 + \beta_1 [\![\texttt{location} = 2]\!] + \beta_2 [\![\texttt{location} = 3]\!] ,$$

our two contrasts are

$$C_1 : \beta_1 = 0 ,$$

$$C_2 : \frac{1}{2}\beta_1 - \beta_2 = 0 .$$

We can describe these contrasts to R by specifying the coefficients on the non-intercept parameters in the model.

```
> airp.lm1 <- lm(pollution~location,airpollution)
# specify contrasts by giving the coefficients
> contr <- rbind(
+          c(0,1,0),
+          c(0,0.5,-1))
# we can give our contrasts custom names if we like
> contr1 <- rbind(
+          "hill - plains" = c(0,1,0),
+          "suburb - urban" = c(0,0.5,-1))
> summary(glht(airp.lm1,contr1))
< 5 lines removed >
Linear Hypotheses:
                    Estimate Std. Error t value Pr(>|t|)
hill - plains == 0     -6.00       8.21   -0.73     0.74
suburb - urban == 0   -18.00       7.11   -2.53     0.15
(Adjusted p values reported -- single-step method)
```

airp-glht1

If we choose a different description of the model, namely

$$E(Y) = \mu_1 [\![\texttt{location} = 1]\!] + \mu_2 [\![\texttt{location} = 2]\!] + \mu_3 [\![\texttt{location} = 3]\!] ,$$

we can use a more natural description of the contrasts.

airp-glht1a

```
# these look nicer if we parameterize differently in the model
> airp.lm2 <- lm(pollution~-1 + location,airpollution)
> contr2 <- rbind(
+         "hill - plains" = c(1,-1,0),
+         "suburb - urban" = c(1,1,-2))
> summary(glht(airp.lm2,contr2))
< 5 lines removed >
Linear Hypotheses:
                   Estimate Std. Error t value Pr(>|t|)
hill - plains == 0     6.00       8.21    0.73     0.74
suburb - urban == 0  -36.00      14.21   -2.53     0.15
(Adjusted p values reported -- single-step method)
```

Alternatively, we can use the mcp() function with the original model.

airp-glht1b

```
# using mcp() to help build the contrasts
> airp.lm3 <- lm(pollution~location,airpollution)
> contr3 <- mcp(location = rbind(
+         "hill - plains" = c(1,-1,0),
+         "suburb - urban" = c(1,1,-2)
+         ))
> summary(glht(airp.lm3,contr3))
< 8 lines removed >
Linear Hypotheses:
                   Estimate Std. Error t value Pr(>|t|)
hill - plains == 0     6.00       8.21    0.73     0.74
suburb - urban == 0  -36.00      14.21   -2.53     0.15
(Adjusted p values reported -- single-step method)
```

The p-values here are larger than they were in Example 7.3.8 because they have been adjusted to account for the multiple (in this case, two) comparisons being done.

airp-glht1c

```
# unadjusted p-values
> 2 * pt(-0.731,df=3)
[1] 0.5177
> 2 * pt(-2.533,df=3)
[1] 0.0852
```

We can also get the unadjusted p-value using glht() with just one contrast.

airp-glht1d

```
> airp.lm4 <- lm(pollution~location,airpollution)
> contr4 <- mcp(location = rbind(
+         "hill - plains" = c(1,-1,0)))
> summary(glht(airp.lm4,contr4))
< 8 lines removed >
Linear Hypotheses:
                   Estimate Std. Error t value Pr(>|t|)
hill - plains == 0     6.00       8.21    0.73     0.52
(Adjusted p values reported -- single-step method)
```

◁

Example 7.3.16. The `cholesterol` data set in the `multcomp` package contains data from a clinical trial testing various drug treatments effectiveness at reducing cholesterol. The five treatments include three treatments that used a new drug (1 dose of 20 mg, 2 doses of 10 mg, or 4 doses of 5 mg daily) and two control treatments that used competing drugs.

We begin by fitting the ANOVA model and checking the diagnostic plots that R produces.

```
> data(cholesterol,package="multcomp")                    cholesterol
> chol.model <- lm(response~trt,cholesterol)
> plot(chol.model)              # diagnostic plots
> summary(chol.model)
< 8 lines removed >
Coefficients:
            Estimate Std. Error t value Pr(>|t|)
(Intercept)     5.78       1.02    5.67 9.8e-07
trt2times       3.44       1.44    2.39    0.021
trt4times       6.59       1.44    4.57 3.8e-05
trtdrugD        9.58       1.44    6.64 3.5e-08
trtdrugE       15.17       1.44   10.51 1.1e-13

Residual standard error: 3.23 on 45 degrees of freedom
Multiple R-squared: 0.742,      Adjusted R-squared: 0.72
F-statistic: 32.4 on 4 and 45 DF,  p-value: 9.82e-13

> anova(chol.model)
Analysis of Variance Table

Response: response
          Df Sum Sq Mean Sq F value  Pr(>F)
trt        4   1351     338    32.4 9.8e-13
Residuals 45    469      10
```

The diagnostic plots give no cause for concern, and the model utility test has a very small p-value, so we proceed with pairwise comparisons.

```
> chol.glht <- confint(glht(chol.model,mcp(trt="Tukey")))   cholesterol-HSD
> summary(chol.glht)
< 8 lines removed >
Linear Hypotheses:
                    Estimate Std. Error t value Pr(>|t|)
2times - 1time == 0     3.44       1.44    2.39   0.1381
4times - 1time == 0     6.59       1.44    4.57   <0.001
drugD - 1time == 0      9.58       1.44    6.64   <0.001
drugE - 1time == 0     15.17       1.44   10.51   <0.001
4times - 2times == 0    3.15       1.44    2.18   0.2050
drugD - 2times == 0     6.14       1.44    4.25   <0.001
drugE - 2times == 0    11.72       1.44    8.12   <0.001
drugD - 4times == 0     2.99       1.44    2.07   0.2512
drugE - 4times == 0     8.57       1.44    5.94   <0.001
drugE - drugD == 0      5.59       1.44    3.87   0.0030
(Adjusted p values reported -- single-step method)
```

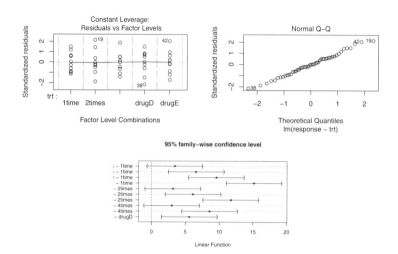

Figure 7.17. Plots of an ANOVA model for the `cholesterol` data set.

```
> plot(confint(chol.glht))
```

Most of the pairwise differences are significant at the familywise error rate of 0.05 as the table of adjusted p-values and plot of 95% simultaneous confidence intervals show. The general story seems to be that the other two drugs do better than the new drug being tested and that the new drug perhaps does somewhat better when used with smaller, more frequent doses.

Had we been primarily interested in just the questions

(1) Does the number of doses of the new drug matter?
(2) Is the new drug different from the old drugs? and
(3) Do the two old drugs differ?

we could test just these questions by setting up the following contrasts:

(1) $\mu_1 - \mu_2$ and $\frac{1}{2}\mu_1 + \frac{1}{2}\mu_2 - \mu_3$,
(2) $\frac{1}{3}\mu_1 + \frac{1}{3}\mu_2 + \frac{1}{3}\mu_3 - \frac{1}{2}\mu_4 - \frac{1}{2}\mu_5$, and
(3) $\mu_4 - \mu_5$.

```
> summary(glht(chol.model, mcp(trt =                          chol-contrasts
+     rbind(
+         "1time - 2times" = c(1,-1,0,0,0),
+         "(1 or 2 times) - 4times" = c(0.5,0.5,-1,0,0),
+         "new - old" = c(2,2,2,-3,-3)/6,
+         "drugD - drugE" = c(0,0,0,1,-1))
+     )))
< 8 lines removed >
Linear Hypotheses:
                        Estimate Std. Error t value Pr(>|t|)
1time - 2times == 0       -3.443      1.443   -2.39   0.0815
```

```
(1 or 2 times) - 4times == 0    -4.871    1.250   -3.90   0.0013
new - old == 0                  -9.027    0.932   -9.69   <1e-04
drugD - drugE == 0              -5.586    1.443   -3.87   0.0014
(Adjusted p values reported -- single-step method)
```

Three of the four hypothesis tests are significant using a familywise error rate of $\alpha = 0.05$. Multiple comparisons are again being handled, but the adjustment is less pronounced now, since we have fewer comparisons. (Compare the results for the contrast $\mu_1 - \mu_2$ here and in the Tukey pairwise comparisons, for example.) Of course, it is not legitimate to first test all pairs (or even to look at the data casually) and then to decide which contrasts are of interest and to report based on the smaller number of comparisons. The decision to restrict attention to only a small set of contrasts must be justified without reference to the data.

Finally, we point out that with a little more effort we can obtain a p-value for the test of $H_0 : \mu_1 = \mu_2 = \mu_3$. See Exercise 7.21. ◁

Dunnet's contrasts

Sometimes, one of the levels of the explanatory variable is a reference or control. In this situation, it makes sense to make all the pairwise comparisons to this reference level but to omit other pairwise comparisons. Dunnet's method makes the appropriate adjustments in this situation, much like Tukey's method did when we wanted all pairwise comparisons. We omit the details of the method this time but provide an example showing how to make the computations using `glht()`.

Example 7.3.17. Returning to Example 7.3.1, suppose that diet A is a standard diet that serves as a reference for the other diets. We can use Dunnet contrasts to compare each of the other diets to diet A.

```
> summary(glht(coag.model, mcp(diet = "Dunnet")))      coag-dunnet
< 8 lines removed >
Linear Hypotheses:
          Estimate Std. Error t value Pr(>|t|)
B - A == 0 5.00e+00  1.53e+00    3.27   0.0097
C - A == 0 7.00e+00  1.53e+00    4.58   <0.001
D - A == 0 2.18e-15  1.45e+00    0.00   1.0000
(Adjusted p values reported -- single-step method)
```

Notice that the values of the test statistics are identical to those on page 448, but the p-values are smaller because we are adjusting for fewer comparisons. ◁

7.3.8. More Examples

Example 7.3.18. Recall Example 6.6.1. In that example we considered a food company that tests flavors by giving the flavor to 50 people (25 men and 25 women) and having them rate the flavor. The result is an overall flavor score based on the 50 individual scores. The samples that are tasted can be prepared in different ways by using, for example, more or less liquid, or by grinding up the ingredients more or less finely. The company is interested in knowing if these different preparation

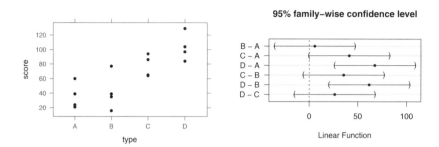

Figure 7.18. Do taste scores depend on preparation methods?

alternatives affect the flavor score. If not, then it really doesn't matter how the flavors are prepared. If so, then they need to control the preparation method carefully if they hope to have comparable results, and they also need to make a decision about which preparation method should be used.

Here are the results of an ANOVA comparing four different preparation methods. The same flavor was used for all 16 groups, but four different preparation methods were employed.

```
> summary(score~type,data=tastetest,                        taste-anova
+           fun=function(x){cbind(mean=mean(x),sd=sd(x))})
score     N=16


+-------+-+--+----------+--------+
|       | |N |mean score|sd score|
+-------+-+--+----------+--------+
|type   |A| 4| 36.000   |17.833  |
|       |B| 4| 41.750   |25.552  |
|       |C| 4| 77.250   |15.086  |
|       |D| 4|103.500   |18.912  |
+-------+-+--+----------+--------+
|Overall| |16| 64.625   |33.388  |
+-------+-+--+----------+--------+
> taste.xy <- xyplot(score~type,data=tastetest)
> taste.lm <- lm(score~type,data=tastetest)
> anova(taste.lm)
Analysis of Variance Table

Response: score
          Df Sum Sq Mean Sq F value Pr(>F)
type       3  12053    4018    10.3 0.0012
Residuals 12   4668     389
> taste.cint <- confint(glht(taste.lm,mcp(type="Tukey"))); taste.cint
< 11 lines removed >


Linear Hypotheses:
           Estimate lwr      upr
B - A == 0    5.750 -35.661  47.161
C - A == 0   41.250  -0.161  82.661
```

Box 7.4. Notation for Two-Way ANOVA

The following notation is used for our discussion of two-way ANOVA:

$$n_{ij} = \text{number of responses with } A = i \text{ and } B = j\,,$$

$$y_{ijk} \text{ or } Y_{ijk} = k\text{th observed response with } A = i \text{ and } B = j\,,$$

$$\mu_{ij} = \text{mean of subpopulation with } A = i \text{ and } B = j\,,$$

$$\bar{y}_{ij\cdot} \text{ or } \overline{Y}_{ij\cdot} = \text{mean of sample responses with } A = i \text{ and } B = j$$

$$= \frac{1}{n_{ij}} \sum_{k=1}^{n_{ij}} y_{ijk}\,.$$

```
D - A == 0   67.500    26.089 108.911
C - B == 0   35.500    -5.911  76.911
D - B == 0   61.750    20.339 103.161
D - C == 0   26.250   -15.161  67.661
```

From this it appears that methods A and B are similar to each other but perhaps different from methods C and D. (We have a pretty small sample and the differences between A and C and between B and C are almost significant at the $\alpha = 0.05$ level, even allowing for the multiple comparisons we are making.) In fact, there is a similarity between methods A and B and between methods C and D. We will return to this data set in the next section. ◁

7.4. Two-Way ANOVA

7.4.1. The Two-Way ANOVA Model

In this section we consider ANOVA models with two categorical predictors. In this situation, there are three important questions:

- What is the effect of factor A?
- What is the effect of factor B?
- Are these effects *additive*?

   That is, if I know the effect of changing factor A and the effect of changing factor B, is the effect of changing both the sum of these two main effects?

Although ANOVA methods can be used for observational studies as well, many of the methods were developed with randomized experiments in mind. In a randomized experiment, the predictors represent treatment conditions that can be combined in all possible ways. In the simplest case there are two categorical variables (A and B) that each have two possible levels (1 and 2). This leads to four possible treatments, which are ideally applied at random and in equal number to the observational units or subjects.

Table 7.5. The taste test data.

	score	scr	liq	type
1	24	coarse	hi	A
2	21	coarse	hi	A
3	39	coarse	hi	A
4	60	coarse	hi	A
5	35	coarse	lo	B
6	39	coarse	lo	B
7	77	coarse	lo	B
8	16	coarse	lo	B
9	65	fine	hi	C
10	94	fine	hi	C
11	86	fine	hi	C
12	64	fine	hi	C
13	104	fine	lo	D
14	129	fine	lo	D
15	97	fine	lo	D
16	84	fine	lo	D

When doing two-way ANOVA, it is common to introduce notation that highlights the structure of the data classified by two categorical variables. Our notation is introduced in Box 7.4. In our taste testing example (Example 7.3.18), there were four observations in each treatment group ($n_{ij} = 4$ for all i and j). The data are displayed in Table 7.5. At the time, we called the four treatments A, B, C, and D. But that hid the fact that this was really a two-way design. The two variables were type of screen (fine or coarse) and amount of liquid (hi or lo) used to prepare the samples for tasting.

Using the one-way ANOVA model of the previous section, we could handle this situation by considering one predictor with four levels. The model would then have four parameters corresponding to the four subpopulation means (μ_{11}, μ_{21}, μ_{12}, μ_{22}). By default, R reparameterizes this as

$$\mu_{11} = \beta_0 \,,$$
$$\mu_{21} = \beta_0 + \beta_1 \,,$$
$$\mu_{12} = \beta_0 + \beta_2 \,,$$
$$\mu_{22} = \beta_0 + \beta_3 \,.$$

This is what we did in Example 7.3.18.

But this approach does not make our primary questions readily answerable. Instead, we prefer a different approach. It is helpful to visualize the four group means involved in the following table:

A \ B	1	2
1	μ_{11}	μ_{12}
2	μ_{21}	μ_{22}

Our three questions correspond to three contrasts:

- $C_1 = \dfrac{\mu_{21} + \mu_{22}}{2} - \dfrac{\mu_{11} + \mu_{12}}{2} = \dfrac{1}{2}\left(-\mu_{11} - \mu_{12} + \mu_{21} + \mu_{22}\right)$ (factor A effect),

- $C_2 = \dfrac{\mu_{21} + \mu_{22}}{2} - \dfrac{\mu_{11} + \mu_{12}}{2} = \dfrac{1}{2}\left(-\mu_{11} + \mu_{12} - \mu_{21} + \mu_{22}\right)$ (factor B effect),

- $C_3 = (\mu_{22} - \mu_{12}) - (\mu_{22} - \mu_{11}) = \mu_{11} - \mu_{12} - \mu_{21} + \mu_{22}$ (interaction effect).

When n_{ij} is the same for each i and j, then these contrasts are orthogonal and span the model space. (When the sample sizes vary by treatment group, these contrasts still span the model space but are not orthogonal.) When these contrasts are orthogonal, we have the following Pythagorean decomposition of SSM:

$$SSM = SS(C_1) + SS(C_2) + SS(C_3)\,,$$

and we can test our three hypotheses of interest with three independent F-tests where

$$F = \frac{SS(C_i)/1}{SSE/(n-4)} = \frac{MS(C_i)}{MSE} \sim \mathsf{F}(1, n-4)\,.$$

As the following example shows, this is exactly what `anova()` will do given the appropriate model formula.

Example 7.4.1. Let's return to our taste testing experiment (Example 7.3.18). The group means can be summarized in the following table.

```
> summary(score~scr+liq, data=tastetest, method="cross")        taste-xtab

 mean by scr, liq

 +-----+
 |N    |
 |score|
 +-----+

 +------+-------+-------+-------+
 |  scr |  hi   |  lo   |  ALL  |
 +------+-------+-------+-------+
 |coarse| 4     | 4     | 8     |
 |      | 36.000| 41.750| 38.875|
 +------+-------+-------+-------+
 |fine  | 4     | 4     | 8     |
 |      | 77.250|103.500| 90.375|
 +------+-------+-------+-------+
 |ALL   | 8     | 8     |16     |
 |      | 56.625| 72.625| 64.625|
 +------+-------+-------+-------+
```

This time we will fit the model using a formula with two predictors. We will say a bit more about the format of the formula in a moment.

```
> taste.lm <- lm(score~scr*liq,data=tastetest)        taste-2way-anova
> anova(taste.lm)
Analysis of Variance Table

Response: score
```

```
          Df Sum Sq Mean Sq F value  Pr(>F)
scr        1  10609   10609   27.27 0.00021
liq        1   1024    1024    2.63 0.13068
scr:liq    1    420     420    1.08 0.31914
Residuals 12   4669     389
```

The first three rows of the ANOVA table correspond to our three contrasts. The fourth row provides MSE for the denominator of our F-statistic. The test for interaction (C_3) has a p-value greater than 0.3, so we proceed to consider the main effects. Of these, only the effect of the screen is significant. This is a small data set, so we don't have much power to detect small differences in taste score, but the effect of texture (fine vs. coarse screen) is large enough to detect even in this small sample.

Notice the labeling of the third row of the ANOVA table: scr:liq. This is R's notation for an interaction. Our model formula score ~ scr * liq is an abbreviation for score ~ scr + liq + scr:liq. See Box 7.6 for a description of model formula syntax in R.   ◁

The previous example illustrates our general approach to this sort of situation.

(1) First test $H_0 : C_3 = 0$ (no interaction).

(2) If we reject $H_0 : C_3 = 0$, then it is unclear what we even mean by the effect of treatment A (or of treatment B) since

$$\mu_{22} - \mu_{12} \neq \mu_{21} - \mu_{11}$$

implies that the response to A depends on the level of B (and vice versa).

(3) If $C_3 = 0$, then C_1 and C_2 are **main effects** for each of our factors and do not depend on the level of the other factor. Thus if we do not reject $H_0 : C_3 = 0$, then we can proceed to evaluate the main effects C_1 and C_2.

Example 7.4.2. It is worthwhile to compare the output from Example 7.4.1 with the following.

taste-2way-lm

```
> taste.lm <- lm(score~scr*liq,data=tastetest)
> summary(taste.lm)
< 9 lines removed >
             Estimate Std. Error t value Pr(>|t|)
(Intercept)     36.00       9.86    3.65   0.0033
scrfine         41.25      13.95    2.96   0.0120
liqlo            5.75      13.95    0.41   0.6874
scrfine:liqlo   20.50      19.72    1.04   0.3191

Residual standard error: 19.7 on 12 degrees of freedom
Multiple R-squared: 0.721,      Adjusted R-squared: 0.651
F-statistic: 10.3 on 3 and 12 DF,  p-value: 0.00121
```

Since only one of the p-values matches our earlier output, it is clear that this is testing different hypotheses. The model being fit here is described in Table 7.7. We

---

Box 7.6. Model Formula Syntax in R

The models fit by the `lm()` and `glm()` functions are specified in a compact symbolic form. An expression of the form `y ~ model` is used to describe how `y` is being modeled by predictors described in `model`. Such a model consists of a series of terms separated by `+` operators. The terms themselves consist of variable and factor names separated by `:` operators. Such a term is interpreted as the interaction of all the variables and factors appearing in the term.

In addition to `+` and `:`, a number of other operators are useful for describing models more succinctly:

* `*` denotes factor crossing: `a*b` is interpreted as `a + b + a:b`.

* `^` indicates crossing to the specified degree. For example `(a+b+c)^2` is identical to `(a+b+c)*(a+b+c)` which in turn expands to a formula containing the main effects for `a`, `b`, and `c` together with their second-order interactions.

* `-` removes the specified terms, so that `(a+b+c)^2 - a:b` is identical to `a + b + c + b:c + a:c`. It can also be used to remove the intercept term: `y ~ x - 1`.

* `0` A model with no intercept can be also specified as `y ~ x + 0` or `y ~ 0 + x`.

* `.` represents "all columns not already in the formula".

Model formulas may also use `%in%` (for nesting) and `offset()` (to include a term with a fixed coefficient of 1).

As we have already seen, formulas may involve arithmetic expressions (e.g., `log(y) ~ a + log(x)`). The `I()` function is used to avoid confusion for the operators listed above, which have special interpretations within model formulas (e.g., `act ~ I(satm + satv)` vs. `act ~ satm + satv`).

For more details, see `?lm()`, `?glm()`, and especially `?formula`.

---

can express the four subpopulation means in terms of the parameters β as follows:

$$\mu_{11} = \beta_0 \,,$$
$$\mu_{21} = \beta_0 + \beta_1 \,,$$
$$\mu_{12} = \beta_0 + \beta_2 \,,$$
$$\mu_{22} = \beta_0 + \beta_1 + \beta_2 + \beta_3 \,.$$

From this it follows that

$$C_1 = \frac{\mu_{21} + \mu_{22}}{2} - \frac{\mu_{11} + \mu_{12}}{2} = \beta_1 + \frac{\beta_3}{2} \,,$$
$$C_2 = \frac{\mu_{12} + \mu_{22}}{2} - \frac{\mu_{11} + \mu_{21}}{2} = \beta_2 + \frac{\beta_3}{2} \,,$$
$$C_3 = (\mu_{22} - \mu_{21}) - (\mu_{12} - \mu_{11}) = \beta_3 \,.$$

Table 7.7. The model matrix and algebraic description of the model fit by `lm(score~scr*liq,taste)`.

	(Intercept)	scrfine	liqlo	scrfine:liqlo
1	1	0	1	0
2	1	0	1	0
3	1	0	1	0
4	1	0	1	0
5	1	1	1	1
6	1	1	1	1
7	1	1	1	1
8	1	1	1	1
9	1	0	0	0
10	1	0	0	0
11	1	0	0	0
12	1	0	0	0
13	1	1	0	0
14	1	1	0	0
15	1	1	0	0
16	1	1	0	0

$$\boldsymbol{y} = \beta_0 \mathbf{1} + \beta_1 [\![\texttt{scr} = \texttt{fine}]\!] + \beta_2 [\![\texttt{liq} = \texttt{lo}]\!] + \beta_3 [\![\texttt{scr} = \texttt{fine and } \texttt{liq} = \texttt{lo}]\!]$$
$$= \beta_0 \mathbf{1} + \beta_1 [\![\texttt{scr} = \texttt{fine}]\!] + \beta_2 [\![\texttt{liq} = \texttt{lo}]\!] + \beta_3 [\![\texttt{scr} = \texttt{fine}]\!] \cdot [\![\texttt{liq} = \texttt{lo}]\!]$$

This explains why the p-values in the last lines of the outputs of `summary(lm())` and `anova(lm())` match – they are testing the same contrast. But since in general $\beta_1 \neq C_1$ and $\beta_2 \neq C_2$, the other two hypotheses being tested are not the same. If $\beta_3 = 0$, then β_1 and β_2 are the two main effects, and these effects are additive. But when $\beta_3 \neq 0$, then β_3 represents an interaction effect and interpreting β_1 and β_2 is a bit more subtle.   ◁

Example 7.4.3.

Q. Fit a model using `summary(lm())` that corresponds precisely to the contrasts C_1, C_2, and C_3 in `anova(lm())` by constructing the model matrix manually.

A. We can build the model matrix and use it as the righthand side of our formula in `lm()`:

taste-2way-matrix

```
> M <- cbind(                                      # model matrix
+          "C1" = rep(c(-1,-1,1,1),each=4)/2,      # C1
+          "C2" = rep(c(-1,1,-1,1),each=4)/2,      # C2
+          "C3" = rep(c(1,-1,-1,1),each=4)/4       # C3
+          )
> taste.lm2 <- lm(score~M,data=tastetest)
> summary(taste.lm2)
< 8 lines removed >
Coefficients:
            Estimate Std. Error t value Pr(>|t|)
(Intercept)    64.62       4.93   13.11  1.8e-08
```

```
MC1              51.50         9.86    5.22  0.00021
MC2              16.00         9.86    1.62  0.13068
MC3              20.50        19.72    1.04  0.31914


Residual standard error: 19.7 on 12 degrees of freedom
Multiple R-squared: 0.721,        Adjusted R-squared: 0.651
F-statistic: 10.3 on 3 and 12 DF,  p-value: 0.00121
```
◁

7.4.2. Model Comparison Tests

We can also obtain the p-values of a two-way ANOVA via model comparison tests, but we have to construct the submodels with some care. Recall that the test for main effects for factors A and B are tests of the null hypotheses

$$H_0 : C_1 = 0 \text{ and}$$
$$H_0 : C_2 = 0 \, ,$$

where

$$C_1 = \frac{\mu_{21} + \mu_{22}}{2} - \frac{\mu_{11} + \mu_{12}}{2} = \beta_1 + \frac{1}{2}\beta_3 \, , \text{ and}$$
$$C_2 = \frac{\mu_{12} + \mu_{22}}{2} - \frac{\mu_{11} + \mu_{21}}{2} = \beta_2 + \frac{1}{2}\beta_3 \, .$$

So the hypotheses for testing main effects have the form

- H_0: $\beta_3 = -2\beta_1$ (no main effect for one factor),
- H_0: $\beta_3 = -2\beta_2$ (no main effect for the other factor).

Example 7.4.4.

Q. Obtain the ANOVA p-values from the taste test data using model comparison tests.

A. The hardest part is coding the models. Here we show two ways to do it. Either we can code the taste data set numerically ourselves or we can copy information from `model.matrix()` which records the coding scheme used in fitting Ω.

```
> ntaste <- data.frame(score = tastetest$score,                    [taste-mct]
+                      scr    = as.numeric(tastetest$scr) - 1,
+                      liq    = as.numeric(tastetest$liq) - 1,
+                      scrliq = ( as.numeric(tastetest$scr) -1 ) *
+                               ( as.numeric(tastetest$liq) -1 )
+                      ); ntaste
   score scr liq scrliq
1     24   0   0      0
2     21   0   0      0
3     39   0   0      0
4     60   0   0      0
5     35   0   1      0
6     39   0   1      0
7     77   0   1      0
8     16   0   1      0
9     65   1   0      0
```

```
10    94    1    0       0
11    86    1    0       0
12    64    1    0       0
13   104    1    1       1
14   129    1    1       1
15    97    1    1       1
16    84    1    1       1
>
> Omega <- lm(score~scr*liq,data=tastetest)
> M <- model.matrix(Omega)
> M2 <- cbind(M[,3], M[,2] - 2 * M[,4])
> M3 <- cbind(M[,2], M[,3] - 2 * M[,4])
>
> omega1 <- lm(score~scr+liq,data=tastetest)
> omega2 <- lm(score~M2,tastetest)
> omega2a <- lm(score~ liq + I(scr - 2 * scrliq),data=ntaste)
> omega3 <- lm(score~M3,tastetest)
> omega3a <- lm(score~ scr + I(liq - 2 * scrliq),data=ntaste)
>
> anova(omega1,Omega)    # test for interaction
Analysis of Variance Table

Model 1: score ~ scr + liq
Model 2: score ~ scr * liq
  Res.Df  RSS Df Sum of Sq     F Pr(>F)
1      13 5089
2      12 4669  1       420 1.08   0.32
# test main effect for scr
# anova(omega2a,Omega)  # this gives the same result as line below
> anova(omega2,Omega)
Analysis of Variance Table

Model 1: score ~ M2
Model 2: score ~ scr * liq
  Res.Df   RSS Df Sum of Sq    F Pr(>F)
1      13 15278
2      12  4669  1     10609 27.3 0.00021
# test main effect for liq
# anova(omega3a,Omega)  # this gives the same result as line below
> anova(omega3,Omega)
Analysis of Variance Table

Model 1: score ~ M3
Model 2: score ~ scr * liq
  Res.Df  RSS Df Sum of Sq     F Pr(>F)
1      13 5692
2      12 4669  1      1024 2.63   0.13
```

Once upon a time, this sort of variable recoding was required in many software packages. Fortunately, this effort is no longer required.   ◁

7.4.3. Interaction Plots

The following example demonstrates an interaction effect which we investigate further using an **interaction plot**.

Example 7.4.5. A study was done to measure the effect of room noise on the performance of second graders on a mathematics test [**ZS80**]. Some of the students were diagnosed as hyperactive; the others were controls. We simulated data that are roughly compatible with the summary statistics presented in the paper analyzing this study.

First we fit an additive model without an interaction term.

```
> model1 <- lm(score~noise+group,mathnoise)                              noise-math1
> anova(model1)
Analysis of Variance Table

Response: score
          Df Sum Sq Mean Sq F value Pr(>F)
noise      1   1440    1440    1.04   0.31
group      1  38440   38440   27.86  6e-06
Residuals 37  51048    1380
> summary(score~group,data=mathnoise,fun=favstats)
score    N=40


+-------+-------+--+---+---+----+------+------+
|       |       |N |min|max|mean|median|sd    |
+-------+-------+--+---+---+----+------+------+
|group  |control|20|116|268|192 |192   |39.792|
|       |hyper  |20| 66|194|130 |130   |34.339|
+-------+-------+--+---+---+----+------+------+
|Overall|       |40| 66|268|161 |161   |48.285|
+-------+-------+--+---+---+----+------+------+
```

Based on this, we might easily have concluded that noise level does not matter but that the controls perform better than the hyperactive children.

Now consider a model with an interaction term.

```
> model2 <- lm(score~noise*group,mathnoise)                              noise-math2
> anova(model2)
Analysis of Variance Table

Response: score
            Df Sum Sq Mean Sq F value  Pr(>F)
noise        1   1440    1440    1.27  0.2672
group        1  38440   38440   33.91 1.2e-06
noise:group  1  10240   10240    9.03  0.0048
Residuals   36  40808    1134
```

There is a clear interaction effect. That is, how (or how much) noise affects performance appears to depend on whether the students are controls or hyperactive. The plots in Figure 7.19 help us visualize the interaction. The plots in the top row are called **interaction plots**. The lines connect group means $\overline{y}_{ij}$ with $\overline{y}_{ij'}$ or

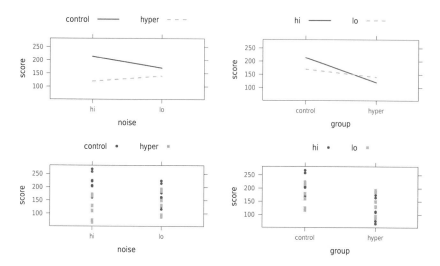

Figure 7.19. Visualizing an interaction effect in the `mathnoise` data.

$\overline{y}_{ij}$ with $\overline{y}_{i'j}$. If there is no interaction, these lines should be parallel because then $\overline{y}_{ij} - \overline{y}_{ij'} = \overline{y}_{i'j} - \overline{y}_{i'j'}$.

The plots were generated with the following code.

```
noise-math-plots
> noise.interaction <- xyplot(score~noise,groups=group,data=mathnoise,
+       type='a',auto.key=list(lines=T,points=F,columns=2))
> noise.xy <- xyplot(score~noise,groups=group,data=mathnoise,
+                 auto.key=list(lines=F,points=T,columns=2))
> noise.interaction2 <- xyplot(score~group,groups=noise,data=mathnoise,
+       type='a',auto.key=list(lines=T,points=F,columns=2))
> noise.xy2 <- xyplot(score~group,groups=noise,data=mathnoise,
+                 auto.key=list(lines=F,points=T,columns=2))
```
◁

7.4.4. Factors with More than Two Levels

When one or both of our predictors have more than two levels, the hypotheses no longer correspond to single contrasts but to combinations of contrasts. For example, if factor A has three levels, then the null hypothesis of interest is

$$H_0 : \mu_{1.} = \mu_{2.} = \mu_{3.} \; ,$$

which is equivalent to

$$H_0 : \mu_{1.} = \mu_{2.} \text{ and } \frac{1}{2} \left(\mu_{1.} + \mu_{2.} \right) = \mu_{3.} \; .$$

This is a combination of two orthogonal contrasts.

In general, if factor A has I levels and factor B has J levels, then the model space will have IJ degrees of freedom because there are IJ group means in the model. Box 7.8 indicates a useful parameterization of this model. Once again, the

---

Box 7.8. Parameterizing the Two-Way ANOVA Model

We can represent the structure of the data for a two-way ANOVA as a rectangle with rows and columns representing the levels of the two factors. When $I = 3$ and $J = 4$, this looks like

	$B = 1$	$B = 2$	$B = 3$
$A = 1$	$\star$	$\circ$	$\circ$
$A = 2$	$\bullet$		
$A = 3$	$\bullet$		
$A = 4$	$\bullet$		

The symbols in the rectangle help us construct a useful parameterization of the model:

group means	$B = 1$	$B = 2$	$B = 3$
$A = 1$	μ	$\mu + \beta_1$	$\mu + \beta_2$
$A = 2$	$\mu + \alpha_1$	$\mu + \alpha_1 + \beta_1 + \gamma_1$	$\mu + \alpha_1 + \beta_2 + \gamma_2$
$A = 3$	$\mu + \alpha_2$	$\mu + \alpha_2 + \beta_1 + \gamma_3$	$\mu + \alpha_2 + \beta_2 + \gamma_4$
$A = 4$	$\mu + \alpha_3$	$\mu + \alpha_3 + \beta_1 + \gamma_5$	$\mu + \alpha_3 + \beta_2 + \gamma_6$

using the following "key":

source	degrees of freedom	label	parameter
intercept	1	$\star$	μ
A	$4 - 1 = 3$	$\bullet$	$\alpha.$
B	$3 - 1 = 2$	$\circ$	$\beta.$
interaction	$(4-1)(3-1) = 6$		$\gamma.$

That is, the $\alpha.$ parameters describe how the means are adjusted as we move from row to row in the first column, the $\beta.$ parameters describe how the means are adjusted as we move from column to column in the first row, and the $\gamma.$ parameters make the "interaction adjustments" required in the remainder of the table.

In general, the degrees of freedom for the two-way ANOVA model can be partitioned as follows:

source	degrees of freedom
intercept	1
A main effects	I-1
B main effects	J-1
interaction effects	(I-1)(J-1)

---

hypotheses of interest may be tested using either contrasts or model comparison tests and the `anova()` function automates this for us.

Example 7.4.6. The `poison` data set contains the results of an experiment in which one of three poisons was given to animals by one of four different treatment methods. The researchers then recorded the length of time (in hours) until the animals died. Several different analyses of these data have been presented, including those in [**BC64**], [**Ait87**], and [**SV99**].

We begin with a simple two-way ANOVA.

```
> poison.lm <- lm(Time~factor(Poison) * factor(Treatment),poison)
> anova(poison.lm)
Analysis of Variance Table

Response: Time
                              Df Sum Sq Mean Sq F value  Pr(>F)
factor(Poison)                 2  103.3    51.7   23.22 3.3e-07
factor(Treatment)              3   92.1    30.7   13.81 3.8e-06
factor(Poison):factor(Treatment)  6   25.0     4.2    1.87    0.11
Residuals                     36   80.1     2.2
```
`poison`

This analysis suggests no interaction effect but clear main effects for both the type of poison and the treatment method. However, a look at the diagnostic plots shows that the assumptions of the ANOVA model are not being met. Figure 7.20 shows that there is much more variability in lifetime (about the model prediction) among the animals who lived the longest.

```
> xplot(poison.lm,w=c(4,2))
```
`poison-resid`

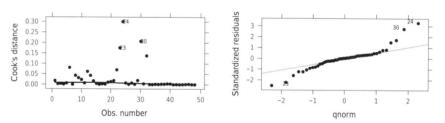

Figure 7.20. The distribution of the residuals in Example 7.4.6 show strong departures from the model assumptions.

Box and Cox [**BC64**] suggested applying a reciprocal transformation to the response variable before fitting the model. This improves the distribution of the residuals greatly (Figure 7.21).

```
> poison.lm2 <- lm(1/Time~factor(Poison) * factor(Treatment),poison)
> xplot(poison.lm2,w=c(4,2))
```
`poison-trans`

In addition to improving the distribution of the residuals, this transformation has a natural interpretation when applied to a variable that measures either a time or a rate – it converts one into the other. So our transformation leads to a model that is predicting the "rate" of death rather than the time until death.

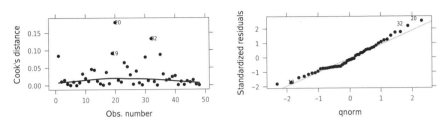

Figure 7.21. Improved residual distribution after transforming the response variable in Example 7.4.6.

In this example, the transformed model presents a similar story, and with stronger evidence. More importantly, the assumptions of the analysis are more nearly satisfied, so we are justified in drawing conclusions from this model.

```
poison-trans-anova
```

```
> anova(poison.lm2)
Analysis of Variance Table

Response: 1/Time
                              Df Sum Sq Mean Sq F value  Pr(>F)
factor(Poison)                 2  0.349  0.1744   72.63 2.3e-13
factor(Treatment)              3  0.204  0.0680   28.34 1.4e-09
factor(Poison):factor(Treatment)  6  0.016  0.0026    1.09    0.39
Residuals                     36  0.086  0.0024
```

It appears that both the type of poison and treatment method matter. We could follow up this analysis by testing a set of contrasts that are of interest – correcting for multiple comparisons, of course.   ◁

7.4.5. Advantages of Two-Way Designs

There are several advantages of two-way designs over two separate one-way designs:

- They use data more efficiently.

  Even if we assume there is no interaction and fit an additive model, for the same total sample size we have more power if we measure both variables on all subjects than if we divide our sample into two subsamples and measure one variable in each sample. In situations where the cost of obtaining an observational unit is high (perhaps a costly experiment must be conducted or perhaps it is difficult to recruit subjects) but the marginal cost of an additional measurement is low (often, but not always the case), studying two factors simultaneously can bring valuable savings.

- We have the potential to detect interaction.

  If we conduct two separate studies, there is no possibility to detect interaction, even if it exists. But by studying two factors simultaneously, we may detect an interaction effect. In some situations, an interaction effect is the most interesting part of the story.

- We can use blocking to reduce the unexplained variation.

Sometimes we are primarily interested in only one factor, but by measuring another factor (the **blocking variable**) that explains some of the variation in our population, we can more easily detect the effect of the variable of interest. The term "block" comes from agriculture where the blocks were rectangular portions of fields. A block design is a generalization of the idea of a paired design. Pairs are blocks of size 2.

There is a old adage in statistics:

Block what you know; randomize the rest.

If there is a factor that plays a role in the variation of the response, it is better to measure it and include it in the model than to leave it in the random (residuals) part of the model. Including such a variable will typically produce smaller estimates for σ and therefore more precise estimates of the model parameters. This benefit is more pronounced when the effect of the blocking variable is large. Of course, if the blocking factor is actually unrelated to the response, we will pay a price in reduced degrees of freedom.

The following example illustrates the power of a blocking variable.

Example 7.4.7. The `palettes` data set contains data from a firm that recycles palettes. Palettes from warehouses are bought, repaired, and resold. (Repairing a palette typically involves replacing one or two boards.) The company has four employees who do the repairs. The employer sampled five days for each employee and recorded the number of palettes repaired. The results are in the table below:

Employee A	123	127	126	125	110
Employee B	118	125	128	125	114
Employee C	116	119	123	119	109
Employee D	113	116	119	115	107

We will perform a number of analyses on this data.

(1) We begin with a one-way ANOVA analysis.

```
> pal.lm1 <- lm(palettes~employee,palettes)        palettes-1way
> anova(pal.lm1)
Analysis of Variance Table

Response: palettes
          Df Sum Sq Mean Sq F value Pr(>F)
employee   3    237    79.0    2.44   0.10
Residuals 16    518    32.4
```

This analysis does not give us any reason to suspect any employees are significantly more or less efficient than the others.

(2) As it turns out, the employer recorded data for each employee on the same five consecutive days. If we do a two-way ANOVA using `day` as a blocking variable, the story changes dramatically.

```
> pal.lm2 <- lm(palettes~employee+day,palettes)     palettes-blocking
> anova(pal.lm2)
Analysis of Variance Table
```

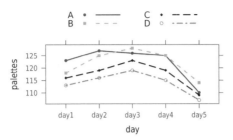

Figure 7.22. The productivity of employees rises and falls together over the course of the week, but their relative productivity remains mostly unchanged.

```
Response: palettes
           Df Sum Sq Mean Sq F value  Pr(>F)
employee    3    237    79.0    24.4 2.1e-05
day         4    479   119.7    37.0 1.2e-06
Residuals  12     39     3.2
```

We now have strong evidence that both employee and the day of the week affect productivity. Although we are not primarily interested in the day-to-day effect, it is important to include this factor in the model. In our first analysis, the presence of a day effect was masking the employee effect. Without the blocking variable, the systematic day-to-day variability was not explainable and simply made each employee look less consistent. Adding in the blocking variable allows us to filter out this effect.

The summary output gives us some indication of what is going on.

```
> summary(pal.lm2)                                      palettes-summary
< 8 lines removed >
Coefficients:
            Estimate Std. Error t value Pr(>|t|)
(Intercept)   120.85       1.14  106.27  < 2e-16
employeeB      -0.20       1.14   -0.18  0.86333
employeeC      -5.00       1.14   -4.40  0.00087
employeeD      -8.20       1.14   -7.21  1.1e-05
dayday2         4.25       1.27    3.34  0.00586
dayday3         6.50       1.27    5.11  0.00026
dayday4         3.50       1.27    2.75  0.01751
dayday5        -7.50       1.27   -5.90  7.3e-05

Residual standard error: 1.8 on 12 degrees of freedom
Multiple R-squared: 0.949,        Adjusted R-squared: 0.919
F-statistic: 31.6 on 7 and 12 DF,  p-value: 8.23e-07
```

The systematic day-to-day variability could have a number of causes, including the state of the palettes to be repaired on a particular day, since palettes that are in worse condition will take longer to repair. There may also be a weekend effect (a tendency to be less productive on days nearer the weekend). See Figure 7.22.

(3) Exercise 7.24 asks you to use contrasts to compare the employees. It would be useful, for example, to know if there is one employee that is especially efficient

or inefficient. The summary output above suggests that employee D is less efficient than the others.   ◁

Blocking can also be used in experimental situations. Suppose, for example, that we are measuring the rate of a chemical reaction with differing amounts of catalyst. If the reaction rate is sensitive to the temperature, we might choose to control the temperature carefully to avoid that source of variability in our results. But then our conclusions would only be directly applicable for reactions at that particular temperature. Alternatively, we could select several different temperatures and repeat each experimental condition (the amount of catalyst) at each of the temperatures.

7.5. Interaction and Higher Order Terms

Many interesting regression models are formed by using transformations or combinations of explanatory variables as predictors. Suppose we have two predictors, x_1 and x_2. Here are some possible models:

- First-order model:
$$Y = \beta_0 + \beta_1 x_1 + \beta_2 x_2 + \varepsilon .$$

- Second-order, no interaction:
$$Y = \beta_0 + \beta_1 x_1 + \beta_2 x_2 + \beta_3 x_1^2 + \beta_4 x_2^2 + \varepsilon .$$

- First-order, plus interaction:
$$Y = \beta_0 + \beta_1 x_1 + \beta_2 x_2 + \beta_3 x_1 x_2 + \varepsilon .$$

- Complete second-order model:
$$Y = \beta_0 + \beta_1 x_1 + \beta_2 x_2 + \beta_3 x_1^2 + \beta_4 x_2^2 + \beta_5 x_1 x_2 + \varepsilon .$$

We can interpret the parameters in these models as follows.

- β_i for $i > 0$ can be thought of as adjustments to the baseline effect given by β_0. (This is especially useful when the predictors are categorical and the intercept has a natural interpretation.)

- When there are no interaction or higher order terms in the model, the parameter β_i can be interpreted as the amount we expect the response to change if we increase x_i by 1 and *leave all other predictors fixed*. The effects due to different predictors are **additive** and do not depend on the values of the other predictors. Graphically this gives us parallel lines or planes.

- With higher order terms in the model, the dependence of the response on one predictor (with all other predictors fixed) may not be linear. Graphically, as we vary one predictor, the response follows a non-linear curve; changing one or more of the other predictors gives rise to a different curve. Importantly, such a curve may be increasing in some regions and decreasing in others.

    Without such higher order terms – even if we use monotonic transformations as discussed in Section 6.5.1 – the model is always monotonic in each predictor. If this agrees with our intuition about the situation, we probably

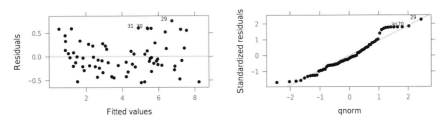

Figure 7.23. Residual plots for a model predicting `thermsPerDay` from `temp`, `kwhpday`, and an interaction term.

have no need for higher order terms. On the other hand, if we expect that our response should reach a maximum (or minimum) for some value of a predictor and fall off from there (or increase from there), this is an indication that we should fit a model with higher order terms.

- With interaction terms, the model is no longer additive: the effect of changing one predictor may depend on the values of the other predictors. Graphically, our lines or curves are no longer parallel.

Example 7.5.1.

Q. In Example 7.2.3 we fit an additive model to the `utilities` data. Is there evidence of an interaction between `temp` and `kwhpday`?

A. Comparing the models with and without an interaction term, we see that adding an interaction term does improve the fit somewhat (adjusted r^2 has increased).

```
> ut.lm3 <- lm(thermsPerDay ~ temp * kwhpday, ut)       utilities-kwh-int
> summary(ut.lm3)
< 8 lines removed >
Coefficients:
             Estimate Std. Error t value Pr(>|t|)
(Intercept) 11.913626   0.491304   24.25  < 2e-16
temp        -0.184309   0.011690  -15.77  < 2e-16
kwhpday     -0.104486   0.018217   -5.74  2.6e-07
temp:kwhpday 0.001564   0.000444    3.52  0.00079

Residual standard error: 0.345 on 66 degrees of freedom
Multiple R-squared: 0.975,       Adjusted R-squared: 0.974
F-statistic:  846 on 3 and 66 DF,  p-value: <2e-16

> ut.plot3 <- xplot(ut.lm3)
```

The residual plots for this model are also slightly improved. (The shape is quite similar, but the residuals are smaller; see Figure 7.23.) In summary, the interaction model is somewhat, but not dramatically, better than the additive model.

But how are we to interpret the interaction model? The coefficient on the interaction term itself seems quite small, but the coefficients on `kwhpday` and `temp` are quite different between the additive and interaction models. This is because they have a different meaning in the new model. The interaction model can be expressed as

$$\texttt{thermsPerDay} = \beta_0 + \beta_1 \texttt{temp} + \beta_2 \texttt{kwhpday} + \beta_3(\texttt{temp} \cdot \texttt{kwhpday})$$
$$= (\beta_0 + \beta_1 \cdot \texttt{temp}) + (\beta_2 + \beta_3 \cdot \texttt{temp})\texttt{kwhpday} ,$$

so for a fixed value of `temp` our model collapses down to a simple linear model. For example, when `temp` $= 36$ (approximately the mean of the temperatures in `ut`), our model equation becomes

$$\texttt{thermsPerDay} = (\beta_0 + \beta_1 \cdot 36) + (\beta_2 + \beta_3 \cdot 36)\texttt{kwhpday} ,$$

and the coefficients can be estimated as follows:

```
> coef(ut.lm3)[1] +  coef(ut.lm3)[2] * 36          utilities-kwh-int2
(Intercept)
     5.2785
> coef(ut.lm3)[3] +  coef(ut.lm3)[4] * 36
  kwhpday
-0.048169
```

This estimated coefficient for `kwhpday` is much closer to the estimate in Example 7.2.3. They would be equal at a temperature of 37.8 degrees:

```
> f <- function(x) {                              utilities-uniroot
+        coef(ut.lm3)[4] * x + coef(ut.lm3)[3] - coef(ut.lm2)[3]
+ }
> uniroot( f, c(20,50) )$root
[1] 37.784
```

The effect of the interaction term is to adjust this coefficient as `temp` changes. In our model with the interaction term, as `temp` increases, the effect of `kwhpday` becomes less pronounced since the coefficients on `kwhpday` and `temp:kwhpday` have opposite signs. Conversely, the electricity use is a more important factor in colder months. A similar analysis could be done reversing the roles of `temp` and `kwhpday`.   ◁

We conclude this section with one more look at the `utilities` data set.

Example 7.5.2.

Q. How well does `month` predict `thermsPerDay` in the `utilities` data set?

A. We would not expect the relationship between `month` and `thermsPerDay` to be monotonic. Let's see if a simple quadratic model provides a reasonable fit.

```
# remove first few observations because of bad meter read     utilities-month
> ut2 <- subset( utilities, subset=(year > 2000 | month > 6) )
> ut2.lm <- lm( thermsPerDay ~ month + I(month^2), ut2 )
> summary(ut2.lm)
< 8 lines removed >
Coefficients:
            Estimate Std. Error t value Pr(>|t|)
(Intercept) 10.82565    0.30767    35.2   <2e-16
month       -2.95419    0.10542   -28.0   <2e-16
I(month^2)   0.20692    0.00776    26.7   <2e-16

Residual standard error: 0.847 on 108 degrees of freedom
```

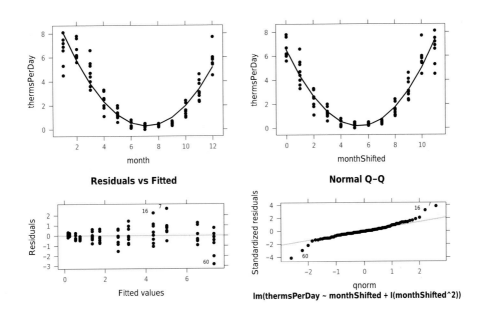

Figure 7.24. Predicting natural gas usage from the month using a second order model. Residual plots are shown for the shifted model where 0 corresponds to February and 11 to January.

```
Multiple R-squared: 0.88,          Adjusted R-squared: 0.878
F-statistic:  397 on 2 and 108 DF,  p-value: <2e-16


> ut2.plot1 <- xyplot(thermsPerDay ~ month, data=ut2,
+                     panel=panel.lm, model=ut2.lm)
> ut2.plot2 <- xplot(ut2.lm, w=1:2)
```

The fit of this model, as measured by adjusted r^2, is not as good as for the models that include `temp` as a predictor. This is to be expected since the effect of `month` on gas usage is primarily through the different average temperatures of the months. On the other hand, if we are trying to estimate future gas usage, this model may be very useful, since it is easier to predict future months than future temperatures.

The fit can be improved by shifting the months so that January and February are at opposite ends of the parabolic fit.

utilities-month2

```
> ut2$monthShifted <- ( ut2$month -2 ) %% 12
> ut2.lm2 <- lm(thermsPerDay ~ monthShifted + I(monthShifted^2), ut2)
> summary(ut2.lm2)
< 8 lines removed >

Coefficients:
                  Estimate Std. Error t value Pr(>|t|)
(Intercept)        6.53083    0.17591    37.1   <2e-16
monthShifted      -2.39239    0.07430   -32.2   <2e-16
I(monthShifted^2)  0.22419    0.00659    34.0   <2e-16
```

```
Residual standard error: 0.713 on 108 degrees of freedom
Multiple R-squared: 0.915,           Adjusted R-squared: 0.913
F-statistic:  581 on 2 and 108 DF,  p-value: <2e-16


> ut2.plot3 <- xyplot(thermsPerDay ~ monthShifted, data=ut2,
+                     panel=panel.lm, model=ut2.lm2)
> ut2.plot4 <- xplot(ut2.lm2, w=1:2)
```

R provides the functions `poly()` and `polym()` for generating models with polynomial predictors. We could also fit our previous model using `poly()` as follows.

```
> ut2.lm3 <- lm(thermsPerDay ~ poly(monthShifted, 2), ut2)    | utilities-month3 |
> summary(ut2.lm3)
< 8 lines removed >
Coefficients:
                        Estimate Std. Error t value Pr(>|t|)
(Intercept)              2.7018     0.0677   39.92   <2e-16
poly(monthShifted, 2)1   1.4922     0.7131    2.09    0.039
poly(monthShifted, 2)2  24.2667     0.7131   34.03   <2e-16

Residual standard error: 0.713 on 108 degrees of freedom
Multiple R-squared: 0.915,           Adjusted R-squared: 0.913
F-statistic:  581 on 2 and 108 DF,  p-value: <2e-16

> quantile( fitted(ut2.lm2) - fitted(ut2.lm3) )
        0%          25%          50%          75%          100%
-1.8652e-14  0.0000e+00  2.2204e-16  8.8818e-16  1.7764e-15
```

Although the fits are the same (up to round-off error), the coefficients of the model are quite different. The `poly()` function generates a set of polynomials of increasing degree $p_1, p_2, \ldots, p_k$ such that the vectors $p_1(\boldsymbol{x}), p_2(\boldsymbol{x}), \ldots, p_k(\boldsymbol{x})$ are orthogonal. The output of `poly()` is the vectors $p_1(\boldsymbol{x}), p_2(\boldsymbol{x}), \ldots, p_k(\boldsymbol{x})$ plus some additional information stored in attributes. Exercise 7.26 explores orthogonal polynomials further.   ◁

7.6. Model Selection

We have now entered a world in which there is (seemingly) no end to the number of models one could choose to fit to a data set. How does one choose which one to use in a particular analysis? In this section we discuss some of the tools available for assisting in model selection.

7.6.1. Model Selection Guidelines

So how do we choose a model? There are no hard and fast rules, but here are some things that play a role:

- A priori theory.

Some models are chosen because there is some scientific theory that predicts a relationship of a certain form. Statistics is used to find the most likely parameters in a model of this form. If there are competing theories, we can fit multiple models and see which seems to fit better.

- Previous experience.

  Models that have worked well in other similar situations may work well again.

- The data.

  Especially in new situations, we may only have the data to go on. Regression diagnostics, adjusted r^2, various hypothesis tests, and other methods like the commonly used information criteria AIC and BIC can help us choose between models. In general, it is good to choose the simplest model that works well.

  In addition to looking at the fit, we must consider the errors in the model. The same sorts of diagnostic checks can be done for multiple regression as for simple linear regression. Most of these involve looking at the residuals. As before, `plot(lm(...))` and `xplot()` will automate generating several types of residual plots. Consideration of both the normality and homoskedasticity assumptions of the error terms in linear models should be a part of selecting a model.

- Interpretability.

  Ideally, we would like a model that can be interpreted, a model that helps us reason about and predict our situation of interest. Simpler models are generally easier to understand and interpret, so "as simple as possible, but no simpler" is a good maxim for selecting interpretable models.

7.6.2. Stepwise Regression

There are a number of methods that have been proposed to automate the process of searching through many models to find the "best" one. One commonly used method is called stepwise regression. Stepwise regression works by repeatedly dropping or adding a single term from the model until there are no such single parameter changes that improve the model (based on some criterion; AIC is the default in R). The function `step()` will do this in R.

If the number of parameters is small enough, one could try all possible subsets of the parameters. This could find a "better" model than the one found by stepwise regression.

Example 7.6.1. Researchers at the University of Minnesota conducted experiments to see whether manipulating the air conditioning fans in the Minneapolis Metrodome can affect the distance traveled by a baseball. Baseballs were launched out of a specially designed canon and several measurements were recorded: the date, an indicator (`Cond`) of whether the air conditioners were blowing in (`Tailwind`) or out (`Headwind`), the angle at which the blowers were directed, the velocity (in feet per second) as the ball left the cannon, the weight (in grams) and diameter (in inches) of the ball, and the distance (in feet) that the ball traveled.

```
> require(alr3); data(domedata)                              domedata
> summary(domedata)
   Cond        Velocity        Angle          BallWt          BallDia
 Head:19   Min.   :149   Min.    :48.3   Min.    :140   Min.    :2.81
 Tail:15   1st Qu.:154   1st Qu.:49.5    1st Qu.:140    1st Qu.:2.81
           Median :156   Median :50.0    Median :141    Median :2.86
           Mean   :155   Mean   :50.0    Mean    :141   Mean    :2.84
           3rd Qu.:156   3rd Qu.:50.6    3rd Qu.:141    3rd Qu.:2.86
           Max.   :161   Max.   :51.0    Max.    :142   Max.    :2.88
      Dist
 Min.   :329
 1st Qu.:348
 Median :352
 Mean   :353
 3rd Qu.:359
 Max.   :374
```

Stepwise regression is one tool that may help us locate a reasonable model from among the many possibilities.

```
                                                        domedata-step
> dome.lm1 <- lm(Dist~Velocity+Angle+BallWt+BallDia+Cond, data=domedata)
> step(dome.lm1,direction="both")
Start:  AIC=135.8
Dist ~ Velocity + Angle + BallWt + BallDia + Cond

            Df Sum of Sq  RSS AIC
- Angle      1        37 1333 135
<none>                   1297 136
- BallWt     1       103 1400 136
- BallDia    1       429 1726 144
- Cond       1       450 1747 144
- Velocity   1       469 1765 144


Step:  AIC=134.75
Dist ~ Velocity + BallWt + BallDia + Cond

            Df Sum of Sq  RSS AIC
<none>                   1333 135
- BallWt     1       111 1444 136
+ Angle      1        37 1297 136
- BallDia    1       409 1742 142
- Velocity   1       481 1814 143
- Cond       1       499 1833 144

Call:
lm(formula = Dist ~ Velocity + BallWt + BallDia + Cond, data = domedata)

Coefficients:
(Intercept)      Velocity        BallWt       BallDia      CondTail
     133.82          1.75         -4.13        184.84          7.99
```

```
> dome.lm2 <- lm(Dist~Velocity+Cond,data=domedata)
> summary(dome.lm2)
< 8 lines removed >
Coefficients:
             Estimate Std. Error t value Pr(>|t|)
(Intercept)    -27.99      83.95   -0.33    0.741
Velocity         2.44       0.54    4.51  8.6e-05
CondTail         6.51       2.59    2.51    0.017

Residual standard error: 7.5 on 31 degrees of freedom
Multiple R-squared: 0.451,       Adjusted R-squared: 0.416
F-statistic: 12.7 on 2 and 31 DF,  p-value: 9.12e-05

> anova(dome.lm2,dome.lm1)
Analysis of Variance Table

Model 1: Dist ~ Velocity + Cond
Model 2: Dist ~ Velocity + Angle + BallWt + BallDia + Cond
  Res.Df  RSS Df Sum of Sq    F Pr(>F)
1     31 1742
2     28 1297  3       446 3.21  0.038
> dome.lm3 <- lm(Dist~Velocity+Cond+BallDia,data=domedata)
> anova(dome.lm2,dome.lm3)
Analysis of Variance Table

Model 1: Dist ~ Velocity + Cond
Model 2: Dist ~ Velocity + Cond + BallDia
  Res.Df  RSS Df Sum of Sq    F Pr(>F)
1     31 1742
2     30 1444  1       298 6.19  0.019
```

◁

7.6.3. Collinearity

Correlated predictors present some additional challenges for regression analysis. We will illustrate them with an example.

Example 7.6.2. Car drivers like to adjust the seat position for their own comfort. Car designers would find it helpful to know where different drivers will position the seat depending on their size and age. Researchers at the HuMoSim laboratory at the University of Michigan collected data on 38 drivers. That data include age (in years), weight (in pounds), height in shoes (in cm), height bare foot (in cm), seated height (in cm), lower arm length (in cm), thigh length (in cm), lower leg length (in cm), and a measurement called hipcenter (the horizontal distance of the midpoint of the hips from a fixed location in the car in mm). Hipcenter is a measure of the position of the seat; the other measurements are descriptions of the driver.

We begin by fitting an additive model using all of the predictors.

seatpos-lm

```
> require(faraway)
> data(seatpos, package="faraway")
> seatpos.lm1=lm(hipcenter ~ ., data=seatpos)
```

```
> summary(seatpos.lm1)
< 8 lines removed >
Coefficients:
            Estimate Std. Error t value Pr(>|t|)
(Intercept) 436.4321   166.5716    2.62    0.014
Age           0.7757     0.5703    1.36    0.184
Weight        0.0263     0.3310    0.08    0.937
HtShoes      -2.6924     9.7530   -0.28    0.784
Ht            0.6013    10.1299    0.06    0.953
Seated        0.5338     3.7619    0.14    0.888
Arm          -1.3281     3.9002   -0.34    0.736
Thigh        -1.1431     2.6600   -0.43    0.671
Leg          -6.4390     4.7139   -1.37    0.182

Residual standard error: 37.7 on 29 degrees of freedom
Multiple R-squared: 0.687,           Adjusted R-squared:  0.6
F-statistic: 7.94 on 8 and 29 DF,  p-value: 1.31e-05
```

Interestingly, although the model utility test shows a small p-value, none of the tests for individual parameters is significant. In fact, most of the p-values are quite large. Recall that these tests are measuring the importance of adding one predictor to a model that includes all the others. When an overall model fit is significant but tests for many of the individual parameters yield large p-values, this is an indicator that there may be **collinearity** among the predictors. In an extreme case, if we included the same predictor in our model twice, leaving out either copy of the predictor would not make the model any worse. Of course, leaving out both copies might have a large impact on the fit. If the correlation between two predictors is high, a similar effect can be observed. In other situations, it may be that linear combinations of some predictors are good proxies for other predictors.

When there is collinearity among the predictors, it is difficult to accurately estimate parameters and interpret the model. If we let r_i^2 be the value of r^2 that results from a linear model that predicts predictor i from all of the other predictors, then it can be shown that

$$\mathrm{Var}(\hat{\beta}_i) = \sigma^2 \cdot \frac{1}{1-r_i^2} \cdot \frac{1}{S_{x_i x_i}}$$

where

$$S_{x_i x_i} = |\boldsymbol{x}_i - \overline{\boldsymbol{x}}_i|^2 .$$

The variance is smaller when r_i^2 is smaller and when $|\boldsymbol{x}_i - \overline{\boldsymbol{x}}_i|^2$ is larger. The latter condition suggests spreading out the values of x_i (if we have control over those values). This cannot be taken to the extreme, however, since placing all the values of x_i at two extremal values would not allow us to verify the linear fit.

The term $\frac{1}{1-r_i^2}$ is called the **variance inflation factor**. When it is large, our estimates of β_i will be very imprecise. One solution to collinearity is to identify and remove unnecessary predictors based on the correlation structure. We can use `cor()` to calculate pairwise correlation or `vif()` from the `faraway` package to calculate the variance inflation factors.

```
> vif(seatpos.lm1)                                              seatpos-vif
     Age  Weight HtShoes      Ht  Seated     Arm   Thigh      Leg
  1.9979  3.6470 307.4294 333.1378  8.9511  4.4964  2.7629   6.6943

> require(car) # for scatterplot.matrix                         seatpos-cor
> scatterplot.matrix(~Age+Arm+hipcenter+Ht+HtShoes+Leg+Seated+Thigh+Weight,
+       reg.line=lm, smooth=TRUE, span=0.5,
+       diagonal = 'density', data=seatpos)
Warning message:
'scatterplot.matrix' is deprecated.
Use 'scatterplotMatrix' instead.
See help("Deprecated") and help("car-deprecated").
> round(cor(seatpos),2)
            Age Weight HtShoes    Ht Seated   Arm Thigh   Leg hipcenter
Age        1.00   0.08   -0.08 -0.09  -0.17  0.36  0.09 -0.04      0.21
Weight     0.08   1.00    0.83  0.83   0.78  0.70  0.57  0.78     -0.64
HtShoes   -0.08   0.83    1.00  1.00   0.93  0.75  0.72  0.91     -0.80
Ht        -0.09   0.83    1.00  1.00   0.93  0.75  0.73  0.91     -0.80
Seated    -0.17   0.78    0.93  0.93   1.00  0.63  0.61  0.81     -0.73
Arm        0.36   0.70    0.75  0.75   0.63  1.00  0.67  0.75     -0.59
Thigh      0.09   0.57    0.72  0.73   0.61  0.67  1.00  0.65     -0.59
Leg       -0.04   0.78    0.91  0.91   0.81  0.75  0.65  1.00     -0.79
hipcenter  0.21  -0.64   -0.80 -0.80  -0.73 -0.59 -0.59 -0.79      1.00
```

The large values of VIF for height and height with shoes is due to the nearly perfect correlation between these measurements. Pairwise correlations can be presented graphically in several ways. See Figure 7.25 for two examples. In this case, we see that the length measurements are highly correlated, so perhaps using just one of them would suffice. Since height is easy to measure, we'll select it first.

```
> seatpos.lm2=lm(hipcenter ~ Age + Weight + Ht, seatpos )     seatpos-lm2
> summary(seatpos.lm2)
< 8 lines removed >
Coefficients:
            Estimate Std. Error t value Pr(>|t|)
(Intercept) 528.29773  135.31295    3.90  0.00043
Age           0.51950    0.40804    1.27  0.21159
Weight        0.00427    0.31172    0.01  0.98915
Ht           -4.21190    0.99906   -4.22  0.00017

Residual standard error: 36.5 on 34 degrees of freedom
Multiple R-squared: 0.656,         Adjusted R-squared: 0.626
F-statistic: 21.6 on 3 and 34 DF,  p-value: 5.13e-08

> vif(seatpos.lm2)
   Age Weight     Ht
1.0930 3.4577 3.4633
```

From the output above, we might expect that dropping weight (and perhaps age) from the model would be a good idea. We could, of course, try others of the length measurements to see which is the most informative predictor.

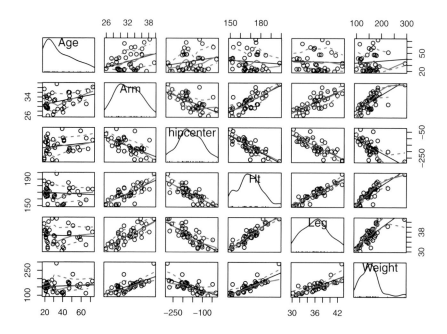

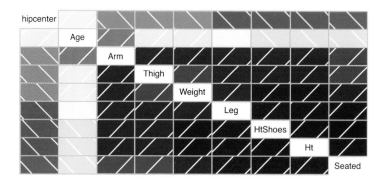

Figure 7.25. Looking at predictors in pairs helps us detect possible collinearity. Some of the variables have been removed from the scatterplot matrix (top panel) for readability.

Alternatively, we could attempt to combine the length measurements into a single measurement. Perhaps the sum of several would make a good predictor.

Principle component analysis is one way of letting the data select linear combinations of the predictors.

```
> pc=with(seatpos,princomp(cbind(HtShoes,Ht,Seated,Arm,Thigh,Leg),   [seatpos-pc]
+     scores=T))
> summary(pc, loadings=T)
Importance of components:
                         Comp.1   Comp.2   Comp.3   Comp.4    Comp.5
Standard deviation      16.93005 2.858520 2.091004 1.543407 1.2087954
Proportion of Variance   0.94527 0.026948 0.014419 0.007856 0.0048189
Cumulative Proportion    0.94527 0.972220 0.986639 0.994495 0.9993141
                          Comp.6
Standard deviation      0.45605740
Proportion of Variance 0.00068593
Cumulative Proportion  1.00000000


Loadings:
        Comp.1 Comp.2 Comp.3 Comp.4 Comp.5 Comp.6
HtShoes  0.649  0.105         0.159 -0.239  0.696
Ht       0.651                0.180 -0.172 -0.714
Seated   0.268  0.385 -0.257 -0.754  0.382
Arm      0.151 -0.463  0.715 -0.487 -0.119
Thigh    0.168 -0.789 -0.560         0.171
Leg      0.181         0.328  0.365  0.851
> seatpos.lmpc <-lm(hipcenter ~ Age + Weight + pc$scores[,1], seatpos )
> summary(seatpos.lmpc)
< 8 lines removed >
Coefficients:
                Estimate Std. Error t value Pr(>|t|)
(Intercept)    -187.0466    47.0013   -3.98  0.00034
Age               0.5713     0.4023    1.42  0.16462
Weight            0.0129     0.3094    0.04  0.96687
pc$scores[, 1]   -2.7625     0.6449   -4.28  0.00014

Residual standard error: 36.3 on 34 degrees of freedom
Multiple R-squared: 0.66,        Adjusted R-squared: 0.63
F-statistic:   22 on 3 and 34 DF,  p-value: 4.26e-08

> vif(seatpos.lmpc)
          Age        Weight pc$scores[, 1]
       1.0741        3.4450         3.4405
```

This method suggests using

```
0.65*HtShoes + 0.65*Ht + 0.27*Seated + 0.15*Arm + 0.17*Thigh + 0.18*Leg
```

as a predictor. This does improve the value of r^2 very slightly, but since this predictor is far from natural, there is no compelling reason to select it.

Yet another option would be to apply stepwise regression and see which predictors remain.

```
# trace=0 turns off intermediate reporting    [seatpos-step]
> seatpos.lmstep<-step(seatpos.lm1, trace=0)
```

```
> summary(seatpos.lmstep)
< 8 lines removed >
Coefficients:
             Estimate Std. Error t value Pr(>|t|)
(Intercept)   456.214    102.808    4.44  9.1e-05
Age             0.600      0.378    1.59    0.122
HtShoes        -2.302      1.245   -1.85    0.073
Leg            -6.830      4.069   -1.68    0.102

Residual standard error: 35.1 on 34 degrees of freedom
Multiple R-squared: 0.681,         Adjusted R-squared: 0.653
F-statistic: 24.2 on 3 and 34 DF,  p-value: 1.44e-08

> vif(seatpos.lmstep)
    Age HtShoes     Leg
 1.0115  5.7778  5.7517
> anova(seatpos.lm1,seatpos.lmstep)
Analysis of Variance Table

Model 1: hipcenter ~ Age + Weight + HtShoes + Ht + Seated + Arm + Thigh +
    Leg
Model 2: hipcenter ~ Age + HtShoes + Leg
  Res.Df   RSS Df Sum of Sq    F Pr(>F)
1     29 41262
2     34 41958 -5      -696  0.1      1
```

The resulting model with three predictors yields a value of r^2 that is nearly as large as for the model with all the predictors. In this particular example, there are quite a number of models with two or three predictors that are very similar in predictive strength. This is the result of the collinearity among the predictors. Final selection in such a situation may be made based on ease of application or interpretation.

There is also a fair amount of variation in seat position that is not explained by any of these measurements. It would be interesting to design an experiment where each subject entered a vehicle and adjusted the seat multiple times to determine how much of this is variation within individual drivers.   ◁

Keep in mind that stepwise regression is a greedy algorithm – at each step it does what looks best locally (by adding or removing one term from the model). This is more efficient than considering all possible models, but it does not always find the best model. Also there may be several models that are nearly equally good based on AIC. A model that is not quite as good as the optimal model may be preferable if it is easier to interpret. So while `step()` can form a good first step in model selection, it is not the end of the story.

7.7. More Examples

We have now covered the major themes of linear models. The various ideas can be "mixed and matched" to give us a large palette of models that can handle a wide range of situations.

Example 7.7.1. The ice data set contains the results of an experiment comparing the efficacy of different forms of dry ice application in reducing the temperature of the calf muscle [**DHM**[+]**09**]. The 12 subjects in this study came three times, at least four days apart, and received one of three ice treatments (cubed ice, crushed ice, or ice mixed with water). In each case, the ice was prepared in a plastic bag and applied dry to the subjects calf muscle.

The temperature measurements were taken on the skin surface and inside the calf muscle (via a 4 cm long probe) every 30 seconds for 20 minutes prior to icing, for 20 minutes during icing, and for 2 hours after the ice had been removed. The temperature measurements are stored in variables that begin with B (baseline), T (treatment), or R (recovery) followed by a numerical code for the elapsed time formed by concatenating the number of minutes and seconds. For example, R1230 contains the temperatures 12 minutes and 30 seconds after the ice had been removed. A number of physiological measurements (sex, weight, height, skinfold thickness, calf diameter, and age) were also recorded for each subject.

The researchers used a **repeated measures ANOVA** to analyze the data. We haven't developed the machinery to do that analysis, but if we select only certain observations or combine multiple measurements into a single value, we can apply some of the analyses that are familiar to us. For this example, we will use $\alpha = 0.10$ as our level of significance because our method is less powerful than the repeated measures analysis. Many of the results suggested by this analysis are confirmed (with stronger evidence) using the repeated measures approach.

Let's begin by checking that the baseline temperatures are comparable for the different treatments by looking at the temperatures just before the ice was applied.

ice-baseline-summary

```
> summary(B1930 ~ Location + Treatment, data=ice,
+                fun=mean, method='cross')
< 7 lines removed >
+-------------+-------+------+------+------+
|   Location  |crushed| cubed|  wet | ALL  |
+-------------+-------+------+------+------+
|intramuscular| 12    |12    |12    |36    |
|             | 35.425|34.800|35.058|35.094|
+-------------+-------+------+------+------+
|surface      | 12    |12    |12    |36    |
|             | 30.983|30.367|30.758|30.703|
+-------------+-------+------+------+------+
|ALL          | 24    |24    |24    |72    |
|             | 33.204|32.583|32.908|32.899|
+-------------+-------+------+------+------+
```

Although we would not expect an interaction between treatment and location, we first fit a model with an interaction term, just to be sure.

ice-baseline-int

```
> base.lmint <- lm(B1930 ~ Location * Treatment, ice)
> anova(base.lmint)
Analysis of Variance Table

Response: B1930
```

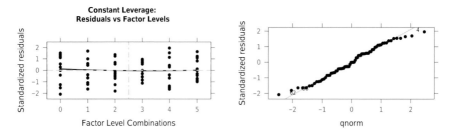

Figure 7.26. Residuals for an additive model at the end of the baseline measurement period.

	Df	Sum Sq	Mean Sq	F value	Pr(>F)
Location	1	347	347	611.91	<2e-16
Treatment	2	5	2	4.08	0.021
Location:Treatment	2	0	0	0.07	0.935
Residuals	66	37	1		

```
> base.lmadd <- lm(B1930 ~ Location + Treatment, ice)      ice-baseline-add
> anova(base.lmadd)
Analysis of Variance Table

Response: B1930
          Df Sum Sq Mean Sq F value Pr(>F)
Location   1    347     347  629.18 <2e-16
Treatment  2      5       2    4.19  0.019
Residuals 68     38       1
> baseplot <- xplot(base.lmadd,w=c(5,2))
> printPlot(xplot(base.lmadd,w=5),size='medium',file='ice-base-resid1')
> printPlot(xplot(base.lmadd,w=2),size='medium',file='ice-base-resid2')
```

As we would expect, there is strong evidence that the surface temperature differs from the intramuscular temperature, and no evidence of an interaction between location and treatment. Figure 7.26 reveals a reasonable distribution of residuals.

It is a bit worrisome that the p-value for `Treatment` is on the small side, since the baseline temperatures are taken before the treatment is applied. A difference in baseline temperatures makes it more difficult to compare the three treatments. Tukey's pairwise comparisons indicate that there may be a difference between baseline temperatures for the cubed ice and crushed ice treatments, with the cubed ice treatments starting from a lower baseline temperature.

```
> require(multcomp)                                         ice-baseline-tukey
> confint(glht(base.lmadd, mcp(Treatment="Tukey")),level=0.9)


        Simultaneous Confidence Intervals


Multiple Comparisons of Means: Tukey Contrasts


Fit: lm(formula = B1930 ~ Location + Treatment, data = ice)


Quantile = 2.088
```

```
90% family-wise confidence level


Linear Hypotheses:
                    Estimate lwr     upr
cubed - crushed == 0 -0.621   -1.068 -0.173
wet - crushed == 0   -0.296   -0.743 0.152
wet - cubed == 0      0.325   -0.123 0.773
```

One potential solution to this difficulty is to model change in temperature relative to the baseline measurement rather than using raw temperature. For example, we can compare the temperature change (relative to baseline) at the end of the treatment phase.

ice-treatment

```
> ice.trt <- lm(T1930 - B1930 ~ Treatment * Location, ice)
> anova(ice.trt)
Analysis of Variance Table

Response: T1930 - B1930
                  Df Sum Sq Mean Sq F value  Pr(>F)
Treatment          2     74      37   13.06 1.7e-05
Location           1   2535    2535  890.10 < 2e-16
Treatment:Location 2     21      10    3.63   0.032
Residuals         66    188       3
```

The output above suggests that there is a difference between the treatments and perhaps an interaction between treatment and location. If we are primarily interested in the intramuscular temperature (since we're trying to cool the muscle, not the skin), we could look only at that subset of the data.

ice-treatment-intr

```
> ice.trt2 <- lm(T1930 - B1930 ~ Treatment, ice,
+                 subset=Location=='intramuscular')
> summary(ice.trt2)
< 8 lines removed >

Coefficients:
               Estimate Std. Error t value Pr(>|t|)
(Intercept)      -2.925      0.507   -5.76  1.9e-06
Treatmentcubed   -0.692      0.718   -0.96    0.342
Treatmentwet     -1.708      0.718   -2.38    0.023

Residual standard error: 1.76 on 33 degrees of freedom
Multiple R-squared: 0.148,        Adjusted R-squared: 0.0964
F-statistic: 2.87 on 2 and 33 DF,  p-value: 0.0711

>
> confint(glht(ice.trt2, mcp(Treatment='Tukey')),level=0.90)
< 12 lines removed >

Linear Hypotheses:
```

```
                   Estimate lwr    upr
cubed - crushed == 0 -0.692  -2.217  0.834
wet - crushed == 0   -1.708  -3.234 -0.183
wet - cubed == 0     -1.017  -2.542  0.509
```

This analysis suggests that wet ice is better than crushed ice, with cubed ice somewhere between (not significantly different from either at this level). Exercise 7.28 asks you to repeat this analysis for surface temperatures.

Alternatively, we might be interested to know whether the difference between surface and intramuscular temperatures is the same for all three treatments. Since skin temperatures are much easier to measure, if skin temperature is a good predictor of intramuscular temperature, it will be much easier to obtain additional subjects for further study.

ice-diff

```
> surface <- ice[ice$Location=='surface',c("Treatment","T1930")]
> intra <- ice[ice$Location=='intramuscular',"T1930"]
> newdata <- cbind(surface,intra)
> names(newdata) <- c('Treatment','SurfTemp','IntraTemp')
> anova(lm(SurfTemp - IntraTemp ~ Treatment, newdata))
Analysis of Variance Table

Response: SurfTemp - IntraTemp
          Df Sum Sq Mean Sq F value Pr(>F)
Treatment  2   38.3    19.1    2.25   0.12
Residuals 33  280.6     8.5
```

Our small data set does not give us enough evidence to reject the null hypothesis that the mean difference between surface and intramuscular temperature is the same for each of the three treatments.

Using the more powerful repeated measures analysis, the researchers were able to conclude that

> As administered in our protocol, wetted ice is superior to cubed or crushed ice at reducing surface temperatures, while both cubed and wetted ice are superior to crushed ice at reducing intramuscular temperatures [DHM+09].

◁

Example 7.7.2. The FUSION study of the genetics of type 2 diabetes was introduced in Exercise 5.28. Now we will analyze the data using a different predictor. Our primary predictor variable will be the number of copies (0, 1, or 2) of the "risk allele" (for a given genetic marker). This number is stored in the Gdose variable in the fusion1 data set. We begin by merging the data in the fusion1 data set (which contains the genotypes) and the pheno data set (which contains the case/control status).

```
# merge fusion1 and pheno keeping only id's that are in both          fusion1-merge
> fusion1m <- merge(fusion1, pheno, by='id', all.x=FALSE, all.y=FALSE)
```

```
> xtabs(~t2d + Gdose, fusion1m)          fusion1-xtabs-dose
        Gdose
t2d          0   1   2
  case      48 375 737
  control   27 309 835
```

Now we can fit our model using logistic regression.

```
> f1.glm1 <-          fusion1-glm1
+     glm( factor(t2d) ~ Gdose, fusion1m, family=binomial)
> f1.glm1
< 4 lines removed >
(Intercept)        Gdose
     -0.532        0.329

Degrees of Freedom: 2330 Total (i.e. Null);  2329 Residual
Null Deviance:            3230
Residual Deviance: 3210        AIC: 3220
```

Recall that the model we are fitting is

$$\log \frac{\pi}{1 - \pi} = \beta_0 + \beta_1 \mathsf{Gdose}$$

or, equivalently,

$$\frac{\pi}{1 - \pi} = e^{\beta_0} \cdot \beta_1^{\mathsf{Gdose}} \ .$$

So the odds of having diabetes depend on the number of risk alleles.

Gdose	log odds of having T2D	odds of having T2D
0	β_0	e^{β_0}
1	$\beta_0 + \beta_1$	$e^{\beta_0} \cdot e^{\beta_1}$
2	$\beta_0 + 2\beta_1$	$e^{\beta_0} \cdot \left(e^{\beta_1}\right)^2$

Because this is a case/control study, the intercept does not have an immediate interpretation since it is sensitive to the ratio of cases to controls enrolled in the study. The odds ratio e^{β_1}, on the other hand, is interpretable as the (multiplicative) increase in the odds of having type 2 diabetes for each copy of the G allele. Our estimated odds ratio is 1.36:

```
> coef(f1.glm1)          fusion1-glm1-or
(Intercept)        Gdose
   -0.53183      0.32930
> exp(coef(f1.glm1))
(Intercept)        Gdose
    0.58753      1.39000
```

Now that we have some idea how to interpret this model, we turn our attention to inference. In particular, we would like to know how strong the evidence is that this particular marker is associated with varying risk for type 2 diabetes.

```
> summary(f1.glm1)                                    fusion1-glm1-sum
< 8 lines removed >
Coefficients:
            Estimate Std. Error z value Pr(>|z|)
(Intercept)  -0.5318     0.1340   -3.97 7.2e-05
Gdose         0.3293     0.0774    4.26 2.1e-05


(Dispersion parameter for binomial family taken to be 1)


    Null deviance: 3231.4  on 2330   degrees of freedom
Residual deviance: 3213.0  on 2329   degrees of freedom
AIC: 3217

Number of Fisher Scoring iterations: 3
```

The deviances in the `summary()` output are likelihood ratio test statistics

$$\text{deviance} = 2 \log \left(\frac{L(\Omega_S)}{L(\Omega)} \right) = 2 \left(\log(\Omega_S) - \log(\Omega) \right) \ ,$$

where Ω_S is the so-called saturated model and Ω is either the null model (only an intercept) or the full model we are fitting (labeled "Residual deviance" in the R output). This means we can easily conduct a model utility test for our model since the likelihood ratio test statistic

$$\begin{aligned}
\lambda &= 2 \left(\log(\Omega_1) - \log(\Omega_2) \right) \\
&= 2 \left(\log(\Omega_1) - \log(\Omega_S) + \log(\Omega_S) - \log(\Omega_2) \right)
\end{aligned}$$

is the difference in the deviances.

```
> 1 - pchisq(3393.6 - 3376.4, df=1)                   fusion1-glm1-mut
[1] 3.3644e-05
```

R provides some utility functions for accessing these values, so there is no need to hand copy them from the output of `summary()`.

```
> f1.glm0 <- glm( factor(t2d) ~ 1, fusion1m, family=binomial)   fusion1-glm1-dev
> deviance(f1.glm0)
[1] 3231.4
> deviance(f1.glm1)
[1] 3213.0
> df1 <- df.residual(f1.glm0) - df.residual(f1.glm1); df1
[1] 1
> 1 - pchisq(deviance(f1.glm0) - deviance(f1.glm1), df=df1)
[1] 1.8370e-05
```

In the case of logistic regression this is not the same test as the z-test in the summary output, although in this case they give very similar results.

This would be strong evidence except for the fact that this test was done in the context of a genomewide association study in which over 300,000 genetic markers were tested. So we would expect more than $300000 \cdot 3.64 \cdot 10^{-5} \approx 10$ tests to have evidence this strong or stronger just by chance even if there are no markers actually associated with type 2 diabetes in the population. ◁

Example 7.7.3.

Q. Does the evidence for association with type 2 diabetes in Example 7.7.2 increase if we allow for the possibility of differences between men and women?

A. As you should expect, there is no reason to limit ourselves to a single predictor when fitting a logistic model. We will include `sex` in our model this time. (In [**SMB**$^+$**07**] additional covariates – including birth province – were also included in the model.)

fusion1-glm2

```
> f1.glm2 <-
+     glm( factor(t2d) ~ Gdose + sex, fusion1m, family=binomial )
> f1.glm2
< 4 lines removed >
(Intercept)        Gdose          sexM
    -0.382        0.337        -0.308

Degrees of Freedom: 2330 Total (i.e. Null);  2328 Residual
Null Deviance:            3230
Residual Deviance: 3200         AIC: 3210
> summary(f1.glm2)
< 8 lines removed >
Coefficients:
           Estimate Std. Error z value Pr(>|z|)
(Intercept)  -0.3816     0.1403    -2.72  0.00652
Gdose         0.3365     0.0777     4.33  1.5e-05
sexM         -0.3083     0.0836    -3.69  0.00023

(Dispersion parameter for binomial family taken to be 1)

    Null deviance: 3231.4  on 2330  degrees of freedom
Residual deviance: 3199.4  on 2328  degrees of freedom
AIC: 3205

Number of Fisher Scoring iterations: 4
```

Since both the number of risk alleles and sex are significant in this model, we might expect that the model utility test will yield a smaller p-value for this model.

fusion1-glm2-dev

```
> deviance(f1.glm0)
[1] 3231.4
> deviance(f1.glm2)
[1] 3199.4
> df2 <- df.residual(f1.glm0) - df.residual(f1.glm2); df2
[1] 2
> 1 - pchisq(deviance(f1.glm0) - deviance(f1.glm2), df=df2)
[1] 1.1263e-07
```

Indeed, the p-value is now sufficiently small to be of genuine interest, even in the context of such a large number of tests.   ◁

Example 7.7.4. An experiment was conducted by students at The Ohio State University in the fall of 1993 to explore the nature of the relationship between a

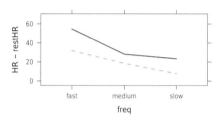

Figure 7.27. An interaction plot for the `step` data in Example 7.7.4.

person's heart rate and the frequency at which that person stepped up and down on steps of various heights. The response variable, heart rate, was measured in beats per minute. There were two different step heights: 5.75 inches (coded as lo), and 11.5 inches (coded as hi). There were three rates of stepping: 14 steps/min. (coded as slow), 21 steps/min. (coded as medium), and 28 steps/min. (coded as fast). This resulted in six possible height/frequency combinations. Each subject performed the activity for three minutes. Subjects were kept on pace by the beat of an electric metronome. One experimenter counted the subject's pulse for 20 seconds before and after each trial. The subject always rested between trials until his or her heart rate returned to close to the beginning rate. Another experimenter kept track of the time spent stepping. Each subject was always measured and timed by the same pair of experimenters to reduce variability in the experiment. Each pair of experimenters was treated as a block.

```
> step.model <- lm(HR - restHR ~ height * freq, step)          step
< 8 lines removed >

Coefficients:
                     Estimate Std. Error t value Pr(>|t|)
(Intercept)             54.60       4.74   11.51  2.9e-11
heightlo               -22.80       6.71   -3.40  0.00236
freqmedium             -26.40       6.71   -3.94  0.00062
freqslow               -31.20       6.71   -4.65  0.00010
heightlo:freqmedium     13.20       9.49    1.39  0.17687
heightlo:freqslow        7.20       9.49    0.76  0.45527

Residual standard error: 10.6 on 24 degrees of freedom
Multiple R-squared: 0.696,        Adjusted R-squared: 0.633
F-statistic:   11 on 5 and 24 DF,  p-value: 1.37e-05


> anova(step.model)
Analysis of Variance Table

Response: HR - restHR
           Df Sum Sq Mean Sq F value  Pr(>F)
height      1   1920    1920   17.07 0.00038
freq        2   4049    2024   17.99 1.7e-05
height:freq 2    218     109    0.97 0.39321
Residuals  24   2700     112
```

```
> interaction.plot <- xyplot(HR - restHR ~ freq, data=step,
+                             groups = height, type='a')
```

The output above and interaction plot in Figure 7.27 reveal no interaction effect. As we would have expected, stepping higher or faster increases the heart rate more. Interval estimates of the magnitude of these effects, adjusting for multiple comparisons, appear below.   ◁

Example 7.7.5. A study was conducted to test the effectiveness of various flavors in attracting rats to poison. Twenty batches of poison were prepared using 5 flavors. The poisons were then placed in one of four locations. After a specified period of time, the researchers returned and weighed the poison to determine how much had been consumed.

Since there is only one observation for each combination of flavor and location, we cannot fit a model that includes interaction. Such a model would have 20 total degrees of freedom: 1 for the intercept, 4 for the main effect of flavor, 3 for the main effect of location, and 12 for interaction effects. That leaves no degrees of freedom to estimate σ, so we have no way to measure the quality of our fit. (In fact, such a model will always fit the data perfectly.)

We can, however, fit an **additive model** that omits the interaction terms. Essentially, this model is assuming *a priori* that there is no interaction between location and flavor.

```
> rat.lm <- lm(consumption~location+flavor,ratpoison)          rat-anova
> anova(rat.lm)
Analysis of Variance Table

Response: consumption
          Df Sum Sq Mean Sq F value  Pr(>F)
location   4    495   123.8   49.93 2.2e-07
flavor     3     56    18.8    7.58  0.0042
Residuals 12     30     2.5
```

The diagnostic plots in Figure 7.28 reveal no major concerns, so we proceed with our ANOVA analysis. Although we cannot formally test for interaction, we can use an interaction plot to give an informal assessment. (See Figure 7.28.) For the most part, this interaction plot looks good, but we see that location A does not fit the pattern as well as the other locations do.

Presumably the investigators are primarily interested in the effect of flavor (since they can market flavor but not location). Location serves as a **blocking factor** in this design. Since each treatment is tested in each block, this design is called a **complete randomized block design**.

It is interesting to compare this analysis with an analysis that ignores the location factor. This is what could have resulted from a **completely randomized design**.

```
> rat.lm1 <- lm(consumption~flavor,ratpoison)                  rat-1way
> anova(rat.lm1)
Analysis of Variance Table
```

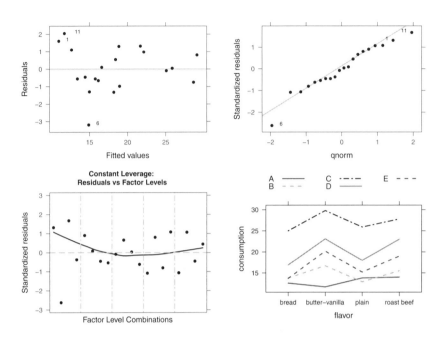

Figure 7.28. Diagnostic plots and an interaction plot for the rat poison data.

```
Response: consumption
         Df Sum Sq Mean Sq F value Pr(>F)
flavor    3    56    18.8    0.57   0.64
Residuals 16   525    32.8
> summary(rat.lm)$sigma
[1] 1.5749
> summary(rat.lm1)$sigma
[1] 5.7287
> (summary(rat.lm1)$sigma/summary(rat.lm)$sigma)^2
[1] 13.232
> summary(rat.lm)
< 8 lines removed >
Coefficients:
                     Estimate Std. Error t value Pr(>|t|)
(Intercept)            10.995      0.996   11.04  1.2e-07
locationB               1.700      1.114    1.53   0.1528
locationC              14.100      1.114   12.66  2.7e-08
locationD               7.225      1.114    6.49  3.0e-05
locationE               4.000      1.114    3.59   0.0037
flavorbutter-vanilla    3.900      0.996    3.92   0.0021
flavorplain             0.760      0.996    0.76   0.4602
flavorroast beef        3.460      0.996    3.47   0.0046

Residual standard error: 1.57 on 12 degrees of freedom
Multiple R-squared: 0.949,        Adjusted R-squared: 0.919
F-statistic: 31.8 on 7 and 12 DF,  p-value: 8.01e-07
```

```
> summary(rat.lm1)
< 8 lines removed >
Coefficients:
                       Estimate Std. Error t value Pr(>|t|)
(Intercept)               16.40       2.56    6.40  8.8e-06
flavorbutter-vanilla       3.90       3.62    1.08     0.30
flavorplain                0.76       3.62    0.21     0.84
flavorroast beef           3.46       3.62    0.95     0.35

Residual standard error: 5.73 on 16 degrees of freedom
Multiple R-squared: 0.097,     Adjusted R-squared: -0.0724
F-statistic: 0.573 on 3 and 16 DF,  p-value: 0.641
```

The variation due to location is large enough in this case to completely mask the effect of flavor, yielding a p-value greater than 0.6. Despite having fewer degrees of freedom for the residuals, the analysis using the block design has an estimated value for σ that is nearly 4 times smaller than the estimate from the one-way design. This, in turn, yields similarly smaller standard errors for the model parameters. Put another way, since the ratio of estimated variances is approximately 13:1, the unblocked design would require sample sizes about 13 times as large to achieve the same power. ◁

7.8. Permutation Tests and Linear Models

We introduced permutation testing with some examples in Chapter 4. In the context of regression, the idea behind this method is the following: Suppose the predictors were unrelated to the response (our null hypothesis). Then the response vector could just as easily have occurred in any permuted order. If we calculate some test statistic (the F-statistic from the model utility test, for example) using *every* permutation of the response, we will determine the sampling distribution for this statistic under this null hypothesis. Variations on this theme can be used to test other hypotheses as well. The only requirement is that the null hypothesis implies that the distributions with and without permutation be identical.

In practice, the number of permutations is often too large for direct computation to be tractable. Sometimes we can analytically derive the sampling distribution and avoid the need for computation. In fact, some of the tests we have already seen are equivalent to permutation tests. Often, however, no such analytical results are known. To use the permutation method in such a situation, we approximate the sampling distribution by computing the test statistic for a large number of random permutations.

Example 7.8.1. Recall Example 7.1.2. There we used normality assumptions to calculate a p-value for the model utility test using the appropriate F-distribution.

```
> require(Devore6); data(xmp13.13,package='Devore6')       concrete-load
Loading required package: Devore6
> concrete <- data.frame(
+     limestone=xmp13.13$x1, water=xmp13.13$x2, strength=xmp13.13$strength)
> concrete.lm1 <- lm(strength ~ limestone * water, concrete)
```

```
> summary(concrete.lm0)                                         concrete-summary
< 8 lines removed >
Coefficients:
             Estimate Std. Error t value Pr(>|t|)
(Intercept)    84.817     12.242    6.93  0.00045
limestone       0.164      0.143    1.15  0.29467
water         -79.667     20.035   -3.98  0.00731

Residual standard error: 3.47 on 6 degrees of freedom
Multiple R-squared: 0.741,          Adjusted R-squared: 0.654
F-statistic: 8.56 on 2 and 6 DF,  p-value: 0.0175
```

Now we will obtain an empirical p-value using 5000 random permutations. (We could increase the accuracy by increasing the number of permutations.)

```
> fstats <- replicate(5000,                                     concrete-perm-f
+     {
+        concrete.lm <- lm(sample(strength) ~ limestone * water, concrete)
+        summary(concrete.lm)$fstat[1]
+     }
+ )
> mean( fstats > summary(concrete.lm1)$fstat[1] )
[1] 0.0086
```

The permutation-based p-value is very close to the p-value from the F-test – certainly close enough for any decision that would be made from this data.   ◁

Example 7.8.2. We can obtain permutation-based p-values for other regression tests as well. Here we compute a permutation-based p-value corresponding to the t-test of $H_0 : \beta_{\text{water}} = 0$.

```
> tstats <- replicate(1000,                                     concrete-perm-t
+     {
+        concrete.lm <- lm(sample(strength) ~ limestone * water, concrete)
+        summary(concrete.lm)$coef[3,3]
+     }
+ )
> mean( abs(tstats) > abs(summary(concrete.lm1)$coef[3,3]) )
[1] 0.324                                                                      ◁
```

It can be shown that the regression F- and t-tests are good approximations to the permutation test when the normality assumptions of the model are met. Since permutation tests are much more computationally demanding – especially to estimate small p-values – we generally prefer other approaches to permutation tests when they are available. On the other hand, since the permutation tests make no specific distributional assumptions (only that when the null hypothesis is true, then the distributions of the response be the same whether permuted or not) yet maintain the correlation structure in the predictors, they can be a valuable tool in situations where the assumptions of an analytically derived test cannot be justified or when there is some potentially important, but unknown, correlation structure.

The method of permutation can be applied in many other situations as well. Fisher's exact test, for example, is a permutation test. In that case, we can derive formulas for the p-values because it is relatively easy to compute the distribution of the tables that result from permutation.

Example 7.8.3. Recall Example 5.5.1. We can do an analysis of this data on smoking habits of students and parents using a permutation test.

smoke-perm

```
> xtabs(~Student+Parents,data=familySmoking) -> smokeTab;
> smokeTab
              Parents
Student          NeitherSmokes OneSmokes BothSmoke
  DoesNotSmoke            1168      1823      1380
  Smokes                   188       416       400
> chisq.test(smokeTab)

        Pearson's Chi-squared test

data:  smokeTab
X-squared = 37.566, df = 2, p-value = 6.96e-09

> observedStat <- chisq.test(smokeTab)$stat
> stats <- replicate(2000,
+     {
+       chisq.test( xtabs(~sample(Student)+Parents,data=familySmoking))$stat
+     }
+ )
> sum( stats > observedStat ) -> x; x / length(stats)   # p-value
[1] 0
> binom.test(x,length(stats),alternative="less")$conf.int
[1] 0.0000000 0.0014967
attr(,"conf.level")
[1] 0.95
```

In a case like this where the p-value is extremely small, a very large number of permutations would be required to estimate the p-value accurately. As we see here, none of our permutations yielded larger values of the test statistic than that observed. From this modest number of permutations we can only conclude that the p-value is likely smaller than 0.0015.   ◁

In applications of permutation testing, the number of permutations done is typically determined by the results of the tests. When p-values are large, an accurate estimate is typically unnecessary, and a modest number of permutations will suffice to convince us that this is the case. But when the p-value is small, it is often important to know how small (at least to within an order of magnitude). This requires permuting until a more extreme test statistic has been observed multiple times.

7.9. Summary

The linear model framework can be extended to handle a wide range of models. Any number of predictor variables, either quantitative or categorical (with any number of levels) may be used. Higher order terms, interaction terms, and transformations further extend the range of available models.

The simplest of these is the **additive model**,

$$\boldsymbol{Y} = \beta_0 + \beta_1 \boldsymbol{x}_1 + \beta_2 \boldsymbol{x}_2 + \cdots + \beta_k \boldsymbol{x}_k + \varepsilon, \qquad \varepsilon \overset{\text{iid}}{\sim} \mathsf{Norm}(0, \sigma), \qquad (7.15)$$

in which the effects of each predictor are added together. In the additive model, β_i can be interpreted as the amount that the mean response changes when x_i is increased by 1 *and all other predictors remain fixed*. The additive model makes the assumption that this value is the same for all combinations of values of the other predictors.

The usual way to accommodate a categorical variable with l levels is to introduce $l - 1$ indicator variables (in addition to the intercept). For a single categorical predictor, the model looks just like (7.15) where

$$x_{ij} = [\![\text{observation } i \text{ is in group } j + 1]\!] \, .$$

In this model, known as the one-way ANOVA model, $\beta_0 = \mu_1$ is the mean of the first group, and for $i > 0$, $\beta_i = \mu_i - \mu_1$ is the difference between the means of group i and group 1. Other coding schemes can be used to describe this model. Two common alternatives are to select the last group as the reference instead of the first or to remove the intercept and fit a separate parameter for the mean of each group.

Interaction allows for the possibility that the effect of one predictor depends on the value of another. For both quantitative and categorical predictors, interactions can be added by including predictors that are formed by taking the product of two other predictors. The null hypothesis that all of the parameters corresponding to interaction terms are zero leads to a test for the presence of an interactions. In a two-way ANOVA, for example, if there is no evidence of interaction, then we can interpret the main effects for each of the categorical predictors as in the additive model. But when there is an interaction effect, it is difficult or impossible to interpret the effect of one predictor without the context of the other predictor.

All of these models can be fit by the method of least squares using `lm()`. (The **generalized linear model** can also be extended to use multiple predictors, both quantitative and categorical. Maximum likelihood fits for these models can be obtained using `glm()`.) Inference for individual parameters of the model as well as a model utility test can be derived using **model comparison tests** just as was the case for simple linear regression in Chapter 6. In the case of ANOVA, a wide variety of hypotheses can be expressed using contrasts, linear combinations of the group means with a coefficient sum of 0. Under the assumptions of the ANOVA model, for any contrast C,

$$\frac{\hat{C} - C}{s/\kappa} \sim \mathsf{t}(\nu) \, ,$$

where ν is the residual degrees of freedom, s^2 is our unbiased estimator for σ^2, κ is a constant such that $E(\boldsymbol{u}_C \cdot \boldsymbol{Y}) = \kappa C$, and $\boldsymbol{u}_C$ is the unit vector associated with C. (See Box 7.3 for details.) The `glht()` (general linear hypothesis test) function in the `multcomp` package can be used to compute p-values and confidence intervals for contrasts. Orthogonal contrasts are easily combined to form more complex hypotheses.

Often we are interested in constructing multiple confidence intervals or testing multiple hypotheses. In an ANOVA situation, for example, we may be interested in comparing each pair of groups or in comparing each treatment group to a common control group. Whenever we make **multiple comparisons** like this, we must be careful to interpret the results in the context of the complete set of confidence intervals or hypothesis tests. A simple, but conservative, adjustment is the **Bonferroni** correction which considers a result among a set of n hypothesis tests to be significant at the level α only if the p-value is less than α/n. Using the duality of confidence intervals and hypothesis tests, we can make a similar adjustment to the width of confidence intervals. Methods of **Tukey** (for make all pairwise comparisons) and **Dunnet** (for comparing all treatment groups to one control group) make less severe corrections that take into account the correlations among the various tests or intervals. The `glht()` function automatically applies a multiple comparisons correction when multiple contrasts are tested simultaneously.

One important use of a categorical predictor is as a **blocking variable**. Blocking can be viewed as a generalization of the ideas of a matched pairs design, which uses blocks of size 2. Typically we are not so much interested in the blocking variable itself but we use it as a way to reduce the variability so that we can more clearly see the impact of other predictors.

The design of experiments and their analysis are intimately connected, and there is a rich literature describing various designs and the analysis appropriate for them going all the way back to Fisher's famous books [**Fis25, Fis35**] in which he discusses important principles of good experimental design, including randomization, replication (so that variability can be estimated), blocking, orthogonality, and various factorial designs that allow for efficient comparison of multiple factors simultaneously. Each of these books has appeared in many editions, and *The Design of Experiments* [**Fis71**] is still an interesting and enjoyable read. Included in it is, of course, Fisher's original discussion of the lady tasting tea. For a more recent treatment focusing on the use of linear models, including many examples of analyses done in R, see the two-volume work of Faraway [**Far05, Far06**]. For those primarily interested in categorical data, the book by Agresti [**Agr90**] is a good reference. For a more thorough treatment of permutation testing see, for example, Good's book on the subject [**Goo05**].

Our treatment of linear models ends with a quotation from the eleventh R. A. Fisher Memorial Lecture, which Samuel Karlin delivered at the Royal Society of London in April, 1983.

> The purpose of models is not to fit the data but to sharpen the questions.
>
> Samuel Karlin

7.9.1. R Commands

Here is a table of important R commands introduced in this chapter. Usage details can be found in the examples and using the R help.

`factor(x)`	Convert x to a factor.
`summary(formula,data,...,` `        fun,method,...)`	Tabulate various summary statistics for a data set [Hmisc].
`favstats(x)`	Compute some basic summary statistics for x [fastR].
`project(y,v,...)`	Project y in the direction of v [fastR].
`dot(x,y)`	Dot product of x and y [fastR].
`vlength(x)`	Vector length (norm) of x [fastR].
`model <- lm(y~x,...)`	Fit a linear model.
`model <- glm(y~x,...)`	Fit a generalized linear model.
`summary(model)`	Print summary of `model`.
`resid(model)`	Residuals of `model`.
`plot(model); xplot(model)`	Diagnostic plots for `model`.
`anova(model)`	Print ANOVA table for `model`.
`anova(model1,model2)`	Print ANOVA table for model comparison test of two nested models.
`confint(model,...)`	Confidence intervals for model parameters.
`predict(model,...)`	Confidence intervals and prediction intervals for the response variable.
`glht(model,...)`	General linear hypothesis tests – p-values and confidence intervals for contrasts with multiple comparisons corrections as needed [multcomp].
`mcp(...)`	Construct sets of contrast for use with `glht()` [multcomp].
`step(model,...)`	Stepwise regression.
`vif(model,...)`	Variance inflation factor [faraway].
`aov(y~x,...)`	Alternative to `lm()` that stores and prints different information.
`TukeyHSD(aov(y~x,...))`	Tukey Honest Significant Differences.
`splom(...)`	Scatterplot matrix [lattice].
`scatterplot.matrix(...)`	Scatterplot matrix [car].

Exercises

7.1. Prove that $v_i \perp v_0$ where $v_0 = 1$ and $v_i = x_i - \overline{x}_i$.

7.2. Our presentation assumes that when there are two predictors x_1 and x_2, then the additive model space is spanned by the three vectors 1, $x_1 - \overline{x}_1$, and $x_2 - \overline{x}_2$. Prove that this is the case.

7.3. Using the small data set below, fit an additive linear model by doing the projections manually. (You may use `lm()` to check your work.)

	y	x_1	x_2
1	0	1	0
2	2	1	1
3	0	2	1
4	2	2	2
5	-1	3	1
6	6	3	3

7.4. Justify the claims in (7.1) and (7.2).

7.5. Justify the claim in (7.5).

7.6. This exercise refers to the data in Example 7.2.4.

a) Derive each p-value that appears in Example 7.2.4 using a model comparison test.

b) Sometimes SAT scores are expressed as a single score which is obtained by adding SATV and SATM. Can we make better predictions of GPA if we know both subscores than if we only know the combined score? Conduct the appropriate model comparison test to see.

c) As you might expect, `predict()` can produce interval estimates for the additive model as well as point estimates. Using an appropriate model, produce a 95% confidence interval for the mean GPA of students who have SATV and SATM scores of 550 and 650, respectively.

d) Many students do not have both ACT and SAT scores, so it can be handy to convert from one to the other for comparison purposes. Provide a formula that an admissions counselor could use to convert from SAT scores to ACT scores and provide some advice about the accuracy of this conversion.

7.7. Use the R output in Example 7.2.5 to calculate p-values for the hypotheses

- $H_0 : \beta_1 = 2$ and
- $H_0 : \beta_2 = 1$.

7.8. The `students` data set contains ACT and SAT scores, as well as high school and graduation GPAs, for a sample of 1000 students from a single undergraduate institution over a number of years.

  a) At this institution, most students take the ACT. For the small number of students who take the SAT but not the ACT, it would be convenient to have a way of estimating what the ACT score would have been.

  Fit a model that can be used to predict ACT scores from SAT scores in the `students` data set.

  b) Why don't the degrees of freedom for your model add up to 999?

  c) Fit a model that can be used to predict ACT scores from SAT scores in the `gpa` data set. (Note: SAT is the sum of SATM and SATV.)

  d) How do these two models compare?

7.9. In Example 7.2.5 we included all of the subjects in a single analysis. Perhaps the fitted model would be different if we restricted the data to include just men or just women. Fit these models and compare them to each other and to the combined analysis.

7.10. In Example 7.2.5 we included all of the subjects in a single analysis. Perhaps the fitted model would be different if we restricted the data to include only people who do not have type 2 diabetes (the controls, indicated by the variable `t2d`) or only people who do have type 2 diabetes (the cases). Fit these models and compare them to each other and to the combined analysis.

7.11. In Example 7.2.5 we fit a model after performing logarithmic transformations to the response and the predictors. Fit an additive model without performing the transformation and compare it to the model in Example 7.2.5.

  a) Use both adjusted r^2 and AIC to compare the models.

  b) Which model has the best residual behavior?

7.12. Using one of the models to predict weight from height and waist measurements based on the `pheno` data set, give a confidence interval and a prediction interval for the average weight of people that have your height and waist measurements.

  a) Is your weight contained in the confidence interval?

  b) Is your weight contained in the prediction interval?

  c) Is a prediction interval legitimate to use in this situation?

7.13. Do the "little matrix algebra" mentioned on page 433.

7.14. Obtain a different Pythagorean decomposition for the data in Example 7.3.5 by using different bases for the model and residual spaces. You may still use $v_0 = 1$.

7.15. A small study was done comparing 3 types of fertilizer to see which gave the greatest yield. A summary of the data appears below.

```
> summary(yield~type,f,fun=favstats)
yield    N=15

+--------+--+--+----+----+-----+------+---------+
|        |  |N |min |max |mean |median|sd       |
+--------+--+--+----+----+-----+------+---------+
|type    |A| 5|17.0|19.2|17.88|17.6  |0.9121403|
|        |B| 5|18.2|19.8|19.24|19.4  |0.6693280|
|        |C| 5|20.0|22.0|20.96|20.8  |0.8294577|
+--------+--+--+----+----+-----+------+---------+
|Overall|  |15|17.0|22.0|19.36|19.4  |1.5046594|
+--------+--+--+----+----+-----+------+---------+
```

Use this information to calculate the F-statistic and p-value for the model utility test. What do you conclude?

7.16. Do pets help relieve stress? Forty-five self-reported dog lovers were recruited for a study [**ABTK91**]. The data are summarized in [**BM09**]. Fifteen subjects were assigned to each of three treatment groups. The subjects' heart rates were monitored while they performed a stressful task either alone (group C), with a good friend in the room (group F), or with their dog in the room (group P). Mean heart rates and treatment are available in the `petstress` data set.

a) Use diagnostics to decide if an ANOVA analysis is appropriate for this data.

b) If it is appropriate, perform the model utility test and interpret the results.

7.17. Determine the unit direction vector in $\mathbb{R}^9$ for testing the (perhaps silly) null hypothesis

$$H_0 : \mu_1 + 2\mu_2 - 3\mu_3$$

in the context of a one-way ANOVA design if the data are from samples of sizes 2, 3, and 4, respectively, from the three groups.

7.18. Suppose you wanted to construct 10 confidence intervals with a familywise error rate of 5% (or confidence intervals with a 95% simultaneous confidence level). Suppose furthermore that the confidence intervals were independent (not the case when testing all pairwise comparisons).

a) Design an adjustment procedure for this situation and explain why it works.

b) Now generalize to the situation where you have n tests and want a familywise error rate of α.

c) How does this method compare with the Bonferroni correction?

   Hint: Binomial distributions.

7.19. Suppose you have money to sample n subjects for the purposes of comparing the means of I groups. Furthermore, assume that both groups have the same standard deviation. By considering the width of the resulting confidence intervals, explain why it is better to obtain equal-sized samples from each group rather than more from one group and fewer from another.

7.20. Suppose you have money to sample n subjects for the purposes of comparing the means of 2 groups. Under what conditions would it be preferable to have unequal sample sizes for the two groups?

7.21. Recall that in Example 7.3.16 we wanted to test the null hypothesis that the three treatment protocols (1, 2, or 4 doses daily) had the same mean effect. Our null hypothesis for this is $H_0 : \mu_1 = \mu_2 = \mu_3$. This is not a test of a single contrast, but it can be viewed as a combination of two orthogonal contrasts.

Suppose C_1 and C_2 are orthogonal contrasts and
$$H_0 : C_1 = 0 \text{ and } C_2 = 0 .$$
If H_0 is true, then $\boldsymbol{u}_{C_1} \cdot \boldsymbol{Y} \sim \mathsf{Norm}(0, \sigma)$, and $\boldsymbol{u}_{C_2} \cdot \boldsymbol{Y} \sim \mathsf{Norm}(0, \sigma)$, so
$$F = \frac{\left((\boldsymbol{u}_{C_1} \cdot \boldsymbol{Y})^2 + (\boldsymbol{u}_{C_2} \cdot \boldsymbol{Y})^2\right)/2}{SSE/(n - I)} \sim \mathsf{F}(2, n - I) .$$
Use the output given in Example 7.3.16 to calculate this F statistic and use `pf()` to obtain a p-value.

7.22. Repeat Exercise 7.21 with a different orthogonal decomposition of H_0 and confirm that the resulting p-value is the same. Why is it important that the two contrasts be orthogonal?

7.23. The `bugs` data set contains data from an experiment to see if insects are more attracted to some colors than to others [**WS67**]. The researchers prepared colored cards with a sticky substance so that insects that landed on them could not escape. The cards were placed in a field of oats in July. Later the researchers returned, collected the cards, and counted the number of cereal leaf beetles trapped on each card.

Use a linear model to give an analysis of this data. In particular:

a) Decide whether you should use a transformation of the response (`NumTrap`). Two candidate transformations to consider are square root and logarithm.

b) Perform the model utility test.

c) Perform whatever follow-up analysis is appropriate.

7.24. This problem continues the investigation of the `palettes` data set from Example 7.4.7.

a) Why is it not possible to test for an interaction effect between employee and day from this data set? What does R report if you try?

b) If you were hired by the employer to analyze these data and provide a report, what would be the main points of that report?

c) Suppose the employer suspected *prior to collecting the data* that employee D was not as productive as the others. What contrast would be reasonable to test in this situation? Do it in both the blocked and unblocked setting and report what you find.

d) Why does it matter whether the employer suspected employee D was not as productive *prior to* collecting the data or *based on looking at* the data?

e) How does the blocking variable `day` affect our impression of employee D?

7.25. Analyze the `palettes` data (Example 7.4.7) again, but this time use daily percent of palettes repaired as the response variable instead of the raw number of palettes repaired. Compare a one-way ANOVA with a two-way ANOVA using `day` as a blocking variable.

7.26. The function `poly()` was introduced in Example 7.5.2 to specify a second order model with one predictor. We said there that the output of

```
px <- poly( ut2$monthShifted, 2); px
```

is the result of applying polynomials p_1 and p_2 to the values in `ut2$monthShifted`.

a) Show that these two vectors `px[,1]` and `px[,2]` are orthogonal (up to round off error) by computing their dot product.

b) Plot `monthShifted` vs. `px[,1]` and `[px[,2]`. What do you notice?

c) Use `lm()` to determine the coefficients of the polynomials p_1 and p_2.

d) Compute the following four models:

```
       model1 <- lm (thermsPerday ~ monthShifted, ut2)
       model2 <- lm (thermsPerday ~ monthShifted + I(monthShifted^2), ut2)
  model1poly <- lm (thermsPerday ~ poly(monthShifted,1),  ut2)
  model2poly <- lm (thermsPerday ~ poly(monthShifted,2),  ut2)
```

and compare the output of `summary()` for each. Identify places where the same values show up for two or more models and explain why those values are the same. (Keep in mind that there may be small round-off errors and that the output may be displayed with different precision.)

7.27. The figures in Example 7.5.2 and the cyclic nature of weather patterns suggest that a model like

$$\text{thermsPerDay} = \beta_0 + \beta_1 \sin \left(\beta_2 + \frac{12 \cdot \text{month}}{2\pi} \right)$$

would be appropriate.

a) Interpret the parameters β_0, β_1, and β_2 in this model.

b) Why can't we fit this model using `lm()`?

c) Fit the model using maximum likelihood assuming that the error term is normally distribution with a mean of 0.

d) How does this model compare to the model we fit in Example 7.5.2?

e) Why might this not be a good model for the error terms? (That is, why would you expect some months to have more variation than others?) Propose a different model for the errors.

7.28. Repeat the analysis on page 486 using surface temperatures instead of intra-muscular temperatures. How do the two analyses compare?

7.29. Use the `ice` data set to see if the intramuscular temperature changes more for people with smaller `Skinfold` measurements. (Skinfold thickness is a measurement of the thickness of a fold of skin and an approximate measure of subcutaneous body fat.)

7.30. Use the B1930 temperature measurements from the `ice` data set to see if there is evidence that any of the following differ between men and women (before ice is applied):

a) surface temperature,

b) intramuscular temperature,

c) the difference between surface and intramuscular temperature.

7.31. In Example 7.7.3 we fit a logistic regression model using genotype and sex as predictors. The `pheno` data set includes several other variables. Use one of these as an additional term in the model and see if it strengthens the evidence that type 2 diabetes is associated with the genetic marker in `fusion1`.

7.32. Does the genetic marker in `fusion1` predict waist-to-hip ratio (`pheno$whr`)? Fit an appropriate model to answer this question. How will you deal with the fact that both men and women were included in this study?

7.33. Rats were given a dose of a drug proportional to their body weight. The rats were then slaughtered and the amount of the drug in the liver (`y`) and the weight of the liver were measured. What is the interpretation of the following output?

`rat-liver`

```
> require(alr3); data(rat)
> rat.lm <- lm(y~BodyWt*LiverWt,rat)
> summary(rat.lm)

Call:
lm(formula = y ~ BodyWt * LiverWt, data = rat)

Residuals:
     Min       1Q   Median       3Q      Max
-0.13399 -0.04337 -0.00723  0.03639  0.18403

Coefficients:
                 Estimate Std. Error t value Pr(>|t|)
(Intercept)       1.57382    1.46842    1.07     0.30
BodyWt           -0.00776    0.00857   -0.91     0.38
LiverWt          -0.17733    0.19820   -0.89     0.39
BodyWt:LiverWt    0.00109    0.00114    0.96     0.35

Residual standard error: 0.0919 on 15 degrees of freedom
Multiple R-squared:  0.1,        Adjusted R-squared: -0.0799
F-statistic: 0.556 on 3 and 15 DF,  p-value: 0.652
```

7.34. Researchers wanted to compare 6 diets to see which one produces the most weight gain in rabbits. The plan was to feed rabbits from the same litter different diets. Unfortunately, the researchers were only able to use three rabbits per litter. So the design consisted of assigning three of the possible treatments to the rabbits in a litter. The table below depicts this design showing which treatments were used with each litter:

	b1	b2	b3	b4	b5	b6	b7	b8	b9	b10
a	–	●	–	●	–	–	●	●	–	●
b	●	●	–	–	–	●	●	–	●	–
c	●	●	●	●	●	–	–	–	–	–
d	–	–	●	–	●	–	●	–	●	●
e	–	–	–	●	●	●	–	●	●	–
f	●	–	●	–	–	●	–	●	–	●

Similar tables can be created using one of the following:

```
data(rabbit, package='faraway')
xtabs(~treat+block,rabbit)
xtabs(gain~treat+block,rabbit)
```

a) This design was carefully chosen. What nice properties does it have? (This design is called an **incomplete block design**.)

b) What is the advantage of using rabbits from the same litter?

c) The data are in the `rabbit` data set in the `faraway` package. Analyze the data.

d) What can you conclude about the different treatments? Use Tukey's honest significant differences to answer.

7.35. Answer the following questions about the `rabbit` experiment in the previous problem.

a) It is not possible to use an interaction term in this situation. Why not?

b) The additive linear model can be fit with the predictors given in either order. Compare the ANOVA tables that result from doing it in each order. How and why do they differ?

c) Explain how to obtain two of the four p-values (two in each ANOVA table) using a model comparison test.

d) Which analysis is the correct analysis for this situation? Most importantly, which p-value should we use to check for a possible treatment effect? Why?

7.36. The data set `eggprod` in the `faraway` package contains the results of a randomized block experiment. Six pullets (young chickens) were placed into each of 12 pens. The 12 pens were divided into four blocks based on location. In each block, one pen was given each of three treatments. The egg production was then recorded for each pen.

• Do the treatments differ? Carry out an appropriate test and evaluate the results.

• Use Tukey's pairwise comparison method to see which treatment pairs differ significantly.

• Did blocking improve the efficiency of this study? How do you know?

7.37. Model the amount of money (in pounds) spent gambling by British teens (in the 1980s) using the other variables in the `teengamb` data set from the `faraway` package as predictors. Which model do you like best? Be sure to investigate possible interactions between sex and other predictors.

7.38. The `uswages` data set in the `faraway` packages contains a number of variables collected on 2000 US male workers as part of the 1988 Current Population Survey. Find a good model for `wage` using the available predictors. What factors appear to have the greatest influence on wages?

7.39. The table `ex14.26` from the `Devore6` package contains a summary table from a plant experiment to determine whether removing leaves from a certain type of plant affects the plant's ability to form mature fruit. What do these data suggest?

7.40. The table below contains summary data from a study comparing the sizes of left and right feet in 127 righthanded individuals [**LL78**]:

	left foot bigger	same size	right foot bigger	sample size
Men	2	10	28	40
Women	55	18	14	87

What do these data say about sex and foot asymmetry?

A Brief Introduction to R

A.1. Getting Up and Running

A.1.1. Installing R

R runs on just about any machine you will be interested in. Versions for Mac, PC, Linux, UNIX, etc., can be downloaded from the CRAN (Comprehensive R Archive Network) website. Follow the directions given there for your specific platform and the most recent release. Installation is straightforward if you are installing on a local, single-user machine for which you have the necessary permissions.

A.1.2. Installing and Using Packages

R is open source software. Its development is supported by a team of core developers and a large community of users. One way that users support R is by providing **packages** that contain data and functions for a wide variety of tasks.

Installing packages from CRAN

If you need to install a package, most likely it will be on CRAN. Before a package can be used, it must be **installed** (once per computer) and **loaded** (once per R session). For example, to use Hmisc:

```
install.packages("Hmisc")     # fetch package from CRAN to local machine.
require(Hmisc)                # load the package so it can be used.
```

If you are running on a machine where you don't have privileges to write to the default library location, you can install a personal copy of a package. If the location of your personal library is first in R_LIBS, this will probably happen automatically. If not, you can specify the location manually:

```
> install.packages("Hmisc",lib="~/R/library")
```

On a networked machine, be sure to use a different local directory for each platform since packages must match the platform.

Installing packages on a Mac or PC is something you might like to do from the GUI since it will provide you with a list of packages from which you can select the ones of interest. Binary packages have been precompiled for a particular platform and are generally faster and easier to set up, if they are available. Source packages need to be compiled and built on your local machine. Usually this happens automatically – provided you have all the necessary tools installed on your machine – so the only disadvantage is the extra time it takes to do the compiling and building.

Installing other packages

Occasionally you might find a package of interest that is not available via a repository like CRAN. Typically, if you find such a package, you will also find instructions on how to install it. If not, you can usually install directly from the zipped up package file.

```
install.packages('some-package.tar.gz',
                 repos=NULL)             # use a file, not a repository
```

Finding packages

There are several ways to find packages

- Ask your friends.

- Google: Put 'cran' in the search.

- Rseek: `http://rseek.org` provides a search engine specifically designed to find information about R.

- CRAN task views.
  A number of folks have put together task views that list a large number of packages and what they are good for. They are organized according to themes. Here are a few examples of available task views:

Bayesian	Bayesian Inference
Econometrics	Computational Econometrics
Finance	Empirical Finance
Genetics	Statistical Genetics
Graphics	Graphic Displays, Dynamic Graphics, Graphic Devices, and Visualization
Multivariate	Multivariate Statistics
SocialSciences	Statistics for the Social Sciences

- Biocunductor (`http://www.bioconductor.org/`) is another source of packages.

- *R News* (available via CRAN) often has articles about new packages and their capabilities.

- Write your own.
  You can write your own packages, and it isn't that hard to do.

```
> require(aplpack)                                        faithful-stemleaf
> stem.leaf(faithful$eruptions, style='bare')
1 | 2: represents 1.2
 leaf unit: 0.1
            n: 272
    12    1 | 667777777777
    51    1 | 888888888888888888888888888899999999999
    71    2 | 00000000000011111111
    87    2 | 2222222222333333
    92    2 | 44444
    94    2 | 66
    97    2 | 889
    98    3 | 0
   102    3 | 3333
   108    3 | 445555
   118    3 | 6666677777
   (16)   3 | 8888888889999999
   138    4 | 0000000000000000111111111111111
   107    4 | 2222222222223333333333333333
    78    4 | 44444444444445555555555555555555555
    43    4 | 66666666666777777777777
    21    4 | 88888888888899999
     4    5 | 0001
```

Figure A.1. Alternative version of stemplot using `stem.leaf()` from `aplpack`.

A.1.3. Some Useful Packages

`lattice`

`lattice` provides an implementation of trellis graphics. This set of high-level plotting functions makes it relatively simple to produce a wide range of useful graphical summaries of data. `lattice` is included with R but must be loaded before it can be used. In Section A.3 we provide a brief introduction to `lattice` graphics.

`grid`

`lattice` is based on a lower level graphics library in the `grid` package. `grid` contains functions to draw lines, symbols, text, etc., on plots. If you ever need to make fancy custom graphics, you will probably want to become familiar with the functions in `grid`.

`vcd`

`vcd` abbreviates 'visualizing categorical data' and provides a number of tools for this task, including `mosaic()` and `structable` (see Chapter 1). Other functions in `vcd` include `pairs()`, which arranges pairwise mosaic plots for a number of different categorical variables into a rectangular grid.

`aplpack`

The `aplpack` package (short for another plotting package) provides a number of plotting functions, including the `stem.leaf()` function as an alternative to `stem()`. See Figure A.1.

`MASS, DAAG, faraway, cars, alr3`

Each of these packages accompanies a book [**VR97, MB03, Far05, Far06, Fox02, Wei05**] and includes a number of data sets that are used in this book. Some of these are frequently used in examples in other R packages. This is especially true of `MASS` which accompanies the classic book by Venables and Ripley.

A.1.4. Setting Up The R Environment

When R launches, there are a number of files it looks at. You can get all the gory details about what happens and in what order by typing

```
?Startup
```

Actually, that will point you to some further documentation if you really want *all* the gory details.

Here is the short version: When R starts up, it reads a number of system and user files (unless you tell it not to – an important feature when preparing scripts for general use). There are two user files that you may want to edit a bit.[1]

.Rprofile

This is where you can put code that you want run each time a default R session is started. Here is a simple example showing how to load your favorite packages each time an R session is started.

```
# always load my favorite packages
require(lattice)
require(grid)
require(Hmisc)
require(fastR)
# adjust lattice settings
trellis.par.set(theme=col.fastR())
```

.Renviron

This is where you put information to help R interact sanely with the rest of your system. In particular, this is where you can tell R how to find packages. Example:

```
R_LIBS=/Library/Frameworks/R.Framework/Resources/library:~/R/library
```

---

[1]The file names used here are the names used on unix-like machines (including Mac OS X and Linux). Slightly different names (that do not start with a 'dot') are used on a PC.

A.1.5. Getting Help

help()

`help()` provides documentation on functions and language features. It can be abbreviated ?.

```
?c
?"for"
```

apropos()

`apropos()` will list all functions in currently loaded packages matching a specified character string. This can be useful if you "sort of" remember the name of a function.

```
> apropos('hist')
 [1] "event.history"        "hist"
 [3] "hist.data.frame"      "hist.default"
 [5] "hist.FD"              "hist.scott"
 [7] "histbackback"         "histogram"
 [9] "history"              "histSpike"
[11] "ldahist"              "loadhistory"
[13] "panel.histogram"      "panel.xhistogram"
[15] "pmfhistogram"         "prepanel.default.histogram"
[17] "savehistory"          "truehist"
[19] "xhistogram"
```

> apropos

args()

`args()` will list all the **arguments** (inputs) to a function.

```
> args(require)
function (package, lib.loc = NULL, quietly = FALSE, warn.conflicts = TRUE,
    keep.source = getOption("keep.source.pkgs"), character.only = FALSE,
    save = FALSE)
NULL
> args(mean)
function (x, ...)
NULL
> args(sum)
function (..., na.rm = FALSE)
NULL
```

> args

Here are some things to note:

- Some of the R GUIs display this information as soon as you type a function name.
- Function arguments can have default values.
- Sometimes you will see ... listed. More on this later.

- You do not need to name arguments when you use a function. Argument matching in R works as follows:
  - First all named arguments are matched (usually a unique prefix suffices).
  - Then unnamed arguments are matched against unused arguments in order.

example()

Many R functions come with examples that show you how to use the functions. Some of the examples are better than others, but the good ones can be very helpful. For example, to see how to use the `histogram()` function (and several other `lattice` functions as well), type

```
example(histogram)
```

A.1.6. The R "GUI"

R is primarily a programming language. (Actually, the language is officially called S). R provides an implementation of an S interpreter and an interface to interact with the interpreter. (So does a commercial product called S-Plus.) We will largely ignore the GUI (graphical user interface) that comes with R since it is (1) minimalistic and straightforward, (2) platform dependent, (3) of limited use for writing and executing scripts. But there are some things that the various R GUIs do well. For example, many of them provide menu-driven interfaces for finding, installing, and loading packages; for managing graphics windows; for loading data; and for creating external files containing R commands and executing them.

In addition to the standard GUI that ships with R, there is an alternative GUI called JGR (Java GUI for R) available at `http://rosuda.org/JGR/`. JGR is written in Java and is available for multiple platforms.

Those looking for a more traditional GUI approach with drop-down menus, checkboxes, and fill-in forms, should give the `Rcmdr` package a try. It provides a menu-driven interface for many of the things covered in this book, and there are several additional `Rcmdr` plug-in packages that extend the scope even further. Additionally, `Rcmdr` shows the R commands that are being generated by the menus, so it can be a good way to learn new R commands.

A.1.7. Some Workflow Suggestions

In short: *Think like a programmer.*

- Use R interactively only to get documentation and for quick one-offs.
- Store your code in a file rather than entering it at the prompt.

  You can execute all the code in a file using
  ```
  source("file.R")
  ```
  If you do work at the prompt and later wish you had been putting your commands into a file, you can save your past commands with
  ```
  savehistory("someRCommandsIalmostLost.R")
  ```

Then you can go back and edit the file.

- Use meaningful names.

- Write reusable functions.
    Learning to write your own functions will greatly increase your efficiency. (Stay tuned for details.)

- Comment your code.
    It's amazing what you can forget. The comment character in R is #.

A.2. Working with Data

A.2.1. Getting Data into R

Data in R packages

Data sets in the `datasets` package or any other loaded package are available via the `data()` function. Usually, the use of `data()` is unnecessary, however, since R will search most loaded packages (they must have been created with the lazy-load option) for data sets without the explicit use of `data()`. The `data()` function can be used to restore data after it has been modified or to control which package is used when data sets with the same name appear in multiple packages.

```
                                                          packageData-iris
# first line only necessary if iris is already in use
> data(iris)
> str(iris)            # get a summary of the data set
'data.frame':         150 obs. of  5 variables:
$ Sepal.Length:num 5.1 4.9 4.7 4.6 5 5.4 4.6 5 4.4 4.9 ...
$ Sepal.Width :num 3.5 3 3.2 3.1 3.6 3.9 3.4 3.4 2.9 3.1 ...
$ Petal.Length:num 1.4 1.4 1.3 1.5 1.4 1.7 1.4 1.5 1.4 1.5 ...
$ Petal.Width :num 0.2 0.2 0.2 0.2 0.2 0.4 0.3 0.2 0.2 0.1 ...
$ Species :Factor w/ 3 levels "setosa","versicolor",..: 1 1 1 1 1 1 1 1 1
    1 ...
> dim(iris)           # just the dimensions
[1] 150   5
```

```
                                                          iris-help
?iris

iris {datasets}                                       R Documentation
                    Edgar Anderson's Iris Data

Description

This famous (Fisher's or Anderson's) iris data set gives the
measurements in centimeters of the variables sepal length and width and
petal length and width, respectively, for 50 flowers from each of 3
species of iris. The species are Iris setosa, versicolor, and virginica.

Usage

iris iris3
```

Format

iris is a data frame with 150 cases (rows) and 5 variables (columns)
named Sepal.Length, Sepal.Width, Petal.Length, Petal.Width, and Species.

iris3 gives the same data arranged as a 3-dimensional array of size 50
by 4 by 3, as represented by S-PLUS. The first dimension gives the case
number within the species subsample, the second the measurements with
names Sepal L., Sepal W., Petal L., and Petal W., and the third the
species.

Source

Fisher, R. A. (1936) The use of multiple measurements in taxonomic
problems. Annals of Eugenics, 7, Part II, 179-188.

The data were collected by Anderson, Edgar (1935). The irises of the
Gaspe Peninsula, Bulletin of the American Iris Society, 59, 25.

References

Becker, R. A., Chambers, J. M. and Wilks, A. R. (1988) The New S
Language. Wadsworth & Brooks/Cole. (has iris3 as iris.)

See Also

matplot some examples of which use iris.

Examples

```
dni3 <- dimnames(iris3)
ii <- data.frame(matrix(aperm(iris3, c(1,3,2)), ncol=4,
                dimnames = list(NULL, sub(" L.",".Length",
                sub(" W.",".Width", dni3[[2]])))),
                Species = gl(3, 50,
                    lab=sub("S", "s", sub("V", "v", dni3[[3]])))))
all.equal(ii, iris) # TRUE
```

Loading data from flat files

R can read data from a number of file formats. The two most useful formats
are .csv (comma separated values) and white space delimited. Excel and most
statistical packages can read and write data in these formats, so these formats
make it easy to transfer data between different software. R provides read.csv()
and read.table() to handle these two situations. They work nearly identically
except for their default settings: read.csv() assumes that the first line of the file
contains the variable names but read.table() assumes that the data begins on
the first line with no names for the variables, and read.table() will ignore lines
that begin with '#' but read.csv() will not.

The default behavior can be overridden for each function, and there are a number of options that make it possible to read other file formats, to omit a specified number of lines at the top of the file, etc. If you are making the file yourself, always include meaningful names in either file format.

It is also possible to read data from a file located on the Internet. Simply replace the file name with a URL. The data read below come from [**Tuf01**].

read-table

```
# need header=TRUE because there is a header line.
# could also use read.file() without header=TRUE
> traffic <-
+     read.table("http://www.calvin.edu/~rpruim/fastR/trafficTufte.txt",
+     header=TRUE)
> traffic
  year cn.deaths   ny   cn   ma   ri
1 1951       265 13.9 13.0 10.2  8.0
2 1952       230 13.8 10.8 10.0  8.5
3 1953       275 14.4 12.8 11.0  8.5
4 1954       240 13.0 10.8 10.5  7.5
5 1955       325 13.5 14.0 11.8 10.0
6 1956       280 13.4 12.1 11.0  8.2
7 1957       273 13.3 11.9 10.2  9.4
8 1958       248 13.0 10.1 11.8  8.6
9 1959       245 12.9 10.0 11.0  9.0
```

Notice the use of <- in the example above. This is the **assignment operator** in R. It can be used in either direction (<- or ->). In the first line of the example above, the results of `read.table()` are stored in a variable called `traffic`. `traffic` is a **data frame**, R's preferred container for data. (More about data types in R as we go along.)

The `na.strings` argument can be used to specify codes for missing values. The following can be useful for SAS output, for example:

read-sas

```
> traffic <-
+     read.csv("http://www.calvin.edu/~rpruim/fastR/trafficTufte.csv",
+     na.strings=c(".","NA",""))
```

For convenience `fastR` provides `read.file()` which uses the file name to determine which of `read.table()` or `read.csv()` to use and sets the defaults to `header=T`, `comment.char="#"`, and `na.strings=c('NA','',' '.','na')` for both files types.

Manually typing in data

If you need to enter a small data set by hand, the `scan()` function is quick and easy. Individual values are separated by white space or new lines. A blank line is used to signal the end of the data. By default, `scan()` is expecting decimal data (which it calls **double**, for double precision), but it is possible to tell `scan()` to expect something else, like **character** data (i.e., text). There are other options for

data types, but numerical and text data will usually suffice for our purposes. See
?scan for more information and examples.

```
> myData1 <- scan()                                            scan
1: 15 18
3: 12
4: 21 23 50 15
8:
Read 7 items
> myData1
[1] 15 18 12 21 23 50 15
>
> myData2 <- scan(what="character")
1: "red" "red" "orange" "green" "blue" "blue" "red"
8:
Read 7 items
> myData2
[1] "red"    "red"    "orange" "green"  "blue"   "blue"   "red"
```

Be sure when using scan() that you remember to save your data somewhere. Otherwise you will have to type it again.

Creating data frames from vectors

The scan() function puts data into a vector, not a data frame. We can build a data
frame for our data as follows.

```
> myDataFrame <- data.frame(color=myData2,number=myData1)     dataframe
> myDataFrame
   color number
1    red     15
2    red     18
3 orange     12
4  green     21
5   blue     23
6   blue     50
7    red     15
```

Getting data from mySQL data bases

The RMySQL package allows direct access to data in MySQL data bases. This can
be very convenient when dealing with subsets of very large data sets.

Generating data

The following code shows a number of ways to generate data systematically.

```
> x <- 5:20; x                      # all integers in a range   generatingData01
 [1]  5  6  7  8  9 10 11 12 13 14 15 16 17 18 19 20
# structured sequences
```

```
> seq(0,50,by=5)
 [1]   0   5 10 15 20 25 30 35 40 45 50
> seq(0,50,length=7)
[1]   0.0000  8.3333 16.6667 25.0000 33.3333 41.6667 50.0000
> rep(1:5,each=3)
 [1] 1 1 1 2 2 2 3 3 3 4 4 4 5 5 5
> rep(1:5,times=3)
 [1] 1 2 3 4 5 1 2 3 4 5 1 2 3 4 5
> c(1:5,10,3:5)                     # c() concatenates vectors
[1]   1   2   3   4   5 10   3   4   5
```

R can also sample from several different distributions.

generatingData02

```
> rnorm(10,mean=10,sd=2)    # random draws from normal distribution
 [1]   7.0524 10.0244 10.9092 12.3793 11.0753  7.7473 12.8116  6.7872
 [9]   8.5278  8.8866
> x <- 5:20                 # all integers in a range
> sample(x,size=5)          # random sample of size 5 from x (no replacement)
[1] 15 10 14 12 13
```

Functions for sampling from other distributions include rbinom(), rchisq(), rt(), rf(), rhyper(), etc. See Chapter 2 for more information.

A.2.2. Saving Data

write.table() and write.csv() can be used to save data from R into delimited flat files.

writingData

```
> args(write.table)
function (x, file = "", append = FALSE, quote = TRUE, sep = " ",
    eol = "\n", na = "NA", dec = ".", row.names = TRUE, col.names = TRUE,
    qmethod = c("escape", "double"))
NULL
> write.table(ddd,"ddd.txt")
> write.csv(ddd,"ddd.csv")
# this system call should work on a Mac or Linux machine
> system("head -20 ddd.txt ddd.csv")
==> ddd.txt <==
"number" "letter"
"Abe" 1 "a"
"Betty" 2 "b"
"Claire" 3 "c"
"Don" 4 "d"
"Ethel" 5 "e"

==> ddd.csv <==
"","number","letter"
"Abe",1,"a"
"Betty",2,"b"
"Claire",3,"c"
```

```
"Don",4,"d"
"Ethel",5,"e"
```

Data can also be saved in native R format. Saving data sets (and other R objects) using `save()` has some advantages over other file formats:

- Complete information about the objects is saved, including attributes.
- Data saved this way takes less space and loads much more quickly.
- Multiple objects can be saved to and loaded from a single file.

The downside is that these files are only readable in R.

savingData

```
> save(ddd,abc,file="ddd.zip")      # saves both objects in a single file
> load("ddd.zip")                   # loads them both
```

For more on importing and exporting data, especially from other formats, see the *R Data Import/Export* manual available on CRAN.

A.2.3. Primary R Data Structures

Modes and other attributes

In R, data is stored in objects. Each object has a *name*, *contents*, and also various *attributes*. Attributes are used to tell R something about the kind of data stored in an object and to store other auxiliary information. Two important attributes shared by all objects are mode and length.

mode01

```
> w <- 2.5; x <- c(1,2); y <- "foo"; z <- TRUE; abc <- letters[1:3]
> mode(w); length(w)
[1] "numeric"
[1] 1
> mode(x); length(x)
[1] "numeric"
[1] 2
> mode(y); length(y)
[1] "character"
[1] 1
> y[1]; y[2]              # not an error to ask for y[2]
[1] "foo"
[1] NA
> mode(z); length(z)
[1] "logical"
[1] 1
> abc
[1] "a" "b" "c"
> mode(abc); length(abc)
[1] "character"
[1] 3
> abc[3]
[1] "c"
```

Each of the objects in the example above is a vector, an ordered container of values that all have the same mode.[2] The c() function concatenates vectors (or lists). Notice that w, y, and z are vectors of length 1. Missing values are coded as NA (not available). Asking for an entry "off the end" of a vector returns NA. Assigning a value "off the end" of a vector results in the vector being lengthened so that the new value can be stored in the appropriate location.

There are important ways that R has been optimized to work with vectors since they correspond to variables (in the sense of statistics). For categorical data, a factor is a special type of vector that includes an additional attribute called *levels*. A factor can be ordered or unordered (which can affect how statistics are done and graphs are made) and its elements can have mode numeric or character.

A list is similar to a vector, but its elements may be of different modes (including list, vector, etc.). A data frame is a list of vectors (or factors), each of the same length, but not necessarily of the same mode. This is R's primary way of storing data sets. An array is a multi-dimensional table of values that all have the same mode. A matrix is a 2-dimensional array.

The access operators ([ ] for vectors, matrices, arrays, and data frames, and [[ ]] for lists) are actually *functions* in R. This has some important consequences:

- Accessing elements is slower than in a language like C/C++ where access is done by pointer arithmetic.

- These functions also have named arguments, so you can see code like

```
> xm <- matrix(1:16, nrow=4); xm                                  access
     [,1] [,2] [,3] [,4]
[1,]   1    5    9   13
[2,]   2    6   10   14
[3,]   3    7   11   15
[4,]   4    8   12   16
> xm[5]
[1] 5
> xm[,2]                      # this is 1 dimensional (a vector)
[1] 5 6 7 8
> xm[,2,drop=FALSE]           # this is 2 dimensional (still a matrix)
     [,1]
[1,]   5
[2,]   6
[3,]   7
[4,]   8
```

Many objects have a dim attribute that stores the dimension of the object. You can change it to change the shape (or even the number of dimensions) of a vector, matrix, or array. You can see all of the non-intrinsic attributes (mode and length are intrinsic) using attributes(), and you can set attributes (including new ones you make up) using attr(). Some attributes, like dimension, have special functions for accessing or setting. The dim() function returns the dimensions of an object

---

[2]There are other modes in addition to the ones shown here, including complex (for complex numbers), function, list, call, and expression.

as a vector. Alternatively the number of rows and columns can be obtained using `nrow()` and `ncol()`.

attributes

```
> ddd <- data.frame(number=1:5,letter=letters[1:5])
> attributes(ddd)
$names
[1] "number" "letter"

$row.names
[1] 1 2 3 4 5

$class
[1] "data.frame"

> dim(ddd)
[1] 5 2
> nrow(ddd)
[1] 5
> ncol(ddd)
[1] 2
> names(ddd)
[1] "number" "letter"
> row.names(ddd)
[1] "1" "2" "3" "4" "5"
> row.names(ddd) <- c("Abe","Betty","Claire","Don","Ethel")
> ddd                    # row.names affects how a data.frame prints
       number letter
Abe         1      a
Betty       2      b
Claire      3      c
Don         4      d
Ethel       5      e
```

What is it?

R provides a number of functions for testing the mode or class of an object.

whatIsThis

```
> xm <- matrix(1:16, nrow=4); xm
     [,1] [,2] [,3] [,4]
[1,]    1    5    9   13
[2,]    2    6   10   14
[3,]    3    7   11   15
[4,]    4    8   12   16
> mode(xm); class(xm)
[1] "numeric"
[1] "matrix"
> c(is.numeric(xm), is.character(xm), is.integer(xm), is.logical(xm))
[1]  TRUE FALSE  TRUE FALSE
> c(is.vector(xm), is.matrix(xm), is.array(xm))
[1] FALSE  TRUE  TRUE
```

Changing modes and attributes

If R is expecting an object of a certain mode or class but gets something else, it will often try to coerce the object to meet its expectations. You can also coerce things manually using one of the many `as.???()` functions.

asYouLikeIt

```
> apropos("^as\\.")[1:10]      # just a small sample
 [1] "as.array"           "as.array.default"
 [3] "as.call"            "as.category"
 [5] "as.character"       "as.character.condition"
 [7] "as.character.Date"  "as.character.default"
 [9] "as.character.error" "as.character.factor"
# convert numbers to strings (this drops attributes)
> as.character(xm)
 [1] "1"  "2"  "3"  "4"  "5"  "6"  "7"  "8"  "9"  "10" "11" "12" "13" "14"
[15] "15" "16"
# convert matrix to vector
> as.vector(xm)
 [1]  1  2  3  4  5  6  7  8  9 10 11 12 13 14 15 16
> as.logical(xm)
 [1] TRUE TRUE TRUE TRUE TRUE TRUE TRUE TRUE TRUE TRUE TRUE TRUE TRUE TRUE
[15] TRUE TRUE
> alpha <- c("a","1","b","0.5")
> mode(alpha)
[1] "character"
> as.numeric(alpha)      # can't do the coercion, so NAs are introduced
[1]  NA 1.0  NA 0.5
Warning message:
NAs introduced by coercion
> as.integer(alpha)      # notice coersion of 0.5 to 0
[1] NA  1 NA  0
Warning message:
NAs introduced by coercion
```

A.2.4. More About Vectors

Vectors are so important in R that they deserve some additional discussion. In Section 1 we learned how to generate some simple vectors. Here we will learn about some of the operations and functions that can be applied to vectors.

Vectorized functions

Many R functions and operations are "vectorized" and can be applied not just to an individual value but to an entire vector, in which case they are applied componentwise and return a vector of transformed values. Most traditional mathematics functions are available and work this way.

vectors01

```
> x <- 1:5; y <- seq(10,60,by=10); z <- rnorm(10); x; y
[1] 1 2 3 4 5
[1] 10 20 30 40 50 60
```

```
> y + 1
[1] 11 21 31 41 51 61
> x * 10
[1] 10 20 30 40 50
> x < 3
[1]  TRUE  TRUE FALSE FALSE FALSE
> x^2
[1]  1  4  9 16 25
> log(x); log(x, base=10)              # natural and base 10 logs
[1] 0.00000 0.69315 1.09861 1.38629 1.60944
[1] 0.00000 0.30103 0.47712 0.60206 0.69897
```

Vectors can be combined into a matrix using rbind() or cbind(). This can facilitate side-by-side comparisons.

```
# compare round() and signif() by binding rowwise into matrix          vectors01a
> rbind(round(z,digits=2), signif(z,digits=2))
     [,1]  [,2] [,3] [,4] [,5]  [,6]  [,7]  [,8]    [,9] [,10]
[1,] 1.26 -0.33 1.33 1.27 0.41 -1.54 -0.93 -0.29 -0.0100   2.4
[2,] 1.30 -0.33 1.30 1.30 0.41 -1.50 -0.93 -0.29 -0.0058   2.4
```

Functions that act on vectors as vectors

Other functions, including many statistical functions, are designed to work on the vector as a vector. Often these return a single value (technically a vector of length 1), but other return types are used as appropriate.

```
> x <- 1:10; z <- rnorm(100)                                          vectors02
> mean(z); sd(z); var(z); median(z)  # basic statistical functions
[1] -0.0091809
[1] 0.86084
[1] 0.74105
[1] -0.055992
> range(z)                           # range returns a vector of length 2
[1] -2.2239  2.4414
```

```
> sum(x); prod(x)                          # sums and products          vectors02a
[1] 55
[1] 3628800
> z <- rnorm(5); z
[1] -0.412520 -0.972287  0.025383  0.027475 -1.680183
> sort(z); rank(z); order(z)               # sort, rank, order
[1] -1.680183 -0.972287 -0.412520  0.025383  0.027475
[1] 3 2 4 5 1
[1] 5 2 1 3 4
> rev(x)                                    # reverse x
 [1] 10  9  8  7  6  5  4  3  2  1
> diff(x)                                    # pairwise differences
[1] 1 1 1 1 1 1 1 1 1
> cumsum(x)                                  # cumulative sum
 [1]  1  3  6 10 15 21 28 36 45 55
```

Table A.1. Some useful R functions.

cumsum() cumprod() cummin() cummax()	Returns vector of cumulative sums, products, minima, or maxima.
pmin(x,y,...) pmax(x,y,...)	Returns vector of parallel minima or maxima where ith element is max or min of x[i], y[i], ...
which(x)	Returns a vector of indices of elements of x that are true. Typical use: which(y > 5) returns the indices where elements of y are larger than 5.
any(x)	Returns a logical indicating whether any elements of x are true. Typical use: if (any(y > 5)) { ...}.
na.omit(x)	Returns a vector with missing values removed.
unique(x)	Returns a vector with repeated values removed.
table(x)	Returns a table of counts of the number of occurrences of each value in x. The table is similar to a vector with names indicating the values, but it is not a vector.
paste(x,y,..., sep=" ")	Pastes x and y together componentwise (as strings) with sep between elements. Recycling applies.

```
> cumprod(x)                            # cumulative product
 [1]       1       2       6      24     120     720    5040   40320
 [9]  362880 3628800
```

Whether a function is vectorized or treats a vector as a unit depends on its implementation. Usually, things are implemented the way you would expect. Occasionally you may discover a function that you wish were vectorized and is not. When writing your own functions, give some thought to whether they should be vectorized, and test them with vectors of length greater than 1 to make sure you get the intended behavior.

Some additional useful functions are included in Table A.1.

Recycling

When vectors operate on each other, the operation is done componentwise, recycling the shorter vector to match the length of the longer.

```
> x <- 1:5; y <- seq(10,70,by=10)          vectors03
> x + y
[1] 11 22 33 44 55 61 72
Warning message:
In x + y : longer object length is not a multiple of shorter object length
```

In fact, this is exactly how things like x + 1 actually work. If x is a vector of length n, then 1 (a vector of length 1) is first recycled into a vector of length n;

then the two vectors are added componentwise. Some vectorized functions that take multiple vectors as arguments will first use recycling to make them the same length.

Accessing elements of vectors

R allows for some very interesting and useful methods for accessing elements of a vector that combine the ideas above. First, recall that the [ ] operator is actually a function. Furthermore, it is vectorized.

vectors04a

```
> x <- seq(2,20,by=2)
> x[1:5]; x[c(1,4,7)]
[1]  2  4  6  8 10
[1]  2  8 14
```

[ ] accepts `logicals` as arguments well. The boolean values (recycled, if necessary) are used to select or deselect elements of the vector.

vectors04b

```
> x <- seq(2,20,by=2)
> x[c(TRUE,TRUE,FALSE)]          # skips every third element (recycling!)
[1]  2  4  8 10 14 16 20
> x[x > 10]                      # more typical use of boolean in selection
[1] 12 14 16 18 20
```

Negative indices are used to omit elements.

vectors04c

```
> x <- 1:10; x[-7]; x[-c(1,2,4,8)]; x[-length(x)]
[1]  1  2  3  4  5  6  8  9 10
[1]  3  5  6  7  9 10
[1] 1 2 3 4 5 6 7 8 9
```

Here are some more examples.

vectors04d

```
> notes <- toupper(letters[1:7]); a <- 1:5; b <- seq(10,100,by=10)
> toupper(letters[5:10])
[1] "E" "F" "G" "H" "I" "J"
> paste(letters[1:5],1:3,sep='-')
[1] "a-1" "b-2" "c-3" "d-1" "e-2"
> a+b
 [1]  11  22  33  44  55  61  72  83  94 105
> (a+b)[ a+b > 50]
[1]  55  61  72  83  94 105
> length((a+b)[a+b > 50])
[1] 6
> table(a+b > 50)

FALSE   TRUE
    4      6
```

A.2.5. Summarizing Data

Individual variables of a data frame are accessible as vectors via the $ operator. This allows us to apply functions to these vectors.

iris-vectors

```
> table(iris$Species)

    setosa versicolor  virginica
        50         50         50
> range(iris$Sepal.Length)
[1] 4.3 7.9
> table(cut(iris$Sepal.Length,seq(4,8,by=0.5)))

(4,4.5] (4.5,5] (5,5.5] (5.5,6] (6,6.5] (6.5,7] (7,7.5] (7.5,8]
      5      27      27      30      31      18       6       6
> mean(iris$Sepal.Length)
[1] 5.8433
> quantile(iris$Sepal.Length)
  0%  25%  50%  75% 100%
 4.3  5.1  5.8  6.4  7.9
```

Since we know that there are three species of irises, it would be better to compute these sorts of summaries separately for each. There are (at least) two ways to do this in R. The first uses the aggregate() function. A much easier way uses the summary() function from the Hmisc package. This function uses the same kind of formula notation that the lattice graphics functions use.

iris-Hmisc-summary-q

```
> require(Hmisc)
> summary(Sepal.Length~Species,data=iris)
Sepal.Length    N=150

+-------+----------+---+------------+
|       |          |N  |Sepal.Length|
+-------+----------+---+------------+
|Species|setosa    | 50|5.0060      |
|       |versicolor| 50|5.9360      |
|       |virginica | 50|6.5880      |
+-------+----------+---+------------+
|Overall|          |150|5.8433      |
+-------+----------+---+------------+
> summary(Sepal.Length~Species,data=iris, fun=quantile)
Sepal.Length    N=150

+-------+----------+---+---+-----+---+---+----+
|       |          |N  |0% |25%  |50%|75%|100%|
+-------+----------+---+---+-----+---+---+----+
|Species|setosa    | 50|4.3|4.800|5.0|5.2|5.8 |
|       |versicolor| 50|4.9|5.600|5.9|6.3|7.0 |
|       |virginica | 50|4.9|6.225|6.5|6.9|7.9 |
+-------+----------+---+---+-----+---+---+----+
|Overall|          |150|4.3|5.100|5.8|6.4|7.9 |
+-------+----------+---+---+-----+---+---+----+
```

User-defined functions (see Section A.4) can be used in place of `quantile()` to get your favorite statistical summaries.

`xtabs()` cross-tabulates data. `iris` doesn't have a second categorical variable, but we can use the result of a `cut` to cross-tabulate species and binned sepal lengths.

```
> xtabs(~Species+cut(Sepal.Length,4:8), data=iris)          iris-xtabs
          cut(Sepal.Length, 4:8)
Species     (4,5] (5,6] (6,7] (7,8]
  setosa       28    22     0     0
  versicolor    3    27    20     0
  virginica     1     8    29    12
```

A.2.6. Manipulating Data

Adding new variables to a data frame

We can add additional variables to an existing data frame by simple assignment. It is an error to add a vector of the wrong length.

```
> iris$SLength <- cut(iris$Sepal.Length,4:8)          adding-variable
```

```
> summary(iris)                                       adding-variable2
  Sepal.Length    Sepal.Width    Petal.Length    Petal.Width
 Min.   :4.30    Min.   :2.00    Min.   :1.00    Min.   :0.1
 1st Qu.:5.10    1st Qu.:2.80    1st Qu.:1.60    1st Qu.:0.3
 Median :5.80    Median :3.00    Median :4.35    Median :1.3
 Mean   :5.84    Mean   :3.06    Mean   :3.76    Mean   :1.2
 3rd Qu.:6.40    3rd Qu.:3.30    3rd Qu.:5.10    3rd Qu.:1.8
 Max.   :7.90    Max.   :4.40    Max.   :6.90    Max.   :2.5
       Species      SLength
 setosa    :50    (4,5]:32
 versicolor:50    (5,6]:57
 virginica :50    (6,7]:49
                  (7,8]:12
```

Slicing and dicing

`reshape()` provides a flexible way to change the arrangement of data. It was designed for converting between long and wide versions of time series data and its arguments are named with that in mind.

A common situation is when we want to convert from a wide form to a long form because of a change in perspective about what a unit of observation is. For example,

```
> traffic                                             traffic-reshape
  year cn.deaths   ny   cn   ma   ri
1 1951       265 13.9 13.0 10.2  8.0
2 1952       230 13.8 10.8 10.0  8.5
3 1953       275 14.4 12.8 11.0  8.5
```

```
4 1954        240 13.0 10.8 10.5  7.5
5 1955        325 13.5 14.0 11.8 10.0
6 1956        280 13.4 12.1 11.0  8.2
7 1957        273 13.3 11.9 10.2  9.4
8 1958        248 13.0 10.1 11.8  8.6
9 1959        245 12.9 10.0 11.0  9.0
>
> reshape(traffic[,-2], idvar="year",ids=row.names(traffic),
+         times=names(traffic)[3:6],timevar="state",
+         varying=list(names(traffic)[3:6]),
+         v.names="deathRate",
+         direction="long") -> longTraffic
> head(longTraffic)
         year state deathRate
1951.ny 1951    ny      13.9
1952.ny 1952    ny      13.8
1953.ny 1953    ny      14.4
1954.ny 1954    ny      13.0
1955.ny 1955    ny      13.5
1956.ny 1956    ny      13.4
```

In simpler cases, `stack()` or `unstack()` may suffice. `Hmisc` also provides `reShape()` as an alternative to `reshape()`.

A.3. Lattice Graphics in R

A.3.1. Out-of-the-Box Plots and the Formula Interface

The `lattice` graphing functions all use a similar formula interface. The generic form of a formula is

- y ˜ x | z

which can often (both for plotting and for linear models) be interpreted as

- "y modeled by x conditioned on z"

For plotting, y will typically contain a variable presented on the vertical axis, and x a variable to be plotted along the horizontal axis. The condition z is a variable that is used to break the data into sections which are plotted in separate panels. When z is categorical, there is one panel for each level of z. Figure A.2 shows the **scatterplot** produced with

```
> p <- xyplot(Sepal.Length~Sepal.Width|Species,data=iris)
> print(p)
```
`iris-xyplot01`

Note the command `print(p);`. When using R interactively, if a lattice plot is not saved in a variable, then this is not needed since by default the console prints the value of each line when it is executed. This is not the case when R code is run as a batch job. In this case, the explicit `print()` is required.

In the case of a **histogram**, the values for the vertical axis are computed from the x variable, so y is omitted. The plots in Figure A.3 were produced by

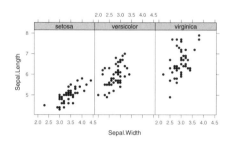

Figure A.2. Scatterplots of sepal length and width.

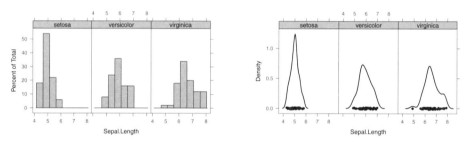

Figure A.3. The output of `histogram(~Sepal.Length|Species,data=iris)` and `densityplot(~Sepal.Length|Species,data=iris)` .

```
> p <- histogram(~Sepal.Length|Species,iris)          iris-histogram01
> print(p)
> p <- histogram(~Sepal.Width|Species,iris)
> print(p)
```

For those who like smoother output, `densityplot()` will draw an empirical density plot using kernel density estimation (see Section 3.5).

When `z` is quantitative, the data is divided into a number of sections based on the values of `z`. This works much like the `cut()` function, but some data may appear in more than one panel. In R terminology each panel represents a shingle of the data. The term shingle is supposed to evoke an image of overlapping coverage like the shingles on a roof. Finer control over the number of panels can be obtained by using `equal.count()` or `co.intervals()` to make the shingles directly (Figure A.4).

```
> p <- histogram(~Sepal.Length|Sepal.Width,data=iris)    iris-histogram02
> print(p)
```

Boxplots are produced using `bwplot()`. The orientation of the boxplots is determined by which variable is on the left (vertical axis) and which is on the right (horizontal axis) in the formula. (See Figure A.5.)

```
> p <- bwplot(Sepal.Length~Species,data=iris)          iris-bwplot01
> print(p)
```

```
> p <- bwplot(Species~Sepal.Length,data=iris)          iris-bwplot02
> print(p)
```

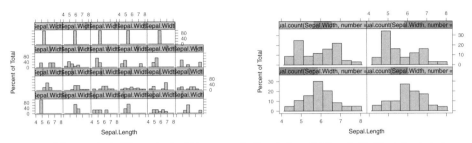

Figure A.4. The output of `histogram(~Sepal.Length|Sepal.Width,data=iris)` and `histogram(~Sepal.Length|equal.count(Sepal.Width,number=4),iris)`.

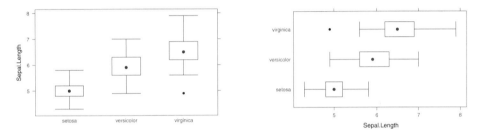

Figure A.5. Boxplots using `bwplot()`.

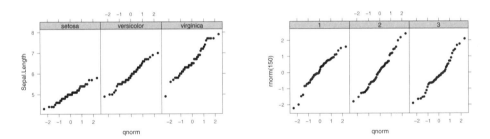

Figure A.6. Normal-quantile plots using `qqmath()`.

Quantile-quantile plots (see Section 3.6) can be made using `qqmath()` By default, the comparison distribution is a standard normal distribution (Figure A.6).

```
> p <- qqmath(~Sepal.Length|Species,data=iris)        qqmath01
> print(p)
> set.seed(1)                        # use fixed random seed
> p <- qqmath(~rnorm(150)|factor(rep(1:3,each=50)))
> print(p)
```

For investigating p-values, we might like to compare with a uniform distribution (Figure A.7, left).

```
                                                      qqmath03
> set.seed(1)                        # use fixed "random" seed
> someData <- data.frame(x=runif(300),group=factor(rep(1:3,each=100)))
```

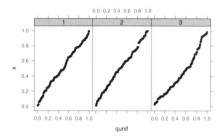 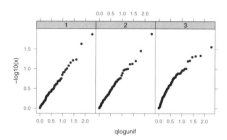

Figure A.7. Quantile-quantile plots using `qqmath()` comparing to the uniform and log-transformed uniform distributions.

```
> p <- qqmath(~x|group, data=someData, distribution=qunif)
> print(p)
```

With just a bit more work we can make a quantile-quantile plot for log-transformed p-values. Of course, this amplifies the discrepancy in the interesting tail of the distribution (Figure A.7, right).

```
> qlogunif <- function(p,a=0,b=1,base=10) {                          qqmath04
+         -log(1-qunif(p,a,b),base)
+ }
> p <- qqmath(~-log10(x) | group, data=someData, distribution=qlogunif)
> print(p)
```

The `lattice` package also provides utilities for comparing several variables at once. For example, a **scatterplot matrix** will make all pairwise scatterplots (Figure A.8).

```
> p <- splom(iris)                                                  iris-splom
> print(p)
```

A.3.2. Customizing Plots via Arguments

The plots we have displayed so far have been pretty basic. In particular, we have accepted most of the default argument values. There are, however, a large number of arguments that allow you to control the look of a `lattice` plot. An extensive list can be found under `?xyplot`, but we will highlight a few of the more important ones here.

- `main = "A title for my plot"` gives your plot a title which appears above the plot.
- `sub = "A subtitle for my plot"` gives your plot a subtitle which appears below the plot.
- `xlab` and `ylab` can be used to change the labels of the axes.
- `xlim = c(low,high)` sets the horizontal range for the viewing rectangle. `ylim` does the same for the vertical range. `c()` concatenates its arguments into a vector and is useful in other contexts as well.

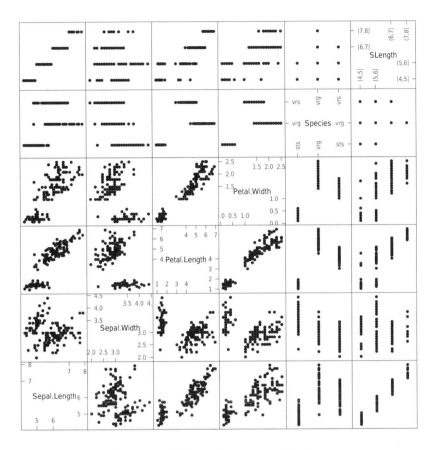

Figure A.8. Scatterplot matrix of iris data.

- `groups` sets a grouping variable. This behaves much like a conditioning variable except that the plots are overlaid in the same panel. Different colors or symbols can be used to differentiate the groups.

- `layout=c(cols,rows)` will arrange the panels in a multi-panel plot into a rectangle with the number of rows and columns specified. (Columns come first because they are along the horizontal axis, which comes first in Cartesian coordinates.) If your rectangle doesn't have enough slots, the plot will be split over multiple "pages". If you have too many slots, some will be left blank.

- `as.table=TRUE` will reverse the vertical ordering of panels. By default, R lays things out with lower values near the bottom and higher values near the top. Sometimes the other order is more natural.

- `cex` controls the size of the symbols displayed in `xyplot()`. The value is a ratio of the displayed size to the default size. For example, to get symbols half the default size, use `cex=0.5`.

- `pch` determines the symbols used in `xyplot()`. The value is an integer, and each integer is mapped to a different symbol or a single character if you want to put letters on your plot. If you provide a vector of numbers, these symbols will be used to distinguish different groups.

- `lwd` and `lty` determine width and type of lines in plots. If you provide a vector of numbers, different line widths will be used to distinguish different groups.

- `col="blue"` will print your symbols in blue. Other color names are available too. (Type `colors()` to see a complete list of named colors.) Again a vector can be used to set colors for multiple groups.

  Note: The `RColorBrewer` package offers a number of interesting color palettes that are intended to work better visually than easy things you might otherwise have tried. The functions `colorRamp()` and `colorRampPalette()` can be used to form sequences of colors that interpolate between a sequence of given colors.

- `fill` sets the interior color for some of the 'open' plot symbols (like 23, an open diamond). In this case `col` sets the outline color.

- `alpha` sets the level of opacity (1 = opaque; 0 = invisible). Using `alpha` values between 0 and 1 produces colors that are partially transparent – allowing colors from below to show through. This can be used for a number of interesting effects. For example, in a scatterplot with lots of data, a low value of `alpha` will make it clear where plotted points are overlapping since each overstrike will darken the dot. Partially transparent rectangles can be used to highlight regions in a plot.

- `auto.key=TRUE` will turn on an automatically generated legend. It isn't perfect but is often sufficient, at least for a rough draft.

- `type` controls which of various flavors of a plot are produced. For example,
  ```
  xyplot( y~x , type='l')
  ```
  will put line segments between data values instead of using plot symbols. `type = 'b'` will do both line segments and points. `type='a'` can be used to make **interaction plots**. For histograms, `type` may be one of `'percent'`, `'density'`, or `'count'` and determines the scale used for the vertical axis.

- `scales` controls the scales for the axes. Values for this argument are nested lists.

- `panel` can be assigned to a function that does the actual plotting in a panel. We'll look at some examples of this in Section A.4.1.

Example A.3.1. Figure A.9 contains an improved scatterplot matrix of the iris data. Now color and shape are used to distinguish the three species. Reducing `alpha` makes it easier to see where multiple observations overlap.

```
> p <- splom(~iris[1:4],data=iris, groups=Species,        iris-splom02
+     pch=c(21,23,25),
+     cex=0.8,
+     alpha=0.7,
+     col='gray50',
+     fill=trellis.par.get('superpose.symbol')$col
+     )
> print(p)                                                              ◁
```

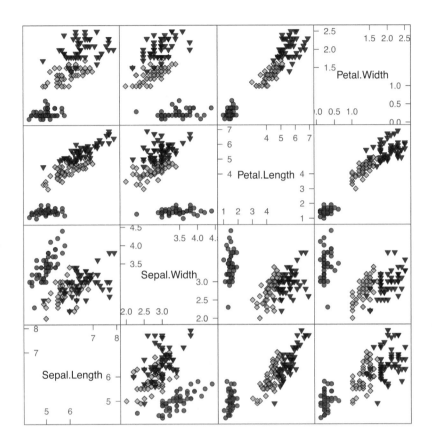

Figure A.9. An improved scatterplot matrix.

Changing lattice defaults with trellis.par.set()

If you tire of adding all these arguments to your plot functions, you can change the defaults of many of these using `trellis.par.set()`. The code below, for example, changes the default plot symbols to use light blue diamonds with black borders.

set-pch

```
> trellis.par.get("plot.symbol")
$alpha
[1] 1

$cex
[1] 0.8

$col
[1] "#0080ff"

$font
[1] 1
```

```
$pch
[1] 1

$fill
[1] "transparent"

# set plot symbol default to be a blue diamond with black border
> trellis.par.set(plot.symbol=list(pch=23,col="black",fill="lightblue"))
> trellis.par.get("plot.symbol")
$alpha
[1] 1

$cex
[1] 0.8

$col
[1] "black"

$font
[1] 1

$pch
[1] 23

$fill
[1] "lightblue"
```

Setting each of the desired defaults individually is unnecessarily tedious; we can use `trellis.par.set()` to change a list of settings all at once. A function can be designed to return the desired list, further simplifying things. Such a list is called a theme. The theme produced by `col.whitebg()`, for example, is an improvement over the defaults that are used in many graphics devices.

```
trellis.par.set(theme=col.whitebg())
```

If you want to design your own themes, you may be interested to know all of the default settings. `lattice` default settings are stored in a nested list structure. That (long) list is returned by

```
trellis.par.get()
```

You can use this as the basis for defining your own themes. As examples, you might look at `col.whitebg()` (in `lattice`) or `col.fastR()` (in `fastR`). Each of these functions returns a list that can be used with `trellis.par.set()`. Most of the plots in this book are produced using

| fastR-theme |

```
> trellis.par.set(theme=col.fastR(bw=TRUE))
```

You can inspect the current `lattice` settings visually with

```
show.settings()
```

A.4. Functions in R

Functions in R have several components:

- a name (like `histogram`)[3]
- an ordered list of named **arguments** that serve as inputs to the function

    These are matched first by name and then by order to the values supplied by the call to the function. This is why we don't always include the argument name in our function calls. On the other hand, the availability of names means that we don't have to remember the order in which arguments are listed.

    Arguments often have **default values** which are used if no value is supplied in the function call.

- a **return value**

    This is the output of the function. It can be assigned to a variable using the assignment operator (`=`, `<-`, or `->`).

- side effects

    A function may do other things (like make a graph or set some preferences) that are not necessarily part of the return value.

When you read the help pages for an R function, you will see that they are organized in sections related to these components. The list of arguments appears in the Usage section along with any default values. Details about how the arguments are used appear in the Arguments section. The return value is listed in the Value section. Any side effects are typically mentioned in the Details section.

Now let's try writing our own function. Suppose you frequently wanted to compute the mean, median, and standard deviation of a distribution. You could make a function to do all three to save some typing. Let's name our function `favstats()`. `favstats()` will have one argument, which we are assuming will be a vector of numeric values.[4] Here is how we could define it:

```
> favstats <- function(x) {                              defFun01
+     mean(x)
+     median(x)
+     sd(x)
+ }
>
> favstats((1:20)^2)
[1] 127.90
```

The first line says that we are defining a function called `favstats()` with one argument, named `x`. The lines surrounded by curly braces give the code to be executed when the function is called. So our function computes the mean, then the median, then the standard deviation of its argument.

But as you see, this doesn't do exactly what we wanted. So what's going on? The value returned by the last line of a function is (by default) returned by the

---

[3]Actually, it is possible to define functions without naming them; and for short functions that are only needed once, this can actually be useful.

[4]There are ways to check the class of an argument to see if it is a data frame, a vector, numeric, etc. A really robust function should check to make sure that the values supplied to the arguments are of appropriate types.

function to its calling environment, where it is (by default) printed to the screen so you can see it. In our case, we computed the mean, median, and standard deviation, but only the standard deviation is being returned by the function and hence displayed. So this function is just an inefficient version of `sd()`. That isn't really what we wanted.

We can use `print()` to print out things along the way if we like.

```
> favstats <- function(x) {                                          defFun02
+     print(mean(x))
+     print(median(x))
+     print(sd(x))
+ }
>
> favstats((1:20)^2)
[1] 143.5
[1] 110.5
[1] 127.90
```

Alternatively, we could use a combination of `cat()` and `paste()`, which would give us more control over how the output is displayed.

```
> altfavstats <- function(x) {                                    defFun02-cat
+     cat(paste("  mean:", format(mean(x),4),"\n"))
+     cat(paste(" edian:", format(median(x),4),"\n"))
+     cat(paste("    sd:", format(sd(x),4),"\n"))
+ }
> altfavstats((1:20)^2)
  mean: 143.5
 edian: 110.5
    sd: 127.90
```

Either of these methods will allow us to see all three values, but if we try to store them ...

```
> temp <- favstats((1:20)^2)                                        defFun02a
[1] 143.5
[1] 110.5
[1] 127.90
> temp
[1] 127.90
```

A function in R can only have one return value, and by default it is the value of the last line in the function. In the preceding example we only get the standard deviation since that is the value we calculated last.

We would really like the function to return all three summary statistics. Our solution will be to store all three in a vector and return the vector.[5]

```
> favstats <- function(x) {                                          defFun03
+         c(mean(x),median(x), sd(x))
+ }
```

---

[5]If the values had not all been of the same mode, we could have used a list instead.

```
> favstats((1:20)^2)
[1] 143.50 110.50 127.90
```

Now the only problem is that we have to remember which number is which. We can fix this by giving names to the slots in our vector. While we're at it, let's add a few more favorites to the list. We'll also add an explicit `return()`.

defFun04

```
> favstats <- function(x) {
+        result <- c(min(x),max(x),mean(x),median(x), sd(x))
+     names(result) <- c("min","max","mean","median","sd")
+     return(result)
+ }
> favstats((1:20)^2)
   min    max    mean median      sd
  1.00 400.00 143.50 110.50 127.90
> summary(Sepal.Length~Species,data=iris,fun=favstats)
Sepal.Length     N=150


+-------+----------+---+---+---+------+------+-------+
|       |          |N  |min|max|mean  |median|sd     |
+-------+----------+---+---+---+------+------+-------+
|Species|setosa    | 50|4.3|5.8|5.0060|5.0   |0.35249|
|       |versicolor| 50|4.9|7.0|5.9360|5.9   |0.51617|
|       |virginica | 50|4.9|7.9|6.5880|6.5   |0.63588|
+-------+----------+---+---+---+------+------+-------+
|Overall|          |150|4.3|7.9|5.8433|5.8   |0.82807|
+-------+----------+---+---+---+------+------+-------+
```

Notice how nicely this works with `summary()` from the `Hmisc` package. You can, of course, define your own favorite function to use with `summary()`. The `favstats()` function in `fastR` includes the quartiles, mean, standard deviation, and variance.

A.4.1. Customizing Lattice Plots via Panel Functions

User-defined functions can be used to modify the behavior of `lattice` graphics plots. Typically this is done by defining a function to replace the default `panel` argument. For example, suppose we wanted to add the diagonal line with equation $y = x$ to a scatterplot so that we can see how similar two variables are for our subjects.

defFun11

```
> x = (0:10)/10
>
> myData=data.frame(x=x, y=sin(x))
>
> panel.xyplotWithDiag <- function(x,y,...) {
+     panel.xyplot(x,y,...)
+     panel.abline(a=0,b=1,col="gray30",lwd=2)
+ }
```

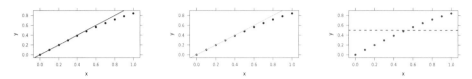

Figure A.10. An illustration of the `panel` argument of `xyplot()`.

There are a few things to notice in this example:

- `panel.xyplot()`

    This is the default panel function for `xyplot()`. It takes the data for a single panel and plots the appropriate symbols or lines. We still want that printed, so we simply call it inside our new panel function.

- `...`

    The `...` argument collects all the arguments supplied by the caller of a function but not listed in the argument list. There are a number of uses of this feature. Here it allows to avoid listing all of the arguments to `panel.xyplot()` (and there are a lot of them) when we make our new function. We simply pass them along using `...`.[6]

- `panel.abline()`

    This function draws a line with a given slope (`b`) and intercept (`a`). (It can also make horizontal and vertical lines using the arguments `h` or `v`.) Its primary purpose is to be added to other panel functions like we have just done. There are a number of additional arguments to `panel.abline()`, so we included `...` here too.

As you can see from Figure A.10, this does what we want. But the current function is not very flexible. With a little more care we can build a function that will plot the line $y = x$ by default but also allow us to do other lines easily.

```
> xyplot(y~x, data=myData,panel=panel.xyplotWithDiag)          defFun12
>
> panel.xyplotWithLine <- function(x,y,intercept=0,slope=1,...) {
+     panel.xyplot(x,y,...)
+     panel.abline(a=intercept,b=slope,...)
+ }
>
> xyplot(y~x, data=myData, panel=panel.xyplotWithLine)
> xyplot(y~x, data=myData,
+     inter=0.5, slope=0, pch=16,
+     lwd=2, col="gray30", lty=2,
+     panel=panel.xyplotWithLine,
+ )
```

Notice that the `col` argument affects both the symbols and the line. This is because both `panel.xyplot()` and `panel.abline()` have a `col` argument (which

---

[6]You can access the `...` list directly via `list(...)`.

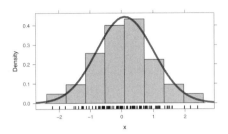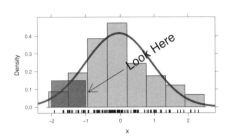

Figure A.11. Two illustrations of the `panel` argument of `histogram()`.

they receive as part of ...). With a little more work we could separate those and provide a separate argument for the color of the line.

Rather than do that, let's do another example that simply combines some of the ingredients that are available in the `lattice` package. A fancy histogram appears in Figure A.11.

> histogram-rug-density

```
> x <- rnorm(100)
> p <- histogram(~x, type='density',
+             panel=function(x,y,...) {
+                 panel.rug(x,...)
+                 panel.histogram(x,...)
+                 panel.mathdensity(
+                     dmath=dnorm, args=list(mean=mean(x),sd=sd(x)),
+                     lwd=5, col="black", lty=1, alpha=0.5,
+                     ...)
+                 }
+         )
> print(p)
```

A.4.2. Adding `grid` Elements to Plots

The `lattice` package provides high-level graphics functions for making various types of statistical plots. The `grid` package provides low-level graphics functions for making things like line segments, rectangles, dots, text, axes, etc., and placing them on a graph. The implementation of `lattice` is based on `grid`, and new plotting functions can be written completely in `grid`. More importantly for the typical user, `grid` functions can also be added to panel functions to make highly customized plots without starting from scratch. This is especially handy because of the preprocessing of the data that is done by `lattice` before the panel function is called.

Grid graphics are based on the notion of a viewport. A viewport is simply a rectangular region on a graphics device (like your screen or a file). Locations within a viewport can be specified in the usual Cartesian coordinate manner, but a number of different units may be used, including those listed in Table A.2. Of these the most important are `npc` and `native`.

As an example, the right-hand plot in Figure A.11 shows a fancy histogram produced using the following code.

histogram-rug-density-grid

```
> x <- rnorm(100)
> p <- histogram(~x, type='density',
+               panel=function(x,y,...) {
+                    panel.rug(x,...)
+                    panel.histogram(x,...)
+                    panel.mathdensity(
+                        dmath=dnorm, args=list(mean=mean(x),sd=sd(x)),
+                        lwd=5, col="black", lty=1, alpha=0.5,
+                        ...)
+                    grid.text("Look Here",
+                        x= 0.5, y=0.48,
+                        just="left",
+                        default.units="npc",
+                        rot=33,
+                        gp=gpar(col="black",cex=2)
+                        )
+                    grid.segments(
+                        x0= 0.48, x1= unit(-0.92,"native"),
+                        y0= 0.45, y1= unit(0.085,"native"),
+                        arrow = arrow(),                    # default arrow
+                        default.units="npc",
+                        gp=gpar(col="black")
+                        )
+                    grid.rect( x = -2, y=0, width=1, height=0.15, #unit(0.15,"npc"),
+                        default.units="native",
+                        just=c("left","bottom"),
+                        gp=gpar(col="black", fill="gray40", alpha=0.6)
+                        )
+               }
+       )
```

The functions with names beginning `grid` are low-level graphics functions. Details for each can be found using R's help utilities. The default unit for these functions is specified with `default.units`, but other units may be used if specified explicitly using `unit()`. The argument `gp` is assigned using the function `gpar()` and is used to set graphical parameters like color that are applicable to a wide range of graphical objects.

A.4.3. Going 3-d

Much data is multi-dimensional. Paper and computer screens (and retinas, too, for that matter) are fundamentally 2-dimensional. This leads to a dilemma: How do we deal with more than two dimensions. We have already seen some ways: We've used multiple panels, multiple colors, multiple shapes, and overlays to view complex data.

Table A.2. Units available in `grid` graphics.

npc	*normalized parent coordinates.* In this measure the lower left corner is (0,0) and the upper right corner is (1,1). This allows positioning things in terms of percentages of the viewport size.
snpc	*square normalized parent coordinates.* This unit is similar to `npc` but relative to the smaller of the width and height. This is useful for making sure that squares are entirely contained in a viewport.
native	All viewports have a native coordinate system (stored as `xscale` and `yscale`). This coordinate system is established when the viewport is created. For viewports designed to display data, this system is typically in the units of the data. In particular, this is the case for the viewports into which the panel functions draw.
char	multiples of the current nominal font size.
lines	multiples of the current height of a line of text.
strwidth, strheight	the width or height a specified text string would take up (need to supply the text string).
grobwidth, grobheight	the width or height a specified graphical object would take up (need to supply the grob).
inches	distance in inches horizontally or vertically from lower left corner.
cm, mm	same, but in cm or mm. (Also available: various printing measures like `points`, `bigpts`, `picas`, `dida`, `cicero`, `scaledpts`.)
null	In some settings, this space-filling unit can be used. This is primarily used to arrange viewports within other viewports. So, for example, it is possible to arrange a viewport for a title that is 3 lines tall and another viewport below it that takes up all of the remaining space by having the first be `unit(3,'lines')` tall and the second be `unit(1,'null')` tall.

We could also attempt to render something that appears 3-dimensional. R has some ability to do this. Take a look at the following, for example:

- `levelplot()`, `contourplot()`, and `wireframe()` (see Figure 5.3),
- `cloud()` (for 3-d scatterplots in `lattice`),
- `scatterplot3d()` (for 3-d scatterplots in the `scatterplot` package).

A.5. Some Extras in the `fastR` Package

The `fastR` package includes a number of functions that begin with the letter x. Each of these functions adds some functionality to a similarly named function without the initial x. Typically, this additional functionality is provided either via additional arguments or by choosing different default values for the existing arguments.

It is instructive to see how some of these functions were written.

Example A.5.1. The `plot()` function can plot a number of different types of objects. If we inspect the `plot()` function, we see that there appears to be very little to it.

```
> plot                                                                 plot
function (x, y, ...)
{
    if (is.function(x) && is.null(attr(x, "class"))) {
        if (missing(y))
            y <- NULL
        hasylab <- function(...) !all(is.na(pmatch(names(list(...)),
            "ylab")))
        if (hasylab(...))
            plot.function(x, y, ...)
        else plot.function(x, y, ylab = paste(deparse(substitute(x)),
            "(x)"), ...)
    }
    else UseMethod("plot")
}
<environment: namespace:graphics>
```

`UseMethod()` tells R to look at the first argument in the function call and to use the `class` attribute of this object to decide what to do. We can see a list of all the available methods for `plot()` using the `methods()` function.

```
> methods(plot)                                                  plot-methods
 [1] plot.acf*            plot.data.frame*    plot.decomposed.ts*
 [4] plot.default         plot.dendrogram*    plot.density
 [7] plot.ecdf            plot.factor*        plot.formula*
[10] plot.hclust*         plot.histogram*     plot.HoltWinters*
[13] plot.isoreg*         plot.lm             plot.medpolish*
[16] plot.mlm             plot.ppr*           plot.prcomp*
[19] plot.princomp*       plot.profile.nls*   plot.shingle*
[22] plot.spec            plot.spec.coherency plot.spec.phase
[25] plot.stepfun         plot.stl*           plot.table*
[28] plot.trellis*        plot.ts             plot.tskernel*
[31] plot.TukeyHSD

   Non-visible functions are asterisked
```

If no specially designed method is found, `plot.default()` is used.

In Chapter 6 we introduce several diagnostic plots for linear models. The `plot()` function can be used to provide a number of diagnostic plots. The `fastR` packages provides a method for `xplot()` that produces similar plots using `lattice` graphics.

xplot

```
> xplot
function (x, ...)
{
    UseMethod("xplot")
}
<environment: namespace:fastR>
> methods(xplot)
[1] xplot.default xplot.lm
> xplot.default
function (...)
{
    plot(...)
}
<environment: namespace:fastR>
```

You can inspect the code of `xplot.lm()` using

```
xplot.lm
```

◁

Example A.5.2. `xqqmath()` is a wrapper around `qqmath()` that calls `qqmath()` with a different default for `panel`.

xqqmath

```
> require(fastR);
> methods(xqqmath)
[1] xqqmath.formula xqqmath.numeric
> xqqmath.formula
function (x, data = NULL, panel = panel.xqqmath, ...)
{
    require(lattice)
    qqmath(x, data = data, panel = panel, ...)
}
<environment: namespace:fastR>
```

All the work is buried in `panel.xqqmath()`, which you can inspect with

```
panel.qqmath
```

The resulting function `xqqmath()` adds a reference line using `panel.qqmathline`. (See Figure A.12.)

xqqmath-example

```
> x = rnorm(100)
> p <- qqmath(~x, main='QQ-plot using qqmath()')
> q <- xqqmath(~x, main='QQ-plot using xqqmath()')
```

◁

In a similar way, `xhistogram()` makes it easy to overlay a density curve on a histogram, and `xpnorm()` adds a graphical display to the output of `pnrom()`. See the `fastR` documentation for additional functions and full documentation.

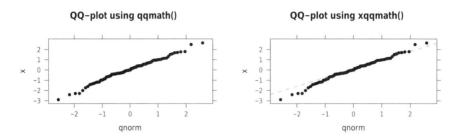

Figure A.12. Comparing `qqmath()` and `xqqmath()`.

A.6. More R Topics

A.6.1. Getting Stuff out of R

R is a great statistical package (and programming language). But if you want to present your statistical results, you probably want to do that using some other tool like LaTeX, WORD, or PowerPoint. To do that, you need to be able to get text and graphics from R into the other software. Grabbing text is easy using copy and paste. You may need to choose a mono-spaced font to make sure things align properly after pasting. Alternatively, you can save R output to a file using the `sink()` function. This can be useful when writing scripts.

Plots made in R can be saved in a number of formats, including pdf, postscript, jpeg, and png. Some of the GUIs contain menu options to help you save graphical output to a file, and you can always grab a screen shot (Command-Shift-4 on a Mac). You can also use R functions to redirect graphical output to a file. For example, to save a plot in pdf format 6 inches wide and 4 inches tall, use

```
myplot <- histogram(y~x,data=data)
pdf(file='myGreatPlot.pdf',width=6,height=4)
trellis.par.set(theme=col.whitebg())
print(myplot)
dev.off()
```

Note that the trellis options are set only for a given device, and each file is a different device, so if you want the settings in your file to match what is on your screen, you need to set them again. Also notice that `dev.off()` is necessary to actually write the file. It also returns output to the previous device (which might be a window on your screen or another file). You can see a list of all the open devices using `dev.list()`. The function used to open a new screen device is machine dependent. For example, on a UNIX or Linux machine you can use `x11()` to open a new X11 window, and on a Mac you can use `quartz()` to get a new quartz window. The GUIs for Mac and PC automate this for you.

For other file formats, `postscript()`, `jpeg()`, and `png()` work similarly, but the width and height are specified in pixels for jpeg and png files. For a full list of arguments and more details, see the built-in help.

A.6.2. Writing Scripts in R

Instead of typing R commands one at a time and executing them, you can save the commands in a file and execute all the commands in the file at once. From the R terminal, use

```
source("myFile.R")
```

to execute the code in a file called myFile.R. You can also do this from the command line using

```
    R CMD BATCH -vanilla myFile.R
```

Output will be stored in myFile.Rout.

For all but the simplest things, saving your code in a file and running it as a batch job is a good idea since it is easier to edit and reuse code that way. Should you be working from the terminal and decide you wish you had been saving your work, there is a way to do this after the fact as well:

```
savehistory(file="myGreatIdeas.Rhistory")
```

will save the commands of your current session to a file. That file can then be edited to select the portion that was of interest. The sink() function can be used to save portions of the output to files of your choice.

For more sophisticated scripting, including a mechanism for handling argument passing, the recently developed Rscript will work on UNIX-like systems (including Mac OS). To use Rscript, simply include the following lines in your script:

```
#! /usr/bin/Rscript --vanilla --default-packages=utils
args <- commandArgs(TRUE)
```

Executing the script will place the command line arguments in the list args and execute the R code in the remainder of the file. The getopt package provides further utilities for processing command line arguments with Rscript.

A.6.3. Advanced Programming in R

To write more complicated functions, you will probably need to learn a bit more about the R programming language. For a brief introduction to programming in R, see Chapter 6 of *R for Beginners* by Emmanuel Paradis [**Par**]. For more detailed information consult one of the manuals or contributed documents available at CRAN.

Exercises

A.1. Try to guess the results of each of the following R commands. Use R to check your guesses.

```
odds <- 1 + 2*(0:4);
primes <- c(2,3,5,7,11,13);
length(odds);
length(primes);
odds + 1;
odds + primes;
odds * primes;
odds > 5;
sum(odds > 5);
sum(primes < 5 | primes > 9);
odds[3];
odds[-3];
primes[odds];
primes[primes >=7];
sum(primes[primes > 5]);
sum(odds[odds > 5]);
```

whats-up

A.2. The data set `ChickWeight` contains data from an experiment on the effect of diet on early growth of chicks.

  a) Which chick was heaviest at the end of the study? How heavy was it? What diet was it on?

  b) Which chick was lightest at the end of the study? How heavy was it? What diet was it on?

A.3. The data set `ChickWeight` contains data from an experiment on the effect of diet on early growth of chicks. First remove from the data any chicks that do not have measurements over the full time span of the study.

  a) For each chick in the study, determine the absolute amount of weight gained.

  b) For each chick in the study, determine the percentage increase in weight.

  c) Using numerical and graphical summaries, compare the weight gained by chicks on the different diets. Does it appear diet has an impact on weight gain?

A.4. Write an R function that computes the third largest value in a vector of numbers. Your function should report an error if there are fewer than three numbers in the vector or if the input is something other than a vector of numbers.

A.5. Write a function `kmax(x,k=1,...)` that returns the kth largest element of x. When k=1, your function should behave just like `max()`.

A.6. Write a function `wmax(x,...)` that returns the index of the largest element of x. If the largest value occurs more than once, report the index of the first occurrence.

R provides the `which.max()` function which will probably be more efficient than the code you write because it uses R's internal representation of vectors.

A.7. The `read.table()` function assumes by default that there is no header line in the file. Write a function called `xread.table()` that assumes by default that there is a header line but otherwise behaves just like `read.table()`.

Some Mathematical Preliminaries

This appendix reviews some mathematical material and establishes notation that will be used throughout this book. When reading and writing mathematics, keep in mind that (in computer science terms) *mathematics is a strongly typed, case-sensitive language that is context-sensitive and allows operator overloading.*

- Strongly typed.

    Every mathematical object has a type. Mathematical objects can be, for example, real numbers or integers or functions or sets. But each object is a *something*. Always be sure you know what type of mathematical objects you are dealing with.

    Mathematicians often declare the type of an object with a phrase like "let x be a _____."

- Case-sensitive.

    Capitalization (and fonts) matter. We use capitalization and font changes to indicate specific things. So x, X, $\boldsymbol{x}$, and $\boldsymbol{X}$ all mean different things; we will use this notation to indicate four different types of objects. This helps us get around limitations of the size of our alphabet and allows us to choose notation that reminds us of the type of object we are representing.

    Frequently we will also employ conventions that designate certain portions of the alphabet for certain types of objects. Thus we may expect f to be a function and x to be an input (a real number, perhaps).

- Context-sensitive.

    To avoid cumbersome notation, we often omit some of the notation when context suffices to fill in the details.

- Operator overloading.

    Mathematical operators mean different things depending on the types of the mathematical objects they operate on. Thus, for example, $a \cdot b$ can mean different things depending on whether a and b are numbers, vectors, or matrices.

This book assumes familiarity with the differential and integral calculus. This appendix provides a brief introduction to some other mathematical preliminaries that may or may not be familiar to you.

B.1. Sets

A **set** is a collection of objects of a certain type. We can have sets of integers, sets of people, sets of functions, even sets of sets. The most important feature of a set is **membership**. Some things belong to a set and other things do not. Often mathematicians use capital letters to denote sets and lowercase letters to denote potential **elements** (members) of a set. So if A is a set and x is a potential element, we write

$$x \in A \qquad \text{if } x \text{ is an element of } A, \text{ and}$$
$$x \notin A \qquad \text{if } x \text{ is not an element of } A.$$

We can describe sets in many ways. If a set is small, we may simply list all of its elements. For example, the set of one-digit prime numbers could be presented as

$$A = \{2, 3, 5, 7\} \, .$$

The order in which the elements are listed is immaterial, so we could describe this same set as

$$A = \{7, 3, 5, 2\} \, .$$

Larger sets may be described in other ways. Often sets are simply described using words. As long as it is clear which things are members of the set and which are not, the set has been adequately defined. Special sets of numbers are traditionally denoted using a special font. For example $\mathbb{R}$ is used to denote the **real numbers** (numbers representable as decimals) and $\mathbb{Z}$ is used to represent the **integers** (positive and negative whole numbers and 0).

Definition B.1.1. A set A is a **subset** of another set B if all members of A are also members of B. This is denoted $A \subseteq B$.

If A is a subset of B, then we say B is a **superset** of A. This is denoted $B \supseteq A$.

Example B.1.1. The following are examples of the subset relationship between sets:

- $\{2, 4, 7\} \subseteq \{7, 2, 5, 4, 8, 1\}$.
- $\mathbb{Z} \subseteq \mathbb{R}$.
- The set of even integers is a subset of the set of integers.   ◁

The set of possible outcomes of a random process is called the **sample space**, and a subset of the sample space is called an **event**. For example, if we flip a coin twice, recording each time whether the result is a head (H) or a tail (T), then the sample space is

$$S = \{HH, HT, TH, TT\}\,.$$

The event

$$E = \{HH, TT\}$$

consists of those outcomes where the two coins match.

We can build new sets from old sets in a variety of ways. A subset is often described using **set-builder notation**. For example,

$$\{0, 1, 2, 3, 4, 5\} = \{x \in \mathbb{Z} \mid 0 \le x \le 5\}\,.$$

This notation indicates a superset from which the elements are drawn and a condition that determines which elements are selected. The general format is

$$\{x \in A \mid \text{condition}\}$$

where A is the superset and all elements of A making the condition true are in the set being defined (and nothing else). The **open intervals** on the real line can be defined using this notation as follows:

$$(a, b) = \{x \in \mathbb{R} \mid a < x < b\}\,.$$

The union, intersection, and complement operators are defined as

$$A \cup B = \{x \mid x \in A \text{ or } x \in B\}\,,$$
$$A \cap B = \{x \mid x \in A \text{ and } x \in B\}\,,$$
$$A^c = \{x \in U \mid x \notin A\}\,.$$

The complement A^c is defined relative to some **universal set** U. Typically this set is not specified but is understood from context. The complement of an event, for example, is relative to the sample space.

These operations and relations are often depicted using a **Venn diagram**. In Figure B.1, A is represented by the striped region, B by the shaded region, $A \cap B$ by the football shaped region that is both shaded and striped, $A \cup B$ by the portion that is either striped or shaded (or both), etc. The universal set U is represented by the outer rectangle. Thus the white portion of the diagram represents $(A \cup B)^c$.

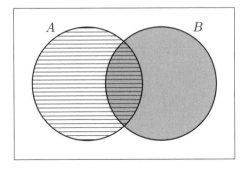

Figure B.1. A Venn diagram with two sets A and B.

If a set A is finite, then the number of elements of the set is denoted $|A|$ and is read "cardinality of A" or more informally "size of A". The Venn diagram in Figure B.1 can be used to justify the following important lemma.

Lemma B.1.2. *If A and B are finite sets, then*

$$|A \cup B| = |A| + |B| - |A \cap B| \,. \qquad \qquad \square$$

The **Cartesian product** of two sets is the set of ordered pairs where the first component comes from the first set in the product and the second component from the second set:

$$A \times B = \{\langle a, b \rangle \mid a \in A \text{ and } b \in B\} \,.$$

This idea can be extended to products of any number of sets.

B.2. Functions

Intuitively, a **function** from A to B is a rule that assigns to each element of A some element of B. More formally, we write $f : A \to B$ if

- $f \subseteq A \times B$, and
- for any $a \in A$, there is exactly one $b \in B$ such that $\langle a, b \rangle \in f$.

The set A is called the **domain** of the function f. The set B is the **codomain**. The set

$$f(A) = \{f(a) \mid a \in A\}$$

of all possible outputs of the function is called the **range** of the function.

Functions can be described with formulas

$$f(a) = a^2$$

or with tables of values

a	$f(a)$
0	0
1	1
2	4
3	9
4	16
$\vdots$	$\vdots$

or in any other means that makes it clear which element of B is assigned to each element of A.

Piecewise defined functions

Frequently we will define functions using different formulas in different regions of the range. For example, the absolute value function may be defined by

$$|x| = \begin{cases} x & \text{if } x \geq 0, \\ -x & \text{if } x < 0. \end{cases}$$

That is, $|\,x\,| = x$ when x is positive or zero, but $|\,x\,| = -x$ when x is negative.

Another important example of a piecewise defined function is an **indicator function** (or **characteristic function**), which often provides a notational convenience in statistics. Indicator functions take on the values 1 and 0 depending on whether the input belongs or does not belong to some set A. Different statisticians use different notation for this. One common notation uses a subscripted I or χ:

$$I_A(x) = \chi_A(x) = \begin{cases} 1 & \text{if } x \in A, \\ 0 & \text{if } x \notin A. \end{cases}$$

We prefer the following variation on **Iverson bracket** notation since it reduces the amount of subscripting required, works well when the set A is described using text, and, in fact, works equally well for any statement, not just for the statement $x \in A$:

$$[\![x \in A]\!] = \begin{cases} 1 & \text{if } x \in A, \\ 0 & \text{if } x \notin A. \end{cases}$$

Indicator functions can be used arithmetically, as in the following example.

Example B.2.1. The absolute value function can be expressed as

$$|\,x\,| = x[\![x > 0]\!] - x[\![x < 0]\!] \,.$$

$\triangleleft$

B.3. Sums and Products

We will frequently have need of sums and products of sets of numbers. Likely you are familiar with notation like

$$\sum_{i=1}^{5} 2i = 2(1) + 2(2) + 2(3) + 2(4) + 2(5) = 30 \,.$$

Here each number from 1 to 5 is substituted for i in the expression $2i$, and the resulting numbers are summed. The set $I = \{1, 2, 3, 4, 5\}$ is called the **index set** for this sum, and we could also have written this sum as

$$\sum_{i \in I} 2i \,.$$

There is no reason that the index set must be a set of consecutive integers. For example, if $A = \{2, 3, 5, 7\}$, then

$$\sum_{a \in A} a = 2 + 3 + 5 + 7 = 17 \,.$$

When the index set is clear from context, we will often omit it and write

$$\sum_{a} a = 2 + 3 + 5 + 7 = 17$$

or even

$$\sum a = 2 + 3 + 5 + 7 = 17 \,.$$

In each case, the sum is over "all a's currently being considered". If this is ever ambiguous, we can return to the more explicit notation for clarity.

Products are handled similarly, but we replace $\sum$ with $\prod$. So, for example, factorials of positive integers can be defined as

$$n! = \prod_{i=1}^{n} i \ .$$

Sums and products need not have finite index sets. Infinite sums (also called **series**) and products are defined in terms of limits. For example, if we let

$$S_n = \sum_{i=1}^{n} a_i \ ,$$

then

$$\sum_{i=1}^{\infty} a_i = \lim_{n \to \infty} S_n \ ,$$

provided this limit exists. Infinite products are defined similarly.

Example B.3.1. A **geometric series** is a series in which consecutive terms differ by a common ratio. For example, the series

$$1 + \frac{1}{2} + \frac{1}{4} + \cdots = \sum_{i=0}^{\infty} \left(\frac{1}{2}\right)^i$$

is a geometric series with ratio $\frac{1}{2}$.

We can evaluate both the partial sums and the infinite sum for this series (and all convergent geometric series). If we let S_n be the sum of the first n terms of the series, then

$$
\begin{aligned}
S_n &= 1 + \tfrac{1}{2} + \tfrac{1}{4} + \cdots + \left(\tfrac{1}{2}\right)^{n-1} , \\
\tfrac{1}{2} S_n &= \quad\ \ \tfrac{1}{2} + \tfrac{1}{4} + \cdots + \left(\tfrac{1}{2}\right)^{n-1} + \left(\tfrac{1}{2}\right)^{n} .
\end{aligned}
$$

So

$$S_n - \frac{1}{2} S_n = \frac{1}{2} S_n = 1 - \left(\frac{1}{2}\right)^n ,$$

and

$$S_n = \frac{1 - \left(\frac{1}{2}\right)^n}{\frac{1}{2}} \ .$$

From this it follows that

$$\sum_{i=0}^{\infty} 2^{-i} = 1 + \frac{1}{2} + \frac{1}{4} + \cdots = \lim_{n \to \infty} \frac{1 - \left(\frac{1}{2}\right)^n}{\frac{1}{2}} = 2 \ . \qquad \triangleleft$$

The reasoning in the previous example extends naturally to any geometric sum.

Lemma B.3.1. *For any real numbers a and $r \notin \{0, 1\}$,*

$$\sum_{i=0}^{n-1} ar^i = \frac{a - ar^n}{1 - r} \ .$$

If in addition $|\,r\,| < 1$, then

$$\sum_{i=0}^{\infty} ar^i = \frac{a}{1 - r} \ .$$

Proof. Exercise B.10.   □

Sums and products of vectors are easily computed in R. See page 522 for examples.

Exercises

B.1. Under what conditions is $A \cap B = A$? Explain.

B.2. Under what conditions is $A \cup B = A$? Explain.

B.3. Under what conditions is $A \cup B = A \cap B$? Explain.

B.4. Use Venn diagrams to show that $(A \cup B)^c = A^c \cap B^c$.

Be sure to accompany your diagrams with sufficient text to make the explanation clear.

B.5. Show that $(A \cap B)^c = A^c \cup B^c$.

B.6. Describe in words the function

$$f(a, b) = (a - b)[\![a > b]\!] \ .$$

B.7. For any sets A and B, we can define the function

$$f(x) = [\![x \in A]\!] \cdot [\![x \in B]\!] \ .$$

  a) For what values of x is $f(x) = 1$?

  b) For what values of x is $f(x) = 0$?

  c) Express $f(x)$ another way.

B.8. Let $f(x)$ be an indicator function. What can you say about $g(x) = (f(x))^2$?

B.9. Evaluate the following sums and products:

  a) $\displaystyle\sum_{i=2}^{5} i^2,$ **b)** $\displaystyle\sum_{n=1}^{4} n,$ **c)** $\displaystyle\sum_{x=1}^{5}(2x - 1),$ **d)** $\displaystyle\prod_{n=2}^{4} n.$

B.10. Prove Lemma B.3.1.

B.11. Suppose S is a subset of the positive integers less than 100. What does $\sum_{i=1}^{99} [\![i \in S]\!]$ tell us about S?

B.12. Generalize Lemma B.3.1 for geometric sums and series with an arbitrary starting point. That is, derive formulas for $\sum_{i=i_0}^{i_0+n-1} ar^i$ and $\sum_{i=i_0}^{\infty} ar^i$.

B.13. Prove that $\sum_{i=1}^{n} i = \dfrac{n(n+1)}{2}$.

B.14. Prove that $\sum_{i=1}^{n} 2i - 1 = n^2$.

B.15. Prove that $\sum_{i=1}^{n} i^2 = \dfrac{n(n+1)(2n+1)}{6}$.

B.16. Prove that $\sum_{i=1}^{n} i^3 = \dfrac{n^2(n+1)^2}{4}$.

B.17. Express the mean of the elements in the set S using sum notation.

B.18. Simplify the following sums:

a) $\sum_{a=1}^{n} (2a - 1)x,$

b) $\sum_{y=1}^{n} xy,$

c) $\sum_{x \in S} (x - m)$, where $|S| = 10,$

d) $\sum_{x \in S} (x - m)$, where $|S| = 10$ and $m = \sum_{x \in S} \dfrac{x}{10}.$

B.19. The function $f : \{0, 1, 2, 3, 4\} \to \mathbb{R}$ is defined in the table below:

x	0	1	2	3	4
$f(x)$	$\frac{1}{6}$	$\frac{1}{3}$	$\frac{1}{4}$	$\frac{1}{6}$	$\frac{1}{12}$

a) What is the value of $\sum_{x} f(x)$?

b) What is the value of $\sum_{x} x f(x)$?

c) What is the value of $\sum_{x} x^2 f(x)$?

B.20. Derive a formula for $\displaystyle\prod_{i=1}^{n} \frac{i}{i+1}$.

Geometry and Linear Algebra Review

While it is possible to learn the statistics in this book without reference to linear algebra, a certain amount of linear algebra serves to motivate some of the methods and simplifies notation and algebraic manipulation at several points. This appendix provides the necessary background in linear algebra. Many of the results presented here can be generalized, but we will focus our attention on vectors in $\mathbb{R}^n$ since such vectors can represent the n values of a variable. Indeed, this is precisely how R codes data in data frames: as a list of vectors each having the same length.

C.1. Vectors, Spans, and Bases

Let's begin by considering the usual Cartesian coordinate system for the real plane ($\mathbb{R}^2$). A pair of real numbers $\boldsymbol{x} = \langle x_1, x_2 \rangle$[1] can be represented as a **point** in the plane that is x_1 units to the right of and x_2 units above the origin or as an **arrow** that extends right x_1 units and up x_2 units. There is, of course, a natural correspondence between these two representations. (See Figure C.1.) There are certain advantages to the arrow representation, however. First, it does not require us to specify an origin. More importantly, the arrow representation provides a natural interpretation for a number of **vector** operations. For example, addition of vectors is done componentwise:

$$\boldsymbol{x} + \boldsymbol{y} = \langle x_1, x_2 \rangle + \langle y_1, y_2 \rangle = \langle x_1 + y_1, x_2 + y_2 \rangle \, .$$

---

[1] We use angle brackets rather than parentheses around vectors to distinguish them from open intervals.

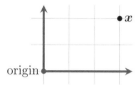
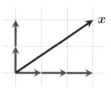

Figure C.1. Representing the vector $\boldsymbol{x} = \langle 3, 2 \rangle$ in Cartesian coordinates as a point (left) and as an arrow (right).

Geometrically, we can represent addition by placing vectors end to end:

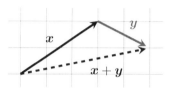

It is immediate from the definition that the order of the addends does not matter. This can be seen geometrically as well:

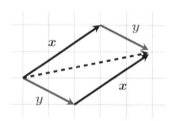

Vectors in higher dimensions work analogously – although the pictures become more difficult to imagine when $n > 3$. We will use boldface type to indicate that a variable is a vector. Subscripts on a non-bolded variable will indicate the ordered components of the vector. For example, we may write $\boldsymbol{x} = \langle x_1, x_2, x_3, x_4 \rangle \in \mathbb{R}^4$. Occasionally we will have need of a sequence of vectors, which we will denote as $\boldsymbol{x}_1, \boldsymbol{x}_2, \boldsymbol{x}_3, \ldots$. In this case, double subscripting is used to indicate components. The third component of the second vector in the preceding sequence is denoted x_{23}. We will let $\boldsymbol{0}$ and $\boldsymbol{1}$ denote the vectors consisting of 0's and 1's in each component. The number of components of these vectors will be clear from context.

When a vector $\boldsymbol{x} \in \mathbb{R}^n$ is multiplied by a real number r (also called a **scalar** in this context), *each* component of $\boldsymbol{x}$ is multiplied by r:

$$r\boldsymbol{x} = \langle rx_1, rx_2, \ldots, rx_n \rangle .$$

This is also what happens to vectors in R:

vec-mult

```
> x <- c(1,2,3)
> 4 * x
[1]  4  8 12
```

The two operations just described – **vector addition** and **scalar multiplication** – are the fundamental operations of a vector space. It is straightforward to show the following lemma.[2]

Lemma C.1.1. *Let u, v, and w be vectors in $\mathbb{R}^n$ and let a and b be real numbers. Then*

(1) $u + v \in \mathbb{R}^n$. $\mathbb{R}^n$ *is closed under vector addition.*

(2) $au \in \mathbb{R}^n$. $\mathbb{R}^n$ *is closed under scalar multiplication.*

(3) $(u + v) + w = u + (v + w)$.   *Vector addition is associative.*

(4) $u + v = v + u$.   *Vector addition is commutative.*

(5) $0 + u = u$. 0 *is an additive identity.*

(6) $1u = u$. 1 *is an identity for scalar multiplication.*

(7) $u + (-1)u = 0$. $(-1)u$ *is an additive inverse for u.*

(8) $a(u + v) = au + av$.   *Scalar multiplication distributes over vector addition.*

(9) $(a + b)u = au + bu$.   *Vector addition distributes over scalar multiplication.*

(10) $(ab)u = a(bu)$.   *This is a form of "mixed associativity".*

                                                                    □

If we let $u = \langle 1, 0 \rangle$ and $v = \langle 0, 1 \rangle$, then $\langle 3, 2 \rangle = 3\langle 1, 0 \rangle + 2\langle 0, 1 \rangle = 3u + 2v$. This is indicated in Figure C.1. It is not a coincidence that this vector can be expressed as a **linear combination** of u and v; the same is true for any vector in $\mathbb{R}^2$ since $\langle a, b \rangle = au + bv$. We will say that the set $\{u, v\}$ **spans** $\mathbb{R}^2$. More generally, if $S = \{v_1, v_2, \ldots, v_k\}$ is a finite set of vectors, then span(S) is the set of all linear combinations of vectors from S:

$$x \in \text{span}(S) \iff x = \sum_i a_i v_i \quad \text{for some real numbers } a_i.$$

As just shown, span$\{\langle 1, 0 \rangle, \langle 0, 1 \rangle\} = \mathbb{R}^2$. If S is an infinite set, then span(S) consists of all possible *finite* linear combinations from S.[3]

A set of vectors is **linearly independent** if none of the vectors is in the span of the others; otherwise we say the set is linearly **dependent**. Note that if a set is linearly independent, then if we left any of the vectors out, the span would get smaller, but if the set is linearly dependent, we can remove at least one vector without changing the span. A set of linearly independent vectors that span all of $\mathbb{R}^n$ is called a **basis** for $\mathbb{R}^n$.

Example C.1.1.

Q. Show that $\{\langle 1, 0 \rangle, \langle 0, 1 \rangle\}$ is a basis for $\mathbb{R}^2$ and find another basis.

A. We have already shown that the span is all of $\mathbb{R}^2$. Now we show independence. $a\langle 1, 0 \rangle = \langle a, 0 \rangle \neq \langle 0, 1 \rangle$, so $\langle 0, 1 \rangle$ is not in the span of $\langle 1, 0 \rangle$. Similarly, $\langle 1, 0 \rangle$ is not in the span of $\langle 0, 1 \rangle$, so the vectors are independent.

---

[2]In fact, the properties in Lemma C.1.1 are essentially the definition of a **vector space**. Any two sets (a set of vectors V, playing the role of $\mathbb{R}^n$, and a set of scalars S, playing the role of $\mathbb{R}$) and two operators (vector addition and scalar multiplication) satisfying the 10 properties above form a **vector space over S**. The only vector spaces we will require are the **Euclidean spaces** $\mathbb{R}^n$ and their subspaces.

[3]Another way to handle spans of infinite sets is to require that all but finitely many of the coefficients a_i of a linear combination be 0.

Now let's find another basis. If we think geometrically, we know that we need two "directions" that can be combined to get anywhere in the plane. Any two non-parallel directions should work. For example, $B = \{\langle 1,1\rangle, \langle 1,-1\rangle\}$. It is clear that neither vector is a multiple of the other, so the vectors in B are independent. To see that B spans $\mathbb{R}^2$, let $m = a + b/2$ and notice that

$$\langle a,b\rangle = \langle m,m\rangle + \langle a-m, m-a\rangle = m\langle 1,1\rangle + (a-m)\langle 1,-1\rangle \, .$$

$\lhd$

The first basis given in Example C.1.1 is called the **standard basis** for $\mathbb{R}^2$. The standard basis for $\mathbb{R}^n$ consists of n vectors that each have exactly one non-zero coordinate, and that coordinate is 1. Before giving an example in a higher dimension, we present the following lemma that gives an alternative characterization of independence.

Lemma C.1.2. *Let $S = \{\boldsymbol{v}_1, \boldsymbol{v}_2, \ldots, \boldsymbol{v}_k\}$ be a set of vectors in $\mathbb{R}^n$. Then S is a **linearly independent** set of vectors if and only if*

$$\sum_i a_i \boldsymbol{v}_i = \boldsymbol{0} \implies \text{ all the } a_i\text{'s are } 0 \, .$$

Proof. Exercise C.1. $\square$

Example C.1.2.

Q. Find two different bases for $\mathbb{R}^3$.

A. By an argument similar to the one just given, $\{\langle 1,0,0\rangle, \langle 0,1,0\rangle, \langle 0,0,1\rangle\}$ is a basis. Here is another basis:

$$B = \{\langle 1,0,0\rangle, \langle 1,1,0\rangle, \langle 1,1,1\rangle\} \, .$$

To see that it has the proper span, observe that

$$\langle a,b,c\rangle = (a-b)\langle 1,0,0\rangle + (b-c)\langle 1,1,0\rangle + c\langle 1,1,1\rangle \, .$$

Now suppose that

$$a_1\langle 1,0,0\rangle + a_2\langle 1,1,0\rangle + a_3\langle 1,1,1\rangle = \langle a_1 + a_2 + a_3, a_1 + a_2, a_3\rangle = \boldsymbol{0} \, .$$

It is immediate that $a_3 = 0$, from which it follows that a_2 and a_1 must also be 0. So by Lemma C.1.2, the vectors in B form a linearly independent set of vectors. $\lhd$

A **subspace** of $\mathbb{R}^n$ is a subset that is closed under addition and scalar multiplication. In other words, a subspace is equal to its own span. The span of any set of vectors is a subspace. More importantly,

Theorem C.1.3 (Dimension Theorem). *Let S be a subspace of $\mathbb{R}^n$. Then:*

(1) *S is the span of a finite set of independent vectors. Such a set is called a **basis** for S.*

(2) *Every basis for S has the same size (number of linearly independent vectors). This size is called the **dimension** of S, denoted $\dim(S)$.* $\square$

Lemma C.1.4. *Let S be a set of vectors from $\mathbb{R}^n$. Then:*

(1) *If S is independent, then $|S| \leq n$.*

(2) *If* $\text{span}(S) = \mathbb{R}^n$, *then* $|S| \geq n$.

(3) *If S is a basis for* $\mathbb{R}^n$, *then* $|S| = n$. $\square$

We will not prove Theorem C.1.3 and Lemma C.1.4 here. Proofs can be found in standard linear algebra textbooks. The notions of basis and dimension will be important in our discussion of linear models. In that context, the dimension of a subspace is usually called **degrees of freedom**.

C.2. Dot Products and Projections

In our description of the vector spaces $\mathbb{R}^n$ we have said nothing about how to multiply two vectors. There are several different multiplication operators that one could consider. Componentwise multiplication is a natural idea and is available in R:

vec-mult2

```
> u <- c(1,2,3)
> v <- c(4,5,6)
> u * v
[1]  4 10 18
```

This operation is not as important, however, as the **dot product** (also called the **inner product**). The dot product $\boldsymbol{a} \cdot \boldsymbol{b}$ is easy to compute: It is the sum of the componentwise products. That is,

$$\boldsymbol{u} \cdot \boldsymbol{v} = \sum_i u_i v_i .$$

The dot product is easily defined in R, and the function `dot()` has been included in `fastR` as a convenience.

dot

```
> dot <- function(x,y) { sum(x*y) }
```

vec-dotprod

```
> dot(u,v)
[1] 32
```

Notice that by the Pythagorean Theorem, the dot product of a vector with itself gives the square of its length:

$$\boldsymbol{v} \cdot \boldsymbol{v} = \sum_i v_i^2 = |\boldsymbol{v}|^2 .$$

The `vlength()` function in `fastR` computes the length of a vector using this identity.

Although it is not obvious from its definition, the dot product has an important geometric interpretation:

Lemma C.2.1. *Let u and v be vectors in $\mathbb{R}^n$. Then*

$$u \cdot v = |u|\,|v|\cos(\theta) ,$$

where θ is the angle between the two vectors.

Proof. First note that it suffices to prove the lemma for unit vectors u and v since $au \cdot bv = (ab)u \cdot v$. In two dimensions ($\mathbb{R}^2$) the lemma follows easily from the identity

$$\cos(A - B) = \cos A \cos B + \sin A \sin B$$

and the following diagram:

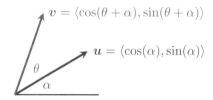

If we let $A = \alpha$ and $B = \theta + \alpha$, then

$$\cos(\theta) = \cos(\theta + \alpha - \alpha)$$
$$= \cos(\theta + \alpha)\cos(\alpha) + \sin(\theta + \alpha)\sin(\alpha) = v_1 u_1 + v_2 u_2 = u \cdot v . \qquad \square$$

Corollary C.2.2 (Projections in $\mathbb{R}^n$). *Let u be a unit vector and let x be any vector in $\mathbb{R}^n$. Then:*

(1) $u \cdot x = |x|\cos(\theta)$.

    This gives the length *of the projection of x in the direction of u. We'll refer to this quantity as a **projection coefficient**. (See Figure C.2.)*

(2) $(u \cdot x)u = u(u \cdot x)$ *is the **projection** of x in the direction u.* $\qquad \square$

Corollary C.2.3. *If $v, x \in \mathbb{R}^n$, then the projection of x in the direction of v is given by*

$$\mathrm{proj}(x \to v) = \frac{v \cdot x}{|v|^2}v .$$

Proof. Exercise C.2. $\qquad \square$

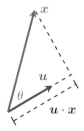

Figure C.2. The projection of x in the direction of a unit vector u.

The `project()` function in `fastR` can be used to compute projection coefficients or projection vectors.

```
> project                                                          vec-proj
function (x, u = rep(1, length(x)), type = c("vector", "length"))
{
    type = match.arg(type)
    switch(type, vector = u * (dot(x, u)/dot(u, u)), length = dot(x,
        u)/sqrt(dot(u, u)), )
}
<environment: namespace:fastR>
> project(u,v)
[1]  1.6623 2.0779 2.4935
> project(u,v,type='length')
[1]  3.6467
```

We can use projections to decompose elements of a vector space into components in the direction of each basis vector.

Lemma C.2.4. *Let* v_1 *and* v_2 *be linearly independent vectors, and let*

$$x = \alpha v_1 + \beta v_2 \ .$$

Then

$$\alpha w_1 = \mathrm{proj}(x \to w_1)\,, \ \text{and} \ \beta w_2 = \mathrm{proj}(x \to w_2) \tag{C.1}$$

where

$$w_1 = v_1 - \mathrm{proj}(v_1 \to v_2) \ \text{and} \ w_2 = v_2 - \mathrm{proj}(v_2 \to v_1) \ .$$

Sketch of proof. Figure C.3 shows the decomposition $x = \alpha v_1 + \beta v_2$ and the vectors w_2 and $\mathrm{proj}(x \to w_2)$ in the case that x is between v_1 and v_2 (so $\alpha > 0$ and $\beta > 0$). Exercise C.5 asks you to explore other configurations.

In every case, by similar triangles,

$$\mid \beta \mid = \frac{|\beta v_2|}{|v_2|} = \frac{|\,\mathrm{proj}(x \to w_2)|}{|w_2|} \ . \tag{C.2}$$

If w_2 and $\mathrm{proj}(x \to w_2)$ point in the same direction (as they do in Figure C.3), then $\beta > 0$; otherwise $\beta < 0$. So β is a kind of quotient of $\mathrm{proj}(x \to w_2)$ and w_2 and the lemma is proved for β. A symmetric argument provides the result for α. □

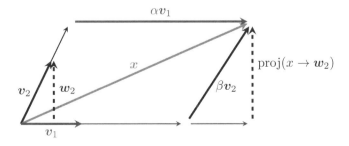

Figure C.3. Decomposing x into non-orthogonal components.

Example C.2.1.

Q. Express $\langle 2, 5 \rangle$ as $\alpha \langle 1, 1 \rangle + \beta \langle 2, 3 \rangle$.

A. First we define a function in R that computes the vector decomposition.

vdecomp

```
> vdecomp <- function(x,v1,v2) {
+     w1 <- v1 - project(v1,v2) ; w2 <- v2 - project(v2,v1)
+     p1 <- project(x,w1,type='v') ; p2 <- project(x,w2,type='v')
+     a  <- sign( dot(w1, p1) ) * vlength(p1) / vlength(w1)
+     b  <- sign( dot(w2, p2) ) * vlength(p2) / vlength(w2)
+     structure( list( coefficients=c(a,b),
+         x.orig = x, x.proj = a * v1 + b * v2
+         ))
+ }
```

Now we use `vdecomp()` to answer the question.

vector-decomp1

```
> v1 <- c(1,1)
> v2 <- c(2,3)                        $x.orig
> x  <- c( 2,5)                       [1] 2 5
> vdecomp(x,v1,v2)
$coefficients                         $x.proj
[1] -4  3                             [1] 2 5
```

So $\langle 2, 5 \rangle = 2 \langle 1, 1 \rangle + 3 \langle 2, 3 \rangle$.   ◁

Example C.2.2.

Q. Find the projection of $x = \langle 2, 3, 5 \rangle$ onto the span of $v_1 = \langle 1, 1, 1 \rangle$ and $v_2 = \langle 1, 2, 3 \rangle$ and express the projection as a linear combination of v_1 and v_2.

A. Let $x = x_1 + h$, where $h \perp v_1$ and $h \perp v_2$. Since all the projections in the proof of Lemma C.2.4 that involve x also involve a vector in $\text{span}(v_1, v_2)$, applying (C.1) to x or x_1 yields the same α and β. Doing so will thus find both the projection x_1 and the coefficients α and β.

vector-decomp2

```
> v1 <- c(1,0,0)                      $x.proj
> v2 <- c(1,1,1)                      [1] 2 4 4
> x  <- c( 2,3,5)
> vdecomp(x,v1,v2)                    > h <- x - vdecomp(x,v1,v2)$x.proj;
$coefficients                        > round(h,8)
[1] -2  4                             [1]  0 -1  1
                                     > round(dot(h,v1),8)
$x.orig                               [1] 0
[1] 2 3 5                            > round(dot(h,v2),8)
                                      [1] 0
```

So the projection of x onto the span of $\{v_1, v_2\}$ is $\langle 2, 4, 4 \rangle = -2v_1 + 4v_2$.   ◁

The vector decomposition process can be iterated to decompose vectors into more than two components.

Example C.2.3.

Q. Let $x = \langle 2, 7, 3 \rangle$. Express x as $x = \alpha_1 v_1 + \alpha_2 v_2 + \alpha_3 v_3$ for the vectors $\{v_1, v_2, v_3\} = \{\langle 1, 0, 0 \rangle, \langle 1, 1, 1 \rangle, \langle 1, 2, 3 \rangle\}$.

A. Let $x = x_1 + h$ where $h \perp v_1$ and $h \perp v_2$. We can use Lemma C.2.4 to find α_1 and β_1 such that $x_1 = \alpha_1 v_1 + \beta_1 v_2$ and then use Lemma C.2.4 again to find α_2 and β_2 such that

$$x = \alpha_2 x_1 + \beta_2 v_3$$
$$= \alpha_2(\alpha_1 v_1 + \beta_1 v_2) + \beta_2 v_3$$
$$= \alpha_2 \alpha_1 v_1 + \alpha_2 \beta_1 v_2 + \beta_2 v_3 \,.$$

Applying this to our particular vectors, we see that

$$\langle 2, 7, 3 \rangle = -9\langle 1, 0, 0 \rangle + 15\langle 1, 1, 1 \rangle - 4\langle 1, 2, 3 \rangle \,.$$

vector-decomp3

```
> v1 <- c(1,0,0)
> v2 <- c(1,1,1)
> v3 <- c(1,2,3)
> x  <- c(2,7,3)
> vdecomp(x,v1,v2)
$coefficients
[1] -3  5

$x.orig
[1] 2 7 3

$x.proj
[1] 2 5 5

> a1 <- vdecomp(x,v1,v2)$coef[1]
> b1 <- vdecomp(x,v1,v2)$coef[2]
> x1 <- a1 * v1 + b1 * v2; x1
[1] 2 5 5
# decompose x into x1 and v3
```

```
> vdecomp(x,x1,v3)
$coefficients
[1]  3 -4

$x.orig
[1] 2 7 3

$x.proj
[1] 2 7 3

> a2 <- vdecomp(x,x1,v3)$coef[1]
> b2 <- vdecomp(x,x1,v3)$coef[2]
# this should equal x
> a2 * (a1 * v1 + b1* v2) + b2 * v3
[1] 2 7 3
# the three coefficients
> c( a2 * a1, a2 * b1, b2)
[1] -9 15 -4
```

◁

C.3. Orthonormal Bases

Definition C.3.1. Two vectors $x, y \in \mathbb{R}^n$ are called **orthogonal** (written $x \perp y$) if $x \cdot y = 0$.

The intuition behind this definition is that if x and y are both non-zero vectors, then $x \perp y$ if and only if the angle between them is 90 degrees. We will frequently have need of a special type of basis called an orthonormal basis.

Definition C.3.2. Let S be a subspace of $\mathbb{R}^n$. An **orthonormal basis** for S is a basis of unit vectors that are pairwise orthogonal.

The standard bases for $\mathbb{R}^n$ are orthonormal, but there are many other orthonormal bases. Given any basis of orthogonal vectors, we can obtain an orthonormal basis by rescaling. So the only challenge in obtaining an orthonormal basis is the orthogonality condition.

Example C.3.1.

Q. Find an orthonormal basis for $\mathbb{R}^3$ that includes the vector $\boldsymbol{u}_1 = \frac{1}{\sqrt{3}}\mathbf{1}$.

A. $\boldsymbol{v} \perp \boldsymbol{u}_1$ if and only if $\sum v_i = 0$, so let $\boldsymbol{v}_2 = \langle 1, -1, 0 \rangle$. Now the first two coordinates of our third vector must be equal, and the sum must be zero, so let $\boldsymbol{v}_3 = \langle 1, 1, -2 \rangle$. Finally let $\boldsymbol{u}_i = \boldsymbol{v}_i/|\boldsymbol{v}_i|$. Then $\{\boldsymbol{u}_1, \boldsymbol{u}_2, \boldsymbol{u}_3\}$ forms the desired basis. $\lhd$

Example C.3.2.

Q. Find an orthonormal basis for $\text{span}\{\langle 1, 1, 1, 1 \rangle, \langle 1, 2, 3, 4 \rangle\}$.

A. Let $\boldsymbol{v}_1 = \langle 1, 1, 1, 1 \rangle$ and let $\boldsymbol{v}_2 = \langle 1, 2, 3, 4 \rangle$. Since $\boldsymbol{v}_1 \cdot \boldsymbol{v}_2 = 10$, these vectors are not orthogonal. But if we let $\boldsymbol{p}_2 = \frac{\boldsymbol{v}_1 \cdot \boldsymbol{v}_2}{|\boldsymbol{v}_1|^2}\boldsymbol{v}_1$ be the projection of $\boldsymbol{v}_2$ in the direction of $\boldsymbol{v}_1$, then $\boldsymbol{w}_2 = \boldsymbol{v}_2 - \boldsymbol{p}_2$ will be orthogonal to $\boldsymbol{v}_1$. This is clear from the geometry (Figure C.2), but it is also easily established algebraically:

$$\boldsymbol{w}_2 \cdot \boldsymbol{v}_1 = (\boldsymbol{v}_2 - \boldsymbol{p}_2) \cdot \boldsymbol{v}_1$$

$$= \left(\boldsymbol{v}_2 - \frac{\boldsymbol{v}_1 \cdot \boldsymbol{v}_2}{|\boldsymbol{v}_1|^2}\boldsymbol{v}_1 \right) \boldsymbol{v}_1$$

$$= \boldsymbol{v}_2 \cdot \boldsymbol{v}_1 - \frac{\boldsymbol{v}_1 \cdot \boldsymbol{v}_2}{|\boldsymbol{v}_1|^2}\boldsymbol{v}_1 \cdot \boldsymbol{v}_1$$

$$= \boldsymbol{v}_2 \cdot \boldsymbol{v}_1 - \frac{\boldsymbol{v}_1 \cdot \boldsymbol{v}_2}{|\boldsymbol{v}_1|^2}|\boldsymbol{v}_1|^2$$

$$= \boldsymbol{v}_2 \cdot \boldsymbol{v}_1 - \boldsymbol{v}_1 \cdot \boldsymbol{v}_2 = 0 \, .$$

An orthonormal basis is obtained by dividing each vector by its length:

$$\boldsymbol{u}_1 = \frac{\boldsymbol{v}_1}{|\boldsymbol{v}_1|} = \langle \frac{1}{2}, \frac{1}{2}, \frac{1}{2}, \frac{1}{2} \rangle \, ,$$

$$\boldsymbol{u}_2 = \frac{\boldsymbol{w}_2}{|\boldsymbol{w}_2|} = \frac{1}{2\sqrt{5}}\langle -3, -1, 1, 3 \rangle \, .$$

$\lhd$

This process can be used iteratively to find an orthonormal basis for any subspace of $\mathbb{R}^n$. The procedure is called **Gram-Schmidt orthonormalization**. One important detail of Gram-Schmidt orthonormalization is that one of the directions may be freely chosen (as long as it lies in the vector space).

When the basis vectors are orthogonal, it is especially easy to decompose a vector into multiples of the basis vectors.

Lemma C.3.3. *If $v_1, v_2, \ldots v_n$ form an orthogonal basis for a subspace S of $\mathbb{R}^m$, then for any $x \in S$,*

$$x = \sum_{i=1}^{n} \operatorname{proj}(x \to v_i),$$

and

$$|x| = \sum_{i=1}^{n} |\operatorname{proj}(x \to v_i)|^2.$$

Proof. Exercise C.8. □

C.4. Matrices

Definition C.4.1. An $m \times n$ (read m-by-n) real **matrix** (plural: matrices) is a table with m rows and n columns of real numbers. The numbers m and n are called the **dimensions** of the matrix. The number of rows is always specified first.

The entry in row i and column j of matrix A is denoted A_{ij}. As we have done in the previous sentence, we will use boldface capitals for matrices but will remove the bold when denoting an individual element of a matrix. The use of capitals is common for matrices. The boldface helps distinguish matrices from random variables. This leaves one point of potential ambiguity: Both a matrix and a vector of random variables will be denoted using boldface capitals. Context should make it clear which type of object is meant.

When we want to display a matrix explicitly, we will surround the entries in square brackets, for example

$$A = \begin{bmatrix} 1 & 2 & 3 \\ 4 & 5 & 6 \end{bmatrix}.$$

A vector of length n can be thought of either as a 1-by-n **row matrix** or as an n-by-1 **column matrix**, and rows and columns of matrices can be thought of as vectors. Thus a matrix can be thought of as a collection of row (or column) matrices. The rows and columns of matrix A are denoted $A_{i\cdot}$ and $A_{\cdot j}$, respectively. When it is clear from context whether we are interested in rows or columns, we sometimes drop the dot and write A_i or A_j.

Definition C.4.2. The product of two matrices A and B is a matrix formed by taking dot products of rows of A with columns of B. That is,

$$[AB]_{ij} = A_{i\cdot} \cdot B_{\cdot j}$$

provided the length of the rows of A is the same as the length of the columns of B.

Example C.4.1.

Q. Let $A = \begin{bmatrix} 1 & 2 & 3 \\ 4 & 5 & 6 \end{bmatrix}$ and let $B = \begin{bmatrix} 5 & 1 \\ 2 & 3 \\ 4 & -1 \end{bmatrix}$. Compute AB and BA.

A. Applying the definition, we have

$$AB = \begin{bmatrix} 1(5) + 2(2) + 3(4) & 1(1) + 2(3) + 3(-1) \\ 4(5) + 5(2) + 6(4) & 4(1) + 5(3) + 6(-1) \end{bmatrix} = \begin{bmatrix} 21 & 4 \\ 54 & 13 \end{bmatrix},$$

$$BA = \begin{bmatrix} 5(1) + 1(4) & 5(2) + 1(5) & 5(3) + 1(6) \\ 2(1) + 3(4) & 2(2) + 3(5) & 2(3) + 3(6) \\ 4(1) - 1(4) & 4(2) - 1(5) & 4(3) - 1(6) \end{bmatrix} = \begin{bmatrix} 9 & 15 & 21 \\ 14 & 19 & 24 \\ 0 & 3 & 6 \end{bmatrix}.$$

◁

Note that matrices can only be multiplied if their dimensions are compatible. The preceding example clearly shows that AB is not, in general, equal to BA. It is possible that one product is defined and the other not or that both are defined but have different dimensions. Even if A and B are compatible square matrices, in which case both products exist and have the same dimensions, the two products may still differ.

Definition C.4.3. The $n \times n$ **identity matrix** I (or I_n) is a square matrix with 1's along the major diagonal and 0's elsewhere. That is,

$$I_{ij} = \begin{cases} 1, & i = j, \\ 0, & i \neq j. \end{cases}$$

The following lemma explains why I is called the identity matrix.

Lemma C.4.4. *For any matrix A, $AI = IA = A$.*

Proof. Exercise C.15   □

Definition C.4.5. Let A be a square matrix. The **inverse** of A, denoted A^{-1}, is a matrix such that

$$AA^{-1} = A^{-1}A = I.$$

There are many applications of matrix multiplication. One that will be important in our study of linear models is the abbreviated notation of certain algebraic equalities.

Example C.4.2.

Q. Express the following system of equations using matrix algebra:

$$\begin{array}{rcrcl} 5x & + & 2y & = & 3, \\ 3x & + & 1y & = & 1. \end{array}$$

A. This is equivalent to

$$\begin{bmatrix} 5 & 2 \\ 3 & 1 \end{bmatrix} \begin{bmatrix} x \\ y \end{bmatrix} = \begin{bmatrix} 3 \\ 1 \end{bmatrix}.$$

◁

Example C.4.3.

Q. Use linear algebra to solve the system in the preceding example.

A. Let $A = \begin{bmatrix} 5 & 2 \\ 3 & 1 \end{bmatrix}$, let $b = \begin{bmatrix} 3 \\ 1 \end{bmatrix}$, and let $x = \begin{bmatrix} x \\ y \end{bmatrix}$. We want to solve

$$Ax = b.$$

(Note that we are using lowercase for the column matrices since they are equivalent to vectors.)

If we knew $\boldsymbol{A}^{-1}$, we could easily obtain a solution since

$$\boldsymbol{A}^{-1}\boldsymbol{A}\boldsymbol{x} = \boldsymbol{A}^{-1}\boldsymbol{b}$$

implies that

$$\boldsymbol{x} = \boldsymbol{A}^{-1}\boldsymbol{b} \,.$$

We will not discuss methods for inverting matrices here, but note that $\boldsymbol{A}^{-1} = \begin{bmatrix} -1 & 2 \\ 3 & 5 \end{bmatrix}$, since

$$\begin{bmatrix} -1 & 2 \\ 3 & 5 \end{bmatrix} \begin{bmatrix} 5 & 2 \\ 3 & 1 \end{bmatrix} = \begin{bmatrix} 5 & 2 \\ 3 & 1 \end{bmatrix} \begin{bmatrix} -1 & 2 \\ 3 & 5 \end{bmatrix} = \begin{bmatrix} 1 & 0 \\ 0 & 1 \end{bmatrix} \,.$$

Thus

$$\boldsymbol{x} = \boldsymbol{A}^{-1}\boldsymbol{b} = \begin{bmatrix} -1 & 2 \\ 3 & 5 \end{bmatrix} \begin{bmatrix} 3 \\ 1 \end{bmatrix} = \begin{bmatrix} -1 \\ 4 \end{bmatrix} \,.$$

It is easily verified that this is indeed a solution.

In Section C.4.1 we show how to use R to compute inverses. It is worth mentioning, however, that there are ways to solve such a system of linear equations (such as Gaussian elimination) that do not require finding an inverse. Typically these methods are preferred because they are more efficient and less prone to round-off error. $\triangleleft$

Example C.4.4.

Q. Express the following system of equations using matrix algebra:

$$\begin{aligned} y_1 &= \alpha + \beta x_1 , \\ y_2 &= \alpha + \beta x_2 , \\ y_3 &= \alpha + \beta x_3 . \end{aligned}$$

A. This is equivalent to

$$\begin{bmatrix} y_1 \\ y_2 \\ y_3 \end{bmatrix} = \begin{bmatrix} 1 & x_1 \\ 1 & x_2 \\ 1 & x_3 \end{bmatrix} \begin{bmatrix} \alpha \\ \beta \end{bmatrix} \,.$$

We can also express this as

$$\boldsymbol{y} = \boldsymbol{1}\alpha + \boldsymbol{x}\beta \,. \qquad \triangleleft$$

Definition C.4.6. The **transpose** of a matrix $\boldsymbol{A}$, denoted $\boldsymbol{A}^\top$, is the matrix that results by reversing the rows and columns. That is,

$$A_{ij}^\top = A_{ji} \,. \qquad \square$$

To save space, we will often write a column matrix as the transpose of a row matrix:

$$\begin{bmatrix} 1 & 2 & 3 & 4 \end{bmatrix}^\top = \begin{bmatrix} 1 \\ 2 \\ 3 \\ 4 \end{bmatrix} \,.$$

C.4.1. Matrices in R

R has a number of utilities for working with matrices. A vector of length mn can be converted to an m-by-n matrix simply by changing its dimensions with the `dim()` function. R arranged the elements column by column.

```
> M = 1:12                                                      mat-dim
> dim(M) = c(3,4)
> M
     [,1] [,2] [,3] [,4]
[1,]   1    4    7   10
[2,]   2    5    8   11
[3,]   3    6    9   12
```

In addition to setting the dimensions of a matrix, `dim()` can also be used to query the dimensions of an existing matrix.

```
> dim(M)                                                   mat-checkdim
[1] 3 4
```

An alternative method for building a matrix from a vector uses the `matrix()` function. This does not change the original vector but builds a new matrix using the values from the vector.

```
> x <- 1:12                                                      matrix
> matrix(x,nr=2)                  # 2 rows, entries columnwise
     [,1] [,2] [,3] [,4] [,5] [,6]
[1,]   1    3    5    7    9   11
[2,]   2    4    6    8   10   12
> matrix(x,nr=3,byrow=TRUE)       # 3 rows, entries rowwise
     [,1] [,2] [,3] [,4]
[1,]   1    2    3    4
[2,]   5    6    7    8
[3,]   9   10   11   12
> matrix(x,nc=3,byrow=TRUE)       # 3 columns, entries rowwise
     [,1] [,2] [,3]
[1,]   1    2    3
[2,]   4    5    6
[3,]   7    8    9
[4,]  10   11   12
> x                               # x is unchanged
 [1]  1  2  3  4  5  6  7  8  9 10 11 12
```

Recycling makes it easy to fill a matrix with a single value or to build a matrix where all the values are "missing".

```
> matrix(1,nr=4,nc=3)             # matrix of all 1's       matrix-recycle
     [,1] [,2] [,3]
[1,]   1    1    1
[2,]   1    1    1
[3,]   1    1    1
[4,]   1    1    1
> matrix(nr=3,nc=2)               # matrix of missing data
```

```
      [,1] [,2]
[1,]    NA   NA
[2,]    NA   NA
[3,]    NA   NA
```

The latter is useful for building a container in which values will be stored as they are computed.

Matrices can also be created by "binding" together the appropriate row or column vectors.

mat-bind

```
> A = rbind(1:3,4:6); A
     [,1] [,2] [,3]
[1,]   1    2    3
[2,]   4    5    6
> B = cbind(c(5,2,4),c(1,3,-1)); B
     [,1] [,2]
[1,]   5    1
[2,]   2    3
[3,]   4   -1
```

R can readily convert a vector into a matrix and will do so in situations where a matrix is required but a vector is specified. The default conversion is to a column matrix, but if a row matrix is required, the vector may be coerced into a row matrix instead. The `as.matrix()` function can be used to explicitly make the conversion.

as-matrix

```
> as.matrix(1:4)
     [,1]
[1,]   1
[2,]   2
[3,]   3
[4,]   4
```

The `t()` function computes the transpose.

matrix-t

```
> t(1:4)                          # transpose column into row
     [,1] [,2] [,3] [,4]
[1,]   1    2    3    4
> M
     [,1] [,2] [,3] [,4]
[1,]   1    4    7   10
[2,]   2    5    8   11
[3,]   3    6    9   12
> t(M)
     [,1] [,2] [,3]
[1,]   1    2    3
[2,]   4    5    6
[3,]   7    8    9
[4,]  10   11   12
```

The operator for matrix multiplication is `%*%`. (The operator `*` performs componentwise multiplication on matrices of the same shape.)

```
> A %*% B              # Note: A*B does not work          mat-mult
      [,1] [,2]
[1,]   21    4
[2,]   54   13
> B %*% A
      [,1] [,2] [,3]
[1,]    9   15   21
[2,]   14   19   24
[3,]    0    3    6
```

Since R coerces vectors to be matrices, we can use matrix multiplication to compute the dot product.

```
> 1:4 %*% 1:4                                            mat-dot
      [,1]
[1,]   30
> sum(1:4 * 1:4)
[1] 30
```

Notice what is happening here. The vectors are converted to matrices. The first vector is converted to a row matrix and the second to a column matrix to make the dimensions compatible for multiplication. So the product being computed is

$$\begin{bmatrix} 1 & 2 & 3 & 4 \end{bmatrix} \begin{bmatrix} 1 \\ 2 \\ 3 \\ 4 \end{bmatrix} = 1 + 4 + 9 + 16 = 30 \ .$$

The product

$$\begin{bmatrix} 1 \\ 2 \\ 3 \\ 4 \end{bmatrix} \begin{bmatrix} 1 & 2 & 3 & 4 \end{bmatrix} = \begin{bmatrix} 1 & 2 & 3 & 4 \\ 2 & 4 & 6 & 8 \\ 3 & 6 & 9 & 12 \\ 4 & 8 & 12 & 16 \end{bmatrix}$$

can be computed with the outer() function.

```
> outer(1:4, 1:4)                                       mat-outer
      [,1] [,2] [,3] [,4]
[1,]    1    2    3    4
[2,]    2    4    6    8
[3,]    3    6    9   12
[4,]    4    8   12   16
```

The function outer() can also be used in conjunction with arbitrary functions to build structured matrices.

```
> outer(1:4, 1:4, FUN=function(x,y) {paste(x,':',y,sep='')})   mat-outer-fun
      [,1]  [,2]  [,3]  [,4]
[1,] "1:1" "1:2" "1:3" "1:4"
[2,] "2:1" "2:2" "2:3" "2:4"
[3,] "3:1" "3:2" "3:3" "3:4"
[4,] "4:1" "4:2" "4:3" "4:4"
```

Matrix inverses can be computed using the `solve()` command. As the following example illustrates, the numerical methods used can lead to some round-off error.

```
> x <- as.matrix(c(3,1))          # vector as column matrix
> A <- rbind(c(5,2),c(3,1))
> Ainv <- solve(A); Ainv          # solve() computes inverse
      [,1] [,2]
[1,]   -1    2
[2,]    3   -5
> A %*% Ainv
           [,1] [,2]
[1,] 1.0000e+00    0
[2,] 4.4409e-16    1
> Ainv %*% A
           [,1] [,2]
[1,] 1.0000e+00    0
[2,] 3.5527e-15    1
> Ainv %*% x                      # solution to system
      [,1]
[1,]   -1
[2,]    4
```

mat-solve

The linear algebra routines used in quality statistical packages are selected for efficiency and to reduce the effects of such round-off error.

Exercises

C.1. Prove Lemma C.1.2.

C.2. Prove Corollary C.2.3.

C.3. Show that for any vectors v, x, and y,

$$v \cdot (x + y) = v \cdot x + v \cdot y, \text{ and}$$
$$v \cdot (x - y) = v \cdot x - v \cdot y.$$

C.4. For each of the following pairs of vectors, determine the projection of x in the direction of v. When the vectors are in $\mathbb{R}^2$, illustrate with a picture.

a) $x = \langle 1, 0 \rangle$; $v = \langle 1, 1 \rangle$.

b) $x = \langle 1, 0 \rangle$; $v = \langle 1, -1 \rangle$.

c) $x = \langle 1, 0 \rangle$; $v = \langle 1, 2 \rangle$.

d) $x = \langle 1, 2, 3 \rangle$; $v = \langle 1, 1, 1 \rangle$.

e) $x = \langle 1, 1, 1 \rangle$; $v = \langle 1, 2, 3 \rangle$.

f) $x = \langle 1, 2, 3 \rangle$; $v = \langle 1, -1, 0 \rangle$.

g) $x = \langle 1, 2, 3, 4 \rangle$; $v = \langle 1, 1, -1, -1 \rangle$.

h) $x = \langle 1, 1, -1, -1 \rangle$; $v = \langle 1, -1, 1, -1 \rangle$.

C.5. In this exercise we explore other cases in the proof of Lemma C.2.4.

    a) Draw a version of Figure C.3 for the case that x is between v_1 and v_2 but the angle between v_1 and v_2 is obtuse. Show that Lemma C.2.4 holds in this case as well.

    b) Draw a version of Figure C.3 for the case that x is between v_1 and $-v_2$ and the angle between v_1 and v_2 is obtuse. Show that Lemma C.2.4 holds in this case as well.

    c) Draw a version of Figure C.3 for the case that x is between $-v_1$ and $-v_2$ and the angle between v_1 and v_2 is obtuse. Show that Lemma C.2.4 holds in this case as well.

C.6. For $k \in \{a, b, c, d, e, f, g, h\}$, let w_k be the answer to part k of Exercise C.4.

    a) Show that $\langle 1, 0 \rangle = w_a + w_b$.

    b) Show that $\langle 1, 0 \rangle \neq w_a + w_c$.

    c) Show that $\langle 1, 2, 3 \rangle \neq w_d + w_f$.

    d) Explain when we get equality and inequality in situations like those above.

    e) Generalize the result illustrated by w_h.

C.7. Let $x \in \mathbb{R}^n$, let w be the projection of x in the direction of $\mathbf{1}$, and let $v = x - w$.

    a) Show that each component of w is equal to $\bar{x} = \sum x_i / n$. (We will usually denote w as $\bar{x}$.)

    b) Show that $|v|^2$ is $(n-1)s^2$ where s^2 is the sample variance. (We will usually denote v as $x - \bar{x}$.)

    c) Show by direct computation of the dot product that $\bar{x} \perp x - \bar{x}$.

    d) Since $\bar{x} \perp x - \bar{x}$, the Pythagorean Theorem implies that

$$|\bar{x}|^2 + |x - \bar{x}|^2 = |x|^2 \,.$$

    Show this by direct algebraic manipulation. That is, express each squared length in terms of x_i's and $\bar{x}$ and show that the left- and right-hand sides are equal. (Isn't linear algebra nice?)

C.8. Prove Lemma C.3.3.

C.9. Find an orthonormal basis for $\text{span}\{\langle 1, 1, 1 \rangle, \langle 1, 1, -2 \rangle\}$.

C.10. Find an orthonormal basis for $\text{span}\{\langle 1, 1, 1 \rangle, \langle 1, 2, 3 \rangle\}$.

C.11. Find an orthonormal basis for $\text{span}\{\langle 1, 1, 1, 1 \rangle, \langle 1, 1, -1, -1 \rangle, \langle 1, -1, -1, -1 \rangle\}$.

C.12. Find an orthonormal basis for $\mathbb{R}^4$ that includes $\frac{1}{\sqrt{10}} \langle 1, 2, 3, 4 \rangle$.

C.13. Write your own version of the function `project(x,v)` that projects x in the direction of v. Do not require that v be a unit vector.

C.14. Develop the Gram-Schmidt orthonormalization procedure and code it as an R function. You can provide input to your function either as a list of vectors or as a matrix. In the latter case, interpret the columns as the input vectors.

C.15. Prove Lemma C.4.4.

C.16. Show that $\left(\boldsymbol{A}^{\top}\right)^{\top} = \boldsymbol{A}$.

C.17. True or false: If $\boldsymbol{A}$ is square, then $(\boldsymbol{A}^{\top})^{-1} = \left(\boldsymbol{A}^{-1}\right)^{\top}$.

C.18. Show that $\boldsymbol{A}\boldsymbol{A}^{\top}$ is always a square matrix and find its dimensions.

Review of Chapters 1–4

This appendix presents, in outline form, an overview of the most important topics and results from the first four chapters.

D.1. R **Infrastructure**

(1) Frequently used packages: `fastR`, `lattice`, `Hmisc`, `MASS`, `faraway`.

(2) Want a GUI? Give `Rcmdr` a try.

(3) R workflow.

    (a) You can tell R to load your favorite packages every time it launches.

    (b) R commands can be (and should be) saved in unformatted ASCII files using your favorite text editor. Use `source()` or the R GUI to execute commands in your file.

    (c) If you like working at the command line, you can also use the following:
R CMD `batch` *filename*.

(4) R output.

    (a) Images can be saved as pdf or png files either using the GUI or using one of

```
pdf(...); ... ; dev.off()
```

    or

```
png(...); ... ; dev.off()
```

    (b) Text output can be copied and pasted into your favorite text editor. Be sure to choose a mono-spaced font.

    (c) You can save text output to a file using

```
sink("filename"); ... ; sink()
```

    If you use R CMD `batch` *filename*, your input and output will be stored in a file named *filename*`.Rout`.

D.2. Data

(1) Organization:
  - R uses data frames to store data.

  - Rows represent observational units (subjects, cases, etc.).

  - Columns represent variables.

  - Use `dataFrame$variable` to access one variable as a vector of values.

(2) We have focused primarily on two simple types of sampling: simple random sampling and iid random sampling).

(3) There are two major types of variables: quantitative and categorical.

(4) Important numerical summaries: mean, median, quantiles (`quantile()`), standard deviation (`sd()`), variance (`var()`).

(5) Important graphical summaries: histogram (`histogram()`), stemplot (`stem()`), boxplot (`bwplot()`), scatterplot (`xyplot()`), quantile-quantile plot (`qqmath()`).

(6) Data can be imported from files: `read.table()`, `read.csv()`.

```
> names(batting)                                      review-data
 [1] "player" "year"    "stint"  "team"   "league" "G"      "AB"
 [8] "R"      "H"       "H2B"    "H3B"    "HR"     "RBI"    "SB"
[15] "CS"     "BB"      "SO"     "IBB"    "HBP"    "SH"     "SF"
[22] "GIDP"
> require(Hmisc)
> summary(HR~team,data=batting,fun=max,
+         subset=(year==2005&league=="AL"),nmin=1)
HR     N=612


+-------+---+---+--+
|       |   |N  |HR|
+-------+---+---+--+
|team   |BAL| 46|27|
|       |BOS| 52|47|
|       |CHA| 38|40|
|       |CLE| 38|33|
|       |DET| 45|21|
|       |KCA| 46|21|
|       |LAA| 40|32|
|       |MIN| 37|23|
|       |NYA| 51|48|
|       |OAK| 41|27|
|       |SEA| 48|39|
|       |TBA| 43|28|
|       |TEX| 50|43|
|       |TOR| 37|28|
+-------+---+---+--+
|Overall|   |612|48|
+-------+---+---+--+
> plot1 <- histogram(~AB, data=batting,subset=year==2005)
```

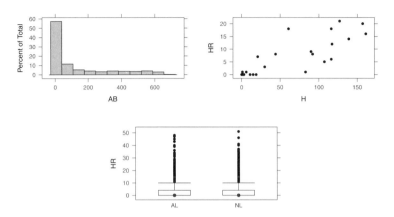

Figure D.1. Some plots showing the distribution of at bats, the relationship between hits and home runs, and home runs by league during the 2005 Major League baseball season.

```
> plot2 <- xyplot(HR~H, subset=(team=="DET" & year==2005), data=batting)
> plot3 <- bwplot(HR~league, data=batting,subset=year==2005)
```

D.3. Probability Basics

(1) Probability assigns a numerical value (between 0 and 1) to measure the like-lihood of an event (set of outcomes) of some random process.

(2) Two big questions to get you going in the right direction:
  (a) What is random here?

  (b) What do I know about its distribution?
      Possible answers: pmf or pdf, cdf, mgf, moments or other statistical measures, how to simulate, etc.

(3) Random variables:
  (a) pmf or pdf, cdf, probability, and quantile calculations,

  (b) palette of familiar distributions (see below) and when they are (poten-tially) good models,

  (c) joint, marginal, and conditional distributions.

D.4. Probability Toolkit

(1) Probability axioms and rules, including:
  (a) $0 \leq \mathrm{P}(A) \leq 1$,

  (b) $\mathrm{P}(S) = 1$ (S is the *sample space*),

  (c) $\mathrm{P}(A \cup B) = \mathrm{P}(A) + \mathrm{P}(B) - \mathrm{P}(A \cap B)$,

  (d) $\mathrm{P}(A \cap B) = \mathrm{P}(A) \cdot \mathrm{P}(B \mid A)$.

(2) Independence.

(3) Counting techniques. [Big idea: Don't double count! Don't skip!]

(4) The cdf method.

    Since the pdf of a continuous random variable does not correspond directly to a probability statement, it is often easier to work with the cdf and then use differentiation to obtain the pdf (if needed).

(5) Computation of moments (including expected value and variance).

    (a) $E(X^k) = \sum x^k f_X(x)$ or $\int x^k f_X(x)\,dx$.

    (Think: sum of value $\times$ probability.)

    (b) Computational formulas:

- $Var(X) = E(X^2) - E(X)^2$.

- $E(aX + b) = a\,E(X) + b$.

- $Var(aX + b) = a^2\,Var(X)$.

- $E(X + Y) = E(X) + E(Y)$.

- $E(XY) = E(X) \cdot E(Y)$, provided X and Y are independent.

- $Var(X + Y) = Var(X) + Var(Y)$, provided X and Y are independent.

- $Var(X + Y) = Var(X) + Var(Y) + 2\,Cov(X, Y)$.

(6) mgf methods:

- computing moments from the mgf via differentiation,

- mgfs of linear transformations and independent sums,

- recognizing distributions by recognizing mgf,

- used in proof of Central Limit Theorem.

(7) Simulation techniques, including `sample()` and `replicate()`.

D.5. Inference

(1) Framework.

    (a) The three distributions: population, sample, sampling distribution.

    (b) Statistic = function : data $\to \mathbb{R}$.

    (c) Estimator = statistic used to estimate a parameter.

    (d) Hypothesis tests (four steps, interpreting p-values, types of error, power).

    (e) Confidence intervals (confidence levels, coverage rate).

    (f) Assumption checking and robustness (simulations).

(2) Applications.

    (a) `binom.test()` for inference about proportions.

    (b) `fisher.test()`, based on hypergeometric distribution, for testing independence of rows and columns in a 2-way table.

    (c) `t.test()` for inference about population mean using sample mean (unknown population variance):

        • paired *t*-test: *t*-test on difference between two other variables for the *same* observational unit. (Example: before/after measurements on same person.)

    (d) *z*-procedures:

        • inference for μ assuming σ is known – rarely used,

        • as approximation to `binom.test()` when $n\pi$ is large enough – `prop.test()`.

    (e) Empirical p-values and permutation tests.

D.6. Important Distributions

Summaries of a number of important distributions appear in Tables A, B, and C on the back inside cover and its facing page and also on page 236.

Exercises

D.1. In each of the following situations, determine a reasonable model (distribution) for the random variables mentioned. List all parameters involved. If possible, determine numerical values for the parameters.

    a) Flip a fair coin 10 times. Let H be the number of heads.

    b) Roll a standard, fair, six-sided die repeatedly until you get a 4. Let R be the number of rolls.

    c) Flip a coin repeatedly until you obtain three tails. Let X be the number of times the coin is flipped.

    d) Sit along a city street and watch the vehicles go by. Let T be the time until a red pick-up truck passes.

    e) Let C be the number of phone calls that arrive at a call center on a Monday morning (8–noon).

    f) Let F be the free-throw percentage of a random basketball player from the Big Ten.

Table D.1. Table for Exercise D.5.

true proportion of heads	probability of rejection ($\alpha = 0.05$)
0%	
10%	
20%	
30%	
40%	
45%	
50%	
55%	
60%	
70%	
80%	
90%	
100%	

D.2. Let $X \sim \mathsf{Binom}(20, .25)$. Determine the mean, standard deviation, and median of X.

Do this two ways: once using R and once using pencil and paper. (Don't just copy from the summary sheet – recalculate these values.)

D.3. Let $X \sim \mathsf{Exp}(\lambda = 2)$. Determine the mean, standard deviation, and median of X.

Do this two ways: once using R and once using pencil and paper. (Don't just copy from the summary sheet – recalculate these values.)

D.4. A coin is tossed 100 times to see if it is a fair coin. The result is 40 heads and 60 tails. What conclusion do your draw from this data? Explain.

D.5. In the situation of the previous problem, what is the probability of correctly identifying a biased coin as biased?

  a) Fill in Table D.1.

  b) Use simulation to check your theoretical calculations.

  c) Use R to make a graph of these results.

Bonus: Try other values of α and produce plots that show the dependence of the results on both the true proportion of heads and the value of α.

D.6. Suppose $\boldsymbol{X} = \langle X_1, X_2, \ldots, X_n \rangle \overset{\text{iid}}{\sim} \mathsf{Unif}(0, 1)$. Let $Y = \min \boldsymbol{X}$.

  a) What is $\mathrm{P}(Y \leq .05)$?

  b) What is the pdf of Y?

  c) Explain how this result is related to conducting multiple hypothesis tests.

Note that answers to the first two questions will depend on n, the length of the vector $\mathbf{X}$.

D.7. Let X and Y be continuous random variables. Define a new random variable Z according to the following process. First flip a biased coin with probability α of obtaining a head. If the result is heads, sample from X; otherwise sample from Y. Such a distribution is called a **mixture** of X and Y.

   a) Derive a formula for the cdf of Z in terms of α, and the cdfs of X and Y.

   b) Derive a formula for the pdf of Z in terms of α, and the pdfs of X and Y.

   c) Let $W \sim 0.3 \cdot \mathsf{Norm}(8, 2) + 0.7 \cdot \mathsf{Norm}(16, 3)$ (that is, a 30-70 mix of two normal distributions). What is $P(W \le 12)$?

   d) Use R to plot the pdf of W.

D.8. If $X = e^Y$ and $Y \sim \mathsf{Norm}(\mu, \sigma)$, then X is said to have a **lognormal distribution**. The name comes from the fact that the logarithm of X has a normal distribution.

   a) Derive the pdf for X. (You may express this in terms of φ, the pdf of the standard normal distribution.)

   b) Write an R function `dlognorm(x,mu=0,sigma=1)` that returns the value of the pdf of X at x.

   c) Plot the pdf of X using R for several values of μ and σ.

   d) Describe the shape of the pdf of a lognormal random variable.

   e) Write an R function `rlognorm(n,mu=0,sigma=1)` that returns n random draws from a lognormal distribution with the given parameters.

   f) Write R functions `plognorm()` and `qlognorm()` that do the obvious things. What arguments do these functions need?

You can use `dlnorm()`, `rlnorm()`, etc., to check your work, but do not use them inside your functions. You may, however, make use of `dnorm()` and friends.

D.9. The R data set `faithful` contains eruption times for 272 eruptions of the Old Faithful Geyser in Yellowstone National Park. Consider these to be a random sample from the population of eruption times.

   a) Compute a 95% confidence interval for the mean eruption time.

   b) Is there any reason to be concerned about the coverage rate of this confidence interval?

   c) Is there any other reason to be concerned about this confidence interval?

D.10. Write an R function `moment()` such that given a vector of probabilities `probs`, a vector of values `vals`, and a positive integer k, `moment(probs, vals, k)` returns the kth moment of the discrete random variable X described by `probs` and `vals`. Test your function by computing some means and variances for which you know the answer.

   Bonus: Add a boolean `centered` so that your function returns moments about the mean when `centered` is true and moments about 0 otherwise.

D.11. Repeat the preceding problem for a continuous random variable. You can describe the random variable in one of two ways:

- a function `density` giving the pdf of X or
- a function `density` and a pair `range` giving an interval on which the pdf is non-zero.

Choose reasonable default values for your argument(s).

    Bonus: Write your function to allow either type of input.

Hints, Answers, and Solutions to Selected Exercises

Additional solutions are available at the companion web site for this book:

http://www.ams.org/bookpages/amstext-13

1.2.

```
> pulsePlot1 <- histogram(~pulse, data=littleSurvey)
> pulsePlot2 <- histogram(~pulse, data=littleSurvey, subset=pulse>30)
> pulseSubset <- littleSurvey$pulse[ littleSurvey$pulse > 30 ]
> mean(pulseSubset)
[1] 70.641
> median(pulseSubset)
[1] 70
```

1.4.

```
> t <- table(littleSurvey$number); t

 1  2  3  4  5  6  7  8  9 10 11 12 13 14 15 16 17 18 19 20 21 22 23
 4  7 13 12  7  3 22  3  6  4  8 13 13  8 13  9 22  4  5  4 11  6 16
24 25 26 27 28 29 30
 7 14  2 20  7  8  8
> plot <- histogram(~number, littleSurvey, breaks=0.5 + 0:30)
```

587

```
> max(t)
[1] 22
> which(t == max(t))
 7 17
 7 17
> which(t == min(t))
26
26
> table(littleSurvey$number %% 2 == 0)

FALSE  TRUE
  182    97
```

1.5.

a) odd case: m is middle value; even case: middle values are $m - d$ and $m + d$ for some d, so m is the median.

b) Suppose L values are less than m and E values equal to m. Then there are also L values greater than m. Since $m - d + m + d = 2m$, the sum of the values is $\sum x_i = \left(\sum_{x_i < m} x_i + \sum_{x_i > m} x_i \right) + \sum_{x_i = m} x_i = 2Lm + Em = (2L + E)m$, so the mean is m.

1.15.   If all the values are the same, then the variance is zero.

1.16.

```
> summary(ERA~lgID, data=pitching2005, subset=GS>=5)     ┌──────────────────┐
ERA     N=217                                            │ pitching2005-era │
                                                         └──────────────────┘

+-------+--+---+------+
|       |  |N  |ERA   |
+-------+--+---+------+
|lgID   |AL| 97|4.6593|
|       |NL|120|4.3798|
+-------+--+---+------+
|Overall|  |217|4.5047|
+-------+--+---+------+
```

2.1.

a) $S = \{\text{HHH}, \text{HHT}, \text{HTH}, \text{HTT}, \text{THH}, \text{THT}, \text{TTH}, \text{TTT}\}$.

b) $A = \{\text{HTT}, \text{THT}, \text{TTH}, \text{TTT}\}$,
$\quad\ B = \{\text{HTT}, \text{TTT}\}$,
$\quad\ C = \{\text{THH}, \text{THT}, \text{TTH}, \text{TTT}\}$.

c) $A^c = \{\text{HHH}, \text{HHT}, \text{HTH}, \text{THH}\}$,
$\quad\ A \cap B = B = \{\text{HTT}, \text{TTT}\}$,
$\quad\ A \cup C = \{\text{HTT}, \text{THH}, \text{THT}, \text{TTH}, \text{TTT}\}$.

2.6.

```
> (choose(13,1)*        # a number to have three of     ┌──────────────────┐
+ choose(4,3)*          # threee of that number         │ cards-full-house │
+ choose(12,1)*         # a different number to have two of └────────────────┘
+ choose(4,2)) /        # two of that numer
```

```
+ choose(52,5)
[1] 0.0014406
```

2.22. The checks for independence show that A and S are very close to being independent. If we think of our table as representing a sample from a larger population, we certainly do not have enough evidence to be convinced that S and A are not independent in the population. We will learn formal procedures for this later.

2.23.

$$
\begin{aligned}
\mathrm{P}(D) &= \mathrm{P}(D \text{ and } AA) + \mathrm{P}(D \text{ and } Aa) + \mathrm{P}(D \text{ and } aa) \\
&= \mathrm{P}(AA) \cdot \mathrm{P}(D|AA) + \mathrm{P}(Aa) \cdot \mathrm{P}(D|Aa) + \mathrm{P}(aa) \cdot \mathrm{P}(D|aa) \\
&= (0.25)(0.01) + (0.5)(0.05) + (0.25)(0.5) = 0.1525 .
\end{aligned}
$$

2.36.

socks

```
> 8*5*4 / choose(17,3)  # 1 sock of each kind means no pairs
[1] 0.23529
> 1 - (8*5*4 / choose(17,3))  # so this is prob of getting a pair
[1] 0.7647
# or do it this way
> ( choose(8,2)*9 + choose(5,2) * 12 + choose(4,2) * 13 +
+    choose(8,3) + choose(5,3) + choose(4,3) ) / choose(17,3)
[1] 0.7647
```

2.42.

value of X	1	2	3	4
probability	0.4	0.3	0.2	0.1

Example calculation: $\mathrm{P}(X = 3) = \frac{3}{5} \cdot \frac{2}{4} \cdot \frac{2}{3}$.

2.49.

playoffs-part

```
### using binomial dist
> 1- pbinom(1,3,0.6)              # win at least 2 of 3
[1] 0.648
### using neg binomial dist
> pnbinom(1,2,0.6)               # lose <= 1 time  before 2 wins
[1] 0.648
```

2.80.

prob-pois-01s

```
>
> dpois(0,6/3)                   # 0 customers in 1/3 hour
[1] 0.13534
> dpois(2,6/3)                   # 2 customers in 1/3 hour
[1] 0.27067
```

3.1. This can easily be done by hand since the kernel is a polynomial, but here is code to do it numerically in R:

prob-dist01

```
> require(MASS)                  # for fractions()
> kernel <- function(x) { (x-2)*(x+2) * as.numeric(x >=-2 & x <= 2) }
> k <- 1 / integrate(kernel,-2,2)$value; k
[1] -0.09375
> f <- function(x) { k * kernel(x) }
> fractions(k)
[1] -3/32
> integrate(f,-2,2)              # check that we have pdf
```

```
1 with absolute error < 1.1e-14
> integrate(f,0,2)
0.5 with absolute error < 5.6e-15
> fractions(integrate(f,0,2)$value)
[1] 1/2
> integrate(f,1,2)
0.15625 with absolute error < 1.7e-15
> fractions(integrate(f,1,2)$value)
[1] 5/32
> integrate(f,-1,1)
0.6875 with absolute error < 7.6e-15
> fractions(integrate(f,-1,1)$value)
[1] 11/16
```

3.10.

a) $P(X \leq 1) = F(1) = 1/4$.

b) $P(0.5 \leq X \leq 1) = F(1) - F(0.5) = 1/4 - 1/16 = 3/16$.

c) $P(X > 1.5) = 1 - F(1.5) = 1 - 9/16 = 7/16$.

d) The median of X: $m^2/4 = 1/2 \implies m = \sqrt{2}$.

e) The pdf of X: $f(x) = x/2$.

```
> f <- function(x) { x/2 }            # define pdf          prob-cdf01
> integrate(f,lower=0,upper=2)        # check it is a pdf
1 with absolute error < 1.1e-14
> xf <- function(x) x * f(x)
> integrate(xf,lower=0,upper=2)       # expected value
1.3333 with absolute error < 1.5e-14
> xxf <- function(x) { x^2 * f(x) }
> integrate(xxf,lower=0,upper=2)      # E(X^2)
2 with absolute error < 2.2e-14
>
# compute the variance using E(X^2) - E(X)^2
> integrate(xxf,lower=0,upper=2)$value  -
+     (integrate(xf,lower=0,upper=2)$value)^2
[1] 0.22222
```

4.7.

```
# This gives the method of moments estimates     miaa-ft-betas
# for the full data set
> beta.mom(miaa05$FTPct)
shape1 shape2
1.7665 1.1337
```

4.16.

```
> x <- c(1,2,4,4,9)                              clt-prob-finite-samples
> mu <- sum(x * 0.2); mu               # population mean
[1] 4
> v <- sum(x^2 *0.2) - mu^2; v         # population variance
[1] 7.6
> pairsums <- outer(x,x,"+")           # compute 25 sums
```

```
> pairmeans <- pairsums/2
>
# sampling distribution with SRS
> srs.means <- as.vector(pairmeans[lower.tri(pairmeans)]); srs.means
 [1] 1.5 2.5 2.5 5.0 3.0 3.0 5.5 4.0 6.5 6.5
> iid.means <- as.vector(pairmeans); iid.means
 [1] 1.0 1.5 2.5 2.5 5.0 1.5 2.0 3.0 3.0 5.5 2.5 3.0 4.0 4.0 6.5 2.5 3.0
[18] 4.0 4.0 6.5 5.0 5.5 6.5 6.5 9.0
>
> srs.mean <- sum(srs.means * 0.1); srs.mean
[1] 4
> srs.var <- sum(srs.means^2 * 0.1) - srs.mean^2; srs.var
[1] 2.85
> v/2 * (5-2) / (5-1)
[1] 2.85
> sqrt(v/2 * (5-2) / (5-1))
[1] 1.6882
>
> var(srs.means)    # N.B: This is the INCORRECT variance
[1] 3.1667
```

4.35.

uvec-prob02

```
> x <- c(3,4,5,8)
> mean(x)
[1] 5
> var(x)
[1] 4.6667
> l <- c(); P<- c()
> for (i in 1:4) {
+     l[i] <- as.numeric(x %*% uvec(i,4))     # proj coefficient
+     P[[i]] <- l[i] * uvec(i,4)              # proj vector
+ }
> l                                           # proj coefficients
[1] 10.0000 -0.7071 -1.2247 -3.4641
> P                                           # proj vectors
[[1]]
[1] 5 5 5 5

[[2]]
[1] -0.5  0.5  0.0  0.0

[[3]]
[1] -0.5 -0.5  1.0  0.0

[[4]]
[1] -1 -1 -1  3

# next two should be the same value
> l[2]^2 + l[3]^2 + l[4]^2
[1] 14
```

```
> 3 * var(x)
[1] 14
```

4.39.

```
> for (species in levels(iris$Species)) {          sepal-width-cint
+         cat (paste('Species:',species,'\n'))
+      print(t.test(iris$Sepal.Width[iris$Species==species])$conf.int)
+ }
Species: setosa
[1] 3.3203 3.5357
attr(,"conf.level")
[1] 0.95
Species: versicolor
[1] 2.6808 2.8592
attr(,"conf.level")
[1] 0.95
Species: virginica
[1] 2.8823 3.0657
attr(,"conf.level")
[1] 0.95
```

Since these confidence intervals do not overlap, it suggests that the mean sepal widths differ from species to species.

4.50.

```
> t.test(endurance$Vitamin,endurance$Placebo,paired=TRUE)    endurance-paired

        Paired t-test

data:  endurance$Vitamin and endurance$Placebo
t = -0.7854, df = 14, p-value = 0.4453
alternative hypothesis: true difference in means is not equal to 0
95 percent confidence interval:
 -180.82    83.89
sample estimates:
mean of the differences
              -48.467

> t.test(endurance$Vitamin-endurance$Placebo) # same as above

        One Sample t-test

data:  endurance$Vitamin - endurance$Placebo
t = -0.7854, df = 14, p-value = 0.4453
alternative hypothesis: true mean is not equal to 0
95 percent confidence interval:
 -180.82    83.89
sample estimates:
mean of x
  -48.467

> t.test(log(endurance$Vitamin),log(endurance$Placebo),paired=TRUE)
```

```
        Paired t-test

data:  log(endurance$Vitamin) and log(endurance$Placebo)
t = -1.8968, df = 14, p-value = 0.07868
alternative hypothesis: true difference in means is not equal to 0
95 percent confidence interval:
 -0.563047  0.034551
sample estimates:
mean of the differences
            -0.26425


> t.test(log(endurance$Vitamin)-log(endurance$Placebo)) # same as above

        One Sample t-test

data:  log(endurance$Vitamin) - log(endurance$Placebo)
t = -1.8968, df = 14, p-value = 0.07868
alternative hypothesis: true mean is not equal to 0
95 percent confidence interval:
 -0.563047  0.034551
sample estimates:
mean of x
 -0.26425


> t.test(log(endurance$Vitamin/endurance$Placebo)) # same as above again

        One Sample t-test

data:  log(endurance$Vitamin/endurance$Placebo)
t = -1.8968, df = 14, p-value = 0.07868
alternative hypothesis: true mean is not equal to 0
95 percent confidence interval:
 -0.563047  0.034551
sample estimates:
mean of x
 -0.26425


> t.test(endurance$Vitamin/endurance$Placebo)

        One Sample t-test

data:  endurance$Vitamin/endurance$Placebo
t = 7.0085, df = 14, p-value = 6.164e-06
alternative hypothesis: true mean is not equal to 0
95 percent confidence interval:
 0.6105 1.1489
sample estimates:
mean of x
  0.87971
```

```
> t.test(1/endurance$Vitamin,1/endurance$Placebo,paired=TRUE)

        Paired t-test

data:  1/endurance$Vitamin and 1/endurance$Placebo
t = 2.4111, df = 14, p-value = 0.03022
alternative hypothesis: true difference in means is not equal to 0
95 percent confidence interval:
 0.00013775 0.00235619
sample estimates:
mean of the differences
             0.0012470

> x <- sum(endurance$Vitamin > endurance$Placebo)
> n <- nrow(endurance)
> binom.test(x,n)

        Exact binomial test

data:  x and n
number of successes = 4, number of trials = 15, p-value = 0.1185
alternative hypothesis: true probability of success is not equal to 0.5
95 percent confidence interval:
 0.077872 0.551003
sample estimates:
probability of success
             0.26667

> prop.test(x,n)

        1-sample proportions test with continuity correction

data:  x out of n, null probability 0.5
X-squared = 2.4, df = 1, p-value = 0.1213
alternative hypothesis: true p is not equal to 0.5
95 percent confidence interval:
 0.089136 0.551675
sample estimates:
      p
0.26667
```

5.2.

- $E(X) = \int_0^1 \theta x^{\theta+1} \, dx = \frac{\theta+1}{\theta+2} = 1 - \frac{1}{\theta+2}$, so we obtain the method of moments estimator by solving

$$1 - \frac{1}{\theta+2} = \overline{x} \,,$$

  which yields $\hat{\theta} = \frac{1}{1-\overline{x}} - 2$.

- The log-likelihood function is

$$l(\theta; \boldsymbol{x}) = n \log(\theta+1) + \sum_i \theta \log(x_i) \,,$$

and

$$\frac{\partial}{\partial \theta} l = \frac{n}{\theta + 1} + \sum_i \log(x_i) \, ,$$

from which we obtain the MLE

$$\hat{\theta} = \frac{-n}{\sum_i \log(x_i)} - 1 = -1 + n \sum_i \log(1/x_i) \, .$$

- Using R, we can evaluate each of these estimators from the data provided.

```
> x <- c(0.90,0.78,0.93,0.64,0.45,0.85,0.75,0.93,0.98,0.78)      prob5-09
> mean(x)
[1] 0.799
> mom <- (1 / (1-mean(x))) - 2; mom
[1] 2.9751
> mle <- ( - length(x) / sum(log(x)) ) - 1; mle
[1] 3.0607
```

5.16.

```
> x <- c(1.00,-1.43,0.62,0.87,-0.66,-0.59,1.30,-1.23,-1.53,-1.94)      lrt-laplace
> loglik1 <- function(theta, x) {
+     m <- theta[1]; lambda <- theta[2]
+     return( sum( log(0.5) + dexp(abs(x-m),rate=lambda, log=T)) )
+ }
> loglik0 <- function(theta, x) {
+     m <- 0; lambda <- theta[1]
+     return( sum( log(0.5) + dexp(abs(x-m),rate=lambda, log=T)) )
+ }
> oldopt <- options(warn=-1)
> free <- nlmax(loglik1,p=c(0,1),x=x)$estimate; free
[1] -0.65815  1.00100
> null <- nlmax(loglik0,p=c(1),x=x)$estimate; null
[1] 0.89525
> stat <- 2 * (loglik1(free,x) - loglik0(null,x)); stat
[1] 2.2329
> 1 - pchisq(stat,df=1)          # p-value based on asymptotic distribution
[1] 0.13510
> options(oldopt)
```

5.36. The kernel of the posterior distribution is given by

$$\text{posterior} = \text{likelihood} \cdot \text{prior}$$

$$= \left(\prod_{i=1}^n \frac{e^{-\lambda} \lambda^{x_i}}{x_i!} \right) \lambda^{\alpha-1} (1-\lambda)^{\beta-1}$$

$$= e^{-n\lambda} \lambda^{n\bar{x}+\alpha-1} (1-\lambda)^{\beta-1} \, .$$

6.16.

a) $\log(y) = \log(a) + x \log(b)$, so $g(y) = \log(y)$, $f(x) = x$, $\beta_0 = \log(a)$, and $\beta_1 = \log(b)$.

b) $\log(y) = \log(a) + b \log(x)$, so $g(y) = \log(y)$, $f(x) = \log(x)$, $\beta_0 = \log(a)$, and $\beta_1 = b$.

6.18. Moving up the ladder will spread the larger values more than the smaller values, resulting in a distribution that is right skewed.

6.24. Hints: (a) Notice that some of the terms in this vector are named. (b) What is the sum of squares of the terms in this vector?

6.35.

```
> oats <- data.frame(                                         oats-variety
+       yield=c(71.2,72.6,47.8,76.9,42.5,49.6,62.8,
+              65.2,60.7,42.8,73.0,41.7,56.6,57.3),
+       variety=c(rep(c("A","B"),each=7)))
> t.test(yield~variety,data=oats)

        Welch Two Sample t-test

data:  yield by variety
t = 0.5527, df = 11.574, p-value = 0.591
alternative hypothesis: true difference in means is not equal to 0
95 percent confidence interval:
 -11.031  18.489
sample estimates:
mean in group A mean in group B
       60.486          56.757
```

The mean yield for variety A is higher in our sample, but the paired *t*-test reveals that this difference is well within what we would expect by random chance even if varieties A and B were equally good.

7.1. $\boldsymbol{v}_0 \cdot \boldsymbol{v}_i = \sum 1 \cdot (x_i - \bar{x}) = \sum x_i - \sum \bar{x} = n\bar{x} - n\bar{x} = 0$, so $\boldsymbol{v}_0 \perp \boldsymbol{v}_i$.

7.6. Here is one example for part (a).

```
# fit some models                                             gpa-mct1
#
> gpa.lm <- lm(gpa~satm+satv+act,gpa)
> gpa.lma <- lm(gpa~ -1 + satm+satv+act,gpa)
#
# model comparison tests for 5 p-values in summary(gpa.lm)
#
> anova(gpa.lma,gpa.lm)
Analysis of Variance Table

Model 1: gpa ~ -1 + satm + satv + act
Model 2: gpa ~ satm + satv + act
  Res.Df  RSS Df Sum of Sq    F  Pr(>F)
1    268 54.5
2    267 47.5  1      6.95 39.0 1.6e-09
```

7.8. The degrees of freedom do not add up to 999 because there is a lot of missing data (especially for SAT scores). The two models are quite similar. The confidence intervals for the parameters show a large overlap.

7.25. Here is one way to get R to calculate the percentages you need.

```
> pal <- palettes$palettes; dim(pal) <- c(5,4); pal     palettes-percs
       [,1] [,2] [,3] [,4]
[1,]   123  118  116  113
[2,]   127  125  119  116
[3,]   126  128  123  119
[4,]   125  125  119  115
[5,]   110  114  109  107
> palperc <- 100 * row.perc(pal); palperc
          [,1]    [,2]    [,3]    [,4]
[1,] 26.170 25.106 24.681 24.043
[2,] 26.078 25.667 24.435 23.819
[3,] 25.403 25.806 24.798 23.992
[4,] 25.826 25.826 24.587 23.760
[5,] 25.000 25.909 24.773 24.318
> palettes$palperc <- as.vector(palperc)
```

B.1. $A \cap B = A$ if and only if $A \subseteq B$.

B.6. $f(a, b)$ is equal to the difference between a and b when a is larger; otherwise it is 0.

B.10. $\sum_{i=2}^{5} i^2 = 2^2 + 3^2 + 4^2 + 5^2 = 54$. We can get R to evaluate this for us as follows:

```
> (2:5)^2                                                sum-ssols
[1]   4   9  16  25
> sum( (2:5)^2 )
[1] 54
```

C.2. The projection of $\boldsymbol{x}$ in the direction of $\boldsymbol{v}$ is the same as the projection in the direction of the unit vector $\frac{\boldsymbol{v}}{|\boldsymbol{v}|}$, so the projection is

$$(\frac{\boldsymbol{v}}{|\boldsymbol{v}|} \cdot \boldsymbol{x}) \frac{\boldsymbol{v}}{|\boldsymbol{v}|} = \frac{\boldsymbol{v} \cdot \boldsymbol{x}}{|\boldsymbol{v}|^2} \boldsymbol{v} \ .$$

C.4.

   a) $\langle 0.5, 0.5 \rangle$.

   b) $\langle 0.5, -0.5 \rangle$.

   c) $\langle 0.2, 0.4 \rangle$.

   d) $\langle 2, 2, 2 \rangle$.

   e) $\langle 3/7, 6/7, 9/7 \rangle$.

   f) $\langle -0.5, 0.5, 0.0 \rangle$.

   g) $\langle -1, -1, 1, 1 \rangle$.

   h) $\langle 0, 0, 0, 0 \rangle$.

C.9.

```
> x <- c(1,1,1)                                          orthonormal-s1
> y <- c(1,1,-2)
> w <- y - project(y,x)
```

```
> dot(x,w)                              # confirm normality
[1] 0
# these two column vectors are orthogonal and have correct span
> cbind( x / vlength(x), w / vlength(w) )
        [,1]      [,2]
[1,] 0.57735  0.40825
[2,] 0.57735  0.40825
[3,] 0.57735 -0.81650
```

Bibliography

[ABTK91] K. M. Allen, J. Blascovich, J. Tomaka, and R. M. Kelsey, *Presence of human friends and pet dogs as moderators of autonomic responses to stress in women*, Journal of Personality and Social Psychology **61** (1991), no. 4, 582–589.

[AC98] A. Agresti and B. A. Coull, *Approximate is better then 'exact' for interval estimation of binomial proportions*, American Statistician **52** (1998), 119–126.

[Agr90] Alan Agresti, *Categorical data analysis*, John Wiley & Sons, New York, 1990.

[Ait87] M. Aitkin, *Modelling variance heterogeneity in normal regression using GLIM*, Applied Statistics **36** (1987), no. 3, 332–339.

[AK66] W. H. Auden and L. Kronenberger (eds.), *The Viking book of aphorisms*, Viking Press, New York, 1966.

[Aka74] H. Akaike, *A new look at the statistical model identification*, IEEE Transactions on Automatic Control **19** (1974), no. 6, 716 – 723.

[AmH82] *American Heritage Dictionary*, Houghton Mifflin Company, 1982.

[And35] Edgar Anderson, *The irises of the Gaspe Peninsula*, Bulletin of the American Iris Society **59** (1935), 2–5.

[Ans73] F. J. Anscombe, *Graphs in statistical analysis*, The American Statistician **27** (1973), no. 1, 17-21.

[BC64] G. E. P. Box and D. R. Cox, *An analysis of transformations (with discussion)*, J. R. Statist. Soc. B **26** (1964), 211–252.

[BD87] George E. P. Box and Norman R. Draper, *Empirical model-building and response surfaces*, John Wiley & Sons, Inc., New York, NY, 1987.

[Ben38] Frank Benford, *The law of anomalous numbers*, Proceedings of the American Philosophical Society **78** (1938), no. 4, 551–572.

[Ber80] Geoffrey C. Berresford, *The uniformity assumption in the birthday problem*, Mathematics Magazine **53** (1980), no. 5, 286–288.

[BHH78] G. E. P. Box, W. G. Hunter, and J. S. Hunter, *Statistics for experimenters*, John Wiley & Sons, New York, 1978.

[BKW80] David A. Belsley, Edwin Kuh, and Roy E. Welsch, *Regression diagnostics: identifying influential data and sources of collinearity*, Wiley series in probability and mathematical statistics, John Wiley & Sons, New York, 1980.

[BM09]   Brigitte Baldi and David S. Moore, *The practice of statistics in the life sciences*, Freeman, 2009.

[Bro98]   Malcolm W. Browne, *Following Benford's law, or looking out for no. 1*, New York Times (1998).

[CB01]   George Casella and Roger L. Berger, *Statistical inference*, 2nd ed., Duxbury Press, June 2001.

[CEF96]   Schooler C., Feighery E., and J. A. Flora, *Seventh graders' self-reported exposure to cigarette marketing and its relationship to their smoking behavior*, American Journal of Public Health **86** (1996), no. 9, 1216–1221.

[CL54]   Herman Chernoff and E. L. Lehmann, *The use of maximum likelihood estimates in χ^2 tests for goodness of fit*, The Annals of Mathematical Statistics **25** (1954), no. 3, 579–586.

[Coo79]   R. D. Cook, *Influential observations in linear regression*, Journal of the American Statistical Association **74** (1979), 169–174.

[Dev03]   J. L. Devore, *Probability and statistics for engineering and the sciences*, 6th ed., Duxbury, 2003.

[DHM07]   Persi Diaconis, Susan Holmes, and Richard Montgomery, *Dynamical bias in the coin toss*, SIAM Review **49** (2007), 211–235.

[DHM$^+$09] J. H. Dykstra, H. M. Hill, M. G. Miller, C. C. Cheatham, T. J. Michael, and R. J. Baker, *Comparisons of cubed ice, crushed ice, and wetted ice on intramuscular and surface temperature changes*, Journal of Athletic Training **44** (2009), no. 2, 136–141.

[Dia77]   Persi Diaconis, *The distribution of leading digits and uniform distribution mod 1*, The Annals of Probability **5** (1977), no. 1, 72–81.

[Dor02]   Neil J. Dorans, *Recentering and realigning the SAT score distributions: How and why*, Journal of Educational Measurement **39** (2002), no. 1, 59–84.

[Doy27]   Arthur Canon Doyle, *A scandal in Bohemia*, The Strand Magazine (1927).

[Far05]   Julian J. Faraway, *Linear models with R*, Chapman & Hall, 2005.

[Far06]   ———, *Extending the linear model with R*, Chapman & Hall, 2006.

[Fis25]   R. A. Fisher, *Statistical methods for research workers*, Oliver & Boyd, 1925.

[Fis35]   ———, *The design of experiments*, Hafner, 1935.

[Fis36]   ———, *The use of multiple measurements in taxonomic problems*, Annals of Eugenics, 7, Part II (1936), 179–188.

[Fis38]   ———, *Presidential address to the first Indian statistical congress*, 1938, pp. 14–17.

[Fis62]   ———, *Confidence limits for a cross-product ratio*, Australian Journal of Statistics **4** (1962).

[Fis70]   ———, *Statistical methods for research workers*, 14th ed., Oliver & Boyd, 1970.

[Fis71]   ———, *The design of experiments*, Hafner, 1971.

[fMA86]   Consortium for Mathematics and Its Applications, *For all practical purposes: Statistics*, Intellimation, 1985–1986.

[fMA89]   ———, *Against all odds: Inside statistics*, Annenberg Media, 1989.

[Fox02]   John Fox, *An R and S-Plus companion to applied regression*, Sage Publications, Thousand Oaks, CA, 2002.

[GCSR03] Andrew Gelman, John B. Carlin, Hal S. Stern, and Donald B. Rubin, *Bayesian data analysis*, 2nd ed., Chapman & Hall, 2003.

[Goo05] Phillip Good, *Permutation, parametric, and bootstrap tests of hypotheses*, 3rd ed., Springer, 2005.

[Hil95a] Theodore P. Hill, *The significant-digit phenomenon*, The American Mathematical Monthly **102** (1995), no. 4, 322–327.

[Hil95b] _____ , *A statistical derivation of the significant-digit law*, Statistical Science **10** (1995), no. 4, 354–363.

[Kal85a] J. G. Kalbfleish, *Probability and statistical inference*, 2nd ed., vol. 2, Springer, 1985.

[Kal85b] _____ , *Probability and statistical inference*, 2nd ed., vol. 1, Springer, 1985.

[Kap09] Daniel T. Kaplan, *Statistical modeling: A fresh approach*, 2009.

[Kec94] Dimitri Kececioglu, *Reliability and life testing handbook*, Prentice Hall, Inc., 1993–1994.

[KM83] R. E. Keith and E. Merrill, *The effects of vitamin C on maximum grip strength and muscular endurance*, Journal of Sports Medicine and Physical Fitness **23** (1983), 253–256.

[Knu97] Donald E. Knuth, *Art of computer programming, volume 2: Seminumerical algorithms (3rd edition)*, Addison-Wesley Professional, November 1997.

[KZC+05] Robert J. Klein, Caroline Zeiss, Emily Y. Chew, Jen-Yue Tsai, Richard S. Sackler, Chad Haynes, Alice K. Henning, John P. SanGiovanni, Shrikant M. Mane, Susan T. Mayne, Michael B. Bracken, Frederick L. Ferris, Jurg Ott, Colin Barnstable, and Josephine Hoh, *Complement factor H polymorphism in age-related macular degeneration*, Science **308** (2005), no. 5720, 385–389.

[Lap20] Pierre S. Laplace, *Théorie analytique des probabilités*, Mme Ve. Courcier, 1820.

[LBC06] Fletcher Lu, J. Efrim Boritz, and H. Dominic Covvey, *Adaptive fraud detection using Benford's law*, Canadian Conference on AI, 2006, pp. 347–358.

[LL78] J. Levy and J. M. Levy, *Human lateralization from head to foot: sex-related factors*, Science **200** (1978), no. 4347, 1291.

[MB03] John Maindonald and John Braun, *Data analysis and graphics using R*, Cambridge University Press, Cambridge, 2003.

[Men65] Gregor Mendel, *Experiments in plant hybridization*, The Origins of Genetics: A Mendel Source Book (C. Stern and E. R. Sherwood, eds.), Freeman, San Francisco, 1865, pp. 1–48.

[MT77] F. Mosteller and J. Tukey, *Data analysis and regression*, Addison-Wesley, 1977.

[New81] Simon Newcomb, *Note on the frequency of use of the different digits in natural numbers*, American Journal of Mathematics **4** (1881), no. 1, 39–40.

[NW72] J. A. Nelder and R. W. M. Wedderburn, *Generalized linear models*, Journal of the Royal Statistical Society. Series A (General) **135** (1972), no. 3, 370–384.

[Par] Emmanuel Paradis, *R for beginners*, available at `http://cran.r-project.org/other-docs.html`.

[Pun83] *The relationship between selected physical performance variables and football punting ability*, Tech. report, Department of Health, Physical Education and Recreation, Virginia Polytechnic Institute and State University, 1983.

[Rad81] M. Radelet, *Racial characteristics and imposition of the death penalty*, American Sociological Review **46** (1981), 918–927.

[Rai76]   Ralph A. Raimi, *The first digit problem*, The American Mathematical Monthly **83** (1976), no. 7, 521–538.

[Rai85]   ———, *The first digit phenomenon again*, Proceedings of the American Philosophical Society **129** (1985), no. 2, 211–219.

[Ric88]   John A. Rice, *Mathematical statistics and data analysis*, 1st ed., Wadworth, 1988.

[Ros88]   N. Rose, *Mathematical maxims and minims*, Rome Press Inc., 1988.

[Sat46]   F. E. Satterthwaite, *An approximate distribution of estimates of variance components*, Biometrics Bulletin **2** (1946), 110–114.

[SC72]   E. Street and M. G. Carroll, *Preliminary evaluation of a food product*, Statistics: A Guide to the Unknown (Judith M. Tanur et al., eds.), Holden-Day, 1972, pp. 220–238.

[SH96]   Z. Sawicz and S. S. Heng, *Durability of concrete with addition of limestone powder*, Magazine of Concrete Research **48** (1996), 131–137.

[SMB+07]   Laura J. Scott, Karen L. Mohlke, Lori L. Bonnycastle, Cristen J. Willer, Yun Li, William L. Duren, Michael R. Erdos, Heather M. Stringham, Peter S. Chines, Anne U. Jackson, Ludmila Prokunina-Olsson, Chia-Jen J. Ding, Amy J. Swift, Narisu Narisu, Tianle Hu, Randall Pruim, Rui Xiao, Xiao-Yi Y. Li, Karen N. Conneely, Nancy L. Riebow, Andrew G. Sprau, Maurine Tong, Peggy P. White, Kurt N. Hetrick, Michael W. Barnhart, Craig W. Bark, Janet L. Goldstein, Lee Watkins, Fang Xiang, Jouko Saramies, Thomas A. Buchanan, Richard M. Watanabe, Timo T. Valle, Leena Kinnunen, Goncalo R. Abecasis, Elizabeth W. Pugh, Kimberly F. Doheny, Richard N. Bergman, Jaakko Tuomilehto, Francis S. Collins, and Michael Boehnke, *A genome-wide association study of type 2 diabetes in Finns detects multiple susceptibility variants*, Science (2007), 2649–2653.

[SMS+04]   K. Silander, K. L. Mohlke, L. J. Scott, E. C. Peck, P. Hollstein, A. D. Skol, A. U. Jackson, P. Deloukas, S. Hunt, G. Stavrides, P. S. Chines, M. R. Erdos, N. Narisu, K. N. Conneely, C. Li, T. E. Fingerlin, S. K. Dhanjal, T. T. Valle, R. N. Bergman, J. Tuomilehto, R. M. Watanabe, M. Boehnke, and F. S. Collins, *Genetic variation near the hepatocyte nuclear factor-4 alpha gene predicts susceptibility to type 2 diabetes*, Diabetes **53** (2004), no. 4, 1141–1149.

[SPSB00]   B. T. Sullivan, E. M. Pettersson, K. C. Seltmann, and C. W. Berisford, *Attraction of the bark beetle parasitoid Roptrocerus xylophagorum (Hymenoptera: Pteromalidae) to host-associated olfactory cues*, Environmental Entomology **29** (2000), 1138–1151.

[SRM53]   R. D. Stichler, G. G. Richey, and J. Mandel, *Measurement of treadware of commercial tires*, Rubber Age **73** (1953), no. 2.

[SS66]   C. Stern and E. R. Sherwood (eds.), *The origins of genetics: A Mendel source book*, Freeman, San Francisco, 1966.

[SV99]   G. K. Smyth and A. P. Verbyla, *Adjusted likelihood methods for modelling dispersion in generalized linear models*, Environmetrics **10** (1999), 696–709.

[SW86]   David. J. Saville and Graham R. Wood, *A method for teaching statistics using n-dimensional geometry*, The American Statistician **40** (1986), no. 3, 205–214.

[SW96]   David J. Saville and Graham R. Wood, *Statistical methods: A geometric primer*, Springer, 1996.

[The89]   The Steering Committee of the Physicians' Health Study Research Group, *Final report on the aspirin component of the ongoing physicians' health study*, New England Journal of Medicine **321** (1989), no. 3, 129–35.

[Tuf01] Edward R. Tufte, *The visual display of quantitative information*, 2nd ed., Graphics Press, Cheshire, CT, 2001.

[Tuk77] John W. Tukey, *Exploratory data analysis*, Addison-Wesley, Reading, MA, 1977.

[Utt05] Jessica M. Utts, *Seeing through statistics*, 3rd ed., Thompson Learning, 2005.

[vB02] Gerald van Belle, *Statistical rules of thumb*, Wiley-Interscience, 2002.

[VR97] W. N. Venables and Brian D. Ripley, *Modern applied statistics with S-Plus*, 2nd ed., Springer-Verlag, Berlin / Heidelberg / London, 1997.

[Wei05] S. Weisberg, *Applied linear regression*, 3rd ed., Wiley, New York, 2005.

[WS67] M. C. Wilson and R. E. Shade, *Relative attractiveness of various luminescent colors to the cereal leaf beetle and the meadow spittlebug*, Journal of Economic Entomology **60** (1967), 578–580.

[Yat34] F. Yates, *Contingency tables involving small numbers and the χ^2 test*, Supplement to the Journal of the Royal Statistical Society **1** (1934), no. 2, 217–235.

[Zag67] S. V. Zagona (ed.), *Studies and issues in smoking behavior*, Universit of Arizona Press, 1967.

[ZS80] Sydney S. Zentall and Jandira H. Shaw, *Effects of classroom noise on performance and activity of second-grade hyperactive and control children*, Journal of Educational Psychology **72** (1980), no. 6, 830–840.

Index to R Functions, Packages, and Data Sets

Index

Page numbers for references to <u>exercises</u> are underlined. Page numbers for references to *examples* are italicized. Page numbers for references to **definitions** are in bold. In *see also* entries, items referred to from Index to R Functions, Packages, and Data Sets are in `typewriter type`.

Titles in This Series